AF179612

Principles of Ecosystem Science

Principles of Ecosystem Science

Edited by Molly Ismay

SYRAWOOD
PUBLISHING HOUSE
New York

Published by Syrawood Publishing House,
750 Third Avenue, 9th Floor,
New York, NY 10017, USA
www.syrawoodpublishinghouse.com

Principles of Ecosystem Science
Edited by Molly Ismay

© 2018 Syrawood Publishing House

International Standard Book Number: 978-1-68286-618-4 (Hardback)

This book contains information obtained from authentic and highly regarded sources. Copyright for all individual chapters remain with the respective authors as indicated. All chapters are published with permission under the Creative Commons Attribution License or equivalent. A wide variety of references are listed. Permission and sources are indicated; for detailed attributions, please refer to the permissions page and list of contributors. Reasonable efforts have been made to publish reliable data and information, but the authors, editors and publisher cannot assume any responsibility for the validity of all materials or the consequences of their use.

The publisher's policy is to use permanent paper from mills that operate a sustainable forestry policy. Furthermore, the publisher ensures that the text paper and cover boards used have met acceptable environmental accreditation standards.

Trademark Notice: Registered trademark of products or corporate names are used only for explanation and identification without intent to infringe.

Cataloging-in-Publication Data

Principles of ecosystem science / edited by Molly Ismay.
p. cm.
Includes bibliographical references and index.
ISBN 978-1-68286-618-4
1. Biotic communities. 2. Ecology. 3. Ecosystem management. I. Ismay, Molly.
QH541 .P75 2018
577.82--dc23

Printed in China.

TABLE OF CONTENTS

PREFACE

The main aim of this book is to educate learners and enhance their research focus by presenting diverse topics covering this vast field. This is an advanced book which compiles significant studies by distinguished experts in the area of analysis. This book addresses successive solutions to the challenges arising in the area of application, along with it; the book provides scope for future developments.

Ecosystem is the community of living organisms and their interaction with their surroundings. They are of various kinds like marine ecosystem, freshwater ecosystem and terrestrial ecosystem. This book presents researches and studies performed by experts across the globe in the field of ecosystem science. While understanding the long-term perspectives of the topics, the book makes an effort in highlighting their impact as a modern tool for the growth of the discipline. It aims to serve as a resource guide for students and experts alike.

It was a great honour to edit this book, though there were challenges, as it involved a lot of communication and networking between me and the editorial team. However, the end result was this all-inclusive book covering diverse themes in the field.

Finally, it is important to acknowledge the efforts of the contributors for their excellent chapters, through which a wide variety of issues have been addressed. I would also like to thank my colleagues for their valuable feedback during the making of this book.

Editor

1

Ecosystem Resilience and Threshold Response in the Galápagos Coastal Zone

Alistair W. R. Seddon[1]*, Cynthia A. Froyd[1,2], Melanie J. Leng[3], Glenn A. Milne[4], Katherine J. Willis[1,5]

1 Long-Term Ecology Laboratory, Biodiversity Institute at the Oxford Martin School, Department of Zoology, University of Oxford, Oxford, United Kingdom, 2 Department of Geography, Swansea University, Swansea, United Kingdom, 3 NERC Isotope Geosciences Laboratory, British Geological Survey, Keyworth, Nottingham, United Kingdom, 4 Department of Earth Sciences, University of Ottawa, Marion Hall, Ottawa, Ontario, Canada, 5 Department of Biology, University of Bergen, Bergen, Norway

Abstract

Background: The Intergovernmental Panel on Climate Change (IPCC) provides a conservative estimate on rates of sea-level rise of 3.8 mm yr^{-1} at the end of the 21st century, which may have a detrimental effect on ecologically important mangrove ecosystems. Understanding factors influencing the long-term resilience of these communities is critical but poorly understood. We investigate ecological resilience in a coastal mangrove community from the Galápagos Islands over the last 2700 years using three research questions: What are the 'fast and slow' processes operating in the coastal zone? Is there evidence for a threshold response? How can the past inform us about the resilience of the modern system?

Methodology/Principal Findings: Palaeoecological methods (AMS radiocarbon dating, stable carbon isotopes ($\delta^{13}C$)) were used to reconstruct sedimentation rates and ecological change over the past 2,700 years at Diablas lagoon, Isabela, Galápagos. Bulk geochemical analysis was also used to determine local environmental changes, and salinity was reconstructed using a diatom transfer function. Changes in relative sea level (RSL) were estimated using a glacio-isostatic adjustment model. Non-linear behaviour was observed in the Diablas mangrove ecosystem as it responded to increased salinities following exposure to tidal inundations. A negative feedback was observed which enabled the mangrove canopy to accrete vertically, but disturbances may have opened up the canopy and contributed to an erosion of resilience over time. A combination of drier climatic conditions and a slight fall in RSL then resulted in a threshold response, from a mangrove community to a microbial mat.

Conclusions/Significance: Palaeoecological records can provide important information on the nature of non-linear behaviour by identifying thresholds within ecological systems, and in outlining responses to 'fast' and 'slow' environmental change between alternative stable states. This study highlights the need to incorporate a long-term ecological perspective when designing strategies for maximizing coastal resilience.

Editor: Jack Anthony Gilbert, Argonne National Laboratory, United States of America

Funding: This research was supported by the UK Natural Environment Research Council, postgraduate studentship (AS) and grant NE/C510667/1 (CF). The funders had no role in study design, data collection and analysis, decision to publish, or preparation of the manuscript.

Competing Interests: The authors have declared that no competing interests exist.

* E-mail: alistair.seddon@zoo.ox.ac.uk

Introduction

Tropical mangrove ecosystems provide essential economic, geomorphological, and ecological ecosystem services by stabilizing eroding coastlines and offering protection from extreme storm surge and tsunami events. They also provide nurseries for economically valuable fishes and crustaceans, and a fuel wood source to local populations [1,2]. Mangroves are well adapted to the dynamic coastal zone; they can respond rapidly to environmental disturbance and tolerate episodic inundations of salt water and sediment movement. However, mangroves are under threat from anthropogenic land degradation and the combined impacts of climate change. Sea level rise is predicted to increase to 3.8 mm yr^{-1} due to the thermal expansion of water and glacial melting and climate change scenarios also predict an increase in the frequency and magnitude of extreme climatic events [3]. As a result, many mangrove ecosystems may be approaching critical thresholds for healthy functioning and it is necessary to understand how they will respond to the impacts of future environmental change [2,4].

Mangrove ecosystems consist of a series of vegetation communities dependent on an elevation gradient, comprised of mangroves, saltmarsh, and cyanobacterial microbial mats which are organized in parallel zones along the coast [5,6]. Because the location of these zones are principally controlled by tidal elevation, any disturbance linked to changes in tidal height or amplitude will result in an adaptation and ecological response [7,8]. The form of ecological response can vary, and can include: (i) increased sediment deposition and a resultant rise in the relative elevation of mangrove stands in response to rising sea levels; (ii) ecological compositional change; and (iii) horizontal migration upslope and inland [8]. Disturbances can arise from either large external perturbations such as tsunami and storm surge events, or from smaller changes such as variations in salinity due to fluctuating tidal levels [9]. Climate change can also alter local hydrological conditions [7], whilst internal competitive relationships within mangrove communities can enhance or reduce the ability of the ecosystem to resist change [10]. These processes are likely to interact on a hierarchy of temporal and spatial scales, leading to non-linear behaviour that is difficult to model or predict [4,11].

The ability of an ecosystem to 'tolerate or adapt to disturbance without collapsing into a different or qualitative state' [12] is an emergent property known as ecosystem resilience [13,14]. Resilience theory provides a framework for scientists to interpret the non-linear dynamics of a natural system and understand the mechanisms of abrupt ecosystem change [15]. Resilient systems require large external perturbations to be driven across a threshold into an alternative stable state. Their high resilience is a result of a negative feedback response to a series of 'fast-small' environmental processes operating at the local scale; these feedbacks maintain the system in equilibrium and allow it to resist change. Non-resilient systems result from the accumulation of 'slow-large' processes, which are influential on larger spatial scales and reduce the resilience over time. In non-resilient systems, thresholds are relatively lower and 'catastrophic regime shifts' occur more readily, resulting in new stable equilibria and new landscape conditions [16].

Resilience is a critical concept in contemporary ecology and has been applied at the local, regional, and global scale [15–17], providing a useful framework for conceptualizing emergent behaviour and in understanding complex responses to environmental change [14,18]. At present, the importance of understanding resilience with respect to future environmental changes in the coastal zone has been identified [2], but there is currently a poor understanding of the long-term (>1000 years) links between resilience and ecological change in mangrove ecosystems. Similarly, although non-linear ecological behaviour in mangroves has been proposed [4], there are few studies which have attempted to identify or explain thresholds of ecological response.

In this study we investigate ecological resilience in a mangrove ecosystem from the Galápagos Islands. We use stable carbon isotopes ($\delta^{13}C$) and AMS radiocarbon dating to examine two key ecological responses in mangrove systems – community compositional change and increasing accumulation rates – to environmental changes over the past 2,700 years. Past changes in salinity are reconstructed using a diatom transfer function in order to estimate short-term tidal disturbances. Geochemical data is used to examine the long-term environmental changes that occurred at the coastal site. and results are also compared to high-resolution palaeoclimatic data from a nearby crater lake [19]. Global scale geophysical modelling is also employed to predict longer term estimates of relative sea-level (RSL) change as a result of post-glacial isostatic adjustment (GIA). In this study we investigate three key questions: What were the 'fast and slow' processes operating in the coastal zone? Is there evidence for a threshold response? How can the past inform us about the resilience of the modern system?

Study Site

Diablas lagoon is a 79 ha coastal lagoon situated on the south coast of the island of Isabela (0°57′6.51″S, 90°59′9.69″W). The lagoon is at sea level, and is separated from the sea by a 6 m high mangrove boundary and sandy beach, 10–300 m in width (Fig. 1). Three mangrove species, *Conocarpus erecta*, *Rhizophora mangle* and *Laguncularia racemosa,* occur around the lagoon. The maximum water depth is 1.5 m along the northern shore, although most of the lagoon is approximately 30 cm deep. The contemporary lagoon sediment is composed of a pink, gelatinous, microbial mat. Salinity varies between 7–9 g L^{-1}, and the pH is 8.4 [20]. Between April and December the Galápagos coastal zone is arid with cooler temperatures due to the cool Humboldt current from the south east. During the wet-season between January and March the southward migration of the ITCZ causes warmer temperatures and intense, daily equatorial showers, whilst El Niño events are an enhancement of the wet season conditions [21]. Due to the negative precipitation/evaporation balance of the Galápagos coastal zone, the low brackish salinities indicate that the lagoon is fed from a freshwater spring with an aperiodic mixing of seawater.

Materials and Methods

Sediment samples

A 4.9 m sediment core was taken in 0.8 m of water in Diablas lagoon, close to the northern fringe of *Rhizophora mangle,* using a Livingstone piston corer [22] (Fig. 1). Samples were wrapped in plastic and foil, transported back to the laboratory, and refrigerated at 4°C to await subsampling. We also collected samples of the possible sources of autochthonous organic material (e.g. mangrove plants, microbial mats) to aid the interpretation of down core variations in carbon isotope composition.

Core chronology and sediment accumulation rates

Age chronologies for each sediment sequence were developed from eight samples using bulk accelerator mass spectrometry (AMS) radiocarbon dating (Table 1). Calibration and age-depth modelling was undertaken in CLAM [23] using the IntCal04 Southern Hemisphere Curve [24]. CLAM is a freely available package written for the R programming language [25]. ^{14}C ages were reported as conventional radiocarbon years BP (AD 1950) and calibrated age ranges were reported as 2σ Weighted Average standard deviations of the calibrated files, rounded to the nearest 5-year interval (Table 1). This calibration assumes no influence from marine carbon due to the reequilibriation of CO_2 from the atmosphere in the shallow lagoon waters (approx 1 m deep) [26] where, as a result, even submerged aquatic lagoon samples would have acquired the bulk of their organic carbon from atmospheric sources. A smooth spline interpolation method was used for age-depth modelling (1000 iterations). Wet sediment accumulation rates without density corrections were estimated by calculating the slope of the curve at the sampling point for each iteration, and errors were calculated at the 95% confidence intervals at the highest and lowest posterior densities.

Carbon isotope ratios

The carbon isotope signature was used to detect major ecological shifts in autochthonous productivity (Fig. 2). Organic composition (both carbon isotope and C/N ratios) are a function of the sources of organic material of the basin, which can result from either aquatic (fresh or brackish) *in situ* productivity or allogenic material originating from the terrestrial or marine environment [27]. In tropical lagoons, thick organic deposits are likely to accrete as a result of the development of mangrove peats [2] or microbial mats [26]. The plants from these two sources of detrital input have different photosynthetic pathways and therefore different isotopic signatures; in general C_3 plants, such as mangroves, are isotopically heavier than microbial mats found in alkaline ponds since they preferentially assimilate the lighter isotope ^{12}C from the atmosphere.

Carbon isotope composition ($\delta^{13}C$) and C/N of the surface and core samples were measured at the NERC Isotope Geoscience Laboratory, British Geological Survey. A sampling resolution of 4 cm was used. Samples were washed in 1 mol HCl to remove carbonates, rinsed with distilled water and sieved under a vacuum using 2 μm filter paper. They were dried, crushed and then

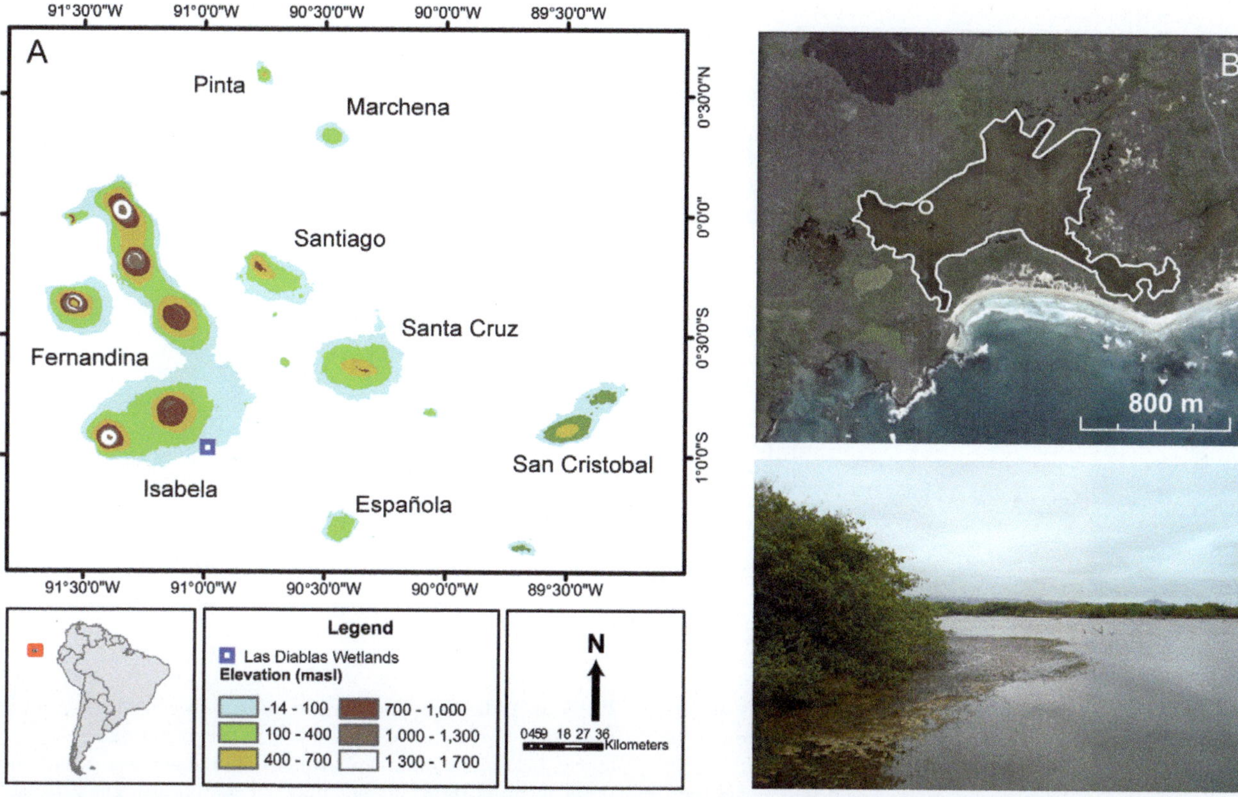

Figure 1. Map satellite image and photograph of the Diablas coastal lagoon. (A) Map of the Galápagos Islands. Diablas lagoon is located in the Diablas wetlands on the south side of Isla Isabela. (B) Satellite image of the Diablas wetlands. The coring location is marked with a circle on both satellite images and the lagoon outline has been traced (*Source: GoogleEarth © 2010*). (C) View from the far-east side of Diablas lagoon, Isabela, looking NW.

weighed into tin capsules. $^{13}C/^{12}C$ analyses were performed by combustion in a Costech Elemental Analyser (EA) on-line to a VG TripleTrap and Optima dual-inlet mass spectrometer, with $\delta^{13}C$ values calculated to the VPDB scale using a within-run laboratory standards calibrated against NBS18, NBS-19 and NBS-22. Replicate analysis of well-mixed samples indicated a precision of $\pm <0.1$‰ (1 SD). C/N ratios can also by measured if required, and these are calibrated against an Acetanilide standard. Replicate analysis of well-mixed samples indicated a precision of $\pm <0.1$.

Stable isotope mixing model

A stable isotope mixing model [28,29] was used to determine the proportion of mangroves vs. microbial mat derived organic carbon in the down core samples based on $\delta^{13}C$ stable isotope measurements (Fig. 3). Surface samples were collected from mangroves and microbial mats growing in the contemporary lagoon environment and measured for their $\delta^{13}C$ signature. After correcting for the anthropogenic ^{13}C Suess effect using mean monthly data (standard curve fitted) from the Mauna Laua

Table 1. AMS radiocarbon calibration results.

Laboratory Code	Depth (cm)	Lab ID	Modern Carbon (%)	$\delta^{13}C$ (‰)	^{14}C age (yr BP)	2σ Cal. Age (yr BP)
OZI797	56	OXR-1	95.7±0.5	−10.7	350±40	285–485
UBA11498	144	ISD_144	87.7±0.2	−12.1	1051±17	860–970
UBA1499	146	ISD_146	877.2±±0.2	−24.5	1098±17	910–990
OZI798	240–242	OXR-2	83.7±0.6	−24.0	1430±60	1165–1405
UBA11500	311	ISD_311	78.9±0.2	−23.5	1908±18	1705–1865
UBA11501	387	ISD_387	76.9±0.2	−28.6	2114±18	1935–2115
OZI1799	440-443	OXR-3	75.7±0.4	−28.6	2240±18	2070–2380
UBA-14628	483	ISD_483	72.34±0.3	−26.4	2601±28	2470–2790

Bulk sediment accelerator mass spectrometry (AMS) radiocarbon results calibrated using the IntCal04 Southern Hemisphere Curve [24]. ^{14}C ages are reported as conventional radiocarbon years BP (AD 1950). Calibrated age ranges are reported as 2σ Weighted Average standard deviations of the calibrated files. These ages are rounded to the nearest 5-year interval. This calibration assumes no influence from marine carbon due to the reequilibriation of CO_2 from the atmosphere in the shallow lagoon water (max 1 m deep) [26]. As a result even submerged aquatic lagoon samples would have acquired the bulk of their organic carbon from atmospheric sources.

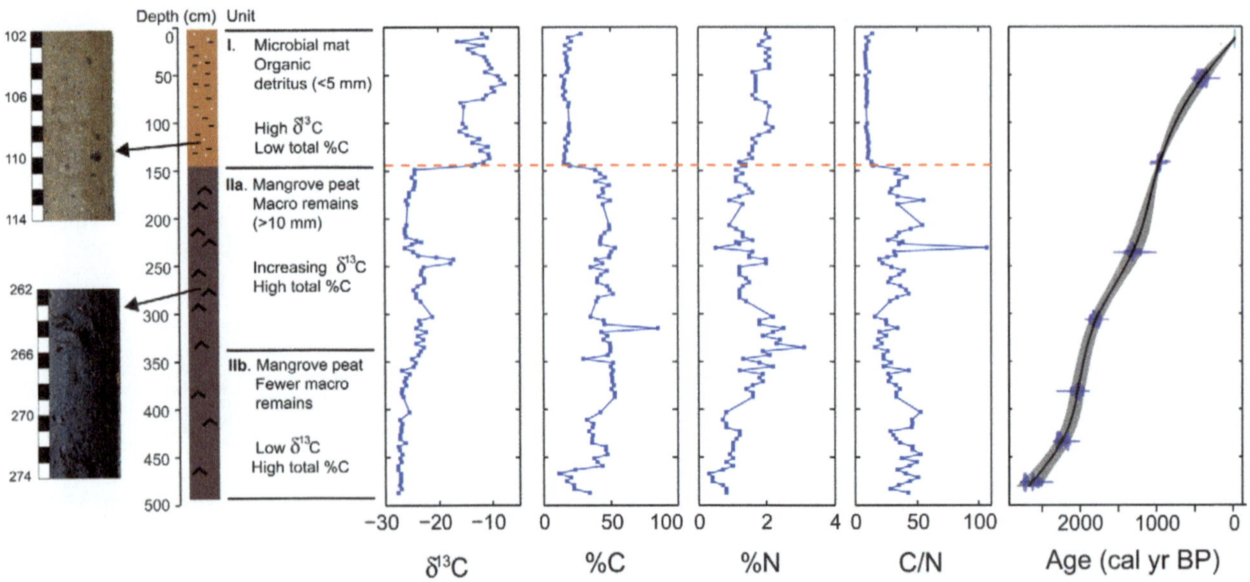

Figure 2. Age-depth, organic and stratigraphic changes observed in the Diablas core.

observatory, Hawaii. [30] and for decompositional processes by −1.9‰ (the middle value of the range cited by Chumra et al 1987), $\delta^{13}C$ values for mangroves were determined to range between −26.4 and −22.9‰ and were assigned a proportion value of 1. Microbial mats contained isotopically heavier carbon, with $\delta^{13}C$ signatures ranging between −16.0 and −10.7‰ and were assigned a proportion value of 0 (Fig. 3b) (mean = −13.2‰).

A logistic curve was fitted to these values using least squares regression (Fig. 3b) ($k = -0.45$, $x^0 = -19.1$, RMSE = 0.038;). Previous studies have used linear mixing models for this purpose [28,29], but are limited by the fact that core samples with values above (below) the maximum (minimum) mean endmember values result in proportions greater than 1 (less than 0) (Fig. 3b, dashed line). The logistic function overcomes this problem by providing an S- shaped curve which maintains all proportional values between 1 and 0.

In this model, the dominant source of organic material is assumed to be autochthonous. Therefore, the model reflects the

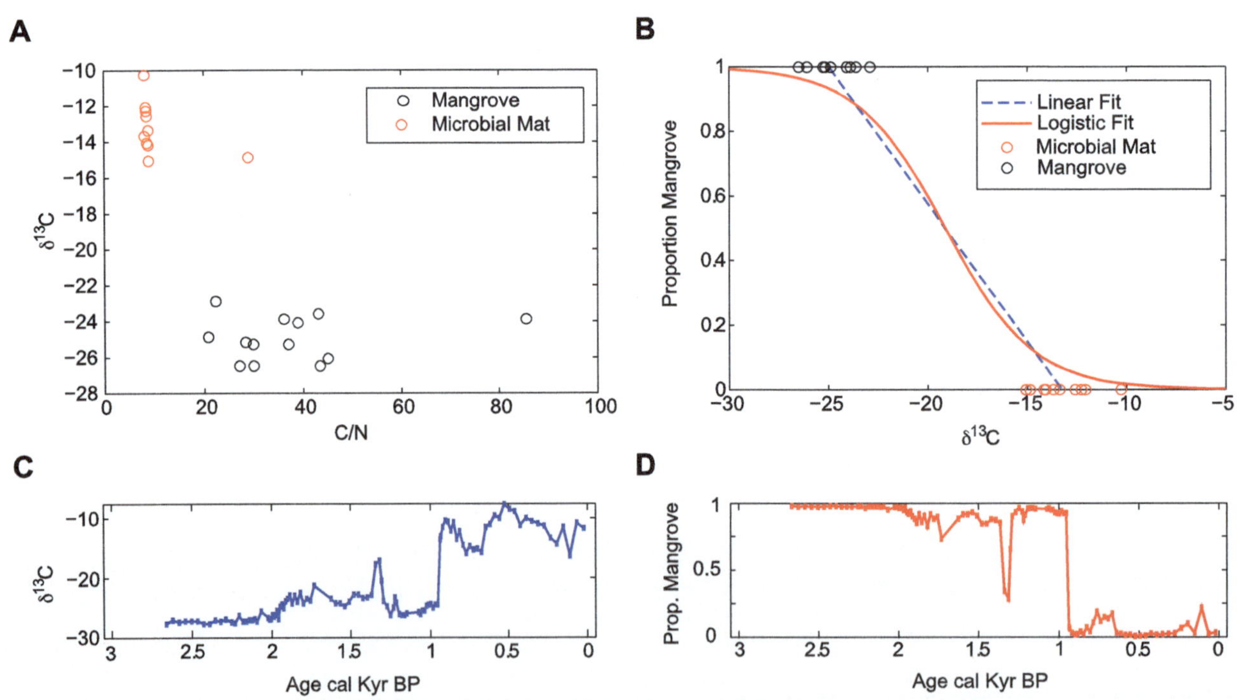

Figure 3. Stable isotope mixing model. (A) C/N and $\delta^{13}C$ results of surface samples of mangroves and microbial mats in Diablas; (B) Comparison between linear fitting (dashed line) and the logistic curve (full line) fitting methodology used in our model. (C) Stable isotope values and (D) mixing model output for Diablas lagoon.

accretion of organic material growing in the lagoon at the sampling point. In terms of biological distributions, an index value of 0 indicates an open water lagoon with no mangroves living at the coring site. Conversely, an index value of 1 indicates that the site was completely covered by mangroves above the surface; i.e an expansion of mangroves growing within the lagoon.

Geochemistry

Changes in the abundance of elements in a sediment profile can be used to infer past changes in palaeoecological and palaeoclimatic conditions (Fig. 4) [31]. Traces of geochemical elements in sediments can be used to detect processes occurring both outside of the lagoon catchment (e.g input of tidal influx) and those occurring as a result of *in situ* chemical processes (e.g. chemical changes affecting elements in the sedimentary column after burial) [32]. These techniques have been applied to interpret past environmental changes in a number of coastal systems [33,34,35].

Traces of geochemical elemental concentrations were measured at 4 cm intervals on 0.2 g of dry sediment using ICP-AES (Inductively Coupled Plasma-Atomic Emission Spectrometer) after applying a bulk sediment digestion technique [36]. This analysis was undertaken on a Perkin Elmer Optima 3300RL operated by the NERC ICP-AES Facility, Department of Geology, Royal Holloway, University of London.

Salinity

Salinity changes were estimated using a Weighted Average (WA) diatom transfer function with downweighting for rare species [37,38] and sample specific errors were estimated by bootstrapping 1,000 samples in RIOJA [39], a freely available package written for in the R programming language [25] (Fig. S1, S2). A total of 40 diatom surface samples from nine Galápagos coastal lagoons were combined with the MOLTEN training set to create a salinity transfer function (Text S1).

One cm^3 diatom samples were taken from both surface material and from sediment cores. The sediment core was subsampled at a resolution of 4 cm. Samples were prepared using the standard diatom digestion preparation procedures [40], the solution was

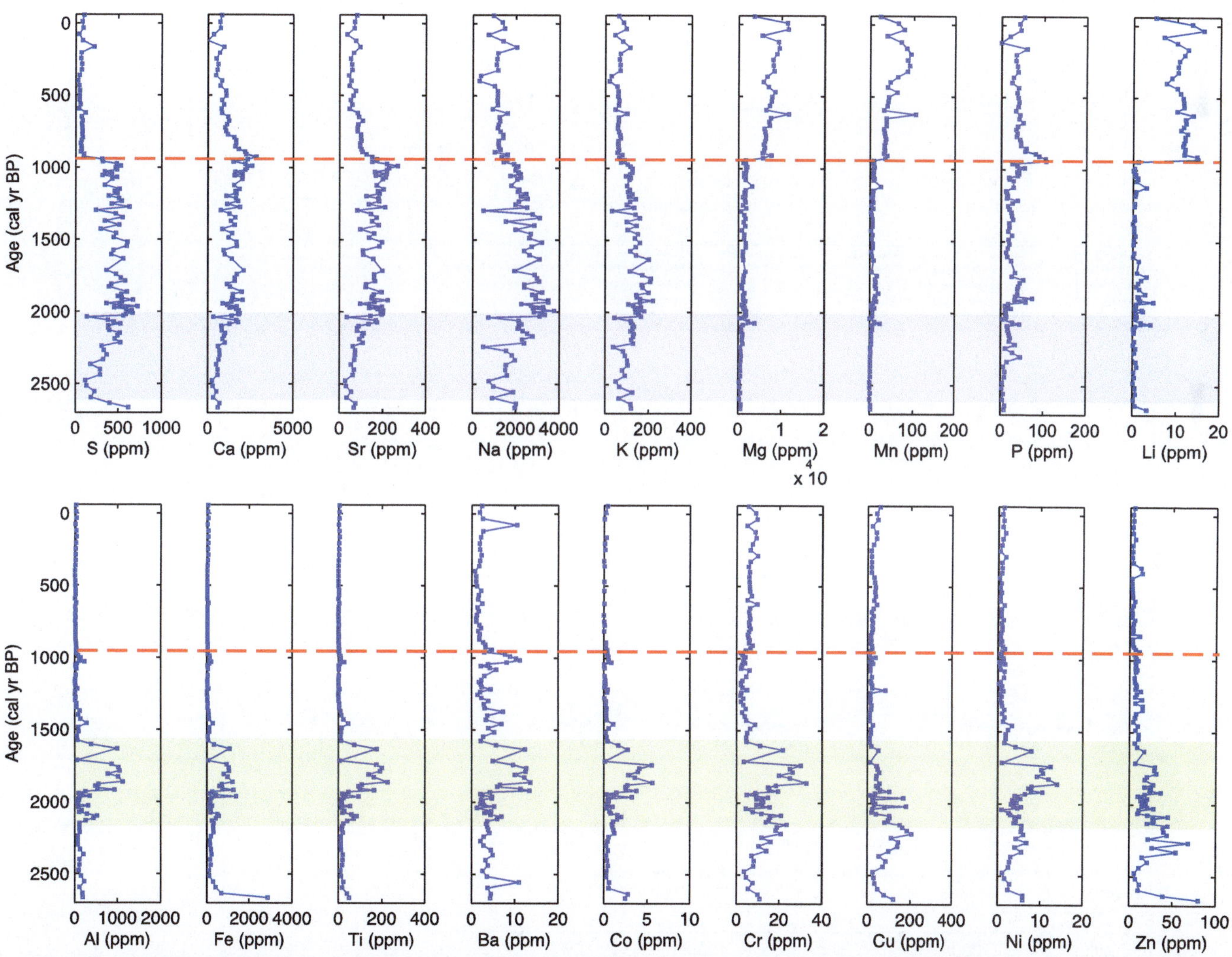

Figure 4. ICP-AES geochemical data from the Diablas lagoon core. Geochemical variations in 18 rare earth elements from the core from the Diablas lagoon. The dashed-red line denotes the major sediment transition identified in figure 2. Blue-shaded area (2700–1990 cal yr BP) indicates the point at which S and associated elements increased, and may indicate increasing tidal influence at the site. Green-shaded area denotes the period (2100–1600 cal yr BP) when geochemical in-wash element indicators (Al, Ti and Fe) were found in higher concentrations in the lagoon and are thought to be indicative of tidal disturbances.

evaporated onto coverslips before being mounted on slides using Naphrax®. Between 300 and 500 valves of all diatom species present were counted.

Glacial Isostatic Adjustment Model

Shorelines in the Galápagos have varied on glacial-interglacial timescales, with sea levels approximately 120 m lower during the Last-Glacial Maximum (LGM) due to an increased proportion of the global water budget locked in terrestrial ice caps [41]. Melting of the ice caps results in changes in relative sea level which is not spatially uniform due to GIA [42]. Output from a GIA model was used to determine the past changes in RSL (Fig. 5). The model has two key components: a model of global ice distribution and a model of Earth properties (e.g. density, elasticity and viscosity). The results shown in Fig. 5 are based on a recently developed ice model [43] and a suite of seven Earth models chosen to sample a range of values for three different parameters often inferred in GIA modelling studies: lithospheric thickness (the thickness of an elastic layer at the model Earth surface), viscosity in the upper mantle (the base of lithosphere to seismic velocity jump at 660 km depth) and viscosity in the lower mantle (660 km to core-mantle boundary). The range of values considered for these parameters were, respectively: 71, 96 & 120 km; 0.1, 0.5 & 1×10^{21} Pas; 1, 10 & 50×10^{21} Pas. These ranges accommodate the majority of plausible values for most locations.

Climate

Climatic conditions in the Galápagos are driven by variability in the El Niño Southern Oscillation (ENSO), which results in a close correlation in weather patterns across different lowland sites (Fig. 5). Data from a high-resolution (~25 years) palaeoclimate record, based on sediment grain size data as a measure of erosional in-wash from a sedimentary sequence from El Junco crater lake, San Cristobal [19] was plotted against the proxy data obtained in this study.

Results

Core stratigraphy

The core is separated into two major stratigraphic sections, an upper section (top-145 cm) of composed of pink, gelatinous sediment which is indicative of a microbial mat (Fig. 2). This top section is homogenous in character, but is mottled with some smaller (approx 1–5 mm) black (organic) detritus and white inorganic particles (<1 mm) which probably are likely to represent gypsum precipitates, commonly observed in shallow water, saline sediments in arid locations [44]. The basal section (145–490 cm) is composed of a dark, organic peat sediment with macroscopic plant remains, roots and shoots, which likely represents a mangrove peat commonly found to accumulate in tropical lagoon locations [45]. This section can be further divided at 323 cm, with fewer large (>10 mm) plant remains in zone 2b at the base. Throughout section 2 there is limited stratigraphic evidence of inorganic material.

Core chronology

Radiocarbon dating results are provided in Table 1, and indicate that the sedimentary sequence spans the last 2,700 years

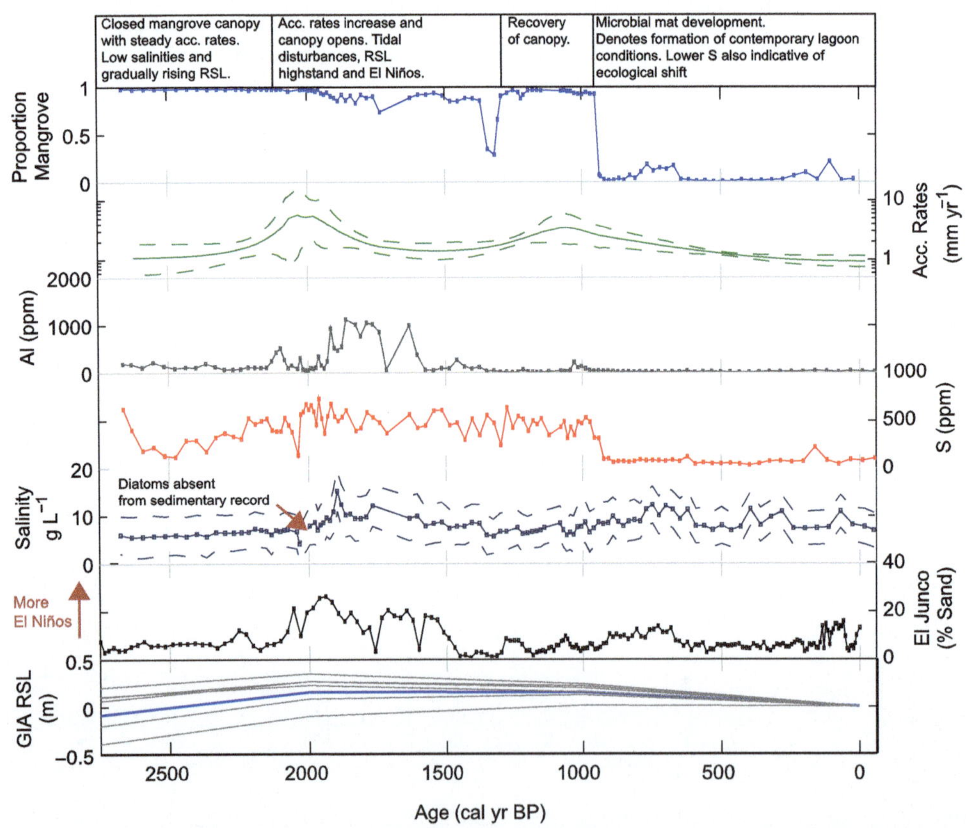

Figure 5. Reconstruction of ecological changes and environmental changes at Diablas lagoon for the past 2,700 years. Note that for the RSL data, seven different modelling parameters are provided (see methods section). The medium parameter model is marked by the thick-blue line. Grain size data is from El Junco crater-lake, San Cristobal (see ref. 19).

(Fig. 2). Between 2670–2200, 1650–1200, and 800 cal yr BP to present, sedimentation rates were low and steady (approximately 1–2 mm yr^{-1}), but the record is punctuated with periods of higher sedimentation rates, peaking at 2050 (5.7 mm yr^{-1}) and 1060 cal yr BP (3.4 mm yr^{-1}). No hiatuses are apparent in the sedimentary record indicating continuous deposition of organic matter through time. In general, these sedimentation rates are in line with other late Holocene mangrove sequences, which accrete between 1.5–2 mm yr^{-1} [45].

Carbon isotope ratios

There is a transition from low (approx − 27‰) to high (approx − 12‰) $\delta^{13}C$ at 940 cal yr BP (Fig 2a). Between 2670 and 940 cal yr BP the core has a low $\delta^{13}C$ and a higher proportion of organic carbon, but between 1960–1270 cal yr BP, deviations towards higher $\delta^{13}C$ occur. The stable isotope mixing model (Fig. 3) indicates four key periods of ecological change at the coring site over the past 2,700 years: (I) a stable mangrove ecosystem between 2670–1960 cal yr BP; (II) slow and steady benthic microbial mat development between 1960–1270 cal yr BP; (III) mangrove ecosystem recovery between 1270–940 cal yr BP; and (IV) an abrupt ecological transition from mangroves to microbial mat (940 cal yr BP), which was dominant for the remainder of the record.

Geochemistry

The major stratigraphic and isotopic changes observed above are easily identifiable in the geochemical signature of the Diablas core (Fig. 4). Na, S, Ca and Sr began to increase in concentrations from 2550 cal yr BP and remained high until 940 cal yr BP (Fig. 4). The enrichment of S and Ca may be due formation of early diagenetic mineral phases in the form of sulphides and carbonates, which have been donated to the sediment by the sea-water and therefore indicate the increasing influence of the tide [46]. Seawater is also enriched in Sr with respect to freshwater, which may explain the coincidence of these elements at this time [47]. The peak in S at 1990 cal yr BP is important, since it coincides with the higher salinity period in the lagoon inferred from the diatom record (see below). S is in high abundance throughout the mangrove peat section of the core since it commonly attaches itself to organic material under anoxic conditions [48].

Mg, Mn, and Li were found in higher abundances at the top of the sequence and were therefore associated with the microbial mat. The reduction of Mn is the first decompositional process used to gain energy from organic matter in anaerobic sediments [31], and, as a result, its elemental concentrations tend to be lower when conditions are most reducing. Conversely, enhanced Mn profiles are common in oxic sediments [31]. Since the microbial mat sediments donate oxygen to the sediment surface, it is likely that the switch from low to high Mn at 940 cal yr BP reflects change in the redox conditions at the site, which are associated with a switch from anoxic to oxidized sediments following the transition of a mangrove derived sediment to a microbial mat.

Between 2020–1600 cal yr BP, there were peaks of Fe, Al, and Ti in the sedimentary sequence (Fig. 4). Covariation of these three elements is unlikely to have resulted from diagenetic effects; instead, these elements are generally assumed to be minerogenic in origin due to erosional processes in the surrounding basin [49]. As a result, these elements are therefore likely to represent either high-energy, above ground, tidal surges, or erosional in-wash due to resulting from the period of stronger, more frequent El Niño events [19].

Salinity

The core is generally composed of benthic brackish water diatoms and experienced a transition of species from the base of the record to the present day (Fig. S2). *Fragilaria* cf. *subsalina* (Grunow) Lange-Bertalot and *Pseudostaurosiropsis geocollegarum* (Witkowski & Lange-Bertalot) Morales were dominant at the base of the record, and indicated stable salinities (approx 3–5 g L^{-1}) for 800 years (Fig. 5). Diatom concentrations suddenly dropped to zero in three samples between 2020–1990 cal yr BP, and were replaced by *Achnanthes submarina* Hustedt. This was the dominant taxon in the lagoon between 1990–1600 cal yr BP, where there was peak in lagoon salinity at 15.1±4 g L^{-1}. A number of marine diatoms (e.g. *Paralia sulcata* (Ehrenberg) Cleve) were also observed in low abundances (<5%, not included in Fig. S2) in the diatom record at this time. Smaller increases in salinity occurred between 800–600 and 400–300 cal yr BP (Fig. 5) and are indicated by the colonization of *Nitzschia palea* (Kützing) W.Smith, followed by *Navicula galapagoensis* Seddon & Witkowski, and *Amphora acutiuscula* Kützing and *A. caroliniana* Giffen. Salinities have varied between 7–11 g L^{-1} for the last 1,100 years.

Relative sea level

Output from the GIA model indicates that sea level increased steadily at our study location, and reached a small highstand of between 0.1–0.4 m between 2000 and 1000 cal yr BP. The predicted amplitude of the highstand is dependent on the adopted Earth viscosity model. This pattern of RSL change was observed in all but one of the model runs (that with a relatively high lower mantle viscosity of 5×10^{22}1Pas), which showed a steadily rising RSL for the past 3,000 years. The predicted change from a RSL rise to fall is due to the reduction in (predominantly Antarctic) ice melt rate during the late Holocene [50]. As the melt rate decreases, the contribution from isostatic processes becomes dominant, which leads to a RSL fall at this locality. The rate of RSL fall is sensitive to the adopted mantle viscosity values. In general, the greater the viscosity, the slower the rate of deformation and the slower the predicted RSL fall, hence providing a possible explanation for the monotonic RSL rise predicted for the Earth model with a relatively high value of lower mantle viscosity.

Discussion

Chronological palaeoenvironmental reconstruction

Sedimentation at the coring site in Diablas began at 2,670 cal yr BP and is probably linked to the fact that relative sea levels were approaching their highest since the Last Glacial Maximum [51]. Salinities were stable, between 3–5 g L^{-1} with no marine diatoms present, but there may have been some indirect marine influences through groundwater infiltrating from below. Lithological evidence, combined with the high%C and stable isotope results, suggests that the base of the core was a mangrove peat comprised of refractory roots and leaf litter, which are typically found beneath the canopy of mangrove forests [52,53,54]. Thus, at this time, there was a dense mangrove canopy living at the coring surface which accreted at a rate of 1–2 mm yr^{-1} for the first 700 years (Fig. 5). This probably represents the first infilling of a basin as RSL rose (Fig. 6).

2020–1600 cal yr BP was an important period of environmental change at Diablas lagoon. Modelling results indicate that sea levels reached a maximum at around this time (between 0.15–0.4 m above present day levels), whilst a number of shorter term, higher energy events were also observed in the record and are the result of a breaching of a seaward barrier (Fig. 6). For example, no diatoms were found in the sediments between 2020–1990 cal yr BP, which likely indicates a high-energy disturbance event, following which salinities in the lagoon increased to 15 g L^{-1} in only 100 years. Elements such as Al, Fe and Ti were also high between 1960 and 1580 cal yr BP and, since marine diatoms were also found in low abundances in the diatom record at this time, may also be indicative of minerogenic deposition as a result of above ground, higher-energy tidal surges (Fig. 6). Since palaeoclimatic data from the nearby highland El Junco crater lake indicates that this was a period of stronger, more frequent El Niño events [19], ENSO variability may have acted as an additional driving force to these disturbances through low pressure ocean surges (Fig. 5). Alternatively, increases in Al, Fe and Ti may have been driven by ENSO variability directly via precipitation driven erosional inwash.

Despite the increasing exposure to tidal surges, mangroves remained the dominant source of biogenic accumulation at the coring site over this 400 year period. An interesting result is the fact that, at the same time that diatoms were removed from the record through tidal disturbance, and when salinities in the lagoon began to increase dramatically, mangrove accumulation rates had peaked at around 5.7 mm yr^{-1}. In this way, the mangroves were able to accrete vertically in line with the tidal inundations, a common response which encourages the stability of the canopy through time (see below).

In addition to this vertical accretion, between 1960–1270 cal yr BP, stable isotope evidence indicates a shift to more variable, heavier isotopes of carbon. Benthic microbial mats are common at the surface of scrub mangrove forests, and it is likely that this represents a reduction or opening of the canopy as following the higher-energy tidal inundations [11]. By 1540 cal yr BP, evidence suggests that the direct effects of the high-energy tidal surges had diminished, and the modelling evidence indicates that RSL was either stable, or had begun to decrease. Sediment accumulation had decreased to 1.4 mm yr^{-1}, and salinities to 7.8 g L^{-1} (Fig. 5). Therefore, we propose that, by then, a barrier had begun to

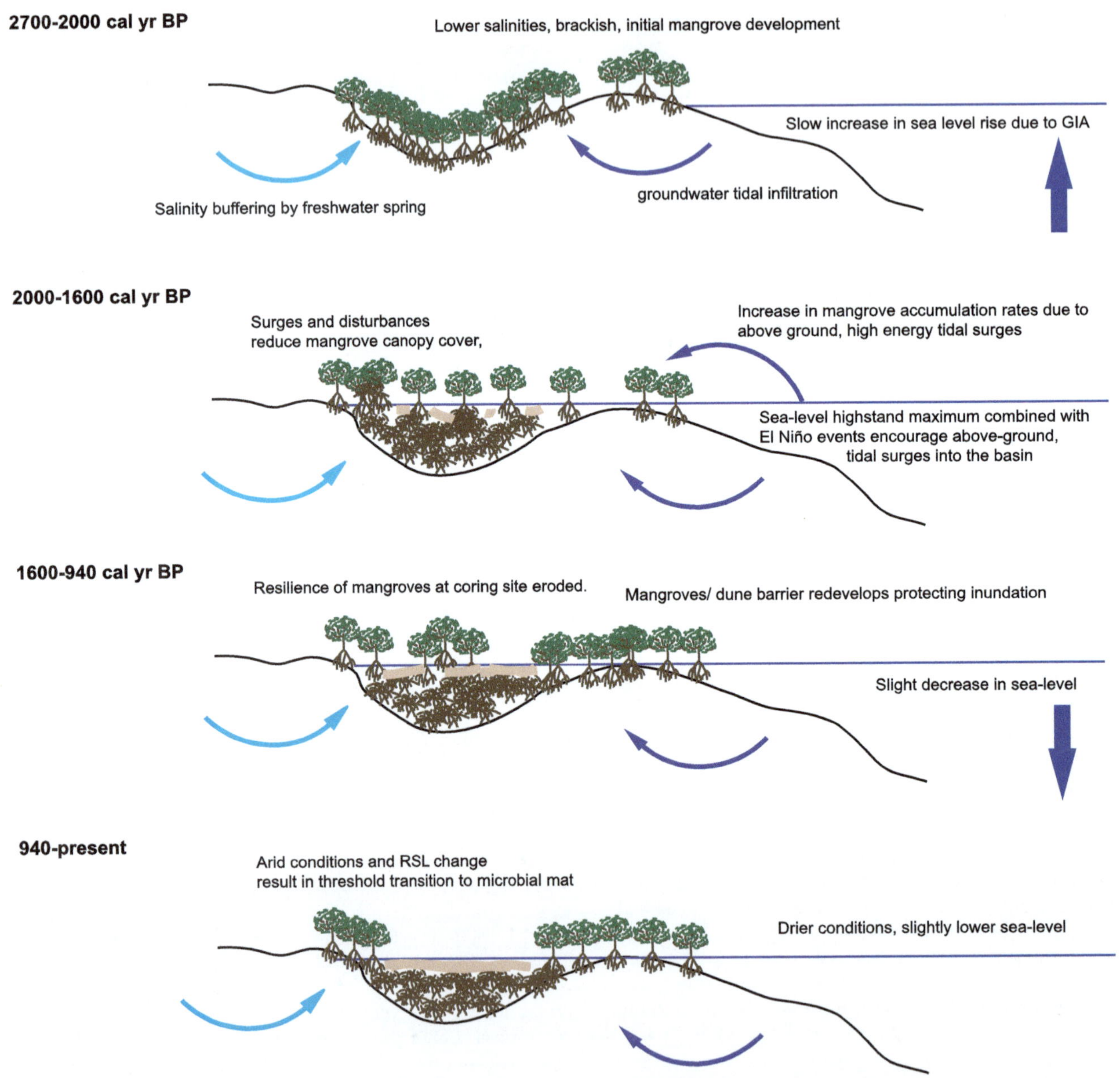

Figure 6. Hypothesised model of environmental changes inferred at Diablas lagoon over the Holocene.

develop at the sea-ward side of the coring site which protected the site from tidal inundation and eventually encouraged mangrove recovery (Fig. 6).

At 940 cal yr BP, there was a rapid ecological transition from mangroves to a microbial mat, as evidenced by both stable isotopic, geochemical, and stratigraphic evidence (Fig. 5). This shift indicates an expansion of the microbial mat dominated lagoonal sediments, and resulted in mangroves only being present at the lagoonal fringes (Fig. 6). Diatom evidence from El Junco crater lake indicates cooler sea-surface temperatures at this time, which would have resulted in drier conditions across the coasts of the Galápagos [55], although studies also indicate that this was a period of stronger, more frequent El Niño events [19,26]. We propose that the combination of drier background climatic conditions and a slightly declining RSL limited the potential for mangrove development within the lagoon basin, and drove the rapid transition to a microbial mat (Fig. 6).

What were the 'fast and slow' processes operating in the coastal zone?

The Diablas wetlands have been highly dynamic for the past 2,700 years experiencing significant environmental and ecological changes linked to GIA, high-energy tidal disturbance events and climatic change. These can be organized into a series of 'fast' (<50 years, the lowest average resolution available between three samples in our record) and 'slow' (>50 years, i.e. decadal-centennial, e.g. [15]) processes and are described briefly below.

Relative Sea Level. Taking the predictions from the GIA model with intermediate parameters, a sea-level highstand of around 0.15 m occurred at 2000 cal yr BP. Since, sea levels were predicted to have changed at an order of magnitude lower than the sedimentation rates observed in our record, the rapid sediment accumulation rates must be associated with other, long-term, geomorphological processes. It is likely that the 4.9 m accumulation resulted from the infilling of the basin as it experienced increasing tidal influence and RSL rise (Fig. 6), whilst the subsequent 'slow' fall in sea level (0.15–0.4 mm yr^{-1}) following the predicted highstand helped drive the transition from mangrove to microbial mat (Fig. 6).

Alternatively, some of the additional sediment accumulation may also have resulted from tectonic subsidence. This is common in volcanic archipelagos such as the Galápagos, where active islands to the west of the archipelago (Isabela) can experience local-regional scale changes in land elevation through gradual subsidence or sudden landslip events [56]. Here, subsidence of the land may have resulted in a more rapid rise in RSL compared to that predicted by the adopted GIA model (which does not account for tectonic processes). However, this possibility is difficult to test with our data.

Tidal Disturbance Events. The system was also punctuated by a series of higher-energy disturbance events associated with above-ground tidal surges (Fig. 5). At around 2000 cal yr BP relative sea levels were at their maximum, disturbances removed diatoms from the record, and lagoon salinities increased to 15 gL^{-1}. Geochemical data may then suggest that the lagoon was subjected to similar high-energy events for approximately the next 400 years. These 'fast' tidal disturbances were also coincident with a period of increased and more frequent El Niño events, a possible driving force behind the tidal surges at the study site (Fig. 6).

Precipitation changes. Palaeoclimatic evidence from El Junco crater lake from the highlands of San Cristobal indicate frequent changes in the strength and magnitude of ENSO over the past 2,700 years, resulting in changes in the precipitation regime on annual-decadal timescales [19]. In the coastal zone, short periods of heavy rainfall (drought) during El Niño (La Niña) events would have reduced (enhanced) lagoon salinities on annual-interannual timescales, but the EEP could also enter El Niño or La Niña phases for hundreds of years. It is likely that the cooler background conditions observed in the EEP between 1000–650 cal yr BP [55] caused increased aridity in the mangrove environment. The accumulation of these drier conditions over time would have increased the aridity stress on the mangroves in the basin, which combined with the slightly decreasing sea level, helped drive the transition from mangrove to microbial mat.

Is there evidence for a threshold response?

Non-linear responses to environmental change occur as a result of negative feedback processes which maintain a system in an equilibrium state. A key negative feedback identified in mangrove ecosystems is the accumulation of organic material in response to a tidal influx [11]. In this way, mangrove communities can accrete vertically in line with tidal inundation linked to rising sea levels or infilling of the lagoon basin. Here, we observed a faster rate of sediment accumulation occurring during a period of disturbances during a small sea level highstand, suggesting that the mechanism of vertical accretion in response to tidal inundation was occurring (Fig. 7). This process would have ensured the general stability of the mangrove canopy over time, maintaining mangroves as the dominant system at the coring site.

According to resilience theory, the accumulation of slow processes can result in an erosion of resilience over time, making a system more susceptible to smaller perturbations and environmental changes. We propose that the historical period of disturbances occurring after 2000 cal yr BP, which had the effect of the opening up the mangrove canopy, caused an erosion of resilience at our study site (Fig. 7). A facilitation effect has been observed in experimental studies on mangroves [10], in which a higher number of mangrove species per unit area encouraged long-term persistence. In an open canopy, the facilitation effect of mangroves would have been reduced [10]. This decrease in resilience would have made it more susceptible to smaller scale environmental perturbations.

Indeed, the transition to a microbial mat in Diablas occurred at a time of drier climatic conditions in the EEP [57], and was also coincident with an apparent decrease in RSL. Thus, the combination of a reduced canopy cover, combined with drier conditions and a slightly decreasing sea level, limited mangrove development in the basin. As a result, a microbial mat replaced mangroves at the coring site, and forced a threshold transition into an alternative stable state (Fig. 7).

How can the past inform us about the resilience of the modern system?

These findings have important consequences in our understanding of the ecological response of mangrove ecosystems to future environmental change. Future predictions on the rates RSL rise linked to climate change are uncertain, but the central estimate of the rate of rise at the end of the 21st century is 3.8 mm yr^{-1} [3]. In mangrove ecosystems which appear to experience low inputs of mineral sediment, organic matter accumulation is key in maintaining their persistence and stability. Studies in Belize and Florida showed that the relative contribution of organic solids to surface accretion was one to three times that of inorganic solids and also indicated biological controls on vertical accretion [58]. Therefore, through the vertical processes described above, mangrove ecological communities were able to accrete at rates comparable to locations with high mineral sedimentation in order to cope with a sea-level rise. At the Diablas wetlands, our data indicates that vertical accretion was a major response to the increased incidence of tidal disturbances, occurring at rates of up to 5.7 mm yr^{-1}.

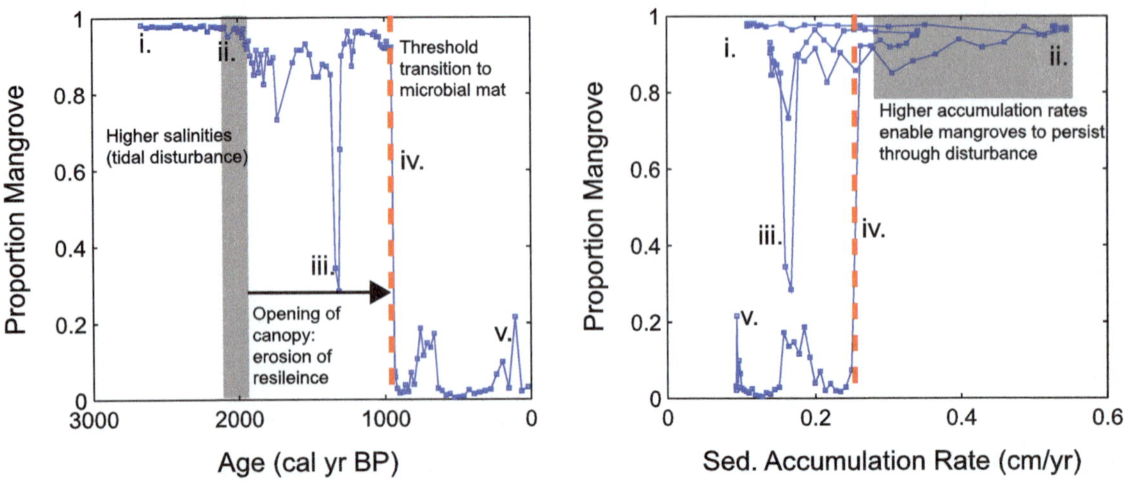

Figure 7. Threshold ecological response at Diablas lagoon. A) Stable isotope mixing model results indicating the main ecological changes at the Diablas lagoon. B) Phase plot to indicate the presence of alternative stable states (mangroves and microbial mat) in the lagoon system. The point at which the threshold between the two states was crossed is marked by a red dashed line. Detailed chronological summary is as follows. (i) 2770 cal yr BP- resilient mangrove stand at the coring site; (ii) 2000 cal yr BP- increase in sediment accumulation rates denotes feedback response of resilient mangroves in response to environmental disturbances; (iii) 1400 cal yr BP- opening of the mangrove canopy due to preceding tidal disturbance events, eroding resilience over time; (iv) (940 cal yr BP) Drier background climatic conditions and slightly decreasing RSL combine, resulting in threshold transition to microbial mat; (v) present day; contemporary lagoon conditions with cyanobacterial mat at the sediment surface.

However, in our study the resilience of mangroves was eroded by a series of historical processes and, as a result, a combination of arid conditions and slow RSL change eventually resulted in a threshold transition to an alternative stable state. Thus, our study implies that in this system, understanding the patterns of past historical processes are essential for determining baselines and for investigating ecological responses. The study highlights the need to incorporate a long-term ecological perspective when designing strategies for maximizing coastal resilience.

Conclusions

Palaeoecology is increasingly being applied as a tool to identify long-term baselines and thresholds in ecosystems at the local-regional scale [59]. In this study, we used palaeoecology to show that tidal inundation, and disturbance can be tolerated by Galápagos coastal communities through complex feedback responses linked to vertical accretion. Sedimentation rates of up to 5.7 mm yr^{-1} were observed in the Diablas ecosystem, and planning procedures which aim to encourage mangrove resilience are essential to ensure the long term provisions of ecosystem services from mangroves in the tropical coastal zone. Paleoecological records can provide important information on the nature of non-linear behaviour, by identifying baselines and thresholds within ecological systems, and in outlining key ecological responses to 'fast' and 'slow' environmental change between alternative stable states.

Supporting Information

Figure S1 Weighted average (WA) diatom transfer function. Plots to indicate the predictive power and residuals from the WA diatom salinity transfer function used in this study.

Figure S2 Diatom stratigraphy and salinity transfer function results. Diatom stratigraphy diagram of key species from both sites and diatom salinity transfer function results. Note that Diablas lagoon was mainly composed of brackish species. As a result, salinity was never greater than approximately 15 g L^{-1}.

Text S1 Weighted average (WA) diatom transfer function. Description and evaluation of the diatom salinity transfer function used in this study.

Acknowledgments

The authors thank Simon Haberle, Iona Flett, Jim Neale, Henk Heijnis, Iain Robertson, Emily Coffey, Benson Schleisser, and Salome Maldonado for their tireless efforts in the field. We thank Keith Bennett for providing the funding for 5 of the 8 radiocarbon dates. The palaeoclimate data was donated by Jessica Conroy (University of Arizona). Thanks also to Kaarina Weckström (University of Copenhagen) for supplying the MOLTEN surface data for the diatom transfer function and Maarten Blaauw (University of Belfast) for modifying the code in his CLAM software in order to estimate sedimentation rates. ICP-AES analyses were performed at the NERC ICP facility based at Royal Holloway University of London. This is publication number 2039 from the Charles Darwin Research Station, Puerto Ayora, Galápagos.

Author Contributions

Conceived and designed the experiments: AWRS CAF MJL KJW. Performed the experiments: AWRS MJL. Analyzed the data: AWRS. Contributed reagents/materials/analysis tools: AWRS CAF MJL GAM. Wrote the paper: AWRS GAM.

References

1. Agardi T, Alder J, Dayton P, Curran S, Kitchingman A, et al. (2005) Coastal Systems. In: Hassan R, Scholes R, Ash N, eds. The Millenium Ecosystem Assessment. Washington, Covelo, London: Island Press. pp 513–550.
2. Alongi DM (2008) Mangrove forests: Resilience, protection from tsunamis, and responses to global climate change. Estuarine Coastal and Shelf Science 76: 1–13.

3. Meehl GA, Stocker TF, Collins WD, Friedlingstein P, Gaye AT, et al. (2007) Chapter 10: Global Climate Projections. In: Solomon SD, Manning Q, M, Chen Z, Marquis M, Averyt KB, et al., editors. Contribution of Working Group I to the Fourth Assessment Report of the Intergovernmental Panel on Climate Change, 2007. Cambrige, UK: Cambridge University Press. pp 748–845.
4. Gilman EL, Ellison J, Duke NC, Field C (2008) Threats to mangroves from climate change and adaptation options: A review. Aquatic Botany 89: 237–250.
5. Ewel KC, Twilley RR, Ong JE (1998) Different kinds of mangrove forests provide different goods and services. Global Ecology and Biogeography 7: 83–94.
6. Adame MF, Neil D, Wright SF, Lovelock CE (2010) Sedimentation within and among mangrove forests along a gradient of geomorphological settings. Estuarine Coastal and Shelf Science 86: 21–30.
7. Rogers K, Saintilan N (2008) Relationships between surface elevation and groundwater in mangrove forests of southeast Australia. Journal of Coastal Research 24: 63–69.
8. Rogers K, Wilton KM, Saintilan N (2006) Vegetation change and surface elevation dynamics in estuarine wetlands of southeast Australia. Estuarine Coastal and Shelf Science 66: 559–569.
9. Feagin RA, Mukherjee N, Shanker K, Baird AH, Cinner J, et al. (2010) Shelter from the storm? Use and misuse of coastal vegetation bioshields for managing natural disasters. Conservation Letters 3: 1–11.
10. Huxham M, Kumara MP, Jayatissa LP, Krauss KW, Kairo J, et al. (2010) Intra- and interspecific facilitation in mangroves may increase resilience to climate change threats. Philosophical Transactions of the Royal Society B-Biological Sciences 365: 2127–2135.
11. Feller IC, Lovelock CE, Berger U, McKee KL, Joye SB, et al. (2010) Biocomplexity in Mangrove Ecosystems. Annual Review of Marine Science 2: 395–417.
12. ResilienceAlliance http://www.resalliance.org/576.php.
13. Holling CS (1986) The resilience of terrestrial ecosystems; local surprise and global change. In: Clark WC, Munn RE, eds. Sustainable development of the biosphere. Cambridge: Cambridge University Press. pp 292–317.
14. Holling CS, Gunderson L, Peterson GD (2002) Sustainability and panarchies. In: Gunderson L, Holling CS, eds. Panarchy. Washington: Island Press. pp 63–102.
15. Dearing JA (2008) Landscape change and resilience theory: a palaeoenvironmental assessment from Yunnan, SW China. Holocene 18: 117–127.
16. Scheffer M, Carpenter SR (2003) Catastrophic regime shifts in ecosystems: linking theory to observation. Trends in Ecology & Evolution 18: 648–656.
17. Rockström J, Steffen W, Noone K, Persson Å, Chapin FS, et al. (2009) Planetary Boundaries: Exploring the Safe Operating Space for Humanity. Ecology and Society 14.
18. Scheffer M, Bascompte J, Brock WA, Brovkin V, Carpenter SR, et al. (2009) Early-warning signals for critical transitions. Nature 461: 53–59.
19. Conroy JL, Overpeck JT, Cole JE, Shanahan TM, Steinitz-Kannan M (2008) Holocene changes in eastern tropical Pacific climate inferred from a Galapagos lake sediment record. Quaternary Science Reviews 27: 1166–1180.
20. Gelin A, Gravez V (2002) Knowledge and Conservation of Galapagos Islands Coastal Lagoons (Ecuador). Basler Stiftung fur Biologische Forschung, Charles Darwin Biological Station and Galapagos National Park. 94 p.
21. Snell H, Rea S (1999) The 1997-98 El Nino in Galapagos: Can 34 years of data estimate 120 years of pattern. Noticas de Galapagos 60: 11–16.
22. Wright Jr HE, Mann DH, Glaser PH (1984) Piston corers for peat and lake sediments. Ecology 65: 657–659.
23. Blauuw M (2010) Methods and code for 'classical' age-modelling of radiocarbon sequences. Quaternary Geochronology 5: 512–518.
24. McCormac FG, Hogg AG, Blackwell PG, Buck CE, Higham TFG, et al. (2004) SHCal04 Southern Hemisphere calibration, 0-11.0 cal kyr BP. Radiocarbon 46: 1087–1092.
25. R Development Core Team (2010) R: A Language and Environment for Statistical Computing, R Foundation for Statistical Computing http://www.R-project.org Vienna, Austria.
26. Sachs JP, Sachse D, Smittenberg RH, Zhang Z, Battisti DS, et al. (2009) Southward movement of the Pacific intertropical convergence zone AD 1400–1850. Nature Geoscience 2: 519–525.
27. Mackie EAV, Leng MJ, Lloyd JM, Arrowsmith C (2005) Bulk organic $\delta^{13}C$ and C/N ratios as palaeosalinity indicators within a Scottish isolation basin. Journal of Quaternary Science 20: 303–312.
28. Phillips DL, Gregg JW (2003) Source partitioning using stable isotopes: coping with too many sources. Oecologia 136: 261–269.
29. Phillips DL, Newsome SD, Gregg JW (2005) Combining sources in stable isotope mixing models: alternative methods. Oecologia 144: 520–527.
30. McCarrol D, Loader NJ (2004) Stable isotopes in tree rings. Quaternary Science Reviews 23: 771–801.
31. Engstrom DR, Wright HE (1984) Chemical stratigraphy in lake sediments as a record of environmental change; Haworth EY, Lund JWG, editors. Leicester, UK: Leicester University Press.
32. Boyle JF (2001) Inorganic geochemical methods in palaeolimnology. In: Last WM, Smol JP, eds. Tracking environmental change using lake sediments Physical and Geochemical Methods. Dordrecht, The Netherlands: Kluwer Adademic Publishers.
33. López-Buendia AM, Bastida J, Querol X, Whateley MKG (1999) Geochemical data as indicators of palaeosalinity in coastal organic-rich sediments. Chemical Geology 157: 235–254.
34. Freitas MC, Andrade C, Cruces A (2002) The geological record of environmental changes in southwestern Portuguese coastal lagoons since the Lateglacial. Quaternary International 93-4: 161–170.
35. Virah-Sawmy M, Willis KJ, Gillson L (2009) Threshold response of Madagascar's littoral forest to sea-level rise. Global Ecology and Biogeography 18: 98–110.
36. Bengtsson L, Enell M (1986) Chemical analysis. In: Berglund BE, ed. Handbook of Holocene palaeoecology and palaeohydrology. Chichester, UK: Wiley. pp 423–451.
37. Birks HJB, Line JM, Juggins S, Stevenson AC, ter Braak CJF (1990) Diatoms and pH reconstruction. Philosophical Transactions of the Royal Society of London, B 327: 263–278.
38. Weckström K, Juggins S (2006) Coastal diatom-environment relationships from the Gulf of Finland, Baltic Sea. Journal of Phycology 42: 21–35.
39. Juggins S (2009) Rioja: Analysis of Quaternary Science Data Package for R: http://www.staff.ncl.ac.uk/staff/stephen.juggins/.
40. Battarbee RW, Jones VJ, Flower RJ, Cameron NG, Bennion H, et al. (2001) Diatoms. In: Smol JP, Birks HJB, Last WM, eds. Tracking Environmental Change Using Lake Sediments. Dordrecht, The Netherlands: Kluwer Academic Publishers. pp 155–202.
41. Siddal M, Rohling EJ, Almogi-Labin A, Melschner D, Schmelzer I, et al. (2003) Sea-level fluctuations during the last glacial cycle. Nature 423: 853–858.
42. Clarke JA, Farrell WE, Peltier WR (1978) Global changes in post glacial sea level: a numerical calculation. Quaternary Research 9: 265–287.
43. Bradley SL, Milne GA, Zhong Y-Q, Horton BP (submitted) Modelling sea-level data from China and Malay-Thailand to develop a new model of Holocene eustatic sea level change. Quaternary Science Reviews).
44. Bauld J (1981) 8. Occurrence of benthic microbial mats in saline lakes. Hydrobiologia 81-82: 87–111.
45. Ellison JC, Stoddart DR (1991) Mangrove ecosystem collapse during predicted sea-level rise - Holocene analogs and implications. Journal of Coastal Research 7: 151–165.
46. Ramirez-Herrera MT, Cundy A, Kostoglodov V, Carranza-Edwards A, Morales E, et al. (2007) Sedimentary record of late-Holocene relative sea-level change and tectonic deformation from the Guerrero Seismic Gap, Mexican Pacific Coast. Holocene 17: 1211–1220.
47. Shen C-C, Liu K-K, Lee M-Y, Lee T, Wang C-H, et al. (2005) A novel method for tracing coastal water masses using Sr/Ca ratios and salinity in Nanwan Bay, southern Taiwan. Estuarine Coastal and Shelf Science 65: 135–142.
48. Lewis JT (1997) The record of deposition and the migration of elements in salt marshes. Southampton: University of Southampton.
49. Shanahan TM, Overpeck JT, Anchukaitis KJ, Beck JW, Cole JE, et al. (2009) Atlantic Forcing of Persistent Drought in West Africa. Science 324: 377–380.
50. Nakada M, Lambeck K (1988) The melting history of the late Pleistocene Antarctic ice sheet. Nature 33: 36–40.
51. Milne GA, Gehrels WR, Hughes CW, Tamisiea ME (2009) Identifying the causes of sea-level change. Nature Geoscience 2: 471–478.
52. Parkinson RW, de Laune RD, White JR (1994) Holocene sea-level rise and the fate of mangrove forests within the wider Caribbean region. Journal of Coastal Research 10: 1077–1086.
53. McKee KL, Faulkner PL (2000) Mangrove peat analysis and reconstruction of vegetation history at the Pelican Cays, Belize. Atoll Research Bulletin 468: 46–58.
54. McKee KL, Cahoon DR, Feller IC (2007) Caribbean mangroves adjust to rising sea level through biotic controls on change in soil elevation. Global Ecology and Biogeography 16: 545–556.
55. Conroy JL, Restrepo J, Overpeck JT, Steinitz-Kannan M, Cole JE, et al. (2009) Unprecedented recent warming of surface temperatures in the eastern tropical Pacific Ocean. Nature Geoscience 2: 46–50.
56. Geist D (1996) On the emergence and submergence of the Galapagos islands. Noticas de Galapagos 56: 5–9.
57. Conroy JL, Overpeck JT, Cole JE, Steinitz-Kannan M (2009) Variable oceanic influences on western North American drought over the last 1200 years. Geophysical Research Letters 36 L17703: 17701–17706.
58. McKee KL (2011) Biophysical controls on accretion and elevation change in Caribbean mangrove ecosystems. Estuarine Coastal and Shelf Science 91: 475–483.
59. Willis KJ, Bailey RM, Bhagwat SA, Birks HJB (2010) Biodiversity baselines, thresholds and resilience: testing predictions and assumptions using palaeoecological data. Trends in Ecology & Evolution 25(10): 583–591.

2

Polychaete Richness and Abundance Enhanced in Anthropogenically Modified Estuaries Despite High Concentrations of Toxic Contaminants

Katherine A. Dafforn[1,2]*, Brendan P. Kelaher[3], Stuart L. Simpson[4], Melinda A. Coleman[5], Pat A. Hutchings[6], Graeme F. Clark[1,2], Nathan A. Knott[5], Martina A. Doblin[2,7], Emma L. Johnston[1,2]

1 Evolution and Ecology Research Centre, School of Biological, Earth and Environmental Sciences, University of New South Wales, Sydney, NSW, Australia, **2** Sydney Institute of Marine Science, Mosman, NSW, Australia, **3** National Marine Science Centre, Southern Cross University, Coffs Harbour, NSW, Australia, **4** Commonwealth Scientific and Industrial Research Organisation, Land and Water, Kirrawee, NSW, Australia, **5** Aquatic Ecosystem Research, New South Wales Department of Primary Industries, Nelson Bay, NSW, Australia, **6** Marine Invertebrates, Australian Museum, Sydney, NSW, Australia, **7** University of Technology Sydney, Broadway, NSW, Australia

Abstract

Ecological communities are increasingly exposed to multiple chemical and physical stressors, but distinguishing anthropogenic impacts from other environmental drivers remains challenging. Rarely are multiple stressors investigated in replicated studies over large spatial scales (>1000 kms) or supported with manipulations that are necessary to interpret ecological patterns. We measured the composition of sediment infaunal communities in relation to anthropogenic and natural stressors at multiple sites within seven estuaries. We observed increases in the richness and abundance of polychaete worms in heavily modified estuaries with severe metal contamination, but no changes in the diversity or abundance of other taxa. Estuaries in which toxic contaminants were elevated also showed evidence of organic enrichment. We hypothesised that the observed response of polychaetes was not a 'positive' response to toxic contamination or a reduction in biotic competition, but due to high levels of nutrients in heavily modified estuaries driving productivity in the water column and enriching the sediment over large spatial scales. We deployed defaunated field-collected sediments from the surveyed estuaries in a small scale experiment, but observed no effects of sediment characteristics (toxic or enriching). Furthermore, invertebrate recruitment instead reflected the low diversity and abundance observed during field surveys of this relatively 'pristine' estuary. This suggests that differences observed in the survey are not a direct consequence of sediment characteristics (even severe metal contamination) but are related to parameters that covary with estuary modification such as enhanced productivity from nutrient inputs and the diversity of the local species pool. This has implications for the interpretation of diversity measures in large-scale monitoring studies in which the observed patterns may be strongly influenced by many factors that covary with anthropogenic modification.

Editor: Martin Solan, University of Southampton, United Kingdom

Funding: This research was primarily supported by the Australian Research Council through an Australian Research Fellowship (DP0990640) awarded to Johnston and a Linkage Grant (LP0990640) awarded to Johnston, Kelaher and Coleman. The funders had no role in study design, data collection and analysis, decision to publish, or preparation of the manuscript.

Competing Interests: The authors have declared that no competing interests exist.

* E-mail: k.dafforn@unsw.edu.au

Introduction

The intense and extensive development by humans across the planet has subjected much of the world's biological diversity to multiple severe anthropogenic stressors [1,2,3] including frequent chemical and physical disturbances, which are often concentrated in urban and industrial areas [4]. Human activity is often concentrated around waterways and associated activities inevitably release contaminants into water bodies and result in other modifications to physico-chemical conditions. As a consequence estuaries in particular have been highly impacted by stressors related to agriculture, industrialisation and urbanisation with almost all estuaries suffering some degree of impact [5]. Apart from physical modification to these systems (e.g. addition of artificial structures), important chemical stressors include toxic contaminants (e.g. metals and hydrophobic organic chemicals such as polycyclic aromatic hydrocarbons (PAHs)) [6], enriching contaminants (e.g. nutrients) [7], and changes to natural conditions such as temperature and salinity that are outside the range of natural variation [8,9]. While much is understood of the effects of individual contaminants [10] and there have been limited comparisons of multiple stressors [7,11,12,13], little is known of whether these effects will extend over large-spatial scales and if the ecology of these systems may influence these effects (e.g. the diversity and resilience of the assemblages, and potential for adaptation over long periods of time).

Environmental monitoring studies often rely on ecological measures of community composition and diversity as indicators of anthropogenic change [14,15]. Such measures are strongly influenced by regional conditions and patterns may not always reflect impacts [16,17]. Recent efforts to monitor ecological impacts have focused on integrating information collected from chemical and ecological monitoring into a more holistic understanding of ecosystem condition [18,19,20]. How this information is collected remains fiercely debated and ranges from laboratory-based experiments, field-based manipulative experiments (generally small-scale), and small- and large-scale observational studies (either snapshot or examining change through time). These each have advantages and disadvantages [21,22], but to distinguish the effects of anthropogenic stressors from other environmental drivers [23] and inform management-oriented biomonitoring requires multiple lines of evidence and experimental studies in order to reliably interpret patterns of community composition and diversity [22,24,25].

Chemical stressors that are released into estuaries accumulate in benthic sediments [13,26] and impact the ecological composition and function of this important habitat [27]. When tested individually and under laboratory conditions, contaminants such as metals and PAHs are known to have toxic effects on aquatic taxa including increased mortality [7,11,12], reduced reproductive potential [13,28] and other sublethal effects [29]. Similarly, field surveys suggest that metal and organic chemical contamination are related to changes to community composition that can include the loss of sensitive species and the increased dominance of more tolerant taxa or individuals. A meta-analysis by [30] determined species richness was reduced by ~40% across contaminated marine communities and the majority of the studies reviewed were conducted in soft sediment communities using field surveys. Contaminants such as dissolved nitrogen and phosphorus also have community-wide effects, but high levels can result in a community composed of very high densities of a few tolerant opportunistic species [31]. In contrast to toxic contaminants, however, such nutrients initially have an enriching effect with the result being observed increases in richness and abundance of primary producers and this has consequences at higher trophic levels [32].

We tested whether impacts of large-scale anthropogenic stress in estuaries could be detected over and above natural spatial variation across regional scales (100 s of kms) using an observational approach to assessing the outcome of contamination on benthic diversity. We surveyed benthic sediments in three heavily modified and four relatively 'pristine' estuaries [33] and we sampled infaunal assemblages, measures of toxic contaminants (metal and PAH concentrations) and indicators of organic enrichment (sediment chlorophyll-a levels, porewater ammonia and total organic carbon) and sediment grain size. We also measured other environmental variables (e.g. temperature and salinity). To aid our understanding of why differences in surveyed infaunal community composition (dominated by polychaetes) occurred in relation to anthropogenic modification, we undertook manipulative field experiments. We redeployed sediments collected from the estuaries with varying degrees of anthropogenic impacts into a single clean estuary and assessed the community composition and diversity of recruiting fauna.

Materials and Methods

Ethics Statement

This study surveyed benthic invertebrate distributions across estuaries and experiments were conducted on invertebrate communities. Sampling in each location was approved and carried out in strict accordance with the New South Wales Department of Primary Industries and New South Wales Marine Parks (Permit No. P09/0072-1.0).

Field surveys of benthic sediments in multiple estuaries

We investigated variation in soft-sediment assemblages in relation to anthropogenic contaminants and other environmental variables at seven sites within seven estuaries. Estuaries were assigned a 'modification category' according to the level of anthropogenic modification and the study was conducted in three heavily modified and four relatively unmodified estuaries stretching across 400 km of the coast of New South Wales, Australia (Figure 1). In this study, heavily modified estuaries represented systems both physically modified and contaminant impacted by anthropogenic activities while relatively unmodified estuaries were systems lacking extensive urbanisation or industry. Port Kembla, Port Jackson and Botany Bay are highly urbanised estuaries with 80–100 years of industrialisation and urbanisation [34,35,36]. Port Hacking, The Clyde, Wagonga Inlet and Jervis Bay are estuaries that are relatively less modified by urbanisation and have no history of major industry. In addition, three of our relatively unmodified estuaries (The Clyde, Wagonga Inlet and Jervis Bay) are within Marine Protected Areas where trawling and commercial fishing are restricted [37]. Seven sites (between 1 and 2 km apart) were sampled in each estuary (Figure 1).

Benthic sediments were collected between February and March 2010. Two replicate sediment grabs and water quality measures were collected at each site from 5 m depth using a Van Veen grab to target 5 kg surficial sediments. Each sediment grab was homogenised in a clean tray and sub-sampled for infauna (250 mL) and measures of toxic contaminants (metal and PAH concentrations) (150 g), organic enrichment indicators (sediment Chl-a levels and organic carbon) (150 g) and sediment grain size (50 g). An additional two sediment grabs were taken from sites in 6 of the surveyed estuaries (Port Kembla, Port Jackson, Botany Bay, Port Hacking, Wagonga Inlet and The Clyde) for the experimental deployment (details below). Plasticware used in sediment collection was previously soaked in 5% HNO_3 for a minimum of 24 h and then rinsed in deionised water (Milli-Q^{TM}). All sediments for chemical analyses were kept in the dark on ice for transport to the laboratory and then frozen at −20°C until analysis.

Temperature, salinity, turbidity, pH, oxygen and Chl-a levels in the water column directly above sediment collection sites were measured using a multiprobe (YSI-Sonde 6600-v2, Yellow Springs, USA) calibrated according to the manufacturer's instructions.

Field experiment sediments deployed in benthic recruitment containers

We used benthic recruitment containers (10 cm diameter ×20 cm height Perspex cylinders) to experimentally investigate the contribution of sediment characteristics to observed patterns by deploying sediments collected during our field survey. Bulk surficial sediment was collected from 5 m depth at 7 sites with unvegetated sediments within 6 estuaries (Port Kembla, Port Jackson, Botany Bay, Port Hacking, Wagonga Inlet and The Clyde). The sediments collected from multiple sites within each estuary were combined and homogenised then sieved (2 mm) to remove shells and larger fauna and flora. This created one sediment homogenate to represent each estuary. The six homogenates represented different sediment characteristics. Homogenate sediments were then divided into 7 replicate subsamples (0.5 kg), which were applied as a top sediment layer (~2 cm) to replicate recruitment containers ($N = 42$). Approximately 1 kg of sand was used to line the bottom of each of the recruitment containers with the homogenate sediments layered on top. Sand was collected from a site in Port Jackson where the chemical composition had been previously characterised as uncontaminated [20]. Recruitment containers were frozen at −20°C for at least 3 mo to defaunate the sediments and remained frozen until deployment where they thawed in situ. For

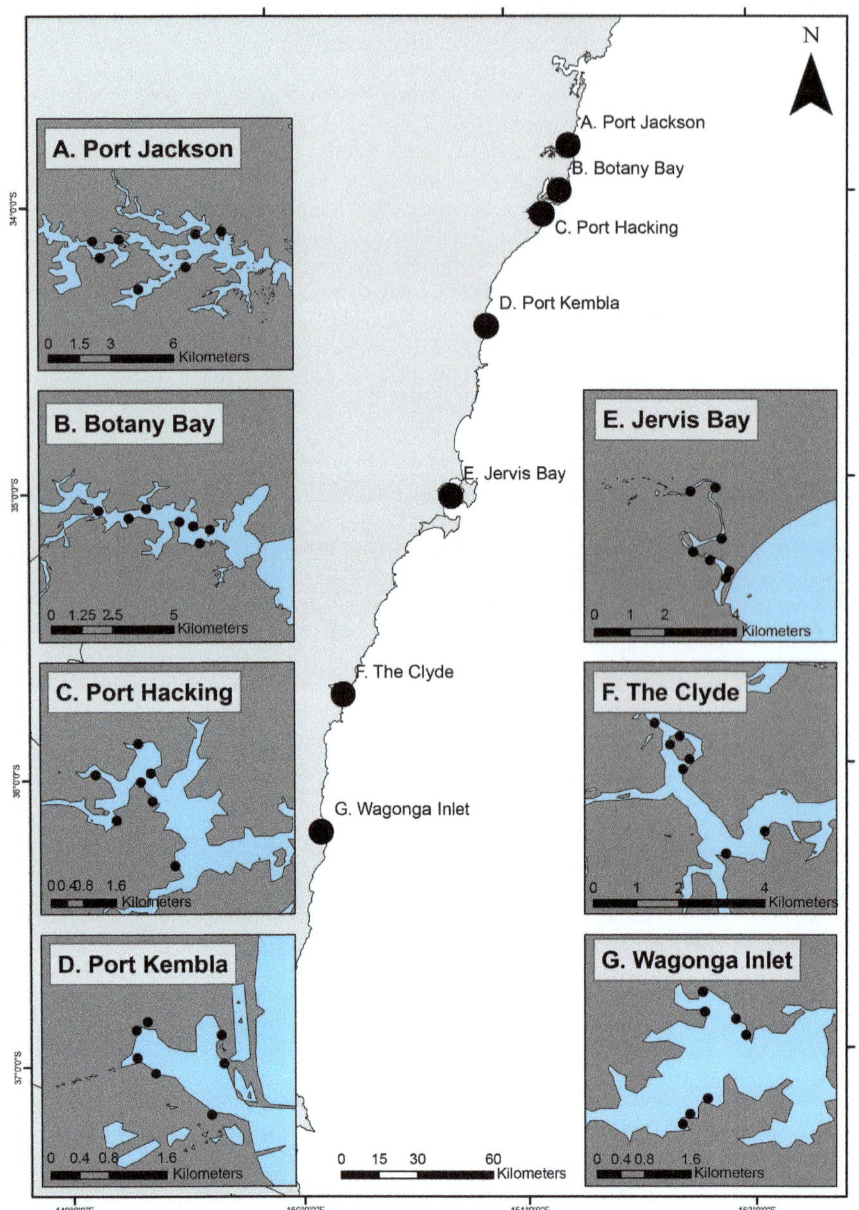

Figure 1. Locations of study sites along the coastline of New South Wales, south-east Australia. A. Port Jackson, B. Botany Bay and D. Port Kembla are heavily modified estuaries. C. Port Hacking, E. Jervis Bay, F. The Clyde and G. Wagonga Inlet are relatively unmodified estuaries.

deployment, the recruitment containers were attached to aluminium frames 2.1 m long and deployed at ~3 m depth (below mean low water spring) in a site previously surveyed within the upper catchment of the Clyde estuary (Figure 1) for 12 weeks (Nov 2010–Feb 2011). Recruitment containers were capped before collection. Surface sediments (~5 cm depth) from each replicate recruitment container were homogenised in a clean tray and sub-sampled for analyses of infauna, metals and grain size. All sediments for chemical analyses were kept in the dark on ice for transport to the laboratory and then frozen until analysis.

Sample processing and analysis

Infaunal sub-samples (125-ml) were stained with Rose Bengal and preserved in a 7% formalin solution then passed through a 2 mm mesh (to remove large debris) and onto a 500-µm sieve. The remaining organisms were sorted with a dissecting microscope and identified to the lowest feasible taxonomic level (mostly order). Polychaetes comprised >70% of all assemblages and were therefore identified to family and we used this higher taxonomic resolution to focus on patterns in polychaete family diversity and polychaete abundance in response to anthropogenic modification. A reference collection was deposited at the Australian Museum.

Sediment characteristics were investigated from benthic sediments collected at each site in the survey and analyses included concentrations of major metal and PAH contaminants, total organic carbon and Chl-a and sediment grain size according to the techniques described in [20]. Sediments deployed in recruitment containers were also analysed for metal contaminants and grain sizes at the start (homogenised mixture) and end (each replicate recruitment container) of the experiment.

Sediment metal analyses were undertaken using a microwave-assisted acid digestion according to Method 3051A [38]. The metal concentrations in acid-digests were analysed using inductively coupled plasma-mass spectrometry (ICP-MS; Perkin Elmer, Optima7300DV, USA). As part of the quality assurance, filter and acid-digest blanks, replicates for 20% of samples, analyte sample-spikes and the certified reference materials (CRMs) were measured. Replicates were within 20% and recoveries for spikes and CRMs, (LGC6137, Graham B. Jackson Pty Ltd, Australia National Research Council Canada, NRCC), were within 85–115% of expected values. Full details of recoveries are included in supplementary materials (Table S1). Metal contaminants were included in further statistical analysis if they exceeded the low trigger values from the interim sediment quality guidelines (SQGs) [39]. Cr, Cu, Ni, Pb, Zn exceeded SQGs in the field surveys and Cu, Pb, Zn in the field experiment.

Sixteen PAHs were analysed including naphthalene, acenaphthylene, acenaphthene, fluorene, phenanthrene, anthracene, fluofranthene, pyrene, benz(a)anthracene, chrysene, benzo(a)pyrene, benzo(b)fluoranthene, benzo(k)fluoranthene, indeno(1,2,3-cd)pyrene, dibenzo(a,h)anthracene and benzo(g,h,i)perylene. Analyses of PAHs in sediments followed Method 8260 [40]. Surrogate PAHs (deuterated internal standards; acenaphthene-d_{10}, phenanthrene-d_{10}, chrysene-d_{12} and perylene-d_{12}) were spiked into all samples and recoveries were $111 \pm 19\%$. Individual PAHs were summed for each sample to give a value of total PAHs [39] and this was used in further statistical analyses.

Sediment Chl-a analysis were undertaken following extractions of pigments in acetone with a spectrophotometer (Lambda 35 UV/Vis) according to Greenberg et al. [41]. Inorganic carbon in benthic sediments was removed by acidification with 2 mL of 1 M HCl overnight [42], and total organic carbon (TOC) was analysed using a Leco CN2000 analyser (Leco Corporation, USA) at a combustion temperature of 1050°C. Sediment grain size analyses were made by wet sieving through graded sieves; gravel (2 mm), sand (2 mm–63 μm), and fines (<63 μm). Samples were then oven dried (24 h at 60°C) and weighed to determine the percentage contribution of each fraction.

Porewaters were extracted from the sediments by centrifugation at 800 g for 5 min and then filtered (0.45 μm, Sartorius Minisart) immediately to minimise exposure to air. Samples were taken for ammonia (stored frozen until analyses). Dissolved ammonia was analysed colorimetrically using a Merck Spectroquant Kit (14752).

Data analysis

In all analyses 'Modification Category' (heavily modified or relatively unmodified) was treated as a fixed orthogonal factor and 'Estuary (within Modification Category)' was a random nested factor. Replicates in each analysis were either sites (survey) or recruitment containers (experiment). Infaunal community composition was compared between modification categories with permutational multivariate analysis of variance (MANOVA) and Principal Components Analysis (PCO). We also investigated changes in taxa richness and abundance between modification categories because these measures are closely linked to contaminants (both inorganic and organic) [31,43], and are robust to taxonomic changes when sampling across large spatial scales e.g. estuaries [44]. Also, many other benthic indices have been developed specifically for European monitoring [45] and require a higher level of taxonomic resolution than is often possible with Australian infauna. Polychaete worms comprised more than 50% of the assemblage on average and were the dominant taxon in all samples (Figure S1). Therefore we limited our univariate analysis of ecological measures (taxa richness and abundance) to polychaete families and polychaete individuals. Sediment characteristics (survey and experiment) and water quality variables (survey only) were also compared between modification categories using these models. Infaunal assemblage data included all taxa identified during the survey or experiment and fourth root transformed resemblance matrices were generated with the Bray Curtis dissimilarity index to reflect changes in relative abundances and composition of assemblages. Univariate measures of polychaete abundance and polychaete family richness were also fourth root transformed and environmental variables were normalised. Resemblance matrices for each dataset were then constructed from Euclidean distances. Permutational analysis of variance (ANOVA) of univariate data and permutational multivariate analysis of variance (MANOVA) were performed using type III sums of squares under the reduced model. Homogeneity of dispersion between groups was tested using PERMDISP and no significant dispersion was detected for 'Modification Category'. Full details of the infauna dataset and analyses are included in supplementary materials (Tables S2, S3, S4, S5, S6).

Principal components analysis (PCA) was used to visualize the entire surveyed infaunal assemblages at each site or in each recruitment container. All analyses were done in PRIMER v6 with PERMANOVA+ [46]. Univariate plots of metals and environmental variables are included in supplementary materials (Figures S2, S3, S4).

Results

Field surveys of sediments in multiple estuaries

Sediments more contaminated in heavily modified estuaries. Benthic sediments had higher metal (Cr, Cu, Ni, Pb and Zn) and PAH concentrations in heavily modified estuaries than relatively unmodified estuaries (Table 1, Figure S2). Many exceeded SQG trigger values (Cr: 80 μg/g; Cu: 65 μg/g; Ni: 21 μg/g; Pb: 50 μg/g; Zn: 200 μg/g) in the heavily modified estuaries and were often above upper limits (Pb: 220 μg/g; Zn: 410 μg/g), but were never above SQG trigger values in the relatively unmodified estuaries. Metal and PAH concentrations also varied among estuaries, with the exception of chromium (Table 1, Figure S2). Sediment chl-a, porewater ammonia and TOC were elevated in heavily modified estuaries suggesting significant inorganic and organic enrichment of these sediments (Table 1, Figure S2). Sediment grain size was finer for benthic sediment from heavily modified estuaries, indicating a greater capacity for contaminant binding and retention and turbidity levels were also highest in heavily modified estuaries (Table 1, Figure S2). Other water quality variables did not differ between modification categories, but varied significantly among estuaries, with the exception of temperature (Table 1, Figure S3).

Infaunal community composition differed in response to modification. The infaunal community composition differed significantly between heavily modified and relatively unmodified estuaries and also among individual estuaries (Table 2; Figure 2). A combination of the two axes of the unconstrained ordination (PCO) emphasised differences among disturbance categories, although a relatively small proportion (39.7%) of the total variation in the community composition was explained. Sites from heavily modified estuaries (positive scores on PCO1 and negative scores on PCO2) were characterised by greater abundances of individuals from the polychaete families Arabellidae, Spionidae, Nephytidae, Cirratulidae, Maldanidae and Capitellidae (Figure 2). Representatives of the family Syllidae were more abundant in relatively unmodified estuaries (negative scores on PCO1 and positive scores on PCO2).

Table 1. Summary results from permutational ANOVA of sediment characteristics and water quality collected in field surveys of benthic sediments.

Variable	Modification category	Estuary(Mo)
Sediment characteristics		
Cr	*	N.S.
Cu	*	**
Ni	*	*
Pb	*	**
Zn	*	**
Total PAHs	**	**
Sediment Chl-a	**	N.S.
Porewater ammonia	**	N.S.
Total organic carbon	N.S.	**
Percent fines	*	N.S.
Water quality		
Turbidity	*	**
Salinity	N.S.	**
pH	N.S.	*
Dissolved oxygen	N.S.	*
Water Chl-a	N.S.	*
Temperature	N.S.	N.S.

Factors include Modification category (Mo; heavily modified or relatively unmodified) and Estuary (Es; nested in Modification category). Sites were the replicates (n = 7).
* = $p<0.05$,
** = $p<0.01$ and N.S. = not significant.

Table 2. Results of permutational MANOVA of (a) the entire infaunal community, and univariate ANOVA of (b) polychaete family richness and (c) polychaete abundance from field surveys of benthic sediments.

Source	df	SS	MS	Pseudo-F	P(perm)
		(a) Infaunal community composition			
Modification category	1	10341	10341	2.04	**0.018**
Estuary (Mo)	5	25372	5074	4.04	**0.000**
Res	39	48951	1255		
		(b) Polychaete family richness			
Modification category	1	391.69	391.69	10.67	**0.044**
Estuary (Mo)	5	183.40	36.68	7.74	**0.001**
Res	39	184.81	4.74		
		(c) Polychaete abundance			
Modification category	1	1610	1610	8.28	**0.040**
Estuary (Mo)	5	971	194	1.47	0.218
Res	39	5144	132		

Factors include Modification category (Mo; heavily modified or relatively unmodified) and Estuary (Es; nested in Modification category). Sites were the replicates (n = 7).

Polychaetes more diverse and abundant in heavily modified estuaries. Polychaete richness (number of families) and abundance (number of individuals) differed significantly between modification categories (Figure 3; Table 2). Heavily modified estuaries supported a more diverse and abundant polychaete community than relatively unmodified estuaries.

In summary, sediments from heavily modified estuaries were more contaminated with metals and PAHs and organically enriched than the relatively unmodified estuaries. These measures of toxic and enriching contamination also varied between estuaries Grain sizes were finer in heavily modified estuaries; conversely most variation in water quality was between estuaries rather than between heavily modified and relatively unmodified estuaries. Therefore differences in the infaunal community and measures of polychaete richness and abundance with modification category are likely to be related to contamination or sediment grain size. Infaunal differences between estuaries could be related to metal concentrations, sediment grain size or environmental variables.

Field experiment sediments deployed in benthic recruitment containers

Characteristics (including metals and grain sizes) of the field-collected sediments differed consistently between modification categories at the start and the end of the field experiment (Figure S4, Table S6). Additionally since the source sediments had increased infaunal abundances (Figure 3; Table 2) they likely experienced more organic loading from the breakdown of these infauna upon deployment. Despite these differences in the sediment characteristics, infaunal community composition did not differ between modification category or source estuary sediment (Figure 4; Table 3). Furthermore, when polychaete ecological measures were analysed with a similar statistical design to the field survey, richness and abundance did not differ between source sediments (Figure 5; Table 3). Patterns of increased

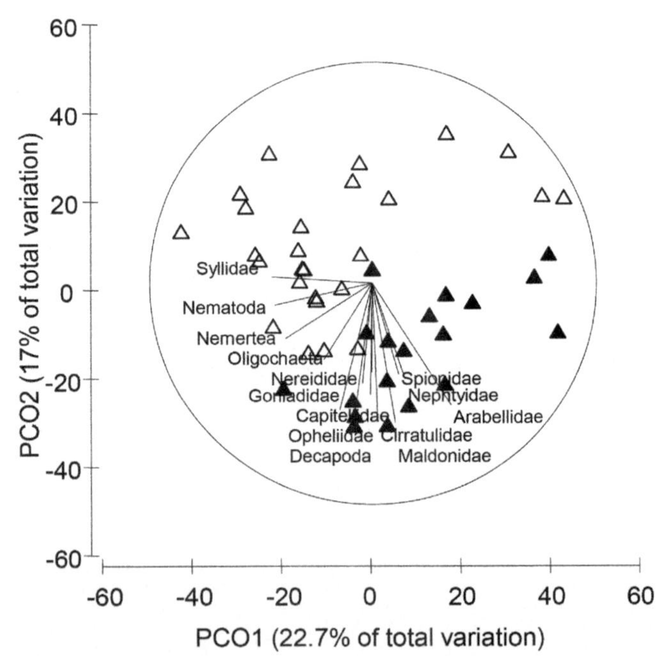

Figure 2. Unconstrained PCA plot of infaunal communities in large-scale field survey. Grabs were sampled from seven heavily modified (filled triangles) or relatively unmodified estuaries (empty triangles). A vector plot of the infaunal community (R>0.4) is also presented to highlight compositional differences among modification categories.

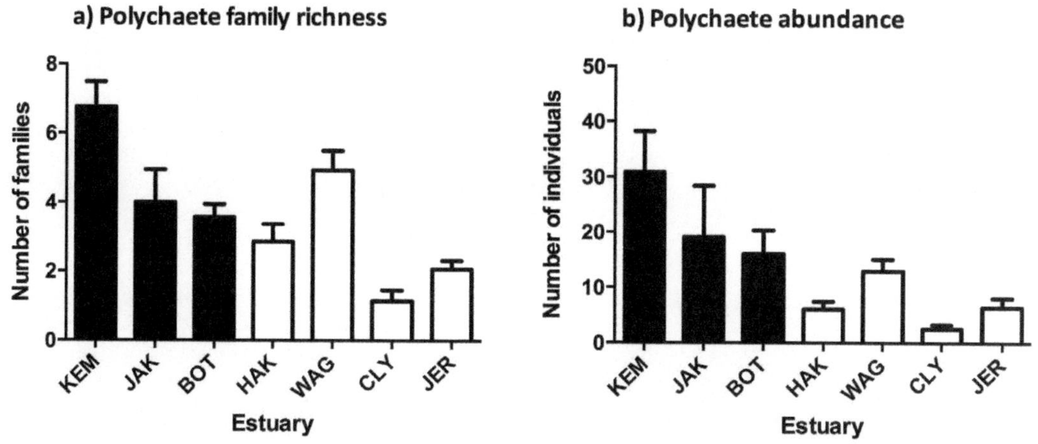

Figure 3. Polychaete family richness and abundance in large-scale field survey. Mean (+S.E.) polychaete (a) family richness and (b) abundance (number of individuals) in benthic sediments collected from seven sites in large-scale field surveys of seven heavily modified (filled bars) or relatively unmodified (open bars) estuaries.

polychaete richness and abundance that were observed in highly contaminated sediments in the field survey were not present in the deployed sediments.

Discussion

Most national biomonitoring guidelines recommend the use of multiple lines of evidence when assessing environmental impacts including the use of experiments to establish causation [39,47,48,49,50]. Environmental monitoring studies, however, rarely use manipulative field experiments to interrogate the patterns found in observational studies. This may result in the misinterpretation of monitoring data and an inability to differentiate the effects of multiple co-varying stressors [26]. We investigated infaunal community composition and contaminant measures in a large-scale survey of benthic sediments. Based on previous research (meta-analysis by [30]) we expected to see clear 'negative' effects of toxic contaminants in heavily modified estuaries (e.g. decreases in abundance and diversity). However, we found that polychaetes in particular were more diverse and abundant in heavily modified estuaries. Sediment Chl-a, porewater ammonia and total organic carbon levels co-varied with toxic contaminants and were also elevated in heavily modified estuaries; we therefore hypothesised that organic enrichment of the sediments may have been driving the observed patterns [31]. Observations from the survey were further investigated with a field experiment that determined the extent to which the patterns were caused by small-scale sediment characteristics or other large-scale environmental conditions that co-vary with anthropogenic modification and disturbance (e.g. organic enrichment and the local species pool). Our field experiment deployed sediments from the surveyed estuaries at one relatively uncontaminated location. Recruitment during the 12 week experiment was subject to local conditions and in this situation we no longer observed elevated polychaete diversity in sediments from heavily modified estuaries. This suggests that estuarine conditions will play a major role in determining benthic community diversity, irrespective of small-scale sediment characteristics.

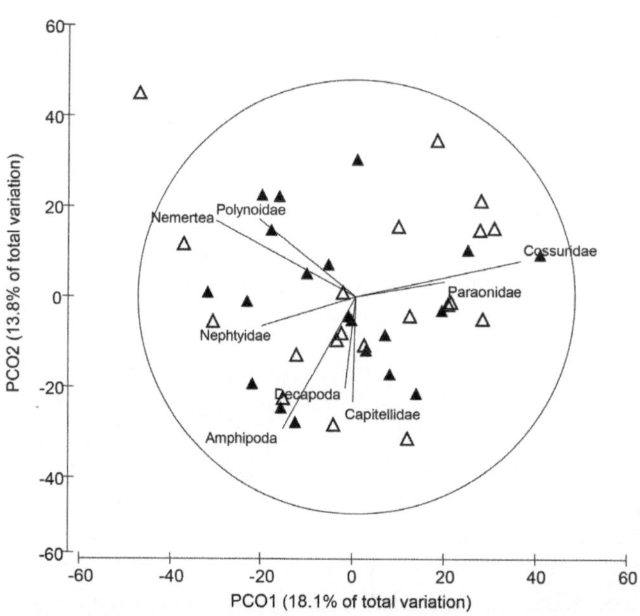

Figure 4. Unconstrained PCA plot of infaunal communities in field manipulated sediment. Sediment was sourced from six estuaries that were either heavily modified (filled triangles) or relatively unmodified estuaries (empty triangles), and deployed in benthic recruitment containers in the Clyde. A vector plot of the infaunal community (R>0.4) is also presented to highlight compositional differences among modification categories.

Habitats that are severely stressed by anthropogenic activities are reported to have substantially reduced species richness (reviewed in [30]) and often support only a few opportunistic or tolerant species [31]. Laboratory and field studies have found infaunal abundances to decrease with increasing metal (Cu) concentrations [51,52] and sediment Cu concentrations measured from estuaries in the current study were among the highest recorded globally in impacted estuaries [53,54,55]. Despite severe metal concentrations far above guideline values up to 1250 ug/g Cu, 550 ug/g Pb and 900 ug/g Zn (Figure S1), we found the opposite for polychaete assemblages, with abundance and family richness higher in heavily modified estuaries. This suggests that toxic effects of contaminants were not significant and may reflect lower bioavailability of contaminants due to increased binding with silt and organic material in heavily modified estuaries [56]. Furthermore, past studies have found some taxa (e.g. polychaetes)

Table 3. Results of permutational MANOVA of (a) the entire infaunal community, and permutational ANOVA of (b) polychaete family richness and (c) polychaete abundance from field transplant experiments of benthic sediments.

Source	df	SS	MS	F	P(perm)
		(a) Infaunal community composition			
Modification category	1	2417.7	2417.7	0.77	0.652
Estuary (Mo)	4	12623	3155.9	1.42	0.061
Res	36	75425	2218.4		
		(b) Polychaete family richness			
Modification category	1	0.02	0.02	0.81	0.423
Estuary (Mo)	4	0.10	0.02	1.24	0.310
Res	36	0.69	0.02		
		(c) Polychaete abundance			
Modification category	1	106.88	106.88	0.72	0.436
Estuary (Mo)	4	590.48	147.62	1.10	0.373
Res	36	4837.40	134.37		

Factors include Modification category (Mo; heavily modified or relatively unmodified), and Estuary (Es; nested in Modification category). BRCs were the replicates (n = 7).

to contain metal tolerant species in both hard substrate and soft sediment communities [51,57,58]. However, more commonly, elevated species richness and abundance in heavily contaminated areas have been associated with organic enrichment [59,60,61].

Most modified estuaries are exposed to a range of inputs (e.g. urban or agricultural runoff or sewage) that would enrich the nutrients within the estuary and may increase the diversity of the local species pool [62]. We found evidence of enhanced sediment Chl-a and high levels of TOC in sediments from heavily modified estuaries suggesting high levels of organic enrichment. Data collected in a separate study also indicated our heavily modified estuaries have higher organic loading than the relatively unmodified estuaries [63]. Organic enrichment from sources such as sewage has been associated with increases in infaunal abundances and diversity [31], which may go some way in explaining the responses we observed in the diversity and abundance of polychaete assemblages from heavily modified estuaries (in particular the surface deposit feeding families; Capitellidae and Cirratulidae). It is important to note that estuaries in NSW are relatively oligotrophic, with nutrient loads an order of magnitude less than estuaries in other parts of the world [64]. Therefore our "heavily modified" estuaries may reflect a state of more moderate pollution where we might expect this increased diversity to be temporary and due to tolerant species [23,31] rather than a "positive" effect of nutrient addition. Estuaries with higher nutrient loads can be subject to severe eutrophication that negatively impacts polychaete diversity and eventually abundance [65,66].

However, large-scale anthropogenic impacts may not be reflected in smaller-scale manipulations [67,68]. In our survey across multiple estuaries we found that polychaete richness and abundance differed among relatively pristine and modified estuaries, however, we found that in experimental deployments of sediment, these ecological measures did not differ between modification category or source estuary. This suggests that there is no small-scale effect of sediment characteristics (even severe metal or organic contamination) because organic enrichment at the scale of an entire estuary is having a greater effect on infaunal diversity and abundance. Our findings agree with recent work [69,70] where continual organic enrichment from farming processes resulted in increased macrofaunal abundances despite expectations of negative impacts from this contamination. However, the apparent lack of sediment metal effects on the experimental community contrasts with recent field mesocosm work [71] in which responses to metal concentrations were evident. Differences in effects may also be related to the scale of the study as their deployments were in multiple estuaries, and reciprocal transplants may be the way forward to optimise the usefulness of manipulative experiments. Temporal scale may have also influenced differences between the results of our experiment and the survey since our surveyed assemblages were well established and our recruitment containers were deployed for only 12 weeks. This period may not have allowed enough time for the initial acclimation of the sediments with respect to their chemistry [72], and subsequent infaunal recruitment before they were collected again and processed. Changes in contaminant concentrations and sediment

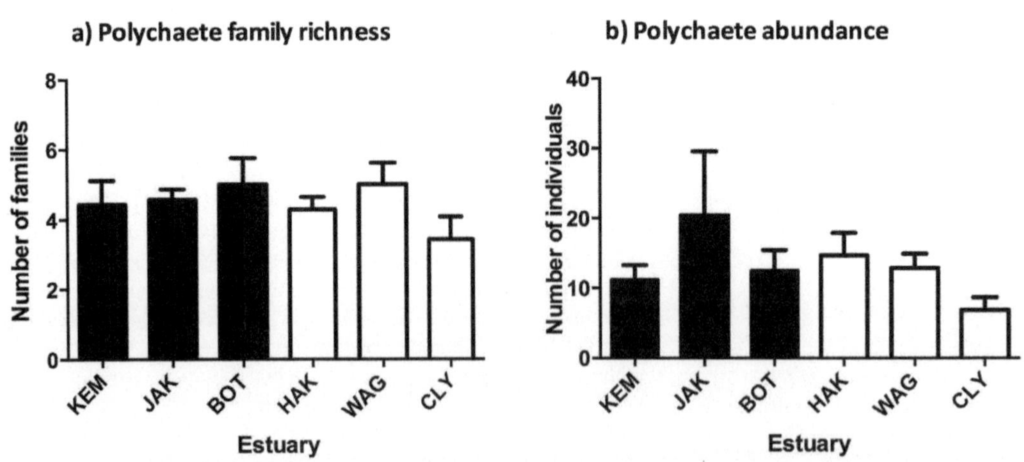

Figure 5. Polychaete family richness and abundance in field manipulated sediments. Mean (+S.E.) polychaete (a) family richness and (b) abundance (number of individuals) in field manipulated sediment sourced from six heavily modified (filled bars) or relatively unmodified (open bars) estuaries and deployed in the Clyde.

loss through resuspension are important considerations for experimental field manipulations of sediments [72]. Metal concentrations decreased during the experiment, but were still highest in the heavily modified sediments at the end of the study. However, studies have demonstrated rapid recolonisation of defaunated sediments [73] and other experimental deployments of sediments have found this period of time to be sufficient to detect effects of similar sediment treatments [74]. Furthermore the number of families and abundances were similar to the observational study results at that site or higher in the experiment indicating that the assemblage had possibly reached equilibrium.

Conclusion

The outcomes of this study have relevance for environmental managers as the patterns we observed across regional scales provided important information about the ecological condition of estuaries and useful directions for future research. The scale of field manipulations may also be an important consideration, as the patterns of increased polychaete richness we observed in heavily modified estuaries across regional scales were not found in our small-scale manipulations within one estuary. This suggests that increases in diversity are related to parameters that co-vary with estuary modification over larger scales such as the productivity of the entire system and the diversity of the regional species pool, and patches of contaminated sediments will not result in similar effects. If environmental managers rely solely on small-scale surveys or manipulations together with meta-analyses from the literature this may lead to a poor understanding of actual impacts over regional scales. Ecological monitoring of human impacts should, as a matter of course, use multiple lines of evidence including field surveys and field experiments in order to reliably identify the drivers of biological patterns and initial observational studies can inform the design of manipulative experiments to further investigate causation.

Supporting Information

Figure S1 Mean (+S.E.) proportional abundances of different taxa analysed from benthic sediment grabs collected in seven estuaries.

Figure S2 Mean (+S.E.) metal and total PAH concentrations (dw), sediment Chl-a, TOC, porewater ammonia and percent fines (<63 μm) analysed from benthic sediment grabs collected in heavily modified (filled bars) and relatively unmodified (open bars) estuaries.

Figure S3 Mean (+S.E) values for water quality variables (salinity, temperature, pH, turbidity, dissolved oxygen and chlorophyll-a) collected in heavily modified (filled bars) and relatively unmodified (open bars) estuaries.

Figure S4 Mean (+S.E.) end metal concentrations (dw) and percent fines (<63 μm) analysed from deployed sediments sourced from heavily modified (filled bars) and relatively unmodified (open bars) estuaries. Starting concentrations are indicated as a single figure above the bars.

Table S1 Details of limits of detection for each of the analytes sampled from sediments.

Table S2 Soft sediment infauna sampled in benthic sediment surveyed from seven sites in seven NSW estuaries that were either heavily modified or relatively unmodified.

Table S3 Soft sediment infauna sampled in field manipulated sediment.

Table S4 Permutational ANOVA results for sediment quality measures in benthic sediment surveyed from seven NSW estuaries.

Table S5 Permutational ANOVA results for water quality variables surveyed from seven NSW estuaries.

Table S6 Permutational ANOVA results for metal contaminants and sediment quality variables measured at the start and end in sediment field experiments deployed in the Clyde estuary.

Acknowledgments

We would like to thank members of the Subtidal Ecology and Ecotoxicology laboratory and volunteers for their assistance. We would also like to thank Bluescope Steel and the NSW Marine Parks Authority for their support. This study was approved and carried out in strict accordance with the New South Wales Department of Primary Industries and New South Wales Marine Parks (Permit No. P09/0072-1.0).

Author Contributions

Conceived and designed the experiments: KAD BPK SLS MAC GFC ELJ. Performed the experiments: KAD GFC BPK MAC ELJ MAD PAH SLS. Analyzed the data: KAD SLS PAH MAD. Contributed reagents/materials/analysis tools: KAD ELJ SLS PAH MAD. Wrote the paper: KAD BPK SLS MAC PAH GFC NAK MAD ELJ.

References

1. Adams SM (2005) Assessing cause and effect of multiple stressors on marine systems. Marine Pollution Bulletin 51: 649–657.
2. Ruiz GM, Fofonoff P, Hines AH, Grosholtz ED (1999) Non-indigenous species as stressors in estuarine and marine communities: assessing invasion impacts and interactions. Limnology and Oceanography 44: 930–972.
3. Vinebrooke RD, Cottingham KL, Norberg J, Scheffer M, Dodson SI, et al. (2004) Impacts of multiple stressors on biodiversity and ecosystem functioning: the role of species co-tolerance. Oikos 104: 451–457.
4. Grimm NB, Foster D, Groffman P, Grove JM, Hopkinson CS, et al. (2008) The changing landscape: ecosystem responses to urbanization and pollution across climatic and societal gradients. Frontiers in Ecology and the Environment 6: 264–272.
5. Lotze HK, Lenihan HS, Bourque BJ, Bradbury RH, Cooke RG, et al. (2006) Depletion, degradation, and recovery potential of estuaries and coastal seas. Science 312: 1806-1809.
6. Birch GF (2000) Marine pollution in Australia, with special emphasis on central New South Wales estuaries and adjacent continental margin. International Journal of Environment and Pollution 13: 573–607.
7. Statham PJ (2012) Nutrients in estuaries – An overview and the potential impacts of climate change. Science of the Total Environment.
8. Rajagopal S, Nair KVK, Azariah A (1995) Response of brown mussel, Perna indica, to elevated temperatures in relation to power plant biofouling control. Journal of Thermal Biology [J 20: 461–467.
9. Roberts DA, Johnston EL, Knott NA (2010) Impacts of desalination plant discharges on the marine environment: A critical review of published studies. Water Research 44: 5117–5128.
10. Strom D, Simpson SL, Batley GE, Jolley DF (2011) Accounting for the influence of sediment particle size and organic carbon on toxicity of copper to benthic invertebrates in oxic/sub-oxic surface sediments. Environmental Toxicology and Chemistry.

11. Trannum HC, Olsgard F, Skei JM, Indrehus J, Overas S, et al. (2004) Effects of copper, cadmium and contaminated harbour sediments on recolonisation of soft-bottom communities. Journal of Experimental Marine Biology and Ecology 310: 87.
12. Martínez-Lladó X, Gibert O, Martí V, Díez S, Romo J, et al. (2007) Distribution of polycyclic aromatic hydrocarbons (PAHs) and tributyltin (TBT) in Barcelona harbour sediments and their impact on benthic communities. Environmental Pollution 149: 104–113.
13. Simpson SL, Spadaro DA (2011) Performance and sensitivity of rapid sublethal sediment toxicity tests with the amphipod Melita plumulosa and copepod Nitocra spinipes. Environmental Toxicology and Chemistry 30: 2326–2334.
14. Hewitt JE, Anderson MJ, Thrush SF (2005) Assessing and monitoring ecological community health in marine systems. Ecological Applications 15: 942–953.
15. Kennish MJ (2002) Environmental threats and environmental future of estuaries. Environmental Conservation 29: 78–107.
16. Keough MJ, Quinn GP (1991) Causality and the choice of measurements for detecting human impacts in marine environments. Marine and Freshwater Research 42: 539–554.
17. Cornell HV, Lawton JH (1992) Species interactions, local and regional processes, and limits to the richness of ecological communities: a theoretical perspective. Journal of Animal Ecology 61: 1–12.
18. Borja A, Bricker SB, Dauer DM, Demetriades NT, Ferreira JG, et al. (2008) Overview of integrative tools and methods in assessing ecological integrity in estuarine and coastal systems worldwide. Marine Pollution Bulletin 56: 1519–1537.
19. Chariton AA, Roach AC, Simpson SL, Batley GE (2010) Influence of the choice of physical and chemistry variables on interpreting patterns of sediment contaminants and their relationships with estuarine macrobenthic communities. Marine and Freshwater Research 61: 1109–1122.
20. Dafforn KA, Simpson SL, Kelaher BP, Clark GF, Komyakova V, et al. (2012) The challenge of choosing environmental indicators of anthropogenic impacts in estuaries. Environmental Pollution 163: 207–217.
21. Chapman PM (2002) Integrating toxicology and ecology: putting the "eco" into ecotoxicology. Marine Pollution Bulletin 44: 7–15.
22. Underwood AJ (1991) The logic of ecological experiments: a case history from studies of the distribution of macro-algae on rocky intertidal shores. Journal of the Marine Biological Association of the UK 71: 841–866.
23. Elliott M, Quintino V (2007) The Estuarine Quality Paradox, Environmental Homeostasis and the difficulty of detecting anthropogenic stress in naturally stressed areas. Marine Pollution Bulletin 54: 640–645.
24. Cole VJ, Glasby TM, Holloway MG (2005) Extending the generality of ecological models to artificial floating habitats. Marine Environmental Research 60: 195–210.
25. Underwood AJ, Chapman MG, Connell SD (2000) Observations in ecology: you can't make progress on processes without understanding the patterns. Journal Of Experimental Marine Biology And Ecology 250: 97–115.
26. Burton GA, Johnston EL (2010) Assessing contaminated sediments in the context of multiple stressors. Environmental Toxicology and Chemistry 29: 2625–2643.
27. Snelgrove PVR (1999) Getting to the bottom of marine biodiversity: sedimentary habitats. Bioscience 49: 129–138.
28. Alquezar R, Markich SJ, Booth DJ (2006) Effects of metals on condition and reproductive output of the smooth toadfish in Sydney estuaries, south-eastern Australia. Environmental Pollution 142: 116–122.
29. Fleeger JW, Carman KR, Nisbet RM (2003) Indirect effects of contaminants in aquatic ecosystems. Science of the Total Environment 317: 207–233.
30. Johnston EL, Roberts DA (2009) Contaminants reduce the richness and evenness of marine communities: A review and meta-analysis. Environmental Pollution 157: 1745–1752.
31. Pearson TH, Rosenberg R (1978) Macrobenthic succession in relation to organic enrichment and pollution of the marine environment. Oceanography and Marine Biology Annual Review 16: 229–311.
32. McKinley A, Johnston EL (2010) Impacts of contaminant sources on marine fish abundance and species richness: a review and meta-analysis of evidence from the field. Marine Ecology Progress Series 420: 175–191.
33. GA (2012) Oz Coasts: Australian online coastal information. Geoscience Australia.
34. Irvine I, Birch GF (1998) Distribution of heavy metals in surficial sediments of Port Jackson, Sydney, New South Wales. Australian Journal Of Earth Sciences 45: 297–304.
35. He Z, Morrison RJ (2001) Changes in the marine environment of Port Kembla Harbour, NSW, Australia, 1975–1995: a review. Marine Pollution Bulletin 42: 193–201.
36. Birch GF, Evenden D, Teutsch ME (1996) Dominance of point source in heavy metal distributions in sediments of a major Sydney estuary (Australia). Environmental Geology 28: 169–174.
37. ANZECC (1999) Strategic plan of action for the national representative system of marine protected areas: a guide for action by Australian governments. Canberra: Australian and New Zealand Environment and Conservation Council Task Force on Marine Protected Areas, Environment Australia.
38. USEPA (2007) Method 3051A Microwave assisted acid digestion of sediments, sludges and oils. Washington: US Environmental Protection Agency.
39. ANZECC/ARMCANZ (2000) Australian and New Zealand Guidelines for Freshwater and Marine Water Quality. Canberra: Australian and New Zealand Environment and Conservation Council and Agriculture and Resource Management of Australia and New Zealand.
40. USEPA (1996) Method 8260B Volatile organic compounds by gas chromatography/mass spectrometry (GC/MS). Washington: US Environmental Protection Agency.
41. Greenberg A, Clesceri L, Eaton A (1992) Standard methods for the examination water and wastewater 18th ed. Maryland, America, American Public Health Association, American Water Works Association, Water Environment Federation.
42. Hedges JI, Stern JH (1984) Carbon and nitrogen determinations of carbonate-containing solids. Limnology and Oceanography 29: 663–666.
43. Moran PJ, Grant TR (1989) The effects of industrial-pollution on the development and succession of marine fouling communities. 1. Analysis of species richness and frequency data. Marine Ecology 10: 231–246.
44. Dauvin J-C (2007) Paradox of estuarine quality: Benthic indicators and indices, consensus or debate for the future. Marine Pollution Bulletin 55: 271–281.
45. Borja A, Barbone E, Basset A, Borgersen G, Brkljacic M, et al. (2011) Response of single benthic metrics and multi-metric methods to anthropogenic pressure gradients, in five distinct European coastal and transitional ecosystems. Marine Pollution Bulletin 62: 499–513.
46. Anderson MJ (2001) A new method for non-parametric multivariate analysis of variance. Austral Ecology 26: 32–46.
47. MacDonald DD, Smith SL, Wong MP, Mudroch P (1992) The development of Canadian marine environmental quality guidelines. Report prepared for the Interdepartmental Working Group on Marine Environmental Quality Guidelines and the Canadian Council of Ministers of the Environment. Ottawa, Environment Canada. 50 pp. +app.
48. CCME (1999) Chapter 6: Protocol for the derivation of Canadian sediment quality guidelines for the protection of aquatic life. Winnipeg: Canadian Council of Ministers of the Environment.
49. Chapman PM (1989) Current approaches to developing sediment quality criteria. Environmental Toxicology and Chemistry 8: 589–599.
50. USEPA (1992) Sediment classification methods compendium. EPA 823-R-92-006. Washington: U.S. Environmental Protection Agency, Office of Water.
51. Olsgard F (1999) Effects of copper contamination on recolonisation of subtidal marine soft sediments - an experimental field study. Marine Pollution Bulletin 38: 448–462.
52. Stark JS (1998) Effects of copper on macrobenthic assemblages in soft sediments: a laboratory experimental study. Ecotoxicology 7: 161–173.
53. Cheung KC, Poon BHT, Lan CY, Wong MH (2003) Assessment of metal and nutrient concentrations in river water and sediment collected from the cities in the Pearl River Delta, South China. Chemosphere 52: 1431–1440.
54. Zhang W, Feng H, Chang J, Qu J, Xie H, et al. (2009) Heavy metal contamination in surface sediments of Yangtze River intertidal zone: An assessment from different indexes. Environmental Pollution 157: 1533–1543.
55. Bryan GW, Langston WJ (1992) Bioavailability, accumulation and effects of heavy metals in sediments with special reference to United Kingdom estuaries: a review. Environmental Pollution 76: 89–131.
56. Chapman PM, Wang F (2001) Assessing sediment contamination in estuaries. Environmental Toxicology and Chemistry 20: 3–22.
57. Dafforn KA, Glasby TM, Johnston EL (2009) Links between estuarine condition and spatial distributions of marine invaders. Diversity and Distributions 15: 807–821.
58. Johnston EL, Keough MJ, Qian P-Y (2002) Maintenance of species dominance through pulse disturbances to a sessile marine invertebrate assemblage in Port Shelter, Hong Kong. Marine Ecology Progress Series 226: 103–114.
59. Abou-Aisha KM, Kobbia IA, El Abyad MS, Shabana EF, Schanz F (1995) Impact of phosphorus loadings on macro-algal communities in the Red Sea coast of Egypt. Water, Air, & Soil Pollution 83: 285–297.
60. Ansari ZA, Ingole BS, Banerjee G, Parulekar AH (1986) Spatial and temporal changes in benthic macrofauna from Mandovi and Zuari estuaries of Goa, West Coast of India. Indian Journal of Sea Research 15: 223–229.
61. López Gappa J, Tablado A, Magaldi NH (1993) Seasonal changes in an intertidal community affected by sewage pollution. Environmental Pollution 82: 157–165.
62. McKinley AC, Miskiewicz A, Taylor MD, Johnston EL (2011) Strong links between metal contamination, habitat modification and estuarine larval fish distributions. Environmental Pollution 159: 1499–1509.
63. Scanes P (2010) NSW Estuarine Catchment Disturbance Ranks. Sydney: NSW Department of Environment, Climate Change, and Water.
64. Scanes P, Coade G, Doherty M, Hill R (2007) Evaluation of the utility of water quality based indicators of estuarine lagoon condition in NSW, Australia. Estuarine, Coastal and Shelf Science 74: 306–319.
65. Diaz RJ, Rosenberg R (2008) Spreading Dead Zones and Consequences for Marine Ecosystems. Science 321: 926–929.
66. Elliott M, de Jonge VN (2002) The management of nutrients and potential eutrophication in estuaries and other restricted water bodies. Hydrobiologica 475/6: 513–524.
67. Nally RM (1997) Scaling artefacts in confinement experiments: a simulation model. Ecological Modelling 99: 229–245.
68. Thrush SF, Whitlatch RB, Pridmore RD, Hewitt JE, Cummings VJ, et al. (1996) Scale-dependent recolonization: the role of sediment stability in a dynamic sandflat habitat. Ecology 77: 2472–2487.

69. Keeley NB, Forrest BM, Macleod CK (2013) Novel observations of benthic enrichment in contrasting flow regimes with implications for marine farm monitoring and management. Marine Pollution Bulletin 66: 105–116.
70. Keeley N, Cromey C, Goodwin E, Gibbs M, Macleod C (2013) Predictive depositional modelling (DEPOMOD) of the interactive effect of current flow and resuspension on ecological impacts beneath salmon farms. Aquaculture Environment Interactions 3: 275–291.
71. O'Brien AL, Keough MJ (2013) Detecting benthic community responses to pollution in estuaries: A field mesocosm approach. Environmental Pollution 175: 45–55.
72. Hill NA, Johnston EL, King CK, Simpson SL (2011) Physico-chemical changes in metal-spiked sediments deployed in the field: Implications for the interpretation of in situ studies. Chemosphere 83: 400–408.
73. Santos SL, Simon JL (1980) Marine soft-bottom community establishment following annual defaunation: larval or adult recruitment? Marine Ecology Progress Series 2: 235–241.
74. Wilkie EM, Roach AC, Micevska T, Kelaher BP, Bishop MJ (2010) Effects of a chelating resin on metal bioavailability and toxicity to estuarine invertebrates: Divergent results of field and laboratory tests. Environmental Pollution 158: 1261–1269.

3

Ecological Succession, Hydrology and Carbon Acquisition of Biological Soil Crusts Measured at the Micro-Scale

Matthew Tighe*, Rebecca E. Haling, Richard J. Flavel, Iain M. Young

School of Environmental and Rural Science, University of New England, Armidale, New South Wales, Australia

Abstract

The hydrological characteristics of biological soil crusts (BSCs) are not well understood. In particular the relationship between runoff and BSC surfaces at relatively large (>1 m^2) scales is ambiguous. Further, there is a dearth of information on small scale (mm to cm) hydrological characterization of crust types which severely limits any interpretation of trends at larger scales. Site differences and broad classifications of BSCs as one soil surface type rather than into functional form exacerbate the problem. This study examines, for the first time, some hydrological characteristics and related surface variables of a range of crust types at one site and at a small scale (sub mm to mm). X-ray tomography and fine scale hydrological measurements were made on intact BSCs, followed by C and C isotopic analyses. A 'hump' shaped relationship was found between the successional stage/sensitivity to physical disturbance classification of BSCs and their hydrophobicity, and a similar but 'inverse hump' relationship exists with hydraulic conductivity. Several bivariate relationships were found between hydrological variables. Hydraulic conductivity and hydrophobicity of BSCs were closely related but this association was confounded by crust type. The surface coverage of crust and the microporosity 0.5 mm below the crust surface were closely associated irrespective of crust type. The $\delta\ ^{13}C$ signatures of the BSCs were also related to hydraulic conductivity, suggesting that the hydrological characteristics of BSCs alter the chemical processes of their immediate surroundings via the physiological response (C acquisition) of the crust itself. These small scale results illustrate the wide range of hydrological properties associated with BSCs, and suggest associations between the ecological successional stage/functional form of BSCs and their ecohydrological role that needs further examination.

Editor: Jordi Moya-Larano, Estacion Experimental de Zonas Áridas (CSIC), Spain

Funding: This research was funded by an internal University of New England staff research support grant. The funder had no role in study design, data collection and analysis, decision to publish, or preparation of the manuscript.

Competing Interests: The authors have declared that no competing interests exist.

* E-mail: mtighe2@une.edu.au

Introduction

Biological soil crusts (BSCs) are widespread in semi-arid and arid zones globally [1,2] and may comprise large areas of the non-vegetated soil surface [3]. Several classifications of BSCs exist which predominantly relate to disturbance sensitivity and/or stages of ecological succession [2,4,5]. These types of classifications are consistent with site observations, and generally follow the sequence of cryptomorphic filamentous cyanobacteria, to early successional/perimorphic bryophytes, followed by the more hypermorphic byrophytes [2]. The relationship between these classifications and the hydrological properties of BSCs are still largely unexplored [2], due in part to the difficulties in quantifying hydrological properties of these surfaces.

Hydrological properties of BSCs are difficult to measure *in-situ.* Studies commonly report on hydrological properties such as runoff at a reasonably large scale (for example, following rainfall simulation over 1 m^2 plots [6]). Runoff values of BSCs at this scale are extremely variable and findings often difficult to place in a wider context [7,8,9]. This is due to several factors; (i) BSCs are often classified and examined as one type of surface cover, irrespective of the range of functional forms present, (ii) there is a lack of information on hydrological properties of different functional crust forms at a small scale (mm) which compound the difficulties in interpreting hydrological responses at larger scales, (iii) previous studies often only compare crusts with soils that have been disturbed (scalped, trampled, burnt, and (iv) different methodologies have been applied in different studies [2,10]. These factors have also caused difficulties in a wider context – with BSCs often covering large areas of the soil surface, the inability to determine hydrological responses of this surface 'type' makes hydrological interpretations of a whole site ambiguous. In particular, BSCs are variably considered as both hydrophilic and hydrophobic and as potentially having high or low porosity, while other hydrological properties such as hydraulic conductivity or potential for waterlogging are only rarely considered [2,11,12].

The variable surface coverage of BSCs at both large and small scales makes hydraulic conductivity of any one type of BSC difficult to quantify, while the relationship of different crust types with inundation or surface saturation may be almost impossible to determine without long term observation of a site or experimental manipulation [11]. Measurement of equivalent pore diameter via permeameter readings is also difficult due to the variable hydrophobic/hydrophilic nature of BSCs [13]. Recently Menon et al. [14] used X-ray tomography to directly measure macro-porosity of crusted surfaces after careful removal and transportation of the intact surfaces. These authors modeled water infiltration and showed how the larger pore sizes (>50 μm) greatly affected water flow in the cyanobacteria and cyanobacteria-lichen dominated crusts they examined. This approach of undertaking a series of measurements on carefully transported, undisturbed BSCs has potential in measuring several hydrological variables under controlled laboratory settings and subsequently determining any relationships these variables have with each other, or with crust type.

In addition to the possibility of moving samples into a controlled laboratory setting, differences in the carbon isotopic signature of BSCs could potentially be used to indicate relative inundation of surfaces. BSCs that have a history of frequent inundation may have an elevated (or more positive) carbon isotope signature than crusts that have been inundated relatively less frequently. However, this potential relationship has not been examined in a semi-arid environment [11,15].

In this study we examine a range of functionally different biological crusts taken from one site in semi-arid south eastern Australia. This study focuses on the porosity and small-scale hydrological properties of different crusted surfaces to determine if there are distinct differences in hydrological properties of the crust types sampled, including C and $\delta^{13}C$ signatures. The aim of this investigation was to ascertain if there were differences in the measured hydrological properties of the different crusts that could account for the wide range of hydrological interpretations made in association with BSCs. A secondary aim of this study was to ascertain if any of these hydrological properties could be directly related to the common classification of BSCs based on disturbance sensitivity/ecological succession.

Methods

Site Description

Biological crusts were collected at one privately owned site approximately 30 km northwest of Nyngan in western NSW, Australia (Figure 1). No permits were necessary for this field study. Permission for the study to be undertaken was granted by the owner of the private land, Mr. A. Gibson. The long-term average annual rainfall at the site is 446 mm, distributed evenly throughout the year. The site was in a mid-slope position and was comprised of a dense *Geijera*, *Dodonaea* and *Eremophila* sp. shrubland with a scattered over-storey of *Eucalytpus populnea* and *E. intertexta*. Soil was a Red Kandosol [16] or a Chromic Luvisol in the FAO classification.

The site had minimal herbaceous plant cover and clumped woody debris and litter cover, and a wide range of biological crust cover occurring both in inter-canopy areas and underneath woody canopies (Figure S1). Grazing history of the site was typical of the region, with set stocking of 0.5–1.5 DSE over 12–18 month periods common. Previous sampling and analysis at the site [9] indicated that surface soil texture was very consistent (sand, silt and clay % means and standard errors of 70.1±0.26, 13.4±0.16, 16.5±0.13; n = 6).

Soil Core Collection

Seven different soil surface types (A through G) were identified and sampled (Table 1) within an area of 50 m×50 m. Three replicates of each surface type were collected, except for sample G where only two examples of the surface type could be found (Table 1, n = 20). The first replicate of each sample was taken as the first of each type encountered during a haphazard search of the area. Subsequent replicates of each type were selected following random distance and orientation walks which were repeated until two more of each surface type were encountered (with the exception of G). Sampling consisted of manually inserting a 5 cm diameter, 10 cm in height zinc alloy cylinder into the soil until the rim of the cylinder was 1–2 mm above the highest point of the soil surface. The intact core and surface was then removed using a hand trowel, wrapped in foam packaging, placed in a foam lined container, and transported to the laboratory. All coring occurred under field moisture conditions (<7% moisture w/w) and samples were left for two weeks to air dry in the laboratory before analysis. Crusts were monitored visually over the whole analysis period to ensure no obvious detrimental effects due to drying or disturbance were observed.

Crust Identification

Crusted surfaces were identified to genus where possible but were placed in a functional group according to life form at a minimum, as per Eldridge and Tozer [5]. Crusts were also classified into groups as related to their sensitivity to disturbance and their approximate appearance in successional progression at a site [2,4,5]. The percent surface area of the biological crust of each sample was calculated using high resolution colour photographs (RGB) and manual adjustment of auto-thresholding in ImageJ following conversion to a binary black and white image, and using the original image as a reference for any manual thresholding adjustment [17]. Cyanobacteria surfaces were inspected both dry and following surface wetting to check for any differences in cover (no differences evident). One replicate of the filamentous cyanobacteria (sample B) was found to be extremely difficult to threshold (dry or wet) and the resulting cover value appeared as an obvious outlier in analyses. This value was omitted from statistical analysis.

3-dimensional Measurements

All intact cores were scanned using a high-resolution Vtomexs X-ray Computed Tomography (CT) system (GE Phoenix). The use of μX-ray tomography for the measurement of soil porosity is now established [18]. The source was a 240 kV 'Direct' tube with a reflection tungsten target and spot size of 4 μm. Preliminary trials indicated optimal energy settings were 130 kV and 110 μA with a 0.5 mm thick copper filter to reduce beam hardening artifacts and all scans were acquired using these settings. A series of 750 angular projections (radiographs) were obtained in a 360° revolution, each the average of 3, 400 ms exposures captured on a 512×512 pixel detector array.

The projections were reconstructed into 16-bit volumes with an isotropic voxel size of 100 μm using GE Phoenix proprietary software on a graphical processor cluster. The volumes were then converted to a stack of.tiff images using VG Studio Max (Version 2.0) and imported into Image J (Version 1.46i for 64-bit Windows 7) for further analysis. The volumes were then converted to 8-bit grey scale values (required for thresholding algorithms), and the centermost 250×250 pixels (equivalent of 25 mm×25 mm) of each slice were selected using a pixel mask. The 'Equalize histogram' algorithm inside the 'Enhance Contrast' tool was used to improve the efficacy and consistency of thresholding. Several representative slices from different depths of the cores were isolated and several 'Automatic Threshold' algorithms were applied which indicated that the 'Yen' automatic threshold [19] was most representative of the pore space in the scans and was consequently applied to all image stacks. The binary images were analysed using SCAMP (Version 1.2) [20,21] to firstly determine the porosity by slice for each stack. Due to the uneven soil surface, the surface was determined to be the first slice where porosity was less than 40%. This selection was visually inspected to confirm the appropriate delineation of the surface. From this digital slice downward porosity, pore space connectivity and pore size distribution measures were obtained for the remaining volume

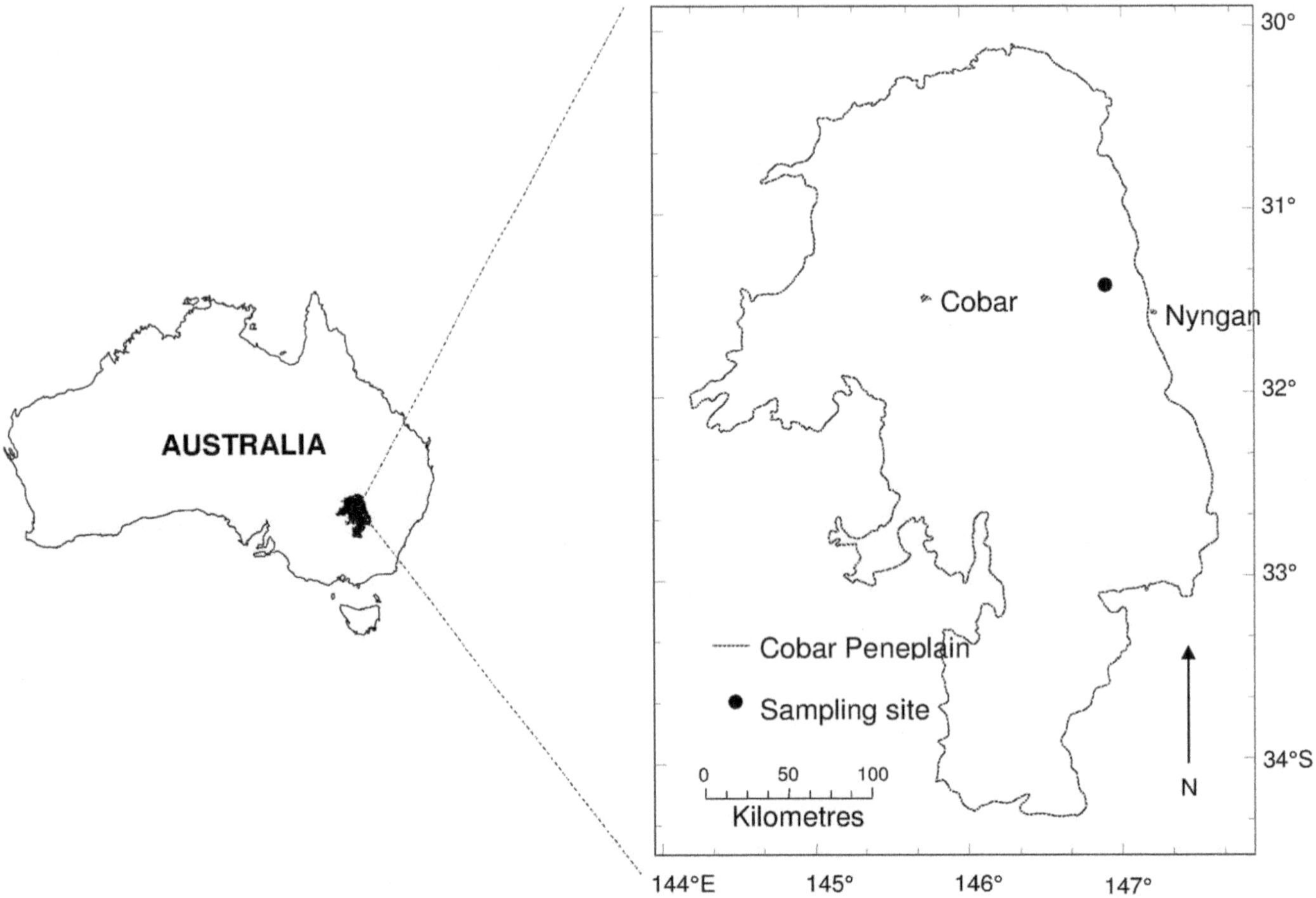

Figure 1. Location of the sampling site in southeastern Australia. The site is located in the Cobar Peneplain bioregion in western NSW, Australia. Nearby regional population centres are also depicted (modified from [38]).

(approx. 430 slices) using SCAMP and the data was summarised using Excel 2010 for Windows 7 (64-bit).

Hydrological Properties

Water repellency of each crusted surface was determined with the water drop penetration test (WDPT) [22]. Six drops were placed on the surface of each intact core using a syringe, ensuring that drops were only placed on the surface of the identified dominant crust. A video recording of each test was made and the time taken for each water drop to penetrate the surface (to the nearest second) was determined from video playback. The average time of penetration was calculated for each sample replicate from the six values. Sample G showed no penetration of water drops before the termination of the test at 1200 s, and the average time for water penetration for sample G was taken as 1200 s.

A custom made microinfiltrometer [23] was used to determine short term (up to 10 min) infiltration of each intact crust surface. The infiltrometer tip (internal diameter 1.61 mm) was placed in contact with the dominant crust identified in each sample replicate and the infiltration (calculated from changes to water weight in a reservoir) was recorded at 10 second intervals. Sorptivity and hydraulic conductivity (the initial, rapid, non-linear rate of water absorbance by a surface and the longer term rate of water movement through a surface respectively) at 0 cm head were estimated using Philip's [24] equation:

$$i = St^{1/2} + At$$

where i = infiltration, S = sorptivity, t = time in seconds, and A = hydraulic conductivity

Infiltration measurements before 50 seconds duration were omitted due to high random variation in early data points, as per the recommendation of Smiles and Knight [25]. Data were graphed as $t^{1/2}$ versus $it^{-1/2}$ with sorptivity and hydraulic conductivity calculated from the intercept and slope of the least squares linear regression. For samples where the average sorptivity of the replicates could not be differentiated from zero a value was not reported. For sample G no infiltration occurred within 10 minutes, with the corresponding sorptivity and hydraulic conductivity values reported as zero. With the exception of sample G, all hydrological calculations were also undertaken using only the infiltration data from after the recorded WDPT for each replicate (as opposed to the 50 s cut off) to determine if the inclusion of earlier data points altered the calculated values. No significant differences were found and the sorptivity and hydraulic conductivity values using the 50 s cut off value were used.

Carbon and $\delta^{13}C$

After infiltration each core was air-dried to constant weight. The surface 5 mm was then manually removed using a finely adjustable sample stand that allowed the sample to be raised 5 mm

Table 1. Dominant crust type and hydrological properties for each sample[1]. Numerical values are arithmetic means ±1 standard error[2].

Variable	Filamentous cyanobacteria – unidentified (A)	Filamentous cyanobacteria – unidentified (B)	Crustose lichen – *Diploschistes* spp. (C)	Liverwort/squamulose lichen mix – *Riccia* spp. and *Cladonia* spp. with minor unidentified filamentous cyanobacteria (D)	Liverwort – *Riccia* spp. (E)	Moss – *Bryum* spp. (F)	Foliose lichen – *Xanthoparmelia* spp. (G)
Sensitivity to disturbance [3]	Low	Low	Mid	Mid to High	High	High	High
Successional stage[3]	Early	Early	Mid	Mid to Late	Late	Late	Late
Percent cover	55.9±5.78bc	24.5±2.71a	82.1±3.16c	34.1±2.36ab	49.8±8.28ab	50.6±a1.44bc	35.2±17.6ab
WDPT (s)	3.6±2.47a	5.9±1.19ab	64.4±16.8b	36.3±29.7b	3.8±1.21a	4.7±1.45a	>1200c
Sorptivity @ 0 cm head (mm $h^{-1/2}$)	NA	0.04±0.008a	0.07±0.03a	0.05±0.03a	0.07±0.03a	NA	NA
Hydraulic conductivity (mm h^{-1})	1.00±0.37bc	0.47±0.07ab	0.09±0.04ab	0.93±0.26bc	2.30±0.83c	4.53±0.43d	0.00d
Porosity 0–5 mm (%)	13.7±0.91a	20.7±3.49a	13.1±1.28a	12.2±1.36a	13.7±2.56a	14.0±3.49a	13.8±0.61a
Porosity 0.5–1 mm (%)	18.1±3.95a	20.5±6.67a	10.5±0.84a	18.9±0.44a	23.7±4.25a	18.2±3.90a	25.1±2.70a
C 0–5 mm (%)	1.94±0.14ab	1.40±0.18a	1.95±0.06ab	1.62±0.10a	2.95±0.45bc	3.29±0.21c	2.54±0.65abc
$\delta^{13}C$ 0–5 mm	−23.4±0.10ab	−22.5±0.23b	−22.5±0.37b	−23.0±0.05ab	−23.2±0.06ab	−24.0±0.23a	−23.2±0.82ab

[1]Means followed by different letters indicate significant differences within a row.
[2]N=3 for each sample except for cover calculation of sample B (n=2) and all variables for sample G (n=2).
[3]As indicated by Belnap [2], Eldridge and Greene [4] and Eldridge and Tozer [5].

above the alloy cylinder. The 0–5 mm section was ground to <0.5 mm and analysed for total C and $\delta^{13}C$ using a Sercon IRMS at the University of New England, Armidale.

Data Analysis

Percent crust cover, WDPT, sorptivity, hydraulic conductivity, porosity (averaged over both the 0–5 and 0.5–1 mm depth increments, see below for explanation of increment choice), C and $\delta^{13}C$ (0–5 mm depth) were analysed for differences between crust types. General linear models were used with distributions and variable transformations specified as appropriate to meet the assumptions of each analysis. Where an analysis indicated a significant overall difference (at $P<0.05$), Tukey's post-hoc tests were applied to determine pairwise differences between crust types.

The above analysis suggested potential associations between crust successional stage and the two variables of WDPT and hydraulic conductivity. Successional stage of crusts were transformed into a rank order variable (1 = Early, 2 = Mid, 3 = Mid to late, 4 = Late) and a general additive mixed model (GAMM) was used to model the association between the rank order of successional stage and both WDPT and hydraulic conductivity. Functional form of crust was specified as a random effect. The association was modeled using a spline with k = 3 degrees of freedom. Significance of the association was determined by plotting the modeled 95% confidence intervals at each successional stage. Summaries of model outputs can be found in Supporting Information (Table S1). Potential associations between pairs of variables were examined using scatterplots, correlations, and linear modelling. Porosity values were also combined into averages across different depth increments (100 μm increments to 5 mm depth) and analysed for significant pairwise associations with hydrological variables using scatterplots and correlation matrices. A strong association was found between hydraulic conductivity and porosity averaged across the 0–1 mm depth increment, and further data exploration indicated this association was due to the strong trend between hydraulic conductivity and the average porosity between 0.5 and 1 mm depth. For each significant association the strength of the association (R^2) and significance (P) were determined from the appropriate linear model, and the results expressed as a linear regression. The potential confounding effect of different functional crust types on the significance of each linear association was additionally tested using a linear mixed model, specifying the crust type as a random effect. The proportional contribution of crust type to total model error was used as an indication of the strength of the effect of crust type in each analysis. Linear associations with 95% confidence intervals, accounting for the random effect off crust type, were produced using the approach of Fox [26]. Summaries of mixed linear model outputs can be found in Supporting Information (Table S2). Hydraulic conductivity and hydrophobicity values for sample G, the foliose lichen, were omitted from the regression analyses due to their extreme nature.

All data analysis was carried out in R version 2.14.1 [27] using the packages mgcv [28], lme4 [29], effects [26], and multcomp [30]. Significance was determined at the 0.05 level, and all assumptions of tests were met.

Results

Hydrological Values of Crust Types

Table 1 indicates the wide range of crust types sampled, both in functional form and using generally accepted classifications of BSC sensitivity to disturbance/ecological succession. There was high variation in WDPT values and hydraulic conductivity values within replicates of each crust type, yet there were still pronounced differences in the hydrophobic nature (as indicated by the WDPT) and the hydraulic conductivity of the crust types. The examined crusts spanned the range of hydrophilic to extremely hydrophobic behaviour according to common classifications of WDPT times [31]. The foliose lichen, considered as highly sensitive to disturbance and a late successional BSC, displayed an extreme degree of hydrophobicity and a corresponding zero value of hydraulic conductivity. The lowest degree of hydrophobicity (high hydrophilic behaviour) was shared between filamentous cyanobacteria, the liverwort, and the moss, spanning the sensitivity to disturbance/successional stage spectrum. The values of hydraulic conductivity ranged across 2 orders of magnitude, and with the exception of the foliose lichen, there was a significant ($P = 0.006$) trend of a 'hump' shaped association between hydrophobicity and crust classification (as sensitivity to disturbance or successional stage), and a similarly significant ($P = 0.0001$), but inverse humped association with hydraulic conductivity (Figure 2).

Sorptivity was low across all crust types (where it could be differentiated from zero), and no significant differences could be detected between the porosity of the crusts, either in the 0–1 or 0–5 mm depth intervals (Table 1). Carbon values in the 0–5 mm increment followed the sensitivity to disturbance classification in that crusts with low sensitivity to disturbance tended to have low C, compared with crusts with high sensitivity to disturbance. The $\delta^{13}C$ values did not show a consistent trend with crust successional stage/sensitivity to disturbance classification (Table 1).

The Relationship between Hydrophobicity and Hydraulic Conductivity

There was some indication of a relationship between hydrophobicity and hydraulic conductivity, but the association was confounded by crust type. Figure 3 shows the exponential decay relationship between hydraulic conductivity and the surface hydrophobicity (WDPT) across all samples, irrespective of crust type. The extremely low hydraulic conductivity values corresponded to WDPT times of greater than 20 seconds. Below this value the slope of the decay curve flattened out significantly, with a wider range of hydraulic conductivity values occurring for WDPT times less than approximately 8 seconds. However, when crust type was included as a random effect in the analysis this relationship was not significant ($P = 0.13$), and the random effect of crust type explained 52% of the total error in the model.

The Relationship between Porosity and Biological Crust Cover

Figure 4 depicts the only significant association detected between porosity and another variable. This association existed irrespective of the type of BSC (i.e. the association remained highly significant when crust type was specified as a random effect in the analysis ($P = 0.0014$). The random effect of crust type accounted for less than 1% of the total error in the model. Thus, irrespective of the type of BSC, as the percent of the sampled surface covered by the crust increased, the porosity averaged across the 0.5–1 mm depth increment decreased. Specifically, as the crust cover increased from approximately 20% of the surface to almost 90%, the porosity decreased by approximately 16%, from 27% to 11% (a relative decrease of 40%). Figure 4 also displays one outlier (open square symbol), which was omitted due to the difficulty in digitally thresholding and estimating the cover of the cyanobacteria in the sample.

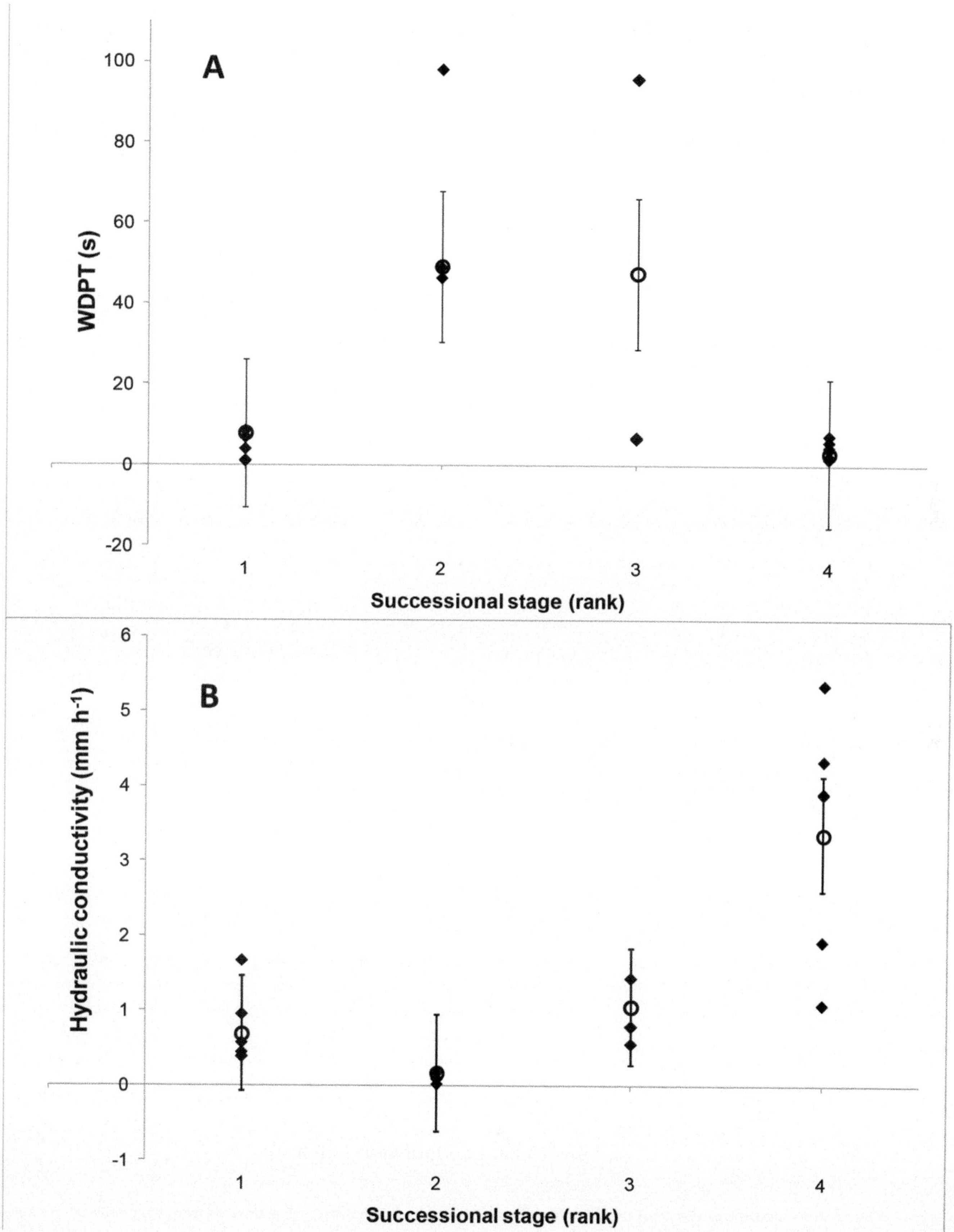

Figure 2. The general trend between successional stage of BSCs and hydrophobicity (as estimated from the results of the water drop penetration test) (A), and hydraulic conductivity (B). Successional stage is presented as a rank order variable (1 = Early, 2 = Mid, 3 = Mid

to Late, 4 = Late). Open circles are predicted response and bars represent 95% confidence intervals following a GAMM analysis with 3 degrees of freedom. Black diamonds are individual data points.

The Relationships between C, δ ^{13}C and Hydraulic Conductivity

Figure 5a depicts the strong and highly significant relationship between C in the 0–5 mm depth increment and hydraulic conductivity, in which C in the crusted surface approximately doubled across the range of hydraulic conductivities measured. The random effect of crust type accounted for 43% of the total error in the model, but even accounting for this effect the association between C and hydraulic conductivity was still highly significant ($P = 0.0005$). As hydraulic conductivity increased and C likewise increased, the δ ^{13}C signature became more negative (Figure 5b). This effect was also highly significant when the random effect of crust type was accounted for ($P = 0.0095$, with crust type accounting for 23% of the total error in the model).

Discussion

There is a long history of conflicting reports of the links between soil surface hydrological behavior and the presence of BSCs. In particular the skew of data towards hot desert environments and the limited number of studies that cover a range of crust types have made it difficult to determine definitive relationships between BSCs and soil surface hydrology [2]. Many past studies have simply referred to BSCs as a single soil surface type, with little to no differentiation of crust types within a study [12]. At the same time, research has identified an ecological succession of BSCs that can occur in any given system. In a recent study Chamizo et al. [32] identified differences in infiltration between functionally different biological crusts types using a rainfall simulator. They found that late successional lichens had low infiltration but conversely moss had a very high infiltration. Our study supports these findings, as well as the existence of several trends suggested in previous meta-analyses [2,7]. This is done by measuring several hydrological properties simultaneously at a very small scale, as opposed to the large scale measured via rainfall simulation.

All crust types were significantly variable in each measure. Thus, it was difficult to detect any general trends across the several crust types. What can be considered the extremes of early successional, low sensitivity to disturbance filamentous cyanobacteria, and the late successional, high sensitivity to disturbance liverwort and moss were easiest to delineate into two groups based on hydrological responses. Warren [7] states that hydrophobicity of BSCs is most commonly reported in relation to sandy soils, and with 70% sand, our findings support this. However, there was an

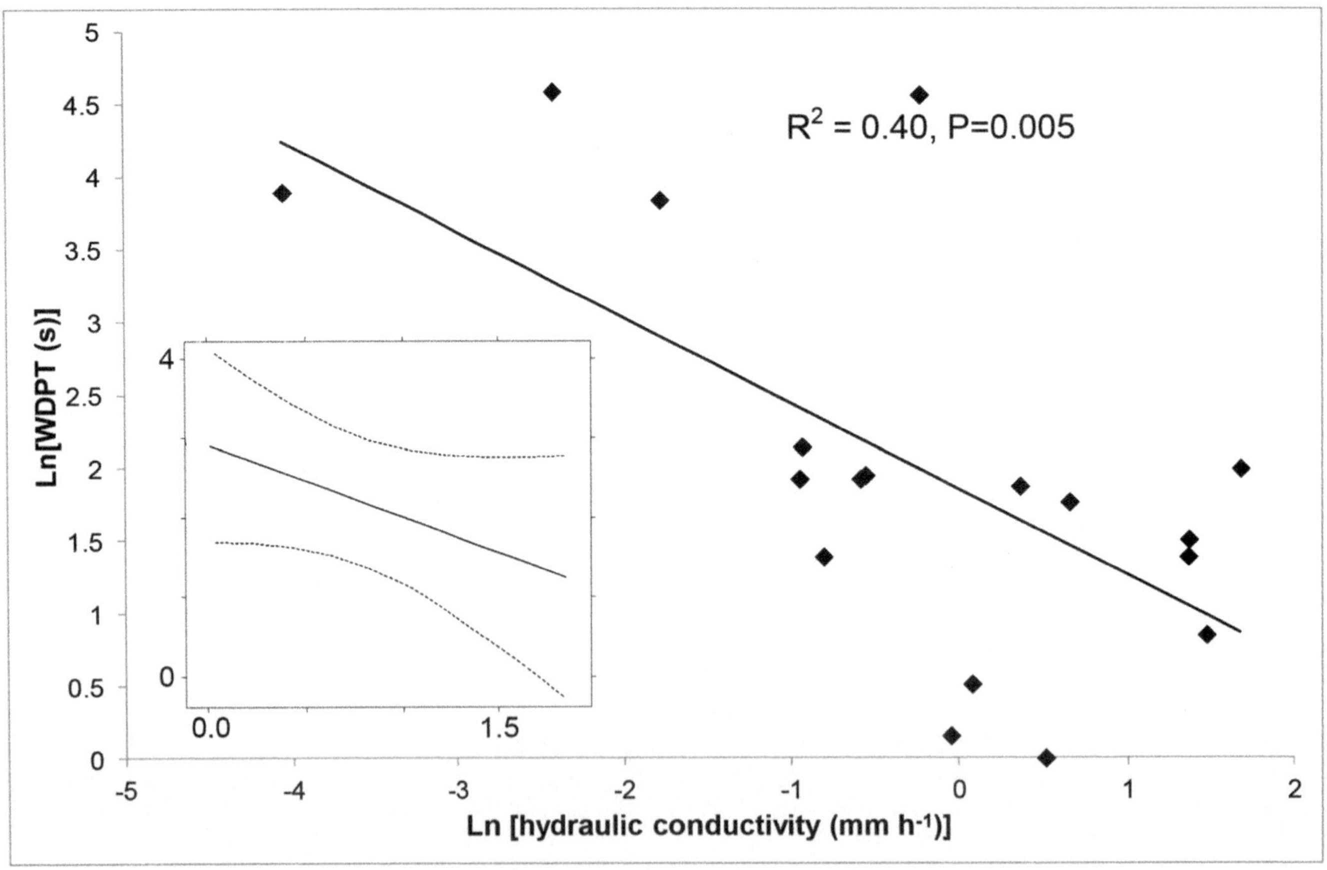

Figure 3. The relationship between hydraulic conductivity (mm h^{-1}) and hydrophobicity (as estimated from the results of the water drop penetration test) of the crust surface. Values are expressed on the natural logarithm scale. INSET – the same association modeled with crust type as a random effect, showing the model prediction and 95% confidence intervals. This association was non-significant ($P = 0.13$). (see text for details). Values from sample G were omitted as outlined in text.

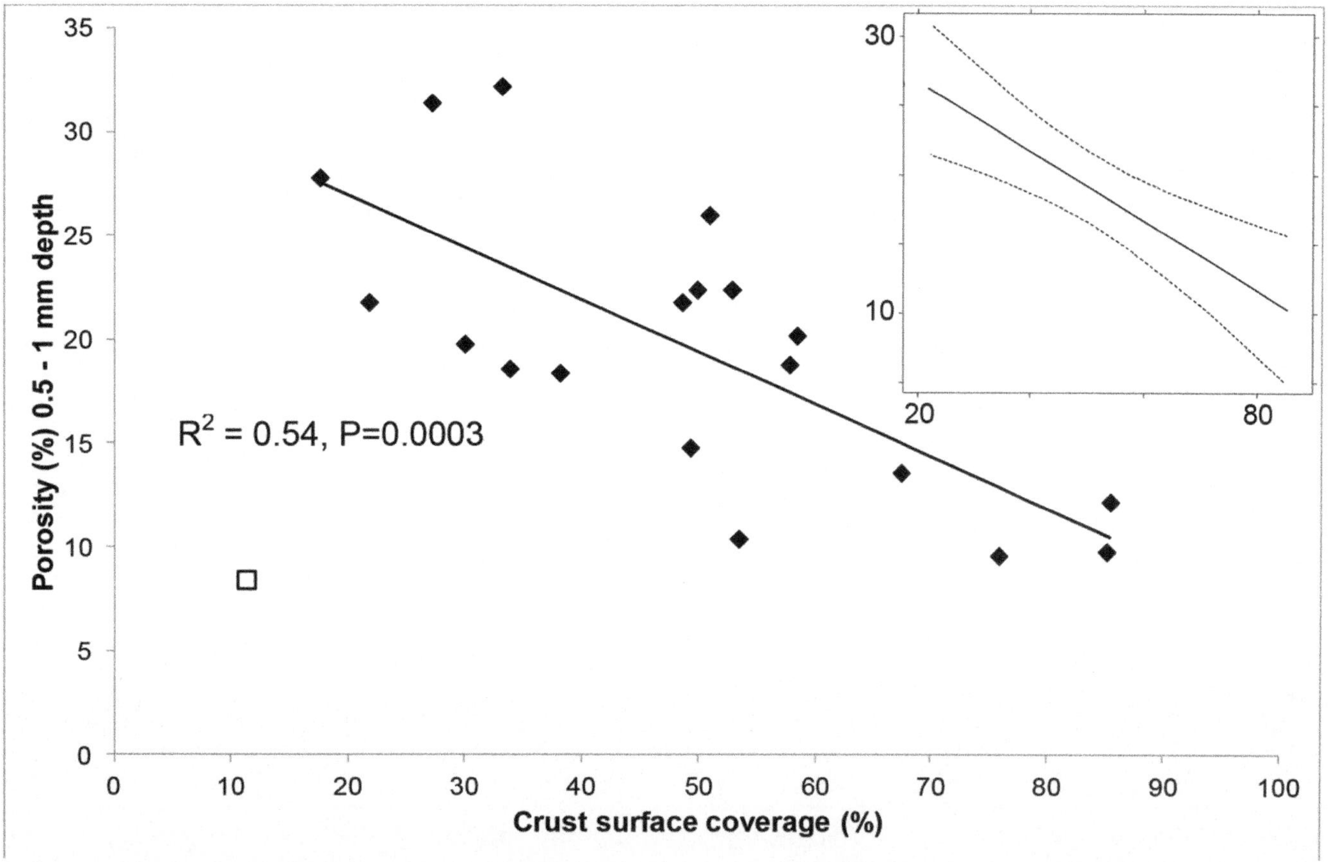

Figure 4. The relationship between surface coverage of BSC and the average porosity of the soil in the 0.5–1 mm depth interval. Open square symbol represents one replicate of sample B (filamentous cyanobacteria) which was omitted from the calculated relationship (see in text for details). The relationship is significant ($R^2 = 0.54$, $P = 0.0003$). INSET – the same association modeled with crust type as a random effect, showing the model prediction and 95% confidence intervals. This association remained significant ($P < 0.0001$) when crust type was included in the analysis as a random effect (see text for details).

extreme range of hydrophobicity between the crust types present at the site we have studied, from 'hydrophilic' to 'severely hydrophobic' in typical water drop penetration nomenclature [31]. The cyanobacteria were hydrophilic with low to medium hydraulic conductivity, and had low to medium C content. The late successional liverwort and moss were also hydrophilic but had relatively high hydraulic conductivities and a correspondingly high C content. The third crust type identified as being highly sensitive to disturbance was the extremely hydrophobic foliose lichen *Xanthoparmelia* spp. This crust was visually very fragile and exhibited a high C content but its hydrological behaviour was markedly different compared with all of the other crusts examined, in that it formed a practically water impermeable seal. These results suggest that hydrophobicity could be both species dependent (i.e. related to organism metabolites/polysaccharide secretions [7]) as well as being conditioned by factors such as functional classifications based on BSC morphology and disturbance response.

The functional crust types classified as having mid range sensitivity to disturbance exhibited a high degree of hydrophobicity and low hydraulic conductivity. Equating sensitivity to disturbance with ecological succession, there is a definite 'hump' shaped response in hydrological behaviour with succession. This response is most obvious in the hydrophilic-hydrophobic-hydrophilic nature of the BSCs, but is also clear (inverted) in the hydraulic conductivities. While these hump shaped associations were significant when the different crust types were included in the analysis as a random effect, our estimates of significance were based on limited degrees of freedom and our results only included a limited range of functional types at each successional stage. The higher hydraulic conductivities and lower hydrophobicities associated with the later successional stages also equate to the BSCs with pronounced above ground leaflet type structures (e.g. *Bryum* sp. (moss) and *Riccia* sp. (Liverwort)). It is possible that the structure of these above ground parts directly affects the hydrological behaviour of the surface at the scale we measured (i.e. water droplet scale). Similarly, the reduced above ground structures of the early to mid successional stage BSCs may account for their lower water affinity. The inclusion of other functional crust types from either the early, mid or late successional stages may alter the association we detected. Thus, while our data suggests a link between successional stage/functional form and hydrological behaviour, the link is not definitive and it is as yet unclear if there is a direct ecological benefit for BSCs in different successional stages or functional attributes to exhibit different hydrological behaviour.

Carbon of the BSCs appeared to lag behind the successional categories, only increasing significantly in the late successional BSCs. Porosity values do not appear to change significantly between crust types, but like other measurements, there was a wide

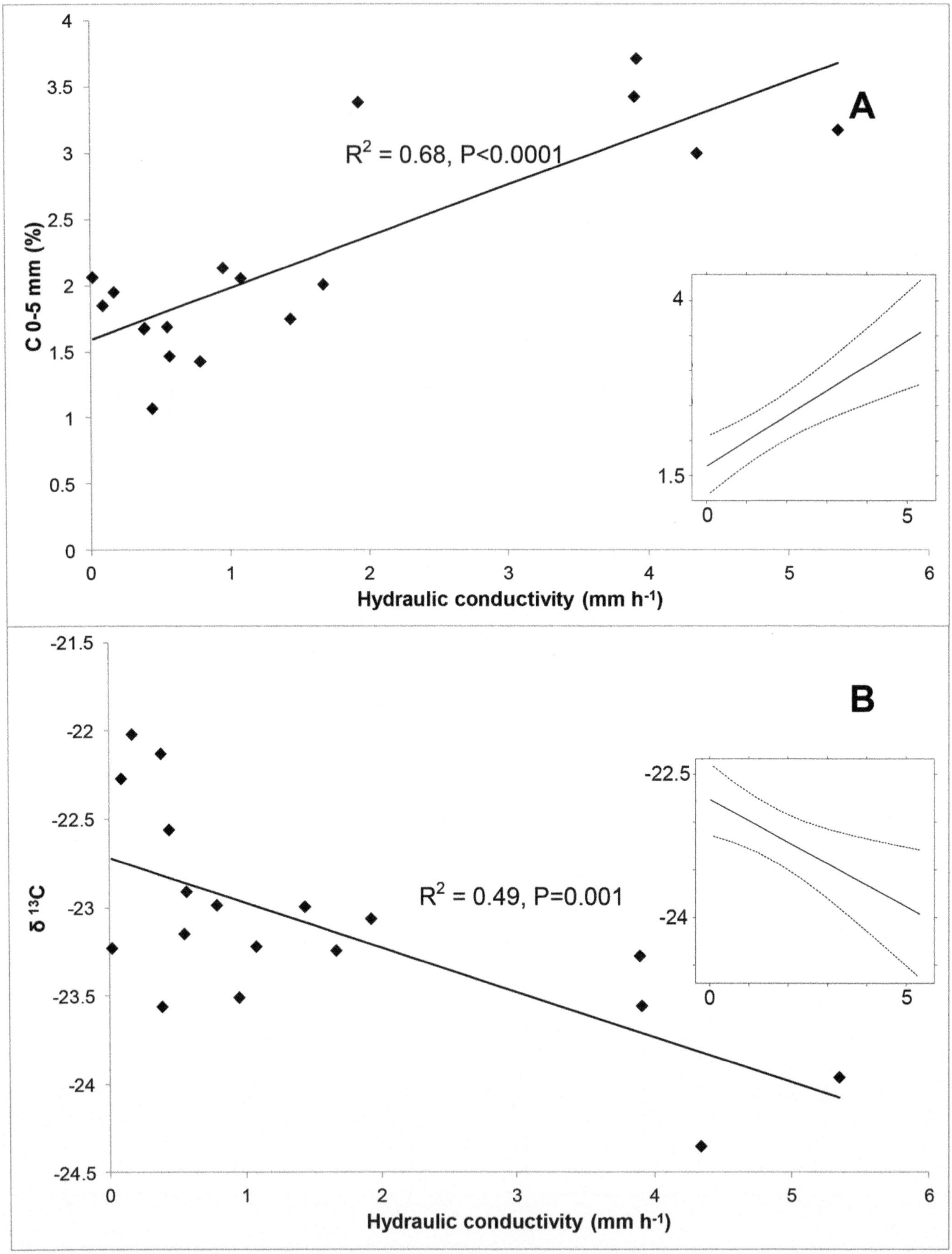
A
R² = 0.68, P<0.0001
C 0-5 mm (%)
Hydraulic conductivity (mm h-1)
B
R² = 0.49, P=0.001
δ 13C
Hydraulic conductivity (mm h-1)

Figure 5. The relationships between C (%) in the 0–5 mm depth interval and hydraulic conductivity (mm h^{-1}) (A), and between $\delta^{13}C$ (0–5 mm) and hydraulic conductivity (mm h^{-1}) (B). Respective relationships are significant ($R^2 = 0.68$, $P < 0.0001$ and $R^2 = 0.49$, $P = 0.001$). INSET – the same associations modeled with crust type as a random effect, showing the model prediction and 95% confidence intervals. These associations remained significant (P = 0.0005 and 0.0095 respectively) when crust type was included in the analyses as a random effect (see text for details). Values from sample G were omitted as outlined in text.

range of variation between replicate measurements. Further exploration of this variation indicated several bivariate relationships that appeared to be independent of crust type.

The hump shaped pattern identified between BSC classification, hydrophobicity and hydraulic conductivities was present as a weak association between hydrophobicity and hydraulic conductivity, but this was confounded by crust type. Hydraulic conductivities of BSCs reported in the literature are often much higher than those reported here, but the difference is consistent with a downscaling effect when smaller samples such as aggregates are measured [33]. The largest average hydraulic conductivity in this study (*Bryum* spp.) was half of the lowest value found in arable Scottish soils using the same custom made infiltrometer. In that study, low values of hydraulic conductivity were considered an indicator of relatively high hydrophobicity. Our results do not directly support this previously found link, but it is probable that the low number of replicates may be compounding the random effect of crust type, effectively masking any existing relationship.

The BSCs with lower hydrophobicity and higher hydraulic conductivity also had the highest C. It is often postulated that high C associated with crusts may retard infiltration as organic detritus and exudates block pores [2]. While we did not find an association between crust coverage and C, we did find that higher crust coverage was associated with lower porosity just beneath the soil surface (i.e. the 0.5–1 mm increment of soil). The coarser scale at which we sampled C (the 0–5 mm increment due to practical limitations) compared to porosity may have masked any existing direct association between C and porosity. Alternatively, a relationship may exist in the pore size range below the resolution of our scans (<100 µm). Miralles-Mellado et al. [34] found several significant associations between porosity and organic C in BSCs from Spain, in which they measured porosity to 50 µm resolution. Thus, both of these possibilities need further examination, particularly as the links between different crust types and micro- and macroporosity do not appear to have been examined in previous BSC studies [14]. Either a smaller volume of the crusted surface could be CT scanned or a larger detector system could be employed to increase the resolution of 3-dimensional scanning and hence the range of pore sizes detected [35]. The latter of these two options is preferable, as selecting a smaller sample volume will decrease the likelihood of calculating a representative porosity for a given increment of soil. Despite these limitations, our results support the hypothesis that higher crust coverage 'clogs' larger pores and reduces associated porosity.

The strong relationship between surface C and hydraulic conductivity may indicate one of two mechanisms; either the increase in hydraulic conductivity associated with the late successional crusts we examined changes wetting and drying cycles and increases C over time, both through crust biomass and exudates/decomposition, or these crusts tend to develop after a surface has increased C to a point that the hydrological properties are altered (i.e. a mid successional crust that has existed for long enough to slowly accrue C to a point that hydraulic conductivity is affected, initiating a change in crust type). Which of these cause and effect relationships dominate needs to be determined to develop our understanding of how, and when, BSCs can be associated with higher surface C. Examining this relationship may also provide information on the ecological benefits both early and late successional BSCs may derive from their relatively hydrophilic nature compared with mid successional crusts.

The link between C and hydrological properties of the BSCs is further evidenced by the association between the isotopic signature of C in the different surfaces and hydraulic conductivity. Biological crusts with the highest hydraulic conductivities also exhibited the most negative $\delta^{13}C$ values. As non-vascular plants are expected to show less C isotopic fractionation with a higher frequency of saturation, this indicates that the BSCs with the lowest inherent ability to conduct water were also those that were saturated with water the longest in this semi-arid environment. While this feature of non-vascular plants has been examined in other environments [11,15], to our knowledge this is the first time this trend has been examined in the type of rangeland system we have investigated. These results suggest that the physiological response (C acquisition) of the different crusts is affected by the hydrological condition of the surface, which in turn alters the chemical signature of the surface. There may be potential to extend the use of C isotopes in this type of research to the occurrence of previous inundation of semi-arid and arid sites, or spatial dynamics of inundation without the need to observe flooding first hand. In addition, given the widespread nature of BSCs in these types of environments, this secondary form of isotopic fractionation may necessitate a re-examination of some model outputs that use $\delta^{13}C$ soil signatures to model shifts between C3 and C4 vegetation in these types of environments [36,37].

Conclusions

This is the first time various hydrological characteristics of a range of crust types have been examined at one site at such a small scale. The high variability found within each crust type and between each crust type helps explain the myriad of different hydrological responses reported in the literature to date. These findings also emphasise the danger of relying on a broad classification such as 'BSC' in which to measure hydrological responses, when a range of crust types, proportional surface coverages, and variable responses may exist at any one site. We did not find any linear relationship between disturbance/successional classification of BSCs and hydrological properties, but there was evidence of a 'hump' shaped association, in that the mid successional/mid range disturbance sensitive crust types were more impermeable to water than either the early pioneering or late successional functional forms, with the exception of *Xanthoparmelia* spp., a highly fragile late successional crust that exhibited higher water impermeability than any other crust type. This trend needs to be confirmed in a wider range of crust types within the spectrum of functional classification.

The small scale at which our measurements were undertaken has allowed us to determine some of the 'inherent' hydrological characteristics of functionally different BSCs, which may serve studies that go on to examine hydrological responses of crusted surfaces at larger scales. For example, the BSCs with the higher hydraulic conductivities and higher values of C in this study are those associated with more pronounced surface relief and tortuosity at larger scales [5], all of which would combine to influence the hydrological responses seen at these larger scales. Similarly, the high runoff associated with smooth, cyanobacteria dominated areas may not be associated with hydrophobicity, but may be related to other factors such as high surface coverage and corresponding low porosity. However, such upscaling will not be simple – BSCs occur as complex assemblages of functional types, and an amalgamated response may be difficult to relate back to small scale measurements, unless only one or two functional forms are dominant at the large as well as small scale. One additional factor related to upscaling responses needs further examination – the influence of hydrological behaviour of crusted surfaces on C isotopic signatures of soil surfaces. As semi-arid and arid rangelands can have large areas dominated by BSCs, the effect of crust isotopic signatures on vegetation change modelling needs to be assessed.

Supporting Information

Figure S1 Image of the study site showing mixed mid-storey of *Geijera*, *Dodonaea* and *Eremophila* sp and minimal herbaceous ground cover.

Table S1 Summarised general additive mixed model (GAMM) outputs for the relationships depicted in Figure 2, including crust type as a random factor, and specifying k = 3 degrees of freedom for the non-parametric spline term of the model.

Table S2 Summarized linear mixed model outputs for bivariate regressions including crust type as a random factor, including MCMC (Markov Chain Monte Carlo) estimated P values.

Acknowledgments

The authors wish to thank Mr. Anthony Gibson for permission to undertake the study on private land, and for providing site information, Ms. Megan Good for assistance in crust identification, and Mrs. Leanne Lisle for assistance with chemical analysis. Comments and suggestions from Dr. Jordi Moya-Laraño and two anonymous reviewers improved this manuscript and are greatly appreciated.

Author Contributions

Conceived and designed the experiments: MT REH RJF IMY. Performed the experiments: MT REH RJF. Analyzed the data: MT REH RJF. Contributed reagents/materials/analysis tools: MT RJF IMY. Wrote the paper: MT REH RJF IMY.

References

1. Evans RD, Johansen JR (1999) Microbiotic crusts and ecosystem processes. Critical Reviews in Plant Sciences 18: 183–225.
2. Belnap J (2006) The potential roles of biological soil crusts in dryland hydrologic cycles. Hydrological Processes 20: 3159–3178.
3. Chamizo S, Stevens A, Canton Y, Miralles I, Domingo F, et al. (2012) Discriminating soil crust type, development stage and degree of disturbance in semiarid environments from their spectral characteristics. European Journal of Soil Science 63: 42–53.
4. Eldridge DJ, Greene RSB (1994) Microbiotic soil crusts: a review of their roles in soil and ecological processes in the rangelands of Australia. Australian Journal of Soil Research 32: 389–415.
5. Eldridge DJ, Tozer ME (1997) A practical guide to soil lichens and bryophytes of Australia's dry country. Sydney, Australia: Department of Land and Water Conservation. 80 p.
6. Tighe M, Muñoz-Robles CA, Reid N, Wilson B, Briggs SV (2012) Hydrological thresholds of soil surface properties identified using conditional inference tree analysis. Earth Surface Processes and Landforms 37: 620–632.
7. Warren SD (2003) Synopsis:Influence of biological soil crusts on arid land hydrology and soil stability. In: Belnap J, Lange OL, editors. Biological Soil Crusts: Structure, Function, and Management. Berlin: Springer-Verlag. 349–360.
8. Greene RSB, Tongway DJ (1989) The significance of (surface) physical and chemical properties in determining soil surface condition of red earths in rangelands. Australian Journal of Soil Research 27: 213–225.
9. Muñoz-Robles CA, Reid NC, Tighe M, Briggs SV, Wilson B (2011) Soil hydrological and erosional responses in patches and inter-patches in vegetation states in semi-arid Australia. Geoderma 160: 524–534.
10. Eldridge DJ (2003) Biological Soil Crusts and Water Relations in Australian Deserts. In: Belnap J, Lange OL, editors. Biological Soil Crusts: Structure, Function, and Management. Berlin: Springer-Verlag. 315–326.
11. Wasley J, Robinson SA, Lovelock CE, Popp M (2006) Some like it wet - biological characteristics underpinning tolerance of extreme ater stress events in Antarctic bryophytes. Functional Plant Biology 33: 443–455.
12. Greene RSB, Chartres CJ, Hodgkinson KC (1990) The effect of fire on the soil in a degraded semi-arid woodland. I. Cryptogam cover and physical and micromorphological properties. Australian Journal of Soil Research 28: 755–777.
13. White I, Sully MJ, Perroux KM (1992) Measurement of Surface - Soil Hydraulic Properties: Disk Permeameters, Tension Infiltrometers, and Other Techniques. In: Topp GC, Reynolds WD, Green RE, editors. Advances in measurement of soil physical properties: Bringing theory into practice: Proceedings of a symposium sponsored by Division S-1 of the Soil Science Society of America. San Antonio, Texas: Soil Science Society of America. 69–103.
14. Menon M, Yuan Q, Jia X, Dougill AJ, Hoon SR, et al. (2011) Assessment of physical and hydrological properties of biological soil crusts using X-ray microtomography and modeling. Journal of Hydrology 397: 47–54.
15. Rice SK, Giles L (1996) The influence of water content and leaf anatomy on carbon isotope discrimination and photosynthesis in *Sphagnum*. Plant, Cell and Environment 19: 118–124.
16. Isbell RF (1996) The Australian soil classification. Melbourne, Australia.: CSIRO.
17. Abramoff MD, Magalhaes PJ, Ram SJ (2004) Image processing with ImageJ. Biophotonics International 11: 36–42.
18. O'Donnell AG, Young IM, Rushton SP, Shirley MA, Crawford JW (2007) Visualization, modelling and prediction in soil microbiology. Nature Reviews Microbiology 5: 689–699.
19. Yen J, Chang F, Chang S (1995) A new criterion for automatic multilevel thresholding. IEEE Transactions on Image Processing 4: 370–378.
20. Crawford J, Deacon L, Grinev D, Harris J, Ritz K, et al. (2012) Microbial diversity affects self-organisation of the soil-microbe system with consequences for function. Journal of the Royal Society Interface 9: 1302–1310.
21. Bird M, Ascough P, Young I, Wood C, Scott A (2008) X-ray tomographic imaging of charcoal. Journal of Archaelogical Science 35: 2698–2706.
22. Doerr SH, Shakesby RA, Walsh RPD (2000) Soil water repellency: its causes, characteristics and hydro-geomorphological significance. Earth-Science Reviews 51: 33–65.
23. Feeney DS, Hallett PD, Rodger S, Bengough AG, White NA, et al. (2006) Impact of fungal and bacterial biocides on microbial induced water repellency in arable soil. Geoderma 135: 72–80.
24. Philip JR (1954) Some recent advances in hydrologic physics. Journal of the Institution of Engineers, Australia 26: 255–259.
25. Smiles DE, Knight JH (1976) A note on the use of the Philip infiltration equation. Australian Journal of Soil Research 14: 103–108.
26. Fox J (2003) Effect displays in R for Generalised Linear Models. Journal of Statistical Software 8: 1–27.
27. R Development Core Team (2011) R: A language and environment for statistical computing. Vienna, Austria.: R Foundation for Statistical Computing.
28. Wood SN (2004) Stable and efficient multiple smoothing parameter estimation for generalized additive models. Journal of the American Statistical Association 99: 673–686.
29. Bates DM (2005) Fitting linear mixed models in R. R News 5: 27–30.
30. Torsten H, Bretz F, Westfall P (2008) Simultaneous Inference in General Parametric Models. Biometrical Journal 50: 346–363.
31. Doerr SH (1998) On standardizing the 'water drop penetration time' and the 'molarity of an ethanol droplet' techniques to classify soil hydrophobicity: a case study using medium textured soils. Earth Surface Processes and Landforms 23: 663–668.
32. Chamizo S, Cantón Y, Lázaro R, Solé-Benet A, Domingo F (2012) Crust composition and disturbance drive infiltration through biological soil crusts in semiarid ecosystems. Ecosystems 15: 148–161.
33. Hallett PD, Young IM (1999) Changes to water repellence of soil aggregates cause by substrate-induced microbial activity. European Journal of Soil Science 50: 35–40.

34. Miralles-Melado I, Cantón Y, Solé-Benet A (2011) Two-dimensional porosity of crusted silty soils: Indicators of soil quality in semiarid rangelands? Soil Science Society of America Journal 75: 1289–1301.
35. Taina IA, Heck RJ, Elliot TR (2008) Application of X-ray computed tomography to soil science: A literature review. Canadian Journal of Soil Science 88: 1–20.
36. Jackson RB, Banner JL, Jobbagy EG, Pockman WT, Wall DH (2002) Ecosystem carbon loss with woody plant invasion of grasslands. Nature 418: 623–626.
37. Wiesmeier M, Steffens M, Kolbl A, Kogel-Knabner I (2009) Degradation and small-scale spatial homogenization of topsoils in intensively-grazed steppes of Northern China. Soil & Tillage Research 104: 299–310.
38. Tighe M, Reid N, Wilson B, Briggs SV (2009) Invasive native scrub and soil condition in semi-arid south-eastern Australia. Agriculture Ecosystems & Environment 132: 212–222.

4

Interacting Coastal Based Ecosystem Services: Recreation and Water Quality in Puget Sound, WA

Jason Kreitler[1]*, Michael Papenfus[2], Kristin Byrd[3], William Labiosa[4]

1 Western Geographic Science Center, U.S. Geological Survey, Boise, Idaho, United States of America, **2** Woods Institute for the Environment, Stanford University, Stanford, California, United States of America, **3** Western Geographic Science Center, U.S. Geological Survey, Menlo Park, California, United States of America, **4** Western Geographic Science Center, U.S. Geological Survey, Seattle, Washington, United States of America

Abstract

Coastal recreation and water quality are major contributors to human well-being in coastal regions. They can also interact, creating opportunities for ecosystem based management, ecological restoration, and water quality improvement that can positively affect people and the environment. Yet the effect of environmental quality on human behavior is often poorly quantified, but commonly assumed in coastal ecosystem service studies. To clarify this effect we investigate a water quality dataset for evidence that environmental condition partially explains variation in recreational visitation, our indicator of human behavior. In Puget Sound, WA, we investigate variation in visitation in both visitation rate and fixed effects (FE) models. The visitation rate model relates the differences in annual recreational visitation among parks to environmental conditions, park characteristics, travel cost, and recreational demand. In our FE model we control for all time-invariant unobserved variables and compare monthly variation at the park level to determine how water quality affects visitation during the summer season. The results of our first model illustrate how visitation relates to various amenities and costs. In the FE analysis, monthly visitation was negatively related to water quality while controlling for monthly visitation trends. This indicates people are responding to changes in water quality, and an improvement would yield an increase in the value of recreation. Together, these results could help in prioritizing water quality improvements, could assist the creation of new parks or the modification of existing recreational infrastructure, and provide quantitative estimates for the expected benefits from potential changes in recreational visitation and water quality improvements. Our results also provide an example of how recreational visitation can be quantified and used in ecosystem service assessments.

Editor: Just Cebrian, MESC; University of South Alabama, United States of America

Funding: This project was supported by Environmental Protection Agency Interagency Agreement IA DW 14-95762701 and the United States Geological Survey Geographical Analysis and Monitoring Program. The funders had no role in study design, data collection and analysis, decision to publish, or preparation of the manuscript.

Competing Interests: The authors have declared that no competing interests exist.

* E-mail: jkreitler@usgs.gov

Introduction

Coastal and marine ecosystem services – the benefits people derive from marine and coastal ecosystems [1,2] – are increasingly used in applied marine ecosystem-based management (EBM) and decision making [1,3,4,5]. This group of ecosystem services can include: coastal protection from reefs, vegetation, and dunes [6,7,8]; provision of marine products like fish and shellfish [9,10]; nutrient cycling and waste filtration [11]; recreational opportunities [12]; and cultural values [11,13]. An understanding of how human actions affect marine ecosystem condition and composition, in the context of EBM and restoration, could guide decisions intended to positively affect the management of ecosystem services and their benefits to human society [5,14].

In Puget Sound, WA, coastal recreational opportunities and water quality are major contributors to human well-being [13,15]. Puget Sound tourism and recreation create annual revenues of over $5 billion and provide 62,000 jobs [16], making them important economic contributors. Shellfishing is an important recreational and commercial activity that is regulated based on water quality. The benefits from improved coastal water quality and increased recreation could also compound to produce opportunities for management and restoration actions that would positively affect people, the environment, and the economy. However, the effect of water quality on recreational behavior in Puget Sound has not been quantified, nor have patterns and variation in recreational use been explored for this region. In Southern California, research has addressed the costs of beach closures [17,18,19], health effects [20,21], and public preferences and the value of recreation [17,18,22,23,24]; these studies provide background for this work, but differences between Puget Sound and Southern California are many, and could lead to different relationships among the factors affecting recreational behavior in Puget Sound.

Studies including recreation as an ecosystem service have used a variety of methods to calculate recreational benefits. Chan et al. [25] used a multicriteria weighted sum to create a GIS value surface of recreational lands, Eigenbrod et al. [26] used survey data to map the popularity of lands for recreational outings, and Hicks et al. [12] and Hein et al. [27] employed the travel cost method (TCM) to value recreation benefits at specific sites. Each technique has benefits and disadvantages. TCM models can value the economic benefits from recreation [28,29,30], but rely on visitation data, which is rare in many areas. Map-based approaches that model the complete coverage of recreational benefits across a study area are intuitive and work well in multi-service assessments, but may not be verifiable. By drawing from both approaches, terrestrial and coastal ecosystem service studies could quantify the factors affecting recreational use, and characterize how land use land cover, demographic, or policy changes may influence multiple ecosystem services [5].

To document and clarify the potential effects of water quality and factors affecting recreational use we investigate two datasets: the first contains recreational visits to Washington State Parks that have access to Puget Sound waters, while the second, a water quality monitoring dataset, is analyzed jointly with visitation to determine if environmental condition partially explains variation in recreational visits. Specifically, we address the following questions:

1. What factors explain variation in the recreational use of State Parks within Puget Sound; and
2. What effect does water quality have on recreational visitation?

The answers to these questions provide support for the ecosystem service objectives of the Puget Sound Nearshore Ecosystem Restoration Project (PSNERP - the proposed large scale estuary restoration in Puget Sound [www.pugetsoundnearshore.org]), the Puget Sound Partnership [31](PSP 2008), and for marine and terrestrial recreation components of ecosystem service models and studies [5,32].

Methods

1.1. Study Area

The Puget Sound region of Washington, USA, (figure 1) is roughly contained by the watersheds draining into Puget Sound from the Cascade and Olympic Mountains, and the Strait of Juan de Fuca between Washington and Vancouver Island, Canada. This land area of approximately 35,500 km^2 has a shoreline of nearly 4,000 km, making the coast a prominent feature on the landscape for the region's 4.4 million inhabitants.

Because much the Puget Sound is deep (450 ft mean depth), the shallow nearshore environment that is most productive and critical for natural ecosystem processes is relatively narrow [16]. The nearshore zone supports many of the marine ecosystem services, including fisheries and recreation, which are characteristic of the Puget Sound region [33]. Historical development patterns have also been centered in this nearshore area due to the maritime roots of Seattle and other nearby communities. Though the region is now a diversified large metropolitan area, the importance of the Sound's ecosystem services are still apparent to residents: over 500,000 boats are registered in the Sound, 280 marinas are operated in the region, the EPA has designated Puget Sound as an Estuary of National Significance, and studies are underway to begin large-scale ecosystem based restoration projects similar in scope to efforts in the Chesapeake Bay and the Florida Everglades [16].

1.2. Data

1.2.1. Visitation. Washington State Parks records visitation numbers based on entrance, camping, and mooring fees. Data are available by month beginning in the late 1980s to mid 1990s to the present, dependent on park. These count data conform to a Poisson distribution through visual inspection, and are simple visitor counts by type to a specific state park and are not distinguished by type of visit (e.g., day use, camping, or mooring). Fifty-seven parks that provide access to the Sound are used in our travel cost method model (table 1); seventeen parks where visitation data and water quality samples were collected concurrently are used for our fixed effects panel model. The requirement of water quality samples and visitation counts leads to an unbalanced panel with 140 observations (Table 2).

1.2.2. Travel distance and demand. Recreational visitation rates are often a function of site amenities and demand from nearby population centers, with demand typically declining with distance [34,35]. We model the relationship between distance and visitation rate using an independent dataset from the Washington State parks reservation system containing a visitor's ZIP code. This dataset is important because it contains the origin location as well as the destination, which the larger visitation dataset lacks. We fit a demand function for visits within 500 miles from Puget Sound parks. Using a road and ferry system dataset and the Network Analyst toolbox in ArcGIS [36], we estimated travel distances for the entire origin/destination matrix. This function relates the visitation rate (# visits/ZIP code population) to the distance traveled ($p<0.001$, $r^2=.67$, figure 2). We then used this relationship to aggregate demand as a function of distance weighted population around each park using the population availability (PA) method of Coombes et al. [37]:

$$PA=\sum_{i=1}^{n}[P_i(a\exp(-bx_i))] \tag{1}$$

where n is the number of US census blocks or Canadian census divisions within the travel distance, i is the census unit, P is the population size of i, a is a constant, b is the decay coefficient, and x is distance. Both a and b are derived from observed relationship in figure 2. The value of PA for each park is then used as an independent variable in explaining park visitation, and is expected to have a positive sign in the visitation model.

1.2.3. BEACH water quality. The Washington State Department of Ecology and Department of Health (DOH) monitor water quality in Puget Sound through the EPA's national Beaches Environmental Assessment and Coastal Health (BEACH) program. *Enterococcus* is tested weekly at coastal recreational swimming beaches during summer months, with water quality results posted on a public website that also notifies users of known pollution events. Beach advisories are issued by counties when bacteria concentrations exceed the EPA threshold, while closures are mainly due to sewage spills or repeated high counts from unknown sources. *Enterococcus* counts exceeded the EPA threshold at seven of the parks we analyzed in this dataset (figure 3).

1.2.4. Other variables. We use several variables to partially explain the variation in visitation. In the fixed effects (FE) model, to control for seasonal, inter-annual, and geographic variations in weather among parks in Puget Sound we use monthly precipitation data by year from the PRISM database [38](PRISM, 2010). In the visitation rate model other explanatory variables included in the model selection process include type of access to the park, park size, beach length, number of listed activities and concessions at each park according to park literature, the number of campsites, and travel time. Travel time to parks from downtown Seattle (which coincidentally is also the population weighted center of the Puget Sound region) was calculated using the same networked travel distance dataset.

1.3. Visitation Rate Model

Recreational demand models typically use the travel cost method (TCM) to explain variation in visitation counts [28,39,40]. These models assume that the visitation rate to a

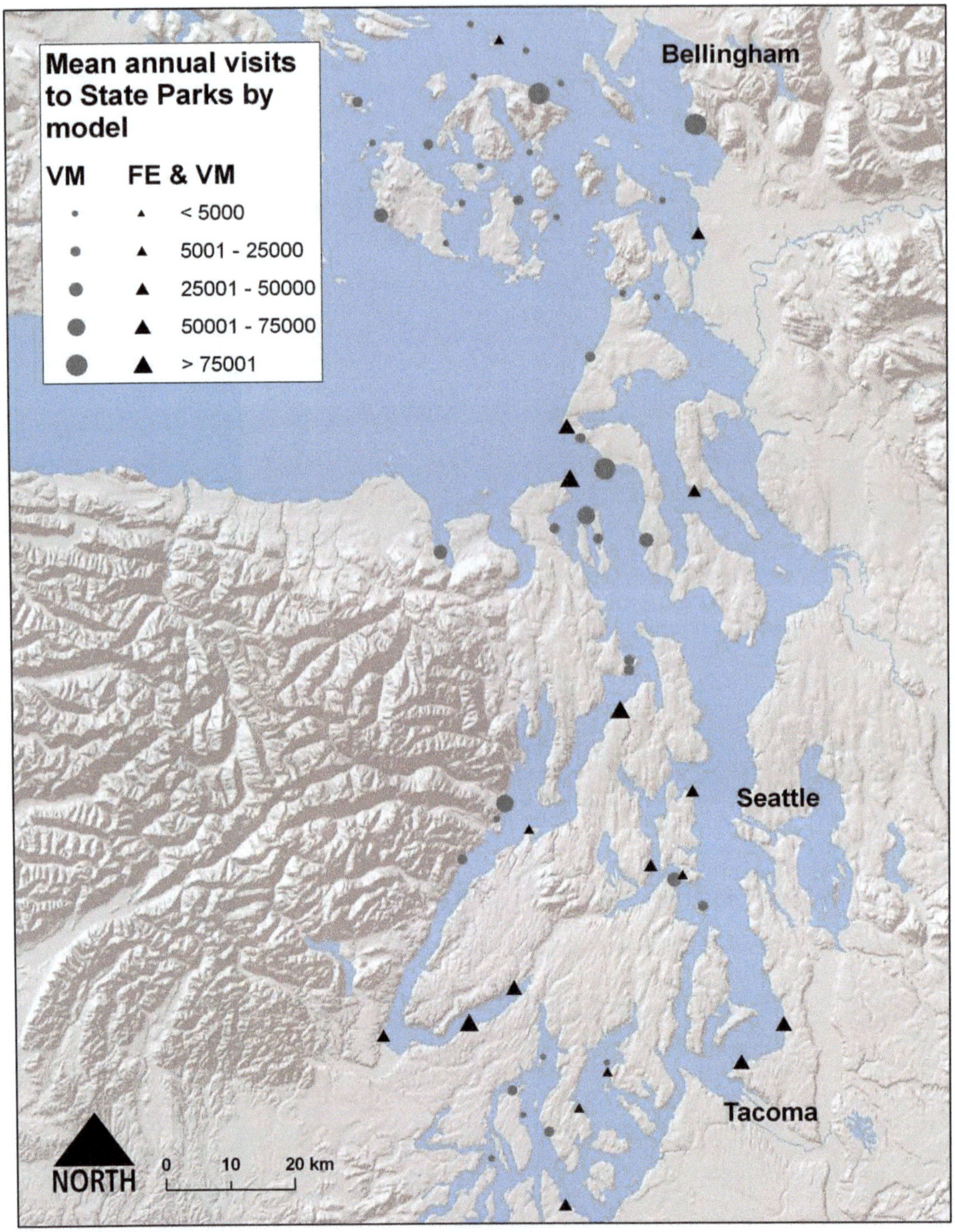

Figure 1. FE and visitation rate model locations and variation in visitors.

location depends on the cost of travel from an origin to the destination, socioeconomic factors, and entrance fees. As observed in figure 2, visitation rate increases as travel cost decreases.

We model recreational visitation as a function of park characteristics, travel cost, access, and population availability (Eq 1). Visits are a count variable modeled using the negative binomial distribution [41]. Our data are overdispersed (variance larger than the mean) and without zero counts, thus two models were tested once specified: the negative binomial (NB) and zero-truncated negative binomial (ZTNB) model. This technique addresses the three main problems associated with truncated count data, that they are non-negative integers, cannot have zero values, and are often over-dispersed [42]. We use an information theoretic approach [43] for model selection, and estimate the models in Stata [44].

The models for mean annual park visitation are estimated as:

$$\ln V_i = \beta_0 + \beta_1 C_i + \beta_2 P_i + \beta_3 D_i \qquad (2)$$

where V is the mean annual count of visitors at park i, C is a vector

Table 1. Summary statistics: visitation rate model data.

Variable	Unit	Mean	Std. Dev.
Annual Visits	# Visitor days/year	32,583	42,862
Campsites	#	38.80	47.40
Camping	Dummy	0.75	0.43
Park size	Acres	284	801
Shore length	meters	2,414	3,241
Population availability	# People (eq 1)	29,214	33,422
Travel time	Minutes	129	63
Travel distance	kilometers	76	35
Ferry	Dummy	0.31	0.46
PWC access	Dummy	0.26	0.44
Activities	#	5.90	3.20
Concessions	#	0.73	1.18
Annual ppt	millimeters	926	335
Summer ppt	millimeters	90	12
Sandy	Dummy	0.31	0.46
Heritage	Dummy	0.10	0.30
Shellfishing	Dummy	0.82	0.38
n = 57			

Table 2. Summary statistics: fixed effect model data.

	Unit	Obs	Month	Mean	Std. Dev.
Visits	Visitor days	46	June	46,235	43,020
		47	July	65,517	50,308
		47	August	57,237	42,375
Bacteria	#/100 ml	46	June	28.1	41.9
		47	July	23.4	29.7
		47	August	17.6	12.4
Precipitation	mm	46	June	35	11
		47	July	22	21
		47	August	33	36
Observations		140			
Parks		17			
Obs/Park	min	3			
	mean	8.2			
	max	12			

for the park's travel cost and access, P is a vector of characteristics of each park, and D is the population availability surrounding each park.

1.4. Fixed Effects (FE) Model

We test the effect of water quality on state park visitation through a fixed-effects panel estimation. The count data are again distributed according to a negative binomial, but in this case we pair repeated monthly visitation counts with *Enterococcus* surveys during summer months to determine the magnitude and direction of the effect of water quality variation. All time invariant heterogeneity among parks is controlled for by this statistical technique [41], leaving changes in visitation to changes in water quality, weather, and time effects. The fixed effects model is estimated as:

$$V_{it} = \beta_0 + \beta_1 E_{it} + \beta_2 M + \beta_3 Y + \beta_4 W_{it} \quad (3)$$

where V is the monthly count of visitors at park i in year t, E is the park's environmental condition as proxied by the *Enterococcus* counts, M is a vector of summer month dummy variables to control for the monthly variation, Y is a vector of year dummy variables to control for interannual variation, and W is the time-variant mean monthly weather conditions. Equation (2) is estimated using the fixed-effects negative binomial panel model in Stata [44], with reference to a June, 2004 baseline (June and 2004 dropped to avoid multicolinearity).

Results

2.1 Visitation Rate Model

Results from our model can be seen in table 3. Using the 57 State Parks in the dataset, our model explains nearly 70% of the null deviance in mean annual park visitation. Out of the variables initially analyzed, six were retained in the best model through an information theoretic approach [43]. The variables that increased visitation include the number of campsites at a park, the park size, and the number of possible activities at the park. Variables that negatively affected visitation include a dummy variable describing accessibility limited to private non-commercial watercraft, the population availability (equation 1), and the travel time to a park from Seattle. The negative binomial and zero-truncated negative binomial performed almost identically, with only small differences among coefficients and between chi^2 statistics.

2.2 FE Model

The fixed-effects model results can be seen in table 4. The coefficient of the natural logarithm of the water quality variable, mean *Enterococcus* counts, indicates a 10% increase will decrease visitation to state parks by 2.5%. The model controls for time-invariant factors that could be affecting visitation counts other than water quality. Month and year effects (except 2005) are significant, while weather effects were insignificant; thus we rule out the potential effect of poor weather contributing to decreased visitation. The month dummy variables for July and August are both positive and have roughly the same coefficient value, indicating they exert a similar increase in visitation in reference to the month of June when conditions are cooler and the summer travel season has just begun. Year effects, with reference to 2004, follow prevailing downward economic conditions, and control for general financial factors among years that may influence visitation.

Discussion

When assessing recreation as an ecosystem service [2,25], visitation is the measure most commonly used to model and quantify the variation of this service. Globally, trends in tourism related to outdoor recreation and wildlife viewing are increasing [45], with recognition that ecosystem services play an important role in generating revenue for conservation and local development [46,47,48]. We used the visitation rate model to understand the factors affecting the regional pattern of recreational visitation, and the FE model to determine how water quality can affect visitation at Puget Sound State Parks. These results could be used in

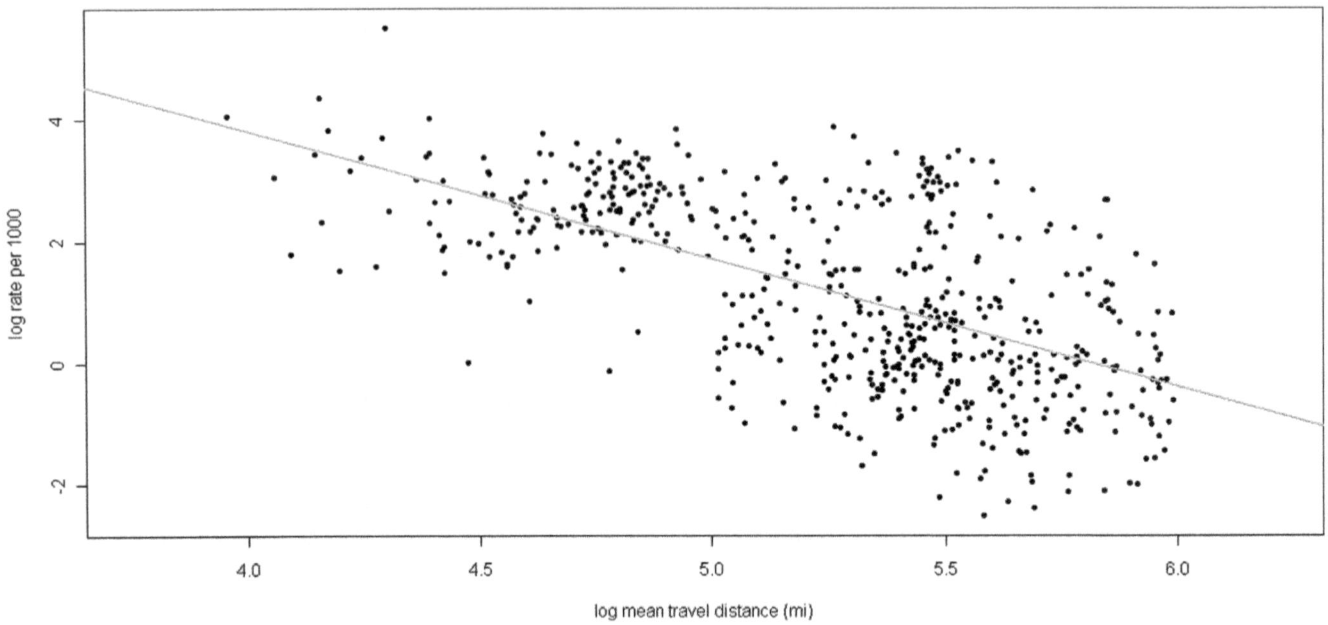

Figure 2. ZIP code visitation rate to Puget Sound State Parks by distance. The natural log of the visitation rate (# visits/1000 population) as a function of the natural log of mean travel distance by zip code ($p<0.001$, $r^2=0.67$).

planning for restoration and water quality improvement decisions in Puget Sound, and give an example of how GIS analysis and visitation modeling can give a thorough understanding of recreational behavior for use in ecosystem service assessments.

The factors affecting visitation to Puget Sound state parks are similar to other studies that have analyzed recreational behavior in coastal [49,50] and terrestrial areas [34,51]. The visitation rate model variables that performed as expected include park size, travel distance, campsites, amenities, and access. All sites within our study had public access, but some require personal watercraft (typically kayaks or small boats) for travel to smaller islands or secluded coastlines without land or ferry access. The size, number

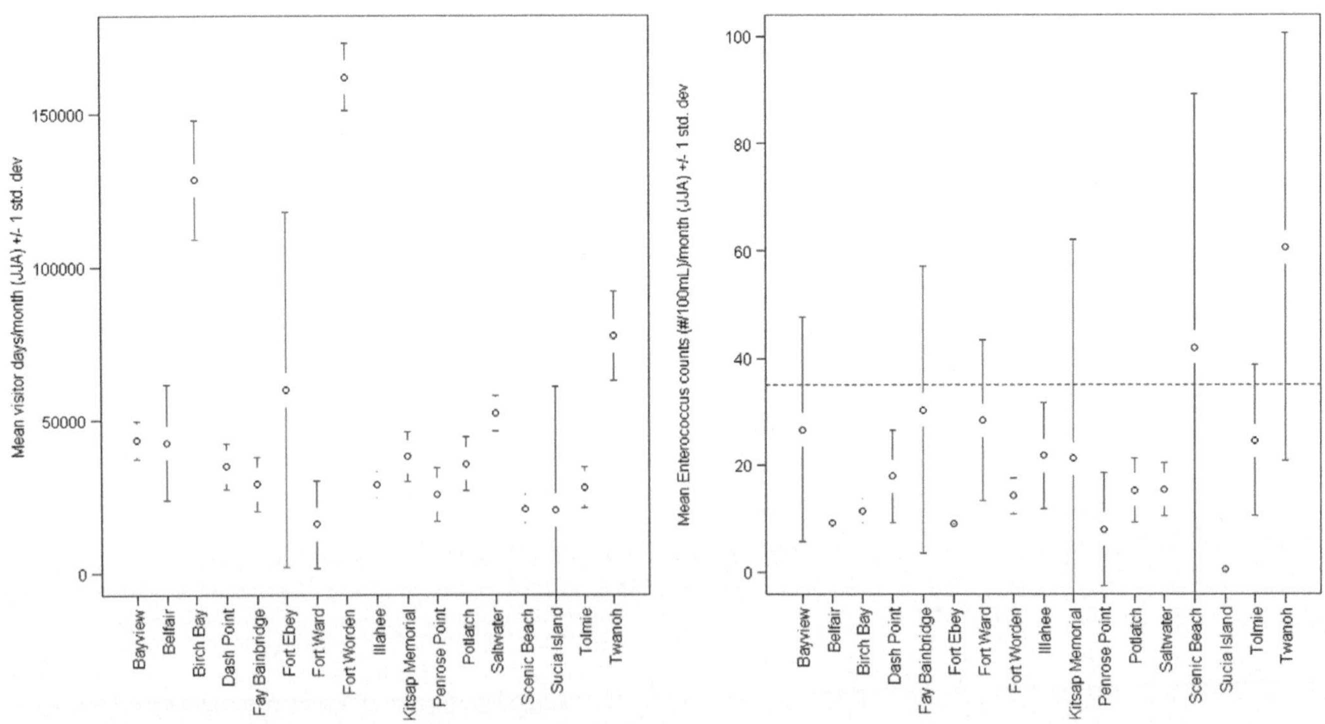

Figure 3. Variation in visitation. Variation in visitation (mean visitor days/month) at the 17 FE model state parks and paired Enterococcus counts from the BEACH program. Dotted line represents the EPA's marine Enterococcus threshold.

Table 3. Visitation rate model results (equation 2).

	NB model				ZTNB model			
Variables	**Coefficients**	**Std. error**	*z*	*p*	**Coefficients**	**Std. error**	*z*	*p*
campsites	0.0078	0.0028	2.75	0.006	0.0078	0.0028	2.75	0.006
ln(acres)	0.2232	0.0717	3.11	0.002	0.2232	0.0717	3.11	0.002
activities (#)	0.1332	0.0450	2.96	0.003	0.1332	0.0450	2.96	0.003
PWC access (dummy)	−1.6565	0.3036	−5.46	0.000	−1.6565	0.3036	−5.46	0.000
ln(PA)	−0.4482	0.1420	−3.16	0.002	−0.4482	0.1420	−3.16	0.002
ln(travel time)	−0.9147	0.3174	−2.88	0.004	−0.9147	0.3174	−2.88	0.004
cons	16.7207	2.6793	6.24	0.000	16.7208	2.6794	6.24	0.000
Dependent variable = mean annual park visitation								
n =	57				57			
LR chi2	75.28				75.08			
prob>chi2	0.000				0.000			

of amenities, and number of campsites are important variables in this context because our data contained general visitation counts that were not stratified by specific recreational activities. Therefore we would expect larger parks with a greater variety of potential activities to attract a larger number of visits. Travel time negatively affected visitation, and was measured from downtown Seattle to each park. We used this as a proxy for actual travel time from origin to park destination because origins were not recorded in the larger State Park dataset. Nonetheless, this travel cost measure likely captures the variation in trip length to parks using the road and ferry network in Puget Sound, and is a large improvement over previous methods [52].

Contrary to our *a priori* expectations, PA (eq 1) had a significant negative relationship with visitation. This is likely due to the activity and purpose of a State Park visit, where a preference for a more natural or semi-natural settings away from urbanized areas could be expected based on local values. If our dataset included frequently used urban parks and coastal access points we might have found less of a negative effect. Therefore the type of recreational activity should be carefully considered when analyzing recreational use, particularly if certain activities may not be desirable in all locations. Similarly, we could be observing visitation displacement (or crowd avoidance, [18,53]) by park visitors. These results indicate a need to collect and analyze visitation data at other points on the Sound to compare recreational use across the entire study area.

Table 4. Response of visits to water quality variation (equation 3).

Variable	Coefficient	Std. error	z	*p*
ln(count)	−0.2565	0.0631	−4.0600	0.0000
ppt	−9.62E−06	1.55E−05	−0.6200	0.5340
June	(dropped)			
July	0.5571	0.1016	5.4800	0.0000
Aug	0.5414	0.0986	5.4900	0.0000
yr2004	(dropped)			
yr2005	−0.1644	0.1095	−1.5000	0.1330
yr2006	−0.2597	0.1207	−2.1500	0.0310
yr2007	−0.4156	0.1223	−3.4000	0.0010
cons	1.9905	0.2307	8.6300	0.0000
Dependent variable = monthly park visitation				
n =	140			
Wald chi^2	87.39			
prob >chi^2	0.0000			

In the fixed-effects model we show that increasing *Enterococcus* counts negatively affect the number of state park visits made in Puget Sound through a revealed preference approach. Swimming, shellfishing, tide pooling, and other recreational activities with water contact are primary activities at these state parks, so it is understandable that decreased water quality would affect visitation. Of the 50 beaches that DOH monitors as part of the BEACH program within Puget Sound, seven failed to meet EPA water quality standards greater than 8% of the time between 2004–2009 [54]. Our sample of parks contains four that failed the BEACH standard greater than 4% of the time, while the standard deviation of *Enterococcus* counts at seven parks overlapped the EPA marine threshold of 35/100 mL (figure 2). Water quality monitoring data has been available since 2003 through a map-based web user interface from the DOH, and has received an average of 170,000, 186,000, and 147,000 web hits for the months of June, July and August, respectively (figure 4). Though survey evidence of actual beach users would be a more definitive source, the number of web hits to the DOH site supports the explanation that variation in visitation could be from recreational avoidance of higher bacterial counts.

The response of recreational behavior to water quality and beach advisories is mixed in the literature. Recreationalists, in general, respond to water quality but not necessarily beach advisories, though there are few studies documenting either. Busch [21] found surfers in California reduced their exposure to poor water quality by following the "72 hour rule" (avoiding water contact for 72 hrs after a rain event), rather than heeding posted beach advisories. Similarly, a study in San Diego, California found beach advisories did not affect recreational site choice, even though survey respondents ranked water quality the highest factor affecting beach experiences [17]. Hanemann et al., [22] and Hanley et al. [50] found a negative response of beach goers to

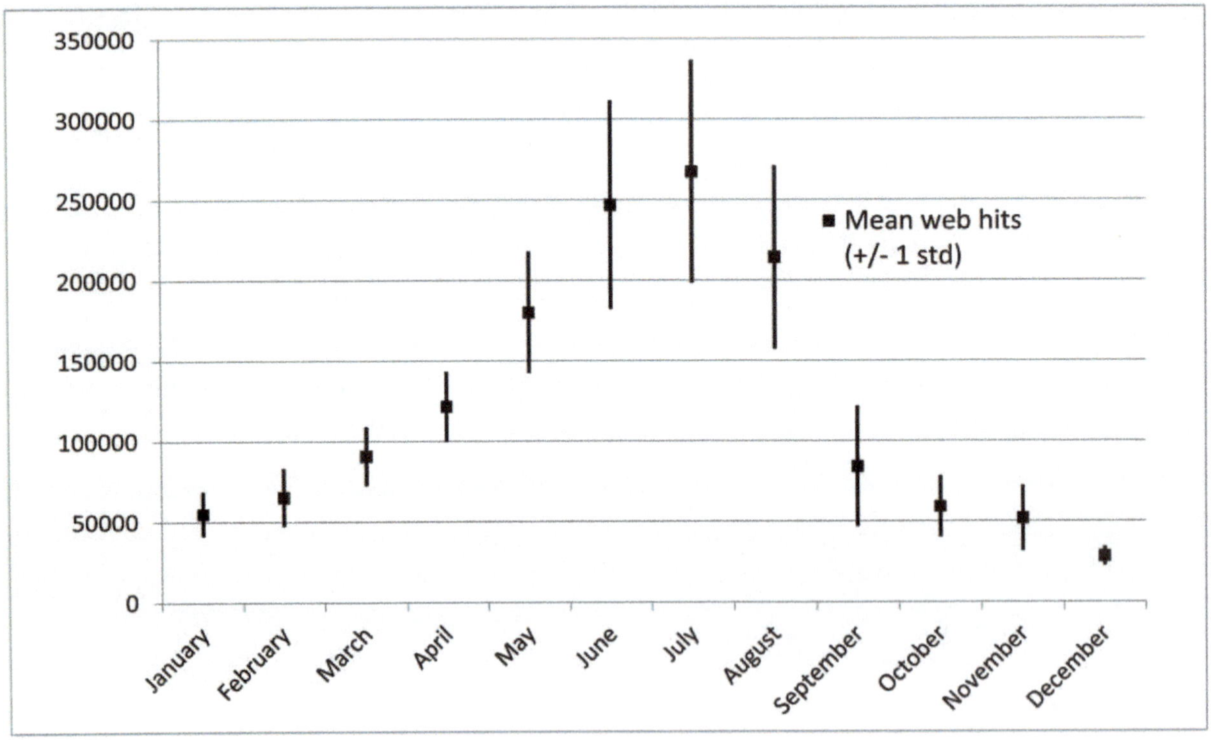

Figure 4. Monthly web hits. Monthly web hits (mean +/−1 st dev) to the WA BEACH program water quality map from 2005–2010.

declining water quality, but only Busch [21] was a true longitudinal revealed preference study. This research adds to the relatively few studies attempting to assess the effect of water quality on recreational visitation, and is the only study available analyzing Puget Sound.

The two models presented in this study were developed for potential Puget Sound wide restoration activities by PSNERP and

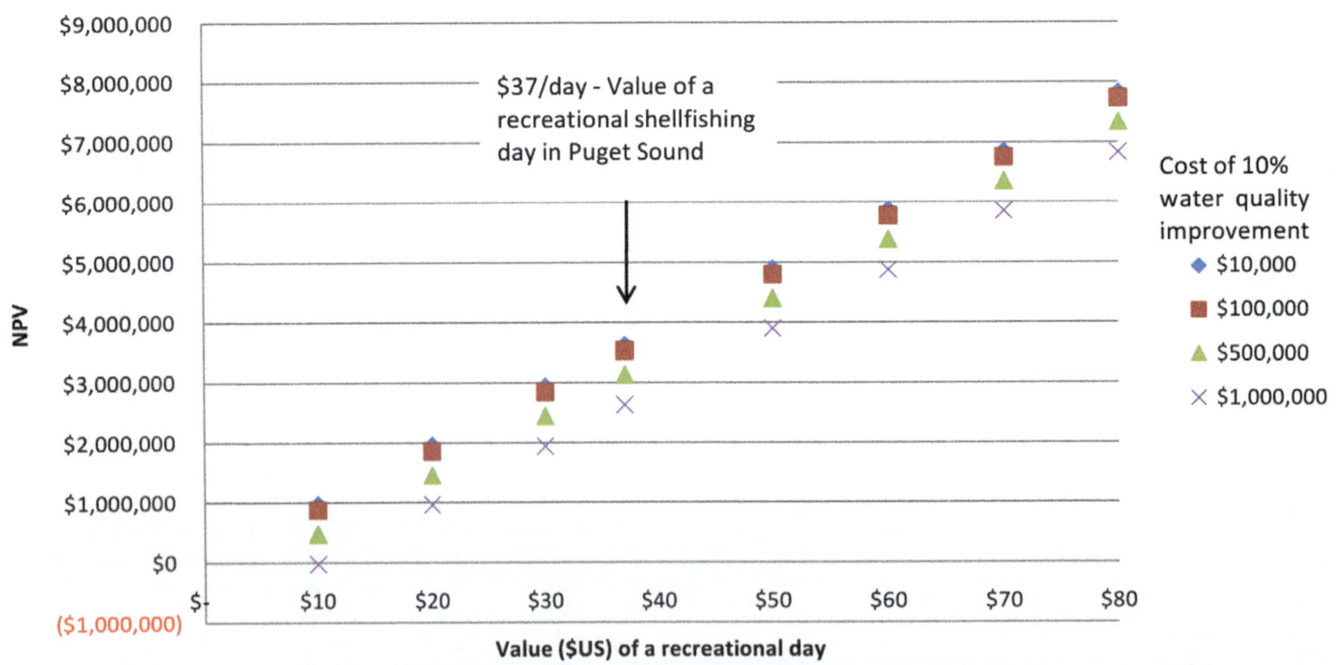

Figure 5. NPV of recreational benefits at Twanoh SP for water quality improvement costs and recreation day values. Net present value calculations assume a 3% discount rate, a 2.5% increase in average summer visitation due to a 10% water quality improvement, and a 25 year time horizon.

the PSP. In Puget Sound there is limited documentation of recreational behavior available, yet these two organizations have objectives that incorporate increasing recreational opportunity. In deciding how to allocation restoration efforts, Puget Sound restoration and cleanup priority would not necessarily proceed from areas with the poorest to best water quality, but rather to areas where the greatest net benefits (increased ecosystem services) would occur after restoration actions. In some areas large restoration costs may be warranted. For example, using the mean monthly Twanoh State Park visitation, a 3% discount rate [55], and a 25 year time horizon, the net present value of recreational benefits after a 10% water quality improvement could yield positive economic benefits for all but one of the improvement cost/recreational day value combinations (figure 5). Using a local estimate for the value of a recreational shellfishing day in Puget Sound ($37 [56]), the benefit/cost ratio of a 10% water quality improvement costing $1 million would be 2.6:1, solely through the value of recreation. Documenting the other ecosystem services is therefore critical to determine where actions could affect bundles of ecosystem services such as commercial and recreational fishing and shellfishing, coastal protection and erosion control, and cultural values [5,57]. The USGS Puget Sound Ecosystem Portfolio Model [58] is a first step in that effort.

Our methods are not without some admitted shortcomings, however. Our visitation data are an aggregated count of visitor days, thus the actual number of trips are likely less than what we report due to multi-day trips. Similarly, people may be taking multi-day trips to multiple parks, which would overestimate our travel cost estimates. These are known critiques of the travel cost method [29,52,59]. We use a single origin (Seattle) to calculate travel costs, which is an acknowledged simplification, due to the absence of origins in our main visitation dataset. Had we employed the reservation dataset (as in the PA variable, figure 2) to estimate a unique travel distance for each park our results might be slightly different. In spite of these small factors we believe this study represents a step forward in quantifying the recreational benefits for ecosystem service studies, particularly for the Puget Sound region.

Ultimately it will likely be most cost-effective to consider the value of ecosystem services when prioritizing restoration actions in Puget Sound. Clean water has clear economic benefits that have been previously addressed [60,61], and in this study we illustrate the positive effect of water quality on the value of recreation. Other values directly tied to water quality, but not present in this study, include commercial and recreational fish and shellfish harvesting, tribal shellfish harvest, residential land values, and other direct and indirect uses [1,4,57]. The ecological functions in Puget Sound have many threats facing them in light of climate and land use change, but linking ecology and human behavior for coastal ecosystem based management [62] is one way restoration could be effective in enhancing human well-being in the Puget Sound region and beyond.

Acknowledgments

We thank Bill Koss and Washington State Parks for sharing recreational visitation data. Jessica Bennet, Jessica Archer, and Jan Jacobs helped by contributing data from the BEACH program and their website traffic. We also thank Mark Plummer, Ken Bagstad, Professor Frank Davis, and two anonymous reviewers for providing helpful and thorough reviews of a previous version of this manuscript.

Product Disclaimer

Mention of trade names or manufacturers does not imply U.S. Government endorsement of commercial products.

Author Contributions

Conceived and designed the experiments: JK MP KB WL. Performed the experiments: JK. Analyzed the data: JK. Contributed reagents/materials/analysis tools: JK MP KB WL. Wrote the paper: JK.

References

1. UNEP (2006) Marine and coastal ecosystems and human well-being: A synthesis report based on the fidings of the Millennium Ecosystem Assessment. Nairobi, Kenya: UNEP. 76 p.
2. Assessment ME (2005) Ecosystem and human well-being: synthesis. Washington, D.C.: Island Press. 155 p.
3. Lester SE, McLeod KL, Tallis H, Ruckelshaus M, Halpern BS, et al. (2010) Science in support of ecosystem-based management for the US West Coast and beyond. Biological Conservation 143: 576–587.
4. McLeod KL, Leslie H, editors (2009) Ecosystem-Based Management for the Oceans. Washington, D.C.: Island Press. 392 p.
5. Chan KMA, Ruckelshaus M (2010) Characterizing changes in marine ecosystem services. F1000 Biology Reports 2: 54. doi:10.3410/B2-54.
6. Barbier EB, Koch EW, Silliman BR, Hacker SD, Wolanski E, et al. (2008) Coastal ecosystem-based management with nonlinear ecological functions and values. Science 319: 321–323.
7. Das S, Vincent JR (2009) Mangroves protected villages and reduced death toll during Indian super cyclone. Proceedings of the National Academy of Sciences of the United States of America 106: 7357–7360.
8. Koch EW, Barbier EB, Silliman BR, Reed DJ, Perillo GME, et al. (2009) Non-linearity in ecosystem services: temporal and spatial variability in coastal protection. Frontiers in Ecology and the Environment 7: 29–37.
9. Aburto-Oropeza O, Ezcurra E, Danemann G, Valdez V, Murray J, et al. (2008) Mangroves in the Gulf of California increase fishery yields. Proceedings of the National Academy of Sciences of the United States of America 105: 10456–10459.
10. Worm B, Barbier EB, Beaumont N, Duffy JE, Folke C, et al. (2006) Impacts of biodiversity loss on ocean ecosystem services. Science 314: 787–790.
11. Beaumont NJ, Austen MC, Atkins JP, Burdon D, Degraer S, et al. (2007) Identification, definition and quantification of goods and services provided by marine biodiversity: Implications for the ecosystem approach. Marine Pollution Bulletin 54: 253–265.
12. Hicks CC, McClanahan TR, Cinner JE, Hills JM (2009) Trade-Offs in Values Assigned to Ecological Goods and Services Associated with Different Coral Reef Management Strategies. Ecology and Society 14(1): 10. [online] Available: http://www.ecologyandsociety.org/vol14/iss1/art10/.
13. Iceland C, Hanson C, Lewis C (2008) Identifying important ecosystem goods and services in Puget Sound. World Resources Institute.
14. Daily GC, Polasky S, Goldstein J, Kareiva PM, Mooney HA, et al. (2009) Ecosystem services in decision making: time to deliver. Frontiers in Ecology and the Environment 7: 21–28.
15. Stinchfield Koontz, Sexton (2009) Social and Economic Considerations for Coastal and Watershed Restoration in the Puget Sound, Washington: A Literature Review. Open File Report: U.S. Geological Survey.
16. Gelfenbaum G, Mumford T, Brennan J, Case H, Deither M, et al. (2006) Coastal habitats in Puget Sound: a research plan in support of the Puget Sound Nearshore Partnership. Seattle.
17. Lew DK, Larson DM (2005) Valuing recreation and amenities at San Diego county beaches. Coastal Management 33: 71–86.
18. Pendleton L (2008) The economics of using ocean observing systems to improve beach closure policy. Coastal Management 36: 165–178.
19. Deacon RT, Kolstad CD (2000) Valuing beach recreation lost in environmental accidents. Journal of Water Resources Planning and Management-Asce 126: 374–381.
20. Given S, Pendleton L, Boehm A (2006) Public Health Costs of Contaminated Coastal Waters: A Case Study of Gastroenteritis at Southern California Beaches. Environmental Science & Technology 40: 4851–4858.
21. Busch J (2009) Surfer and Beachgoer Responsiveness to Coastal Water Quality Warnings. Coastal Management 37: 529–549.
22. Haneman M, Pendleton L, Mohn C, Hilger J, Kurisawa K, et al. (2004) Using revealed preference models to estimate the effect of coastal water quality on beach choice in Southern California.
23. Nelsen C, Pendleton L, Vaughn R (2007) A Socio-economic study of surfers at Trestle's Beach. Shore and Beach 75: 32–37.
24. Pendleton L, Martin N, Webster DG (2001) Public perceptions of environmental quality: A survey study of beach use and perceptions in Los Angeles County. Marine Pollution Bulletin 42: 1155–1160.
25. Chan KMA, Shaw MR, Cameron DR, Underwood EC, Daily GC (2006) Conservation planning for ecosystem services. Plos Biology 4: 2138–2152.
26. Eigenbrod F, Anderson BJ, Armsworth PR, Heinemeyer A, Jackson SF, et al. (2009) Ecosystem service benefits of contrasting conservation strategies in a

human-dominated region. Proceedings of the Royal Society B-Biological Sciences 276: 2903–2911.
27. Hein L, van Koppen K, de Groot RS, van Ierland EC (2006) Spatial scales, stakeholders and the valuation of ecosystem services. Ecological Economics 57: 209–228.
28. Parsons GR (2003) The Travel Cost Model. a Primer for Nonmarket Valuation. 263–329.
29. Mendelsohn R, Olmstead S (2009) The Economic Valuation of Environmental Amenities and Disamenities: Methods and Applications. Annual Review of Environment and Resources 34: 325–347.
30. National Research Council (2005) Valuing ecosystem services: toward better environmental decision-making. Washington, D.C.: National Academies Press.
31. Puget Sound Partnership (2008) Draft 2020 Action Agenda for Puget Sound. Seattle, WA. 96.
32. Nelson E, Daily GC (2010) Modeling ecosystem services in terrestrial systems. F1000 Biology Reports 2: 53. doi:10.3410/B2-53.
33. Leschine TM, Peterson AW (2007) Valuing Puget Sounds's valued ecosystem components. Seattle, WA. Puget Sound Nearshore Partnership report no. 2007-7 Puget Sound Nearshore Partnership report no. 2007-7.
34. Bateman IJ, Lovett AA, Brainard JS (1999) Developing a methodology for benefit transfers using geographical information systems: Modelling demand for woodland recreation. Regional Studies 33: 191–205.
35. Brainard J, Bateman I, Lovett A (2001) Modelling demand for recreation in English woodlands. Forestry 74: 423–438.
36. ESRI (2008) ArcGIS Desktop. 9.3 ed. Redlands, CA: Environmental Systems Research Institute.
37. Coombes EG, Jones AP, Bateman IJ, Tratalos JA, Gill JA, et al. (2009) Spatial and Temporal Modeling of Beach Use: A Case Study of East Anglia, UK. Coastal Management 37: 94–115.
38. PRISM Climate Group (2010) PRISM. Oregon State University.
39. Clawson M (1959) Methods of measuring the deman and value of outdoor recreation. Washington, D.C.
40. Knetsch JL (1963) Outdoor Recreation Demands and Benefits. Land Economics 39: 387–396.
41. Hilbe JM (2008) Negative Binomial Regression. Cambridge: Cambridge University Press.
42. Englin J, Shonkwiler JS (1995) Estimating Social-Welfare Using Count Data Models - An Application to Long-Run Recreation Demand Under Conditions of Endogenous Stratification and Truncation. Review of Economics and Statistics 77: 104–112.
43. Burnham KP, Anderson DR (1998) Model Selection and Multimodel Inference: A Practical Information Theoretic Approach, 2nd ed.: Springer-Verlag.
44. StataCorp (2009) Stata Statistical Software. Release 11 ed. College Station, TX: StataCorp, LP.
45. Balmford A, Beresford J, Green J, Naidoo R, Walpole M, et al. (2009) A Global Perspective on Trends in Nature-Based Tourism. Plos Biology 7.
46. Gossling S (1999) Ecotourism: a means to safeguard biodiversity and ecosystem functions? Ecological Economics 29: 303–320.
47. Naidoo R, Adamowicz WL (2005) Economic benefits of biodiversity exceed costs of conservation at an African rainforest reserve. Proceedings of the National Academy of Sciences of the United States of America 102: 16712–16716.
48. Naidoo R, Adamowicz WL (2005) Biodiversity and nature-based tourism at forest reserves in Uganda. Environment and Development Economics 10: 159–178.
49. Coombes EG, Jones AP (2010) Assessing the impact of climate change on visitor behaviour and habitat use at the coast: A UK case study. Global Environmental Change-Human and Policy Dimensions 20: 303–313.
50. Hanley N, Bell D, Alvarez-Farizo B (2003) Valuing the benefits of coastal water quality improvements using contingent and real behaviour. Environmental & Resource Economics 24: 273–285.
51. Termansen M, McClean CJ, Skov-Petersen H (2004) Recreational site choice modelling using high-resolution spatial data. Environment and Planning A 36: 1085–1099.
52. Bateman IJ, Garrod GD, Brainard JS, Lovett AA (1996) Measurement issues in the travel cost method: A geographical information systems approach. Journal of Agricultural Economics 47: 191–205.
53. Arnberger A, Haider W (2007) Would you displace? It depends! A multivariate visual approach to intended displacement from an urban forest trail. Journal of Leisure Research 39: 345–365.
54. Puget Sound Partnership (2009) State of the Sound. Seattle, WA.
55. National Center for Environmental Economics (2010) Guidelines for preparing economic analyses. In: Policy Oo, editor. Washington, D.C.: U.S. Environmental Protection Agency.
56. TCW Economics (2008) Economic analysis of the non-treaty commercial and recreational fisheries in Washington State. Sacramento, CA.
57. Guerry AD, Plummer ML, Ruckelshaus M, Harvey CJ (2011) Ecosystem service assessments for marine conservation. In: Kareiva P, Daily GC, Ricketts TH, Tallis H, Polasky S, editors. The Theory & Practice of Ecosystem Service Valuation in Conservation.
58. Byrd K, Kreitler J, Labiosa W (2011) Tools for Evaluating and Refining Alternative Futures for Coastal Ecosystem Management: the Puget Sound Ecosystem Portfolio Model US Geological Survye Open File Report.
59. Randall A (1994) A Difficulty with the Travel Cost Method. Land Economics 70: 88–96.
60. Wilson MA, Carpenter SR (1999) Economic valuation of freshwater ecosystem services in the United States: 1971–1997. Ecological Applications 9: 772–783.
61. Leggett CG, Bockstael NE (2000) Evidence of the effects of water quality on residential land prices. Journal of Environmental Economics and Management 39: 121–144.
62. Farber S, Costanza R, Childers DL, Erickson J, Gross K, et al. (2006) Linking ecology and economics for ecosystem management. Bioscience 56: 121–133.

5

Nitrogen Cycling Responses to Mountain Pine Beetle Disturbance in a High Elevation Whitebark Pine Ecosystem

Megan P. Keville[1], Sasha C. Reed[2], Cory C. Cleveland[1]*

1 Department of Ecosystem and Conservation Sciences, University of Montana, Missoula, Montana, United States of America, **2** U.S. Geological Survey, Southwest Biological Science Center, Moab, Utah, United States of America

Abstract

Ecological disturbances can significantly affect biogeochemical cycles in terrestrial ecosystems, but the biogeochemical consequences of the extensive mountain pine beetle outbreak in high elevation whitebark pine (WbP) (*Pinus albicaulis*) ecosystems of western North America have not been previously investigated. Mountain pine beetle attack has driven widespread WbP mortality, which could drive shifts in both the pools and fluxes of nitrogen (N) within these ecosystems. Because N availability can limit forest regrowth, understanding how beetle-induced mortality affects N cycling in WbP stands may be critical to understanding the trajectory of ecosystem recovery. Thus, we measured above- and belowground N pools and fluxes for trees representing three different times since beetle attack, including unattacked trees. Litterfall N inputs were more than ten times higher under recently attacked trees compared to unattacked trees. Soil inorganic N concentrations also increased following beetle attack, potentially driven by a more than two-fold increase in ammonium (NH_4^+) concentrations in the surface soil organic horizon. However, there were no significant differences in mineral soil inorganic N or soil microbial biomass N concentrations between attacked and unattacked trees, implying that short-term changes in N cycling in response to the initial stages of WbP attack were restricted to the organic horizon. Our results suggest that while mountain pine beetle attack drives a pulse of N from the canopy to the forest floor, changes in litterfall quality and quantity do not have profound effects on soil biogeochemical cycling, at least in the short-term. However, continuous observation of these important ecosystems will be crucial to determining the long-term biogeochemical effects of mountain pine beetle outbreaks.

Editor: Ben Bond-Lamberty, DOE Pacific Northwest National Laboratory, United States of America

Funding: Funding sources include a United States Department of Agriculture National Needs Fellowship (USDA-CSREES 2008-38420-19524)(http://www.csrees.usda.gov/fo/nationalneedsgraduatefellowships.cfm) and a grant from the McIntire-Stennis Cooperative Forestry Research Program. The funders had no role in study design, data collection and analysis, decision to publish, or preparation of the manuscript.

Competing Interests: The authors have declared that no competing interests exist.

* E-mail: Cory.Cleveland@umontana.edu

Introduction

Whitebark pine (*Pinus albicaulis*; hereafter referred to as WbP) – a coniferous tree species common in western North America – is considered a keystone species in subalpine ecosystems [1], [2]. Recently, WbP has been the subject of both considerable research and media attention because this important species is facing a number of threats to its existence. WbP has already experienced significant population declines across its distribution [3], and if widespread mortality of WbP continues, the species is in danger of becoming functionally extinct [4]. Among the most serious threats to WbP is the current mountain pine beetle (*Dendroctonus ponderosae*, MPB) outbreak, unprecedented in terms of its extent, particularly in WbP ecosystems [5]. Unfortunately, it is still unclear how this disturbance may affect many ecosystem processes – including nutrient cycling – in WbP dominated ecosystems. Nevertheless, understanding the potential effects on biogeochemistry are important, because nutrient cycling could play a critical role in determining the trajectory of recovery and regeneration of these ecologically important ecosystems following disturbance [6].

Changes in biogeochemical cycling following ecological disturbances have been documented in a wide range of ecosystems [7–10], and nitrogen (N) cycling, in particular, has been shown to change dramatically and rapidly following disturbance [11–13]. Such changes can be ecologically significant for several reasons. For example, N often limits ecosystem processes, particularly in high elevation temperate forests [14]. Thus changes in N pools and fluxes may impact a range of ecosystem characteristics [15–17]. Because N is relatively mobile and can be easily lost from ecosystems (e.g., via leaching; [18–20]), disturbances that mobilize N can further deplete the availability of an already potentially limiting nutrient. In general, disturbance-induced forest mortality can initiate rapid increases in forest floor N due to both increases in dead biomass inputs and reduced plant N demand and uptake. This N can then be quickly lost, as large, episodic inputs of organic N are typically followed by increases in N mineralization and subsequent N losses through NO_3^- leaching and denitrification [10]. However, while some studies have observed significant long-term shifts in N pools and fluxes following disturbance [18], [21], [22], other studies have shown that effects can be modest and short-lived [23], [24], [25]. This suggests that the magnitude and longevity of biogeochemical responses to disturbances such as insect-induced plant mortality are dependent upon characteristics of the affected ecosystem and/or the outbreak itself.

The mountain pine beetle is a phloem-feeding insect that typically undergoes outbreaks in pine forests of the western U.S. and Canada over 30–40 year cycles, and some of the largest WbP populations in the Greater Yellowstone Ecosystem experienced a 320-fold increase in the number of MPB-infested trees between 1999–2007 [4], [26]. During an attack, beetles deposit their eggs in the phloem of a host pine tree where the larvae develop and feed on the phloem [27]. As the larvae feed, they effectively girdle the tree, often leading to mortality within two weeks of a successful attack [5]. For approximately five years following attack, infested stands undergo a number of changes that have the potential to influence biogeochemical cycling. For example, shortly following beetle attack, nutrient and water uptake by host trees ceases, potentially altering soil moisture and soil nutrient pools [13], [28], [29]. Within two years of attack, trees enter the "red" stage of beetle infestation, when needles typically turn red and begin falling to the ground. In other host tree species where litterfall nutrient content has been analyzed, attacked tree litterfall typically has higher N concentrations than normally senescing litterfall, because the attacked trees do not resorb nutrients from their needles before they fall [13], [20], [30]. This large, relatively rapid pulse of needlefall during the red stage provides substantial inputs of carbon (C) and N to the forest floor [29] several years after attack. Five years after attack, trees have typically lost all their needles to the forest floor and reach the "gray" stage.

We took advantage of an ongoing "natural experiment" initiated by the current beetle outbreak to investigate the effects of tree mortality as a result of beetle attack on N cycling in a WbP pine ecosystem. We measured a suite of N cycling metrics under WbP trees occupying three different stages of beetle attack in the Pioneer Mountains of southwestern Montana: 'green' (unattacked) stage, 'red' stage, and 'gray' stage. Given observed changes in N cycling following beetle disturbance in other types of forest ecosystems [13], [29], [30], we began with several hypotheses. First, we hypothesized that litterfall mass would be significantly higher under red stage trees compared to green and gray stage trees, and that litterfall C:N ratios would be lower under infested trees (reflecting minimal foliar nutrient resorption in attacked trees; [29], [30]). Second, we hypothesized that soil inorganic (NH_4^++NO_3^-) and organic N concentrations would be higher under red and gray stage trees compared to green trees. An increase in available soil N could have many possible fates, including microbial immobilization, uptake by understory plants, and/or conversion of NH_4^+ to NO_3^- with the potential for loss from the ecosystem through leaching [18], [21], [22], [28]. Thus, in response to increased N inputs under attacked trees, we hypothesized increases in soil N mineralization and nitrification rates and understory plant foliar N concentrations.

Methods

2.1. Study Site

The study was conducted at Vipond Park (45.697° N, −112.910° W) in the Pioneer Mountains in the Beaverhead-Deerlodge National Forest of southwestern Montana, USA (Fig. 1). All necessary field permits were obtained from the Beaverhead-Deer Lodge National Forest. The site elevation is 2500 m and average temperatures in the region range from −9°C in January to 13°C in July (SNOTEL site 656, 2530 m, 1979–2009 average). Mean annual precipitation is approximately 770 mm falling mostly as snow (SNOTEL site 656, 2530 m 1979–2009 average), and snow covers the ground for approximately 8 months per year. Soils in the area consist of Typic calcicryepts (Inceptisols) and Eutic haplocryalfs (Alfisols) derived from limestone colluvium parent materials (USDA Natural Resources Conservation Service, Web Soil Survey; http://websoilsurvey.nrcs.usda.gov/app/HomePage.htm). The site contains an open canopy forest with whitebark pine (*Pinus albicaulis*) as the dominant canopy tree species co-occurring with occasional lodgepole (*Pinus contorta*) and limber pine (*Pinus flexilis*). The understory is sparse, consisting primarily of perennial grasses and forbs. The current mountain pine beetle outbreak at Vipond Park was first observed in 2005 and, at the time of the study, had progressed to the point where over 70% of the WbP trees are at some stage of beetle infestation (Diana Six, personal communication).

2.2. Sampling Design

WbP at the study site does not grow in a stand per se, but in "clumps" or "patches" [31] consisting of several individuals. Thus, to assess the influence of WbP mortality resulting from beetle kill on N cycling, we established ten 4×4 m (16 m^2) plots centering around individual "focal trees" [32]. This sampling design allowed for the isolation of tree-level effects of beetle infestation at a site that is relatively heterogeneous (as most WbP stands are) in terms of both canopy cover and beetle attack. Three common stages of beetle infestation were investigated in this study: 1) *'green'*, uninfested WbP; 2) *'red'*, recently infested WbP (within 2 years) with needles that had turned red but had not fallen; and 3) *'gray'*, WbP trees infested more than two years ago and that lacked needles. These three categories reflect approximate times since tree death and depict significant changes in terms of potential N fluxes. Ten WbP pine individuals at each of the three infestation stages (30 trees total) were used as focal trees. Focal tree locations were selected based on where patches of trees of a particular beetle infestation stage occurred (i.e., focal trees were adjacent to trees of the same infestation stage/mortality state). Each focal tree was at least 30 m from every other focal tree. In addition to beetle infestation stage, focal trees were chosen based on a >15 cm minimum diameter at breast height (DBH) threshold, as this is considered to be the cutoff size that is conducive to beetle attack. In addition, all focal trees were relatively similar in size, ranging from 15–40 cm DBH.

At each focal tree, soil samples were collected at four different points between the bole and the crown drip line of the tree, in each of four cardinal directions (90° apart) starting 0.5 m from the base of the tree and rotating outward in 0.5 m increments. Soil organic horizon samples were collected to a depth of approximately 5 cm at each sampling point. Mineral soil samples were collected directly below the organic horizon cores to 15 cm depth using an eight cm diameter hand corer. In cases where rocks obstructed the mineral cores, samples were taken adjacent to the organic sampling point after removing the organic material from the mineral soil surface. During each sampling event, soil samples from each of the two depths were composited by tree (i.e., for each three there was an organic and mineral soil horizon sample), placed in coolers and transported to the laboratory at the University of Montana for analysis. Within 48 h, soil samples were sieved (to 2 mm) and subsampled for physical and chemical analyses including gravimetric moisture content, inorganic N concentrations and microbial biomass analysis (see below), and the remaining soil was air-dried for pH and dried at 105°C for total C and N analyses. Soil pH was determined at 2:1 (water: soil) (Beckman Instruments, Fullerton, CA).

Focal tree, soil, and understory plant sampling were done in three phases. First, WbP focal tree foliage, litterfall and standing

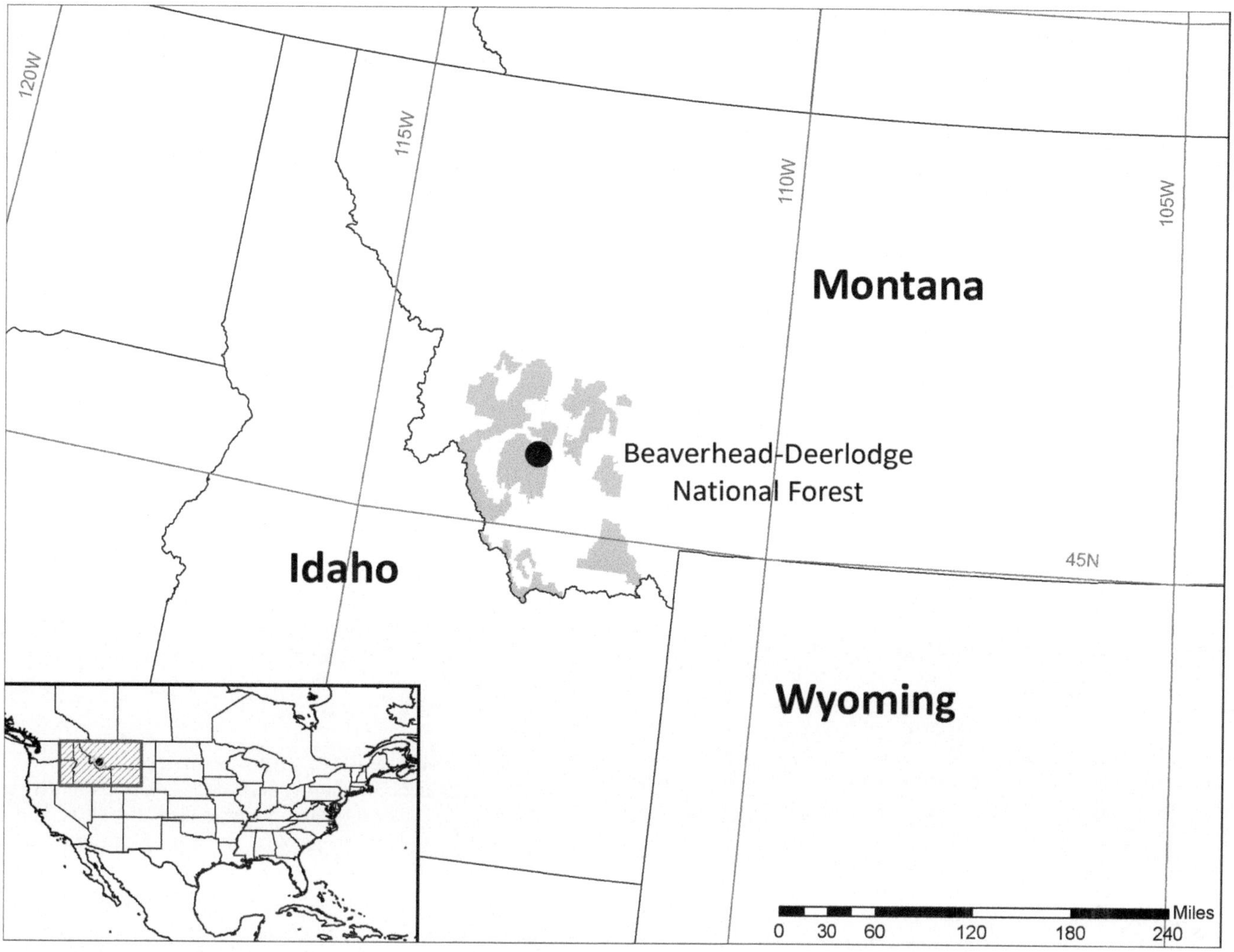

Figure 1. Map of study site. Location of study site, Vipond Park (filled circle) in the Beaverhead-Deer Lodge National Forest (gray shading) in southwestern Montana.

litter assessments were performed in July 2010. Soil samples were collected in September and October 2010; in September, soil inorganic N concentrations and potential N mineralization rates were assayed, and in October soil microbial biomass and total soil C and N concentrations were assessed. Understory cover and foliar chemistry was determined in July of 2011.

2.3. Foliage and Litter Quantity and Quality

One 0.25 m^2 litter trap was placed 0.5 m from the bole of each focal tree in July 2010. Litter was collected after one month to obtain an index of relative litterfall among beetle infestation stages and to perform litter C and nutrient analyses. Following collection, litter was dried at 70°C for 48 hours and subsequently weighed. Oven-dried subsamples from each litter trap were ground with a Wiley Mill (20-mesh screen), weighed into tin capsules (approximately 4 mg sample), and combusted on a CHNS-O elemental analyzer (CE Instruments EA 1110, Thermo Fisher Scientific, Waltham, MA, USA) to determine total litter C and N concentrations (University of Montana Environmental Geochemistry Laboratory, University of Montana, Missoula, MT).

Canopy foliage was collected in July 2010 from green and red stage trees. Samples were taken approximately 2.5 m from the ground at various points around each tree. All green and red foliage was collected near the branch tips in order to maintain consistency with respect to needle cohort. Foliage was analyzed for total C and N concentration using the above protocols. Standing litter was sampled from below each focal tree in July 2010. A 0.25 m^2 sampling quadrat was placed 0.5 m away from each tree and all intact standing litter in the quadrat was collected down to the organic horizon, dried at 70°C for 48 hours and weighed. Subsamples were ground, weighed into tin capsules (approximately 4 mg sample), and combusted on a CHNS-O elemental analyzer (CE Instruments) for total standing litter C and N.

2.4. Understory Vegetation and Foliar N

Understory vegetation under each focal tree was measured in July 2011. Two 0.25 m^2 quadrats were placed 0.5 m from the bole of each tree oriented toward the north and south. Percent ground cover for all understory species in each quadrat was assessed using ocular estimation according to plant functional group (grass, forb, sedge, shrub, seedling). Foliar samples from two common understory species, *Viola praemorsa* (Canary violet) and *Oxytropis sericea* (Silky locoweed), were collected from the quadrats and composited by focal tree and species. Samples were oven-dried at

70°C for 48 hours, ground as described above, weighed into tin capsules (approximately 4 mg sample), and combusted on a Carlo Erba elemental analyzer for total C and N concentration.

2.5. Soil N Pools

Within 24 hours of collection, soil inorganic N (ammonium [NH_4^+] and nitrate [NO_3^-]) was extracted from the September 2010 soil samples. Ten grams of each soil sample were extracted in 2 M KCl for 18 hours, vacuum filtered through 11 µm Whatman (Grade 1) filter paper, and stored at 4°C until analysis. Extracts were analyzed colorimetrically for inorganic N (NH_4^+ and NO_3^-) using a Synergy 2 Microplate Reader (BioTek, USA) after Weatherburn [33] and Doane and Horwath [34], for NH_4^+ and NO_3^-, respectively.

Soil microbial biomass C and N concentrations were analyzed using the chloroform fumigation-extraction method [35]. Briefly, ten grams each of chloroform-fumigated and unfumigated samples from the October 2010 collection (organic and mineral samples composited) were extracted with 0.5 M K_2SO_4 for one hour and vacuum-filtered through 11 µm Whatman (Grade 1) filter paper. Organic C and total N in extracts were analyzed using a Shimadzu TOC-V CPN/TNM-1 analyzer (Shimadzu, Inc, Kyoto, Japan). Microbial biomass C was determined as the difference between extractable organic C in fumigated and unfumigated samples using a proportionality constant (K_c) of 0.45 [36]. Microbial biomass N was determined the same way except using a correction factor (K_n) of 0.54 [35]. Finally, composited organic and mineral soil samples from the October 2010 soil collection were ground, weighed into tin capsules (approximately 7 mg samples), and combusted on a CHNS-O elemental analyzer for total soil C and N determination.

2.6. Soil N Transformations and Fluxes

2.6.1. N mineralization/net nitrification rates. We conducted a 28-day laboratory incubation of soil samples collected in September 2010 (organic and mineral horizons kept separate) to assess net N mineralization rates. Ten grams of field moist, sieved soil were mixed and weighed into plastic vials, covered with perforated plastic wrap, and incubated in the dark at room temperature (22°C) for 28 days. Vials were reweighed weekly and deionized water was added to maintain field moisture of each sample. After 28 days, samples were analyzed for NH_4^+ and NO_3^- as described above (after Weatherburn [33] and Doane and Horwath [34], respectively.) Net N mineralization rates were calculated by subtracting the pre-incubation September 2010 inorganic N values from the final (post-incubation) values, and dividing by the number of incubation days (i.e., 28). Net nitrification rates were obtained in the same way, using NO_3^- concentrations instead of inorganic N (NH_4^++NO_3^-) concentrations.

2.6.2. Soil soluble N fluxes. Soil N fluxes were assessed using ion-exchange resin capsules (Unibest, Inc, Bozeman, MT, USA). Capsules were deployed in the field twice, first capsules remained in soil from July – October 2010, and second capsules remained in soil from October – July 2011, following snowmelt. Resin capsules were inserted in soil beneath each focal tree to 10–15 cm depth by carefully creating a slit in the soil with a hand trowel, inserting the capsule, and carefully removing the blade to minimize disturbance. Once the capsules were collected from the field, resin-exchanged inorganic N (NH_4^++NO_3^-) concentrations were determined following extraction in 2 M KCl and colorimetric analysis.

2.7. Statistical Analyses

All statistical analyses were conducted using SPSS v. 19 (IBM SPSS, Inc., Chicago, IL, USA). Analysis of variance (ANOVA) was used to determine significant differences between the three beetle infestation stages for litterfall mass, standing litter mass, litter C and N concentrations, soil C and N pools and fluxes, and understory vegetation cover and nutrient concentrations. The assumptions of normality and homogeneity of variances were met for all data. Repeated measures ANOVA was used to test for the effects of both time and beetle infestation stage on soil moisture. The normality, homogeneity of variance, and sphericity assumptions were all satisfied. For *post-hoc* analyses, Tukey's HSD test was used to determine differences between stages when ANOVAs were significant ($\alpha = 0.05$).

Results

3.1. Soil Abiotic Characteristics

Soil pH did not vary significantly with beetle infestation stage (Table 1), but soil moisture was higher under both the red and gray stage trees ($P = 0.03$ and $P = 0.04$, respectively) compared with soil moisture soil under the green stage trees. The repeated measures ANOVA indicated that sampling time was a significant factor in explaining differences in soil moisture over the course of the growing season ($P < 0.001$; Table 1). There was not a significant time × infestation stage interaction. However, infestation stage alone explained a significant proportion of the variation in soil moisture among treatments (Table 1).

3.2. Litterfall: Relative Rates and Chemistry

In July 2010, litterfall inputs were more than an order of magnitude higher under the red stage trees (278.38±44.08 g/m^2) than under either green stage or gray trees (11±5 g/m^2 and 77±21 g/m^2 respectively; $P < 0.001$ for both). However, July 2010

Table 1. Mean (±1 SE) soil characteristics under green, red and gray stage whitebark pine trees at Vipond Park in southwestern Montana.

	Infestation Stage		
	Green	**Red**	**Gray**
Soil moisture (%)			
Organic horizon			
July 2010	37.43 (7.71) **a** *	49.32 (11.48) **b** *	45.82 (7.58) **b** *
August 2010	22.77 (10.39) **a** **	34.30 (13.73) **b** **	32.80 (14.79) **b** **
September 2010	37.44 (9.22) **a** *	44.93 (9.52) **b** *	47.93 (12.80) **b** *
Mineral horizon			
July 2010	26.77 (7.05) a *	24.29 (3.03) a *	27.34 (4.42) a *
August 2010	8.31 (6.67) **a** **	12.66 (4.85) **ab** **	15.05 (6.51) **b** **
September 2010	19.93 (3.45) **a** ***	24.22 (2.55) **b** *	24.38 (4.41) **b** *
Bulk soil pH (0–15 cm)	6.17 (0.22) a	6.44 (0.21) a	6.30 (0.19) a
Bulk soil C (%)	7.43 (2.66) a	8.48 (3.23) a	9.72 (4.73) a
Bulk Soil N (%)	0.34 (0.09) a	0.38 (0.14) a	0.41 (0.18) a
Bulk soil C:N	21.25 (1.24) a	22.51 (1.01) a	23.32 (1.35) a

Green stage represents unattacked trees, red stage, trees attacked within the last five years, and gray stage, trees attacked>five years ago. Lower case letters denote significant differences among stages (rows) and number of stars denotes significant differences among sampling dates (columns). Letters and stars denote significant differences at $\alpha = 0.05$.

litterfall inputs did not vary significantly between the gray and green stages (Table 2). Total litterfall N concentrations were significantly higher in gray and red stage litterfall (1.45±0.05% and 1.16±0.07%, respectively) compared to green stage litterfall (1.06±0.04%) ($P=0.001$; $P<0.001$, respectively). This was reflected in the significantly higher litterfall N pools under the red and gray stages compared to the green stage ($P<0.001$ for both) (Table 2). Litterfall C:N ratios varied between stages, with the red and gray stages having significantly lower C:N ratios than the green stage needles ($P=0.046$; $P<0.001$ respectively; Table 2). Foliar C:N ratios varied as well, with the green stage trees having approximately double the foliar C:N (109.88±2.35) of the red stage trees (55.66±2.91; $P<0.001$). There were no significant differences in standing litter mass among infestation stages, although the red and silver stages tended to be higher than the green stage (Table 2). Similarly, standing litter N and C:N did not vary among infestation stages.

3.3. Soil N Pools

Soil inorganic N concentrations (NH_4^++NO_3^-) were higher ($P<0.001$) in the organic horizon compared to the mineral horizon for all stages. Across infestation stages, the organic horizon under red stage trees had higher extractable inorganic N concentrations than the organic horizon of either the green or gray stage trees ($P=0.005$, $P=0.071$) (Table 2). These differences in inorganic N were driven by changes in NH_4^+ across infestation stages. In the mineral horizon, extractable inorganic N did not vary across infestation stages (Table 2) and soil NO_3^- concentrations were very low (i.e., near detection limits) and highly variable for all mineral soils measured. There were no differences in NO_3^- levels across infestation stages in either soil horizon, and there were no significant differences in either microbial biomass N concentrations or total soil N concentrations across infestation stages in either the organic or mineral soil horizons (Tables 1 and 3). However, soil microbial biomass C:N ratios were significantly lower under the red stage trees compared to the gray stage trees ($P=0.01$; Table 3).

3.4. Soil N Cycling

Net soil N mineralization rates were higher in both organic and mineral soils under gray trees, followed by red and then green trees (Table 3), but the differences among stages were not significant.

Table 3. Mean (±1 SE) N cycling characteristics under green, red and gray stage whitebark pine trees at Vipond park in southwestern Montana.

	Infestation Stage		
	Green	**Red**	**Gray**
N mineralization (μg N/g soil/day)			
Organic horizon	1.51 (0.19) a	1.53 (0.44) a	1.65 (0.36) a
Mineral horizon	0.61 (0.21) a	0.63 (0.19) a	1.05 (0.33) a
Microbial biomass C:N	8.53 (0.40) **ab**	7.65 (0.16) **a**	8.98 (0.30) **b**
Resin Capsule Flux (μg N/g soil)	0.46 (0.17) a	1.13 (0.49) a	0.45 (0.22) a
Understory foliar C:N			
Silky Locoweed	8.87 (0.28) a	8.57 (0.17) a	8.35 (0.22) a
Canary Violet	11.14 (0.24) a	11.87 (0.83) a	11.27 (0.34) a

Green stage represents unattacked trees, red stage, trees attacked within the last five years, and gray stage, trees attacked>five years ago. Letters denote significant differences between stages. Letters denote significant differences at $\alpha=0.05$.

Soil inorganic N fluxes as measured with resin capsules did not vary across stages of beetle infestation (Table 2).

3.5. Understory Plant Responses

Understory ground cover did not significantly vary across the three stages of beetle infestation, although grasses and forbs displayed an increasing trend moving from the green to red and gray stages (data not shown). Neither of the two understory forb species (*Viola praemorsa* and *Oxytropis sericea*) exhibited significantly different C:N values across the three stages of beetle infestation (Table 3).

Discussion

The widespread, unprecedented WbP mortality occurring as a result of the current MPB outbreak has profoundly affected WbP ecosystems, and some research has even suggested that WbP may become "functionally extinct" as a result of the current

Table 2. Mean (±1 SE) N input characteristics under green, red and gray stage whitebark pine trees at Vipond Park in southwestern Montana.

	Infestation Stage		
	Green	**Red**	**Gray**
Litterfall Mass (g/m^2)	11.05 (5.23) **a**	278.38 (44.08) **b**	77.48 (21.31) **a**
Litterfall N (%)	1.06 (0.12) **a**	1.16 (0.21) **b**	1.45 (0.15) **b**
Litterfall C:N	48.10 (5.38) **a**	41.25 (7.97) **b**	34.67 (4.34) **b**
Litter Mass (g/m^2)	1938.99 (261.79) a	2994.08 (451.98) a	2655.96 (517.99) a
Litter N (%)	1.48 (0.31) a	1.56 (0.30) a	1.63 (0.20) a
Litter C:N	31.66 (5.69) a	29.25 (5.98) a	27.17 (2.45) a
Organic soil inorganic N (μg N/g soil)	14.96 (6.97) **a**	33.60 (15.76) **b**	21.19 (11.58) **ab**
Mineral soil inorganic N (μg N/g soil)	4.86 (2.96) a	4.73 (2.93) a	6.32 (3.89) a

'Green' stage represents unattacked trees, 'red' stage, trees attacked within the last five years, and 'gray' stage, trees attacked>five years ago. Letters denote significant differences between stages at $\alpha=0.05$.

disturbance [4]. However, what this massive mortality event means for WbP ecosystem biogeochemical cycling remains unknown, and to our knowledge this study is the first to assess how beetle kill alters N cycling at the individual tree level in these high elevation ecosystems. We recognize that our sampling design is effectively a chronosequence, which limits the strength with which we can assert direct causality to the disturbance in the absence of any pre-disturbance data [37]. However, there is no evidence or previous research indicating that factors other than the mountain pine beetle outbreak would have influenced N-cycling changes in our study. We hypothesized that this disturbance would lead to short-term changes in ecosystem N pools and fluxes underneath beetle-infested trees. However, despite some modest changes in soil N cycling in response to the MPB outbreak, overall our results indicate that short-term changes to the N cycle were relatively subtle. This suggests that while the current MPB outbreak may drive large changes in other WbP ecosystem characteristics and ecosystem processes (e.g., succession or disturbance dynamics [4]), changes to the N cycle are relatively small in the short-term, at least compared to the profound effect of the disturbance on other ecosystem characteristics.

Soil moisture varied with beetle infestation stage, such that the red stage trees had the highest soil moisture. There are several possible explanations for this, including changes in transpiration [13], [29], [38], [39], and/or changes in evaporation under green, red or gray trees. For example, increased radiation due to complete loss of canopy cover could promote evaporation, perhaps explaining (at least in part) why soil moisture under gray stage trees was lower than soil moisture under red stage trees. In ponderosa pine stands, Morehouse et al. [30] measured significantly higher photosynthetically active radiation (PAR) reaching the forest floor under beetle-killed trees compared to unattacked trees. Reduced canopy cover of the gray stage trees may partially counteract increases in soil moisture stemming from reduced transpiration. Such changes in soil moisture can have important implications for a number of microbial processes, including N mineralization, respiration and decomposition [40], [41].

Not surprisingly, litterfall inputs were significantly higher under red stage trees compared to green and gray stage trees, reflecting the pulse of litterfall often observed soon after tree mortality following beetle attack. However, increased litterfall (as seen in July) did not translate into differences in standing litter biomass under infested trees. Griffin [13] observed similar patterns in MPB-attacked lodgepole pine stands and attributed this to decomposition breaking down the gradual (3–4 year) influx of litter at a pace that did not allow buildup of a litter layer. Alternatively, the canopy cover patchiness of WbP, as well as the movement of litter during snowmelt or wind and rain events, may contribute to the high variability of standing litter mass, thereby masking detectable shifts among infestation stages.

Litterfall nutrient chemistry varied with infestation stage. Litterfall from gray stage trees had significantly lower C:N ratios than green stage trees, largely driven by higher N concentrations in the needles falling from the infested trees. This result matches the findings of studies conducted in both MPB-attacked lodgepole and ponderosa pine ecosystems [13], [29], [30]. Differences in litterfall nutrient content following MPB attack most likely reflect limited N resorption prior to senescence [30], [42]. However, while we expected these changes in litterfall quantity and quality to translate to subsequent shifts in N cycling as they reached the forest floor and became available for microbial processing and decomposition [12], [20], the shifts we observed for WbP were relatively subtle.

Once increased litterfall N inputs reach the forest floor, the N has multiple possible fates, one of which is uptake by surviving understory vegetation. This could manifest itself through increases in understory net primary productivity, or increases in foliar N concentration of understory plants after the disturbance [43], [44], or both. For example, in lodgepole pine stands attacked by MPB, Griffin et al. [13] observed higher foliar N concentrations in the understory sedge, *Carex geyerii*, in beetle-infested stands compared to uninfested stands, but did not measure higher sedge productivity in infested stands. Other studies of post-beetle stands have observed increases in understory growth in a number of forest types [43], [44], [45]. Increases in available N after disturbance may be responsible for some of this enhanced growth, but other abiotic factors such as increased light and moisture are also likely contributors, particularly in the longer term. We observed no significant differences across infestation stage in the foliar N or C:N ratio of two common understory species (*Viola praemorsa* and *Oxytropis sericea*). There was also no measurable increase in understory ground cover under beetle-attacked trees. Taken together, these data suggest that biogeochemical cycles in different forests may respond differently to attack.

As hypothesized, the highest soil inorganic N values were observed in the organic horizon under red stage trees, and the differences in inorganic N concentrations observed were due to shifts in soil NH_4^+ concentrations, not to changes in NO_3^- concentrations. The increase in soil inorganic N under red stage trees could be explained by one or a combination of factors. For example, higher N in litterfall, reduced plant uptake [10], higher microbial N mineralization [7], [38], [30], lower microbial immobilization of N, or transport of N from the N-enriched litter layer [20] could all contribute to the increase in soil inorganic N in the organic horizon in red stage trees. Other studies have reported similar results, with infested stands having significantly more extractable soil NH_4^+ than uninfested stands [38], [12], [29]. In beetle-infested vs. uninfested ponderosa pine stands, however, there were no significant differences measured in soil inorganic N pools [30]. In our study, inorganic N concentrations were significantly different and correlated with lower soil microbial C:N ratios. Thus, while the N cycling response is not dramatic and was not observable in our assessment of understory plants, there is the potential for beetle mortality to affect soil N cycling in ways that could feedback to influence soil N availability and loss.

In contrast to the differences in inorganic N in the organic (surface) soil horizons, inorganic N concentrations in the mineral soil horizons did not vary among infestation stages. This may suggest that sufficient time has not elapsed for the increased inputs of N to be transported into the lower mineral horizons. Leaching of inorganic N from the organic horizon and litter layers requires time and movement of water through the soil profile [46]. Alternatively, the increased inorganic N in the organic horizon may be immobilized by microbes, lost via denitrification, or taken up by plants before it can move into the mineral horizon.

While we predicted that the relatively low C:N ratios in litter from beetle-infested trees would drive increases in soil N pools, not all forms of inorganic N shifted following beetle attack. Soil NO_3^- concentrations in both the organic and mineral horizons were consistently very low, highly variable and did not differ among beetle infestation stages. Low soil NO_3^- levels are often measured in high elevation, N-limited ecosystems, as available N is often rapidly taken up by plants or immobilized by microbes [47]. Increases in soil NO_3^- may also have been lost from the ecosystem quickly through leaching during rain events or snowmelt [46]. In other disturbed systems, large increases in soil NO_3^- have been observed, but usually involved nearly 100% vegetation mortality, or took place in systems with substantial N deposition [17], [19]. For example, Orwig [48] measured soil NO_3^- in hemlock stands that had been infested by the insect defoliator, hemlock woody adelgid, and observed significantly higher levels in attacked vs. unattacked stands. These forests were nearly pure hemlock and experienced almost complete mortality as the outbreak moved through the system. Soil texture and the timing of snowmelt or rain events also influence the magnitude of N losses from an ecosystem [6]. Ecosystem characteristics such as climate and soil texture, as well as the extent of disturbance-induced mortality interact to determine how mobilized N behaves following the disturbance, particularly with respect to N losses.

Microbial biomass C:N ratios were lower under red stage trees compared to gray stage trees, but surprisingly red stage trees did not differ from green stage trees. Microbial immobilization of N is one pathway mineralized N may take after a disturbance, particularly if N is limiting [49]. In such cases, lower microbial biomass C:N should reflect this, indicating uptake of N. Potential net N mineralization rates in incubated soils did not vary across beetle infestation stage and this result is not consistent with other insect outbreak studies, which found significantly higher net N mineralization potential in soil from attacked stands [12], [30], [38]. Our results appear to argue against increased microbial N mineralization as the driver behind the observed increase in soil inorganic N in the organic soil horizon. One possibility is that reduced plant uptake resulted in the increase in inorganic N rather than increased N mineralization. Alternatively, the timing of sampling may have missed an initial pulse of mineralization following tree mortality. High elevation WbP ecosystems may undergo the majority of their microbial processing of organic matter and decomposition during spring snowmelt, a time at which our study site is inaccessible.

Inorganic N fluxes through the soil profile (as measured with resin capsules) also did not vary with beetle infestation stage in our site. Nitrogen losses are one of the most commonly measured biogeochemical variables in disturbance studies, and increased NO_3^- levels have been observed in a number of insect defoliation studies (often measured in streams draining the disturbed ecosystem; [7], [18], [21], [22]). For example, a spruce bark beetle outbreak in Germany resulted in elevated N concentrations 40 cm below the soil surface for five years following the outbreak [28]. The fact that no measurable changes to N fluxes were observed in this WbP ecosystem is not entirely surprising considering the very low NO_3^- levels measured in the soil profile.

Overall, the results of this study suggest that, from an N cycling perspective, WbP ecosystems experience relatively subtle changes in the years immediately following beetle attack. While N inputs under focal trees changed significantly following tree mortality, they were accompanied by few shifts in internal cycling and pools of N, at least for the variables we measured. In addition, the internal N cycling characteristics that did change with beetle infestation stage varied in the upper, organic soil horizon, but did not appear in the mineral soil horizon, and there was no evidence suggesting significant shifts in N losses from the system. We would expect gaseous N losses to be very low at this site given the coarse texture of the soil and the dry environment [50]. This does not mean that the differences we did observe are not important, as N could strongly limit ecosystem processes in the system, but overall we did not observe the large N cycling responses we expected based on the dramatic mortality and our focal tree sampling design. Given the lack of strong responses at the level of individual trees, we would also predict that the stand-to-ecosystem scale responses would be similarly subtle, but future research is necessary to test the effects of widespread WbP mortality on N cycling at larger scales. However, it is worth noting that this study indicates that WbP ecosystems react differently to MPB attack compared to studies from lower elevation ecosystems [30], [39], which displayed either stronger or more rapid biogeochemical responses to bark beetle attack.

Considering the extent of WbP mortality at the site, the lack of many significant biogeochemical responses to the increased litter C and N inputs was unexpected. The characteristics of WbP ecosystems, however, provide some insight into potential lags in response time to the disturbance, which may ultimately allow regeneration to occur before any longer-term nutrient shifts occur. For example, the very short growing season and harsh climatic conditions that exist in WbP ecosystems most likely cause many microbial processes to progress more slowly than in other ecosystems. From a biogeochemical perspective, climatic constraints on N cycling could have some benefit. Following this large-scale, high mortality disturbance, major shifts in C and N do not appear to occur over the timeline of a few years, but instead may play out over longer timescales that ultimately do not lead to dramatic N losses. Longer-term monitoring at a number of sites is required to determine whether biogeochemical changes are indeed more subtle in WbP ecosystems, or whether they merely take longer to manifest themselves. Thus far the data suggest that, from an N cycling perspective, the prospects for the future regeneration of WbP may be relatively promising.

Acknowledgments

We thank J. Aylward and T. Dietzler for field and laboratory assistance, and R. Rawlinson for help producing the figure. Any use of trade names is for descriptive purposes only and does not imply endorsement by the U.S. Government.

Author Contributions

Conceived and designed the experiments: MPK SCR CCC. Performed the experiments: MPK CCC. Analyzed the data: MPK CCC. Contributed reagents/materials/analysis tools: CCC. Wrote the paper: MPK SCR CCC.

References

1. Farnes PE (1990) SNOTEL and snow course data describing the hydrology of whitebark pine ecosystems. In: Schmidt WC, McDonald KJ, editors. Proceedings- Symposium on whitebark pine ecosystems: Ecology and management of a high-mountain resource. Ogden, Utah: USDA Forest Service General Technical Report INT-270. 302–304.
2. Lanner RM (1980) Avian seed dispersal as a factor in the ecology and evolution of limber and whitebark pine. In: Dancik BP, Higginbotham KO, editors. Proceedings of 6th North American Forest Biology Workshop. Edmonton: University of Alberta. 15–48.
3. Tomback DF (2001) Clark's Nutcracker: Agent of regeneration. In: Tomback DF, Arno SF, Keane RE, editors. Whitebark pine communities: ecology and restoration. Washington DC: Island Press. 88–104.
4. Logan JA, McFarlane WW, Willcox L (2010) Whitebark vulnerability of climate change induced mountain pine beetle disturbance in the Greater Yellowstone Ecosystem. Ecol Appl 20: 895–902.
5. Raffa KF, Aukema BH, Bentz BJ, Carroll AL, Hicke JA, et al. (2008) Cross-scale drivers of natural disturbances prone to anthropogenic amplification: the dynamics of bark beetle eruptions. Bioscience 58: 501–518.
6. Vitousek PM, Gosz JR, Grier CC, Melillo JM, Reiners WA (1982) A comparative analysis of potential nitrification and nitrate mobility in forest ecosystems. Ecol Monogr 52: 155–177.
7. Swank WT, Waide JB, Crossley DA, Todd RL (1981) Insect defoliation enhances nitrate export from forest ecosystems. Oecologia 51: 297–299.
8. Frost CJ, Hunter MD (2004) Insect canopy herbivory and frass deposition affect soil nutrient dynamics and export in oak mesocosms. Ecology 85: 3335–3347.

9. Lovett GM, Canham CD, Arthur MA, Weathers KC, Fitzhugh RD (2008) Forest ecosystem responses to exotic pests and pathogens in eastern North America. Bioscience 56: 395–405.
10. Reynolds BC, Hunter MD, Crossley DA (2000) Effects of canopy herbivory on nutrient cycling in a northern hardwood forest in western North Carolina. Selbyana 21: 74–78.
11. Bormann FH, Likens GE, Eaton JS (1969) Biotic regulation of particulate and solution losses from a forest ecosystem. Bioscience 19: 600–610.
12. Lovett GM, Arthur MA, Weathers KC, Griffin JM (2010) Long-term changes in forest carbon and nitrogen cycling caused by an introduced pest/pathogen complex. Ecosystems 13: 1188–1200.
13. Griffin JM, Turner MG, Simard M (2011) Nitrogen cycling following mountain pine beetle disturbance in lodgepole pine forests of Greater Yellowstone. For Ecol Manage 261: 1077–1089.
14. Gough L, Osenberg CW, Gross KL, Collins SL (2000) Fertilization effects of species density and primary productivity in herbaceous plant communities. Oikos 89: 428–439.
15. Gutschik VP (1981) Evolved strategies in nitrogen acquisition by plants. Am Nat 118: 607–637.
16. Vitousek PM, Howarth RW (1991) Nitrogen limitation on land and in the sea: How can it occur? Biogeochemistry 13: 87–115.
17. Bormann FH, Likens GE (1994) Pattern and process in a forested ecosystem: disturbance, development and the steady state based on the Hubbard Brook Ecosystem Study. New York: Springer. 272 p.
18. Eshleman KN, Morgan RP, Webb JR, Deviney FA, Galloway JN (1998) Temporal patterns of nitrogen leakage from mid-Appalachian forested watersheds: Role of insect defoliation. Water Resour Res 34: 2005–2116.
19. Aber JD, Ollinger SV, Driscoll CT, Likens GE, Holmes RT, et al. (2002) Inorganic nitrogen losses from a forested ecosystem in response to physical, chemical, biotic and climatic perturbations. Ecosystems 5: 648–658.
20. Riscassi AL, Scanlon TM (2009) Nitrate availability in hydrological flow paths for three mid-Appalachian forested watersheds following a large-scale defoliation. J Geophys Res 114: 1–11.
21. Webb JR, Cosby BJ, Deviney FA, Eshleman KN, Galloway JN (1995) Change in the acid-base status of an Appalachian catchment following forest defoliation by the gypsy moth. Water Air Soil Pollut 85: 535–540.
22. Houle D, Duchesne L, Boutin R (2009) Effects of spruce budworm outbreak on element export below the rooting zone: a case study for a balsam fir forest Ann For Sci 66: 707.
23. Lovett GM, Christenson LM, Groffman PM, Jones CG, Hart JE, et al. (2002) Insect defoliation and nitrogen cycling in forests. Bioscience 52: 335–341.
24. Le Mellec A, Michalzik B (2008) Impact of a pine lappet *(Dendrolimus pini)* mass outbreak on C and N fluxes to the forest floor and soil microbial properties in a Scots pine forest in Ger many. Can J For Res 38: 1829–1841.
25. Nave LE, Gough CM, Maurer KD, Bohrer G, Hardiman BS, et al. (2011) Disturbance and the resilience of coupled carbon and nitrogen cycling in a north temperate forest. J Geophys Res 116: DOI: 10.1029/2011JG001758.
26. Logan JA, McFarlane WW, Willcox L (2009) Effective monitoring as a basis for adaptive management: a case history of mountain pine beetle in Greater Yellowstone Ecosystem whitebark pine. iForest-Biogeosciences and Forestry 2: 19–22.
27. Safranyik L, Barclay H, Thomson A, Riel W (1999) A population dynamics model for the mountain pine beetle, *Dendroctonus ponderosae* Hopk. (Coleoptera: Scolytidae). Canadian Forest Service, Information Report BC-X-386 p.35.
28. Huber C (2005) Long lasting nitrate leaching after bark beetle attack in the highlands of the Bavarian Forest National Park. J Environ Qual 34: 1772–1779.
29. Clow DW, Rhoades C, Briggs J, Caldwell M, Lewis Jr WM (2011) Responses of water chemistry to mountain pine beetle induced tree mortality in Grand County, Colorado, USA. Appl Geochem 26: S174–S178.
30. Morehouse K, Johns T, Kaye J, Kaye M (2008) Carbon and nitrogen cycling immediately following bark beetle outbreaks in southwestern ponderosa forests. For Ecol Manage 255: 2698–2708.
31. Arno SF (2001) Community types and natural disturbance processes. In: Tomback DF, Arno SF, Keane RE, editors. Whitebark pine communities: ecology and restoration. Washington DC: Island Press. 74–88.
32. Zinke PJ (1962) The pattern of influence of individual forest trees on soil properties. Ecology 43: 130–133.
33. Weatherburn MW (1967) Phenol-hypochlorite reaction for determination of ammonia. Anal Chem 39: 971–974.
34. Doane TA, Horwath WR (2003) Spectrophotometric determination of nitrate with a single reagent. Anal Lett 36: 2718–2722.
35. Brookes PC, Landman A, Pruden G, Jenkinson DS (1985) Chloroform fumigation and the release of soil nitrogen: a rapid direct extraction method to measure microbial biomass nitrogen in soil. Soil Biol Biochem 17: 837–842.
36. Vance ED, Brookes PC, Jenkinson D (1987) An extraction method for measuring soil microbial biomass C. Soil Biol Biochem 19: 703–707.
37. Johnson EA, Miyanishi K (2008) Testing the assumptions of chronosequences in succession. Ecology Letter 11: 419–431.
38. Jenkins JC, Aber JD, Canham CD (1999) Hemlock woolly adelgid impacts on community structure and N cycling rates in eastern hemlock forests. Can J For Res 29: 630–645.
39. Griffiths RP, Gray AN, Spies TA (2010) Soil properties in old-growth Douglas-fir forest gaps in the western Cascade Mountains of Oregon. Northwest Sci 84: 33–45.
40. Concilio A, Ma S, Li Q, LeMoine J, Chen J, et al. (2005) Soil respiration response to prescribed burning and thinning in mixed-conifer and hardwood forests. Can J For Res 35: 1581–1591.
41. Davidson EA, Janssens IA (2005) Temperature sensitivity of soil carbon decomposition and feedbacks to climate change. Nature 440: 165–173.
42. Stone JE, Kolb TE, Covington W (1999) Effects of restoration thinning on presettlement *Pinus ponderosa* in northern Arizona. Restor Ecol 7: 172–182.
43. Stone WE, Wolfe ML (1996) Response of understory vegetation to variable tree mortality following a mountain pine beetle epidemic in lodgepole pine stands in northern Utah. Plant Ecol 122: 1–12.
44. Metzger KL, Romme WH, Turner MG (2006) Foliar nitrogen patterns following stand-replacing fire in lodgepole pine (*Pinus contorta* var. *latifolia*) forests of the Rocky Mountains, USA. For Ecol Manage 227: 22–30.
45. McCambridge WF, Morris MJ, Edminster CB (1982) Herbage production under ponderosa pine killed by the mountain pine beetle in Colorado. USDA For Serv Int Res (RM-416).
46. Vitousek PM, Gosz JR, Grier CC, Melillo JM, Reiners WA, et al. (1979) Nitrate losses from disturbed ecosystems. Science 204: 469–474.
47. Chapin FS, Matson PA, Mooney HA (2002) Principles of terrestrial ecosystem ecology. New York: Springer. 529 p.
48. Orwig DA, Cobb RC, D'Amato AW, Kizlinski ML, Foster DR (2008) Multi-year ecosystem response to hemlock woolly adelgid infestation in southern New England forests. Can J For Res 38: 834–843.
49. Aber JD, Melillo JM, McClaugherty CA, Eshleman KN (1982) Potential sinks for mineralized nitrogen following disturbance in forest ecosystems. Environ Biogeochem 35: 179–192.
50. Davidson EA, Swank WT (1986) Environmental parameters regulating gaseous nitrogen losses from two forested ecosystems via nitrification and denitrification. Appl Environ Microbiol 52: 1287–1292.

6

Development of an RNA Interference Tool, Characterization of Its Target, and an Ecological Test of Caste Differentiation in the Eusocial Wasp *Polistes*

James H. Hunt[1*◑], Navdeep S. Mutti[2◑], Heli Havukainen[3,4], Michael T. Henshaw[5], Gro V. Amdam[2,3]

1 Departments of Biology and Entomology and W. M. Keck Center for Behavioral Biology, North Carolina State University, Raleigh, North Carolina, United States of America, **2** School of Life Sciences, Arizona State University, Tempe, Arizona, United States of America, **3** Department of Chemistry, Biotechnology and Food Science, Norwegian University of Life Sciences, Aas, Norway, **4** Department of Biomedicine, University of Bergen, Bergen, Norway, **5** Department of Biology, Grand Valley State University, Allendale, Michigan, United States of America

Abstract

Recent advancements in genomics provide new tools for evolutionary ecological research. The paper wasp genus *Polistes* is a model for social insect evolution and behavioral ecology. We developed RNA interference (RNAi)-mediated gene silencing to explore proposed connections between expression of hexameric storage proteins and worker vs. gyne (potential future foundress) castes in naturally-founded colonies of *P. metricus*. We extended four fragments of putative hexamerin-encoding *P. metricus* transcripts acquired from a previous study and fully sequenced a gene that encodes Hexamerin 2, one of two proposed hexameric storage proteins of *P. metricus*. MALDI-TOF/TOF, LC-MSMS, deglycosylation, and detection of phosphorylation assays showed that the two putative hexamerins diverge in peptide sequence and biochemistry. We targeted the *hexamerin 2* gene in 5th (last)-instar larvae by feeding RNAi-inducing double-stranded *hexamerin 2* RNA directly to larvae in naturally-founded colonies in the field. Larval development and adult traits were not significantly altered in *hexamerin 2* knockdowns, but there were suggestive trends toward increased developmental time and less developed ovaries, which are gyne characteristics. By demonstrating how data acquisition from 454/Roche pyrosequencing can be combined with biochemical and proteomics assays and how RNAi can be deployed successfully in field experiments on *Polistes*, our results pave the way for functional genomic research that can contribute significantly to learning the interactions of environment, development, and the roles they play in paper wasp evolution and behavioral ecology.

Editor: Peter K. Dearden, University of Otago, New Zealand

Funding: Research at NCSU was supported by NSF grant #0819807 to JHH and C.M. Grozinger. PROBE received support from the National Program for Research in Functional Genomics, which is funded by the Research Council of Norway (RCN). GVA was supported by RCN 180504, 185306, and 191699 and the PEW Charitable Trust. The funders had no role in study design, data collection and analysis, decision to publish, or preparation of the manuscript.

Competing Interests: The authors have declared that no competing interests exist.

* E-mail: jim_hunt@ncsu.edu

◑ These authors contributed equally to this work.

Introduction

Experimental studies conducted in natural environments are important for understanding ecological patterns and processes [1]. Experimental studies conducted on laboratory model organisms are important for understanding the selective basis of evolution [2]. *Polistes* paper wasps offer an opportunity to pursue a third realm of experimental studies now in its infancy – the implementation of genomics-based methods to experimentally study evolutionary mechanisms in natural environments.

The paper wasp genus *Polistes* has long been an important taxon for field studies of its ecology [3] and behavior [4,5,6]. It is an especially important taxon for understanding the evolution of insect sociality [7,8] and as a model system in behavioral ecology [9,10]. The paper wasp *Polistes metricus* Say, distributed in the Midwestern and Southeastern U. S., has for several decades been the specific focus of diverse studies on aspects of *Polistes* biology. Early work includes physiology [11], inclusive fitness [12], behavior [13], and sex ratio [14]. To extend and expand these and other research areas at both the species and genus levels, *P. metricus* was the first *Polistes* species subjected to analysis of its expressed sequence tags (ESTs – its expressed mRNA) [15], which then were annotated by reference to the genome of a related species, the honey bee *Apis mellifera* [16]. Using this EST resource, Toth et al. [15] documented differences in gene expression among the four categories of adult females that occur in an annual colony cycle of *P. metricus* (Fig. 1). Similarly, Hunt et al. [17] used the EST resource to document differences in gene expression between larvae destined to become workers and larvae destined to become gynes (pre-queens). These studies were 'natural experiments' that were based on collection of specimens from natural environments at particular points in the colony cycle, but they were not manipulative experiments designed to test causal connections.

The manipulative experimental approach of RNA interference (RNAi)-mediated gene targeting occurs post-transcriptionally and involves specific degradation of target mRNA. RNAi is triggered by introduction of double stranded RNA (dsRNA) and can be a near-universal method of gene silencing [18,19]. Among social insects, RNAi has been reported for dsRNA feeding in larval honey bees *A. mellifera* [20,21] and nymphs of the termite

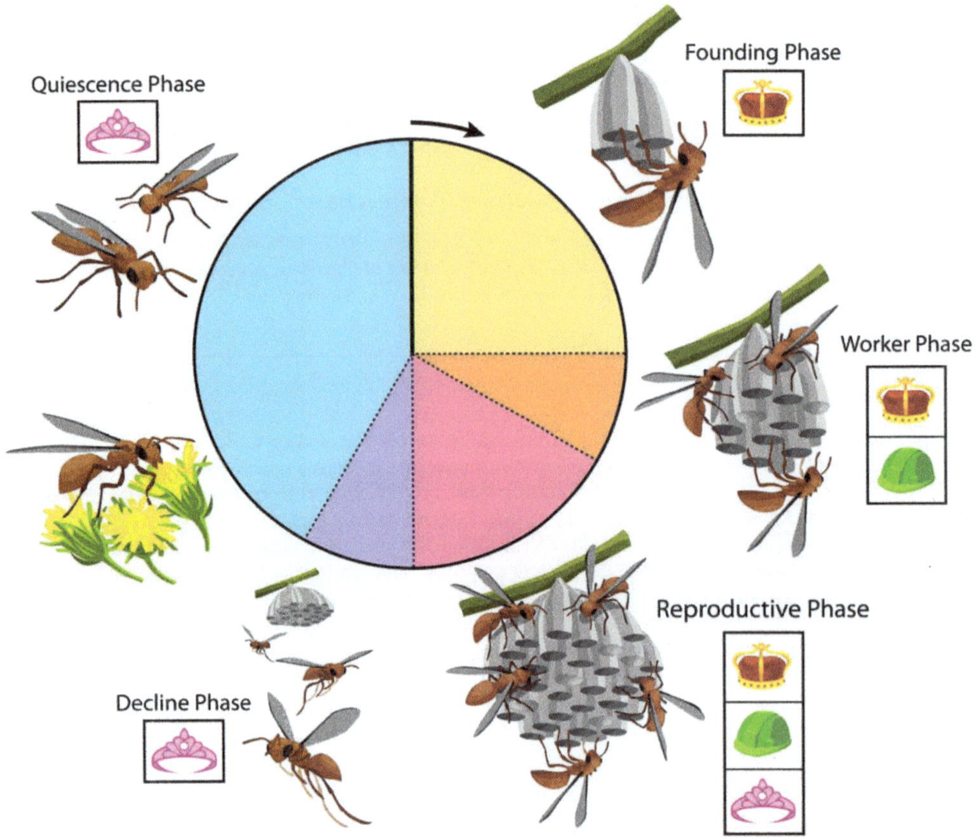

Figure 1. Colony cycle of an annual *Polistes* paper wasp, exemplified by *P. metricus*. Foundress Phase. The annual cycle begins (arrow) when a female emerges from quiescence with undeveloped ovaries. Feeding at flowers enables ovary development, which precedes nest founding. Oviposition occurs immediately upon initiation of each nest cell by the now-maternal foundress (crown icon) that performs all maternal behaviors in addition to oviposition, including foraging, feeding larvae, and nest construction. Worker Phase. The first female offspring to emerge in early summer are workers (hard hat icon) that forage, feed larvae, and construct the expanding nest. The foundress transitions into a true social queen that limits her activities to oviposition and feeding larvae using foods brought to the nest by workers. Reproductive Phase. Female offspring emerging in mid-summer are non-working future foundresses called gynes (tiara icon). Males, not shown, emerge synchronously with gynes. Decline Phase. In late summer the queen and workers die. Gynes and males depart the nest, which now is empty and unattended. Quiescence Phase. Gynes and males mate and feed at fall flowers. With the decline in flower availability and onset of cooler weather, males die, and gynes enter a torpor-like quiescence in sheltered concealment. Workers live only from their emergence until the end of the reproductive phase. A gyne from the reproductive phase of one year has the potential to become a foundress and then queen the following year, thus the tiara to crown transition can represent single individuals that pass from gyne to foundress to queen. Larval treatments and sampling reported in this experiment were conducted in the Reproductive Phase of the colony cycle.

Reticulitermes flavipes [22], and by dsRNA microinjection in adult honey bees [23,24], the red imported fire ant *Solenopsis invicta* Buren [25], and the termite *Cryptotermes secundus* [26]. The majority of these studies were initiated in a laboratory situation. To develop RNAi for use in natural settings could advance the research utility of RNAi for learning mechanisms that underlie important features of social insects' behavioral ecology.

P. metricus may provide an ideal opportunity for this initiative. ESTs for *P. metricus* are now available, and paper wasps' ecology with easily observed nests has made them the subject of extensive natural history studies [8] and a smaller number of experimental studies in captive rearing (e.g. [27]), laboratory settings (e.g. [28]), and natural environments (e.g. [29,30]). Adults feed nectar directly to larvae [8], and this can be mimicked by feeding nectar-like liquids to larvae, providing a vehicle for dsRNA. Specific larvae can be tracked through time by means of 'nest maps', enabling individual marking of newly-emerged adults. These adults can then be tracked on their nests in field situations or collected for subsequent analyses.

Here, we develop RNAi for *P. metricus* for experimental applications in natural environments. This approach enables testing of causal relationships between candidate genes (genotype) and their association with behavior and/or caste ecology (phenotype). We exemplify this potential by exploring a central developmental question with significance in behavioral ecology: the role of specific storage proteins in developmental bias toward workers and gynes (future queens of the next generation). Storage proteins are ecologically important gene products that play a role in diapause not only in *Polistes* [31] but in a broad range of insects [32,33]. *Polistes metricus* has two putative hexameric storage proteins, Hexamerin 1 and Hexamerin 2. Of these, Hexamerin 1 levels in late-larval and early-pupal stages provide a significant predictor of caste bias in adults, whereas Hexamerin 2 does not [31], leaving the roles(s) of the latter protein obscure. Fragments of putatively Hexamerin-encoding transcripts are available as ESTs, but it has been unclear how these relate to the two largely uncharacterized *Polistes* proteins.

We report successful cloning of the Hexamerin 2 encoding gene of *P. metricus* from its available ESTs, and proteomic analyses of the *P. metricus* Hexamerin 1 and Hexamerin 2 proteins. The proteins were identified as separate products with largely diverging peptide sequence and different biochemical properties. Subsequently, dsRNA was synthesized for *hexamerin* 2 and a control gene (*gfp – green fluorescent protein*), and these were used in a larval feeding experiment with naturally-founded *P. metricus* colonies in the field. We show that larval Hexamerin 2 can be successfully suppressed by RNAi and that the adult caste phenotype of Hexamerin 2 knockdowns is not significantly altered. These results as well as suggestive trends in our data support the proposition of diverging roles of the Hexamerin 1 and Hexamerin 2 products in caste development.

Results

Sequence of *hexamerin 2* transcript and protein

The *P. metricus* EST database contains four partial sequences, Contig45616, Contig45913, Contig42334, and Contig40833, that share homology with *A. mellifera* hexamerins. We used RACE to complete the cDNA ends, resulting in a single cDNA (Fig. 2) that overlaps all four contigs. The N-terminal sequence of the encoded protein constitutes a putative secretion signal peptide for an extracellular protein (www.cbs.dtu.dk/services/SignalP/), with cleavage predicted between residues 18 and 19 (amino acids VSA-EY, Fig. 2). The predicted mass of the mature protein is ~68.5 kDa, which is slightly lower than observed on SDS-PAGE gel. This apparent difference in size could be due to post-translational modifications attached to the protein sequence. Accordingly, we tested for N- and O-glycosylation and found potential N-glycosylation sites at asparagine residues at positions 201 and 344 (www.cbs.dtu.dk/services/NetNGlyc/).

The cloned gene belongs to the hemocyanin family and has three conserved domains: hemocyanin_N (an all-alpha domain; residues 29–151), hemocyanin_M (which binds to copper and is necessary for the oxygen transport in the hemolymph; residues 158–415), and hemocyanin_C (an ig-like domain; residues 435–519) [34]. The gene shows highest identity to *A. mellifera* hexamerin 70A (64% identity), while the degree of shared identity to other hexamerins is 55% to *Nasonia vitripennis*, 47% to *Camponotus festinatus*, and 44% to three mosquito sequences: *Aedes aegypti*, *Culex quinquefasciatus*, and *Anopheles gambiae* (Fig. S1).

Mass-spectrometry

MALDI-TOF/TOF and LC-MSMS experiments revealed that Hexamerin 2 corresponds to the cloned gene based on 30%–40% sequence coverage, respectively (Fig. 3C). Hexamerin 1 had four hits against the sequence (YRGEEYLYGHK, LSNGLEK-VEDFDWSK, RIIYYYYR and EASYIIKAR; 8% sequence coverage), but got no hits in the subsequent LC-MSMS analysis. Both protein samples also contained unidentified non-trypsin, non-keratin peptides. MASCOT search score [35] for Hexamerin 2 LC-MSMS result against the cloned sequence was 207, with a sequence-based emPAI [36] score of 1.20. The peptide LTTYFDQFDTSLNNGLVVESQK that spanned the region of the dsRNA was, in particular, identified for Hexamerin 2 (Fig. S2).

Deglycosylation and phosphorylation

The two hexamerins have glycosylation in common, but they differ in the case of phosphorylation. Both Hexamerin 1 and Hexamerin 2 show reduced molecular weight on SDS-PAGE after treatment with PNGase F that removes N-linked glycogroups (Fig. 4A). The estimated size-shift was approximately 1.6 kDa for Hexamerin 1 and 1.8 kDa for Hexamerin 2, likely corresponding to the size of the glycogroup or total glycogroups attached to these proteins. Staining with phosphorylation-specific stain showed that of the two putative hexamerins only Hexamerin 2 was phosphorylated, and it was the only major phosphorylated protein in addition to a putative vitellogenin in *P. metricus* hemolymph (Fig. 4B). This difference at the post-translational level might reflect the divergent effects of the proteins on *P. metricus*.

Verification of hexamerin 2 down-regulation

Compared to controls, larvae fed *hexamerin 2* dsRNA (*hexamerin 2* RNAi) showed a significant reduction in *hexamerin 2* transcript levels (main affects ANOVA: $F_{1,32} = 6.258$, $p = 0.018$) (Fig. 5).

RNAi phenotype and trait associations

Most of the 128 treated larvae (Table S2) could not be included in the final analysis. Forty-seven treated larvae were males, and 31 were cannibalized. Cannibalism occurs naturally and increases in frequency as colonies advance toward the end of their cycle [37]. Two colonies were destroyed (one by a mowing machine; one by *Solenopsis invicta* fire ants) for a loss of 10 specimens (6 *gfp*; 4 *hexamerin 2*). Eight larvae failed to feed or to spin cocoons. Because this is the first experiment of its kind we do not know if this was in response to treatments *per se*, disturbance during treatments, or some other cause. Seven larvae spun cocoons before receiving 4 treatment feedings (3 *gfp*; 4 *hexamerin 2*), and three adult wasps escaped. The remaining 22 larvae received 4 treatment feedings, spun cocoons, emerged as adult females, and were placed in the bioassay (11 total *gfp* from 5 nests; 11 total *hexamerin 2* from 4 nests). We used these 22 specimens (Table S3) for all treatment analyses.

Study site did not affect the wasps' phenotype ($F_{1,16} = 1.000$, $P = 0.500$, $N = 5$, 15). Treatment was also not a significant factor in this corrected model that blocked for the effect of location on phenotypic traits ($F_{1,16} = 0.040$, $P = 0.874$, $N = 10$). No interaction was detected between study site and treatment ($F_{1,16} = 0.042$, $P = 0.840$). Next, we took into account that the total nine nests were split between the treatment categories. Using a nested design ANOVA to control for effects of nest within treatment, we found that neither treatment ($F_{1,\ 13} = 0.008$, $P = 0.929$, $N = 11$) nor nest ($F_{7,13} = 0.528$, $P = 0.798$, $N = 4$, 5) was significant. Thus, although the phenotypes showed some variability between nests (one-way ANOVA: $F_{4,\ 20} = 34.116$, $P = 0.023$, $N = 6$, excluding three nests with only 1 individual observation), the treatment groups were equally affected by inter-individual heterogeneity. We concluded that putative effects of treatment were not confounded by environmental influences or heterogeneity. To conserve degrees of freedom, therefore, factors location and nest were not used for blocking in subsequent analyses.

We initially used a full, parametric model to analyze the phenotypic dataset. Total developmental time, ovary development, forewing length, and adult feeding were included as dependent variables in a one-way analysis of variance (ANOVA). Total developmental time was summed over the measurements of days to cocoon and days cocooned to avoid a colinearity problem caused by the two variables being significantly negatively associated in the dataset (Pearson correlation, $R = -0.538$, $P = 0.001$, $N = 22$). Egg number was not included due to insufficient data and variability (1 egg was laid by a *gfp* control wasp). The ANOVA suggested that *hexamerin 2* RNAi did not alter the *P. metricus* phenotype. This result was independent of whether ovary development was coded as a binary variable (i.e., ovary being active or not: $F_{4,17} = 2.259$, $P = 0.105$, $N = 11$) or as 4 developmental classes ($F_{4,17} = 1.961$, $P = 0.146$, $N = 11$). The analysis conformed to assumptions of ANOVA, as determined by Bartlett's test and Levene's test on variances and normality plots of residuals.

Next, we looked at each variable using a non-parametric approach. We found that days cocooned, ovary development (4 classes), forewing length, and adult feeding were similar between

```
                              cgcggatccagacgctgcgtttgctggcttttatt
gaaaatagaattggtggaggaaagacacccggttactctgacagtctgttaaactaatc
ATGTACAGGAACTTGGTAGTTCTGGCCGTCCTGGCCTTCTGCCTTGTCTCTGCTGAATAT    60
 M  Y  R  N  L  V  V  L  A  V  L  A  F  C  L  V  S  A▲E  Y
TACACCGTTCAAACGGCTGACAAAGTGTTTTTATTAAAACAAAAGAAGATATACAATCTC   120
 Y  T  V  Q  T  A  D  K  V  F  L  L  K  Q  K  K  I  Y  N  L
TTGTATCACGTTTCCCAACCAGACATCGTCAATCCTGATTTATATACCGAGGGTAAACAA   180
 L  Y  H  V  S  Q  P  D  I  V  N  P  D  L  Y  T  E  G  K  Q
TATAGTATTGAAGCCAACATCGATTCCTATACCAATCAAGATGCTGTGAAAGAATTCCTA   240
 Y  S  I  E  A  N  I  D  S  Y  T  N  Q  D  A  V  K  E  F  L
TATCTATACAAGCATGGAATGTTACCACGTAACAGTGTTTTCTCAATCTATTATCCTAAA   300
 Y  L  Y  K  H  G  M  L  P  R  N  S  V  F  S  I  Y  Y  P  K
ATCCTAAAAGAATCGATAGCTTTGTTCCATATGTTTTATTATGCCAAAGATTTTGATGTT   360
 I  L  K  E  S  I  A  L  F  H  M  F  Y  Y  A  K  D  F  D  V
TTCTACAAAACAGCATTATGGGCTCGTATTTATATGAATGAAGGACAATTCATTTTTGCT   420
 F  Y  K  T  A  L  W  A  R  I  Y  M  N  E  G  Q  F  I  F  A
TTTTTCAATGCTGTCGTTCAACGTCCAGATACTATGTATATTCAACTTCCACCTATCTAT   480
 F  F  N  A  V  V  Q  R  P  D  T  M  Y  I  Q  L  P  P  I  Y
GAGCTTTATCCGTATGGATTTTTCAATTCTGAAGCTCTTCAGAAGGCAAATCACGCAAAG   540
 E  L  Y  P  Y  G  F  F  N  S  E  A  L  Q  K  A  N  H  A  K
ATCTTTGGTAAACTAGATTCGCAAAAGTCAGCTGGATATGACACATACATTATTCCAGCT   600
 I  F  G  K  L  D  S  Q  K  S  A  G  Y  D  T  Y  I  I  P  A
AACTATTCTGGCTGGTATATAAACCACGAATATGATCGTGAGAGACAATTAAATTACTTC   660
 N  Y  S  G  W  Y  I  N  H  E  Y  D  R  E  R  Q  L  N  Y  F
ACTGAAGATATCGGCTTGAACTTGTATTACTTCTACTTCAGATATCAATACCCCTTCTGG   720
 T  E  D  I  G  L  N  L  Y  Y  F  Y  F  R  Y  Q  Y  P  F  W
ATGAAAGGCGAAGAATTTAAATTCCCGAAATATCGTGGGGAAGAATACTTATACGGACAC   780
 M  K  G  E  E  F  K  F  P  K  Y  R  G  E  E  Y  L  Y  G  H
AAACAACTTATGACACGTTACCACTTGGAACGTTTATCAAATGGTCTTGAGAAGGTAGAA   840
 K  Q  L  M  T  R  Y  H  L  E  R  L  S  N  G  L  E  K  V  E
GATTTTGACTGGTCCAAAAAAATTCTATCCTGGATATTATCCAACCATGAATTATCACAAT   900
 D  F  D  W  S  K  K  F  Y  P  G  Y  Y  P  T  M  N  Y  H  N
GGGTTGCCATTTGTACAACGTCCATGCTTCAGTATATTCCCATATTACAAATACAAATAT   960
 G  L  P  F  V  Q  R  P  C  F  S  I  F  P  Y  Y  K  Y  K  Y
ATTAGGGATGTAAACGAGATGGAATCGCGTATCACTGGTGCCATCGATTCTGGTCTTGTC  1020
 I  R  D  V  N  E  M  E  S  R  I  T  G  A  I  D  S  G  L  V
ATGGATAAAAATGCCACATTGATCAATATTCATACTCCAGATGGTATTAATATTCTTGGA  1080
 M  D  K  N  A  T  L  I  N  I  H  T  P  D  G  I  N  I  L  G
AATATAATTGAAGGTAATGTGGATTCATATAACTATGATTTCTACGGAAGCTTAGACTAT  1140

 N  I  I  E  G  N  V  D  S  Y  N  Y  D  F  Y  G  S  L  D  Y
TATGCTAGAAAGATTCTTGGATATAATATGGAACCCGCAACACCATATCAAATCACACCA  1200
 Y  A  R  K  I  L  G  Y  N  M  E  P  A  T  P  Y  Q  I  T  P
AGTGCTCTTGAACATTATTCAACATCTTTAAGAGATCCAGCTTTTTATCGGCTCTACAAG  1260
 S  A  L  E  H  Y  S  T  S  L  R  D  P  A  F  Y  R  L  Y  K
AGGATAATCTATTATTACTACAGATACAAAGTACGTCAGGAGCCTTATACCAAGGATATG  1320
 R  I  I  Y  Y  Y  Y  R  Y  K  V  R  Q  E  P  Y  T  K  D  M
GTTGTTTTCCCAAGTCTCAAAGTTGAATCCTTTGCTGTGGACAAACTGACAACTTACTTC  1380
 V  V  F  P  S  L  K  V  E  S  F  A  V  D  K  L  T  T  Y  F
GATCAATTTGATACATCCCTTAACAACGGTTTGGTAGTTGAGAGCCAAAAAGAAGCTGAG  1440
 D  Q  F  D  T  S  L  N  N  G  L  V  V  E  S  Q  K  E  A  E
AGCTACATTATCAAAGCAAGACAATATCGTCTCAATCACAAACCATTTAATTTGCATATA  1500
 S  Y  I  I  K  A  R  Q  Y  R  L  N  H  K  P  F  N  L  H  I
ACCATTAATTCTGAAAAAAGCACTAAAGTAGCTATTCGCATTTTCCTCGGACCCAAATAC  1560
 T  I  N  S  E  K  S  T  K  V  A  I  R  I  F  L  G  P  K  Y
GACGCACTTCATAGATTACTTAACTTTGAAGAGAACTTCAAATATTTCTAT           1610
 D  A  L  H  R  L  L  N  F  E  E  N  F  K  Y  F  Y
```

Figure 2. Nucleotide sequence of *hexamerin* 2 cDNA clone from *P. metricus*, along with the deduced protein sequence. The arrowhead indicates the signal cleavage site as predicted by SignalP (www.cbs.dtu.dk/services/SignalP). The underlined portion of the amino acid sequence corresponds to the dsRNA used for down-regulating *hexamerin 2*. The region corresponding to the 5′ UTR (Untranslated Region) is shown with lowercase letters. The *hexamerin 2* sequence shown is partial and does not include the C-terminal end (~50 amino acids). The GenBank accession number for the *hexamerin 2* sequence is HQ661804.

hexamerin 2 knockdowns and *gfp* controls (Mann-Whitney U-test, P>0.05, N = 11, Fig. 6B, D–F). In contrast, days to cocoon and ovary development score (binary) were significantly different and suggestively different, respectively (Mann-Whitney U-test, P = 0.017; P = 0.094, N = 11, Fig. 6A, C). These results suggest that *hexamerin 2* down-regulation could extend larval development and tentatively reduce ovarian development in adults. However, when corrected for multiple comparisons (Dunn-Sidak Bonferroni, alpha<0.009), neither effect could be considered as statistically reliable.

Finally, we studied the phenotypic trait associations independent of treatment group identity. Adding to the significant negative correlation between days to cocoon and days cocooned (above), the data showed a significant negative relationship between total developmental time and ovary development (Pearson correlation, R = −0.526, P = 0.012, N = 22). This finding indicated that, in the dataset overall, individuals with a shorter total developmental time were more likely to have developed ovaries when scored as adults.

Discussion

Here we have reported the successful development of an RNA interference tool and characterization of its target protein for the primitively eusocial paper wasp *Polistes metricus*. Our most significant contribution to RNAi studies is that we then successfully deployed the RNAi tool in a field setting – that is, in a natural setting outside the laboratory. We did so in the context of an ecological test of caste differentiation in the eusocial wasp *Polistes*, which is a model system for evolutionary studies of insect sociality and behavioral ecology. The foundational aspects of our work included sequencing the *hexamerin 2* gene, matching the cloned gene to the Hexamerin 2 protein product, and identification of the correspondence between the cloned gene and Hexamerin 2 by mass-spectrometry. Our principal biological finding – that Hexamerin 2 plays no role in worker/gyne differentiation – coincides with findings from previous studies using other investigative methodologies.

Our research underscores that *Polistes* is a robust model system, having a rare convergence of assets. Paper wasps, with their 2 dimensional, bare nests allow access for observation and manipulation in their natural environment. As a result, they are a model that can be used to address questions regarding the structure and evolution of insect sociality. Most other social insects have been studied in artificial habitats. Ants and termites nest underground or within solid structures, thus making in-nest

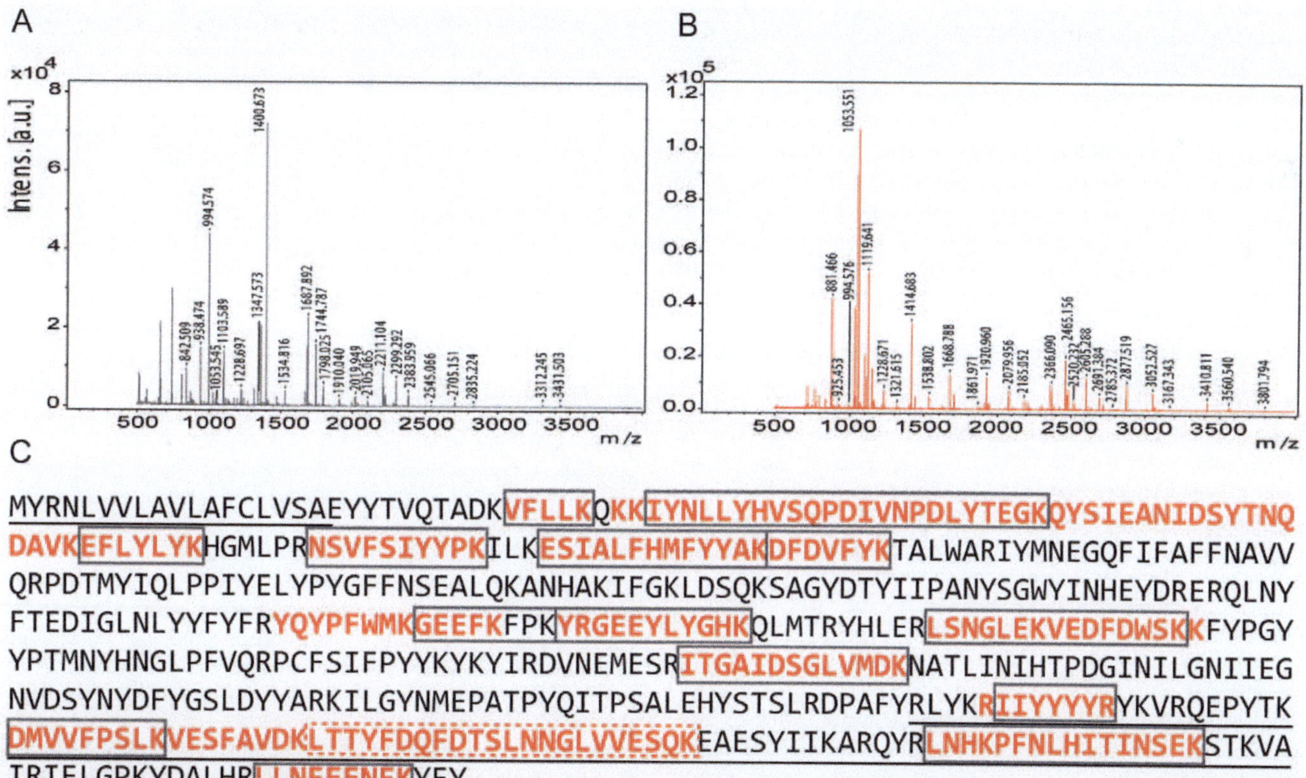

Figure 3. Mass-spectrometric spectra of Hexamerin 1 and Hexamerin 2, and the peptide hits of Hexamerin 2 against hexamerin sequence. The signal sequence in the N-terminus and the dsRNA in the C-terminus are underlined. A: MALDI-TOF/TOF spectra for Hexamerin 1. B: MALDI-TOF/TOF spectra for Hexamerin 2. C: MALDI-TOF/TOF peptide hits (grey boxes) and LC-MSMS hits (red text) for Hexamerin 2. A specifically inspected LC-MSMS peptide found in Hexamerin 2 data, is boxed with a red dashed line.

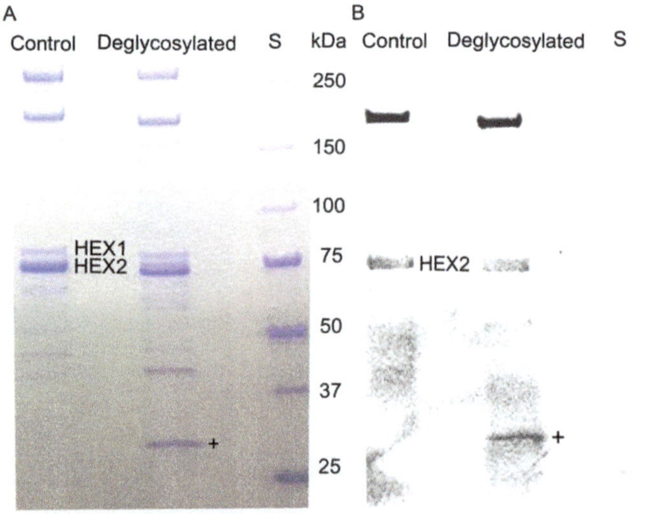

Figure 4. Glycosylation and phosphorylation of *P. metricus* (wild type) hexamerins. Both hexamerins are glycosylated, but only Hexamerin 2 is phosphorylated. **A:** Both hexamerins shift in size toward lower molecular weight after deglycosylation with PNGase F. **B:** The same gel stained with phosphorylation specific Pro-Q diamond stain (Invitrogen). Hexamerin 2 and vitellogenin (between 150 and 250 kDa standard bands) appear to be phosphorylated in *P. metricus* hemolymph. The + sign marks the deglycosylation enzyme PNGaseF.

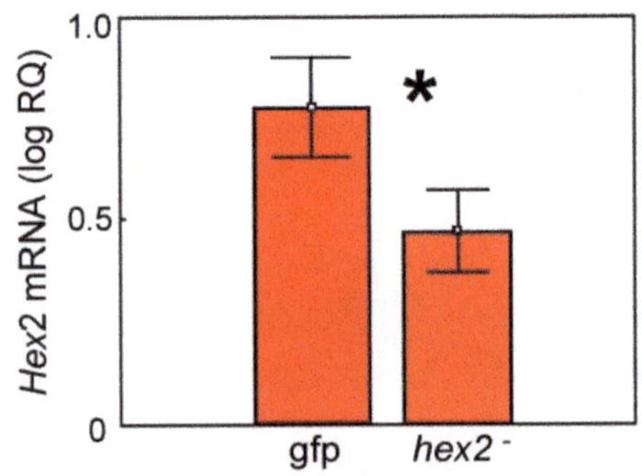

Figure 5. Validation of *hexamerin 2* knockdown after larval feedings. RNA isolated from whole larvae (n = 18 per treatment type) was used for qRT-PCR. Compared to controls, *hexamerin 2* dsRNA feeding elicited a significant reduction in *hexamerin 2* transcript. Bars represent means ± s.e.; asterisk = $P<0.05$.

observations difficult or impossible in the field. Honey bees, although easily observed and the target of the first insect genome project, are domesticated animals and have been selected for docility and honey production. Our work combines the assets of the *Polistes* model system with tools of genomics deployed in a field setting, augmenting the studies of Toth et al. and Hunt et al. [17,31,38,39] and further advancing the *Polistes* model system.

Our small final sample sizes (11 *hexamerin 2* and 11 *gfp*) could reflect several variables that can be addressed in future RNAi experiments. In order to provide maximum assurance that we treated only gyne-destined larvae, we conducted our experiment in late July and August, corresponding to the mid to late phase of gyne production. Because nest mortality is a normal demographic occurrence, a higher sample size could be obtained by beginning an experiment earlier in the phase of gyne production. To increase the number of nest boxes at the start of a season could also increase the final sample size. The frequency of nest foundings in our nest boxes by *P. metricus* (47/300) was unexpectedly lower than that of previous studies conducted in Missouri, USA.

The four putative Hexamerin 1 peptide hits (8% sequence coverage) toward the Hexamerin 2 sequence in the MALDI experiment could be explained by some level of contamination or by sequence identity in these regions. Because the gene that encodes Hexamerin 1 is not identified, precise sequence similarity estimation between two hexamerins is currently infeasible. Previous studies of other hymenopteran species with sequenced genomes have reported relatively high sequence similarity of multiple hexamerins: up to 42% in *A. mellifera* [40] and 79% in the wasp *Nasonia vitripennis* [41].

We found that the two putative hexamerins of *P. metricus* are glycosylated. In Hexamerin 2, the putative N-glycosylation sites are located at asparagines 201 and 344. Glycosylation has been shown to have important stabilizing and structural roles in a related storage protein, arylphorin, in *Antheraea pernyi* silk moth [42]. Based on SDS-PAGE size-estimation, Hexamerin 1 (~77.7 kDa) lies within the typical hexamerin size scale of 75–90 kDa, whereas Hexamerin 2 is smaller than an average hexamerin with its molecular weight of ~68.5 kDa [40,41]. In addition to differences in size, only Hexamerin 2 appears to be phosphorylated. Phosphorylation can be critical for events ranging from signaling to subcellular protein localization [43,44]. Hexamerin-receptor interaction is regulated by receptor phosphorylation in rice moth *Corcyra cephalonica* [45], but effect of phosphorylation of hexamerin itself are currently unclear. The differences that we found between Hexamerins 1 and 2 in both size and phosphorylation suggest different roles for the two proteins.

In a previous study using gel band quantification, Hexamerin 1 was significantly differentially expressed between worker-destined and gyne-destined larvae, but that experiment indicated no specific caste-related role for Hexamerin 2 [31]. In the present study we used dsRNA for the *hexamerin 2* gene to specifically down-regulate production of Hexamerin 2 protein. The absence of a significant caste-related response to RNAi down-regulation of Hexamerin 2 confirms the inference of the previous study. However, our detailed statistical analysis suggests possible trends toward extended larval development time and reduced ovary development in knockdowns, which are characteristics of gynes [31,39]. The detailed statistical analysis is warranted, even though the statistical results *per se* are negative, because both positive and negative statistical outcomes are meaningful for interpretation of the study.

Whereas dsRNA injection studies achieve significant knock-down of the target gene at low dosage [46,47], feeding studies generally require a higher dosage to achieve efficient knockdown. Effective knockdown in *Polistes* was achieved using a dosage of dsRNA consistent with feeding studies in honey bees [21,48] and termites [22]. Control (*gfp*) and *hexamerin 2* larvae were fed equally with dsRNA without adverse phenotypic effects, thereby suggesting that the system will be compatible with future RNAi experiments. A next step, therefore, would be to continue investigation of roles played by hexamerins in *P. metricus* worker/queen differentiation using *hexamerin 1* RNAi in a full replication of the study we report here, once transcripts for the *P. metricus hexamerin 1* gene have been identified. Such an investigation could test the possibilities of compensatory or opportunistic upregulation of Hexamerin 1 by Hexamerin 2. Compensation would be physiological – to have Hexamerin 1 substitute for Hexamerin 2 in functions that are not yet known. 'Opportunistic' would be that Hexamerin 1 increases because more building blocks are available if less Hexamerin 2 is reduced. In addition, methods such as microarray [38] and RNAseq can be employed to identify

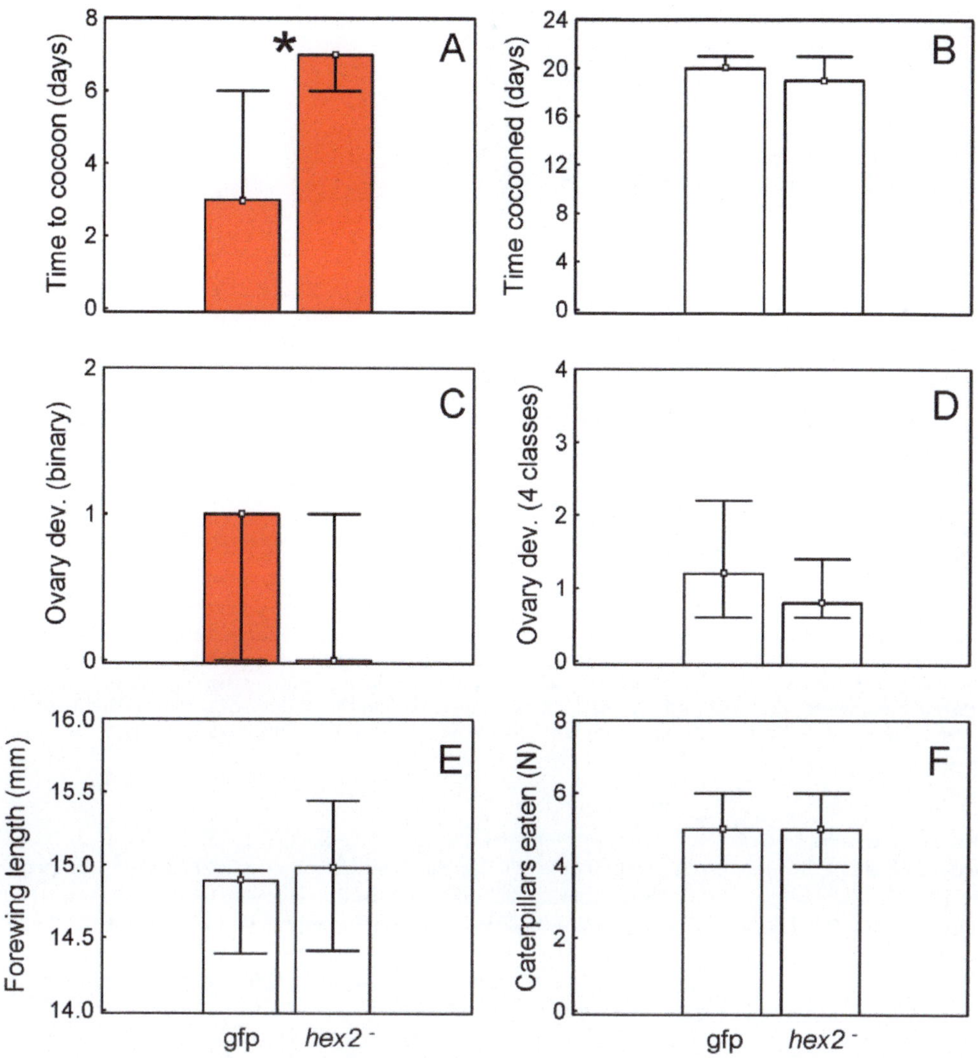

Figure 6. Median and 25–75% interquartile ranges for RNAi phenotypes. *P. metricus* larvae were assigned to a control group that received dsRNA for green fluorescent protein (*gfp*) or to a treatment group that received dsRNA against the *hexamerin 2* gene to trigger RNA interference (*hexamerin2* ⁻). The wasps were monitored during post-treatment development and isolated for bioassays on the day they emerged as adults. **A:** The number of days from the start of treatment with dsRNA until cocoon spinning (time to cocoon). The significance indicator denotes Mann-Whitney U-test, $P = 0.017$, $N = 11$. This test statistic is significant only if multiple testing is not accounted for. **B:** The number of days between cocooning and adult emergence (time cocooned). **C, D:** Ovary development in 13 day-old adults as (C) a binary score (active or not – categories 1 and 2 of Fig. S3) and (D) scored as the 4 developmental classes of Figure S3. **E:** Length in mm of adult forewings from the proximal convergence of the C cell to the distal posterior corner of the 2Rs cell. **F:** Number of caterpillars eaten during the post-emergence isolation period. Control *gfp* and treatment *hexamerin2* ⁻ phenotypes differed significantly for time to cocoon.

candidate genes for other ecologically and evolutionarily important traits such as behavioral dominance, seasonal and nutritional affects on the colony cycle, photoperiodicity, and diapause. The genetic foundations of these traits can then be pursued using RNAi in experiments such as we report here. Ever-lower costs of whole genome sequencing using high throughput sequencing techniques, combined with improvements in annotation and comparative analysis methods, now allow genomic tools to be developed for many species. Efforts are currently underway to sequence genomes of two *Polistes* species: *P. dominulus* and *P. canadensis* (A. L. Toth, personal communication). Thus, soon there will be a substantial increase in genome sequence information for *Polistes*. The RNAi tool that we present here, therefore, is an extremely timely step forward that can enhance the use of paper wasps as model systems for the greater synthesis among gene expression, behavioral ecology, and evolutionary biology of social insects. In addition, although there are unique features of *Polistes* biology that make it particularly suitable for RNAi experiments in the field, our work suggests the possibility that RNAi is a tool that can be adapted for experiments on other species in natural environments.

Methods

Hexamerin sequencing

Four partial EST sequences for hexamerin (contigs 45616, 45913, 42334 and 40833: http://stan.cropsci.uiuc.edu/454/blast/waspblast.html) were available from the *Polistes* wasp EST data set to design primers. Total RNA isolated from 5th instar larvae was treated with DNaseI (Ambion), and 5′ and 3′ RACE were carried out using the GeneRacer Kit (Ambion). For cloning, we used different combinations of primers for 5′ and 3′ RACE (Table S1) in conjunction with the RACE primers provided in the

Figure 7. Field methods and dsRNA delivery. A: "Wasp boxes" have lift-off lids and wire mesh across the open bottom. Boxes were put in place in springtime prior to natural nest establishment on the undersides of lids by *P. metricus* foundresses. The early foundress phase nest in this box is indicated by the arrow. **B:** A reproductive phase study nest on the underside of a box lid during the experiments. Newly-spun cocoons of larvae that had received dsRNA were marked with dots of paint (inset). **C:** dsRNA and sucrose solutions were carried to the field on ice and mixed immediately prior to larval treatments. **D:** dsRNA/sucrose solution was fed directly to the mouthparts of large 5th-instar larvae via micropipette. Larvae consumed the sample quickly, generally in ~1 sec.

kit. Following PCR, the amplified products were separated on 1% agarose gel, and DNA was extracted using the QIAquick Gel Extraction Kit (Qiagen). Thereafter, the PCR products were cloned into pCR® 4-TOPO® vector using the TOPO TA cloning kit (Invitrogen). Several clones were randomly picked and verified by sequencing.

MALDI-TOF/TOF

We identified the cloned hexamerin using peptide mass fingerprinting. To do so we ran dilution series of 0.2–0.8 µl *P. metricus* hemolymph from an adult gyne, a rich source of hexameric storage proteins [49], on a 4–20% gradient SDS-PAGE gel (Bio-Rad) followed by staining proteins with Coomassie blue (Bio-Rad). We then excised protein bands corresponding to Hexamerin 1 and Hexamerin 2 as identified by molecular weight [49] plus a clear gel region between the two proteins as a blank (control) sample. The samples were digested with trypsin and prepared for Ultraflex MALDI-TOF/TOF (Bruker Daltonics) analysis according to existing protocols [50,51] using nC18 StageTip microcolumns [36]. We then processed spectra with FlexAnalysis (Bruker Daltonics) and targeted the annotated peaks for peptide mass fingerprint analyses with Mascot (Matrix Science; in house Mascot server) using all the NCBI databases and the *P. metricus* hexamerin sequence information we obtained by RACE.

LC-MSMS

We excised gel bands of Hexamerinl, Hexamerin 2 (4 gel pieces), and a blank sample as above. Peptides were prepared as above and separated by Ultimate 3000 nano-HPLC (Dionex Corporation, USA) using a nC18 reverse phase column coupled to a 4000 Q-Trap mass spectrometer (Applied Biosystems/MDS SCIEX, Concord, ON, Canada) and eluted as described elsewhere [51]. We used Analyst (AB SCIEX) for data processing and Mascot for data analysis. Mascot search was done against the *P. metricus* hexamerin sequence embedded in all the common NCBI databases. We set the significance threshold at 0.05, ion score minimum at 20, and included only hits with expected values <0.1. A putative peptide corresponding to the size of the sequence LTTYFDQFDTSLNNGLVVESQK, located at the dsRNA region, was inspected in detail in the elution pattern of the Hexamerin 2 peptides.

Deglycosylation and staining of phosphorylated proteins

We tested Hexamerin 1 and Hexamerin 2 for post-translational modifications, glycosylation and phosphorylation, to gain protein-level insight and to make a comparison between the hexamerins. In order to test the presence of N-linked glycogroups, we treated 0.2 µl *P. metricus* hemolymph for 1 h at 37°C with 1 µl PNGase F and kit buffers (New England BioLabs), followed by a 10 min denaturation at 95°C. We treated a control sample similarly but with no PNGase F. We then ran samples on a 4–20% gradient gel (Bio-Rad) and stained them with Pro-Q Diamond (Invitrogen), which exclusively stains phosphorylated proteins. Finally, we visualized the gel in trans-UV light with Gel Doc Imager (Bio-Rad) prior to staining it with Coomassie blue (Bio-Rad) for detection of all proteins.

dsRNA synthesis

dsRNA was synthesized as described by Amdam et al. [24] using the AmpliScribe T7 transcription kit (Epicentre Biotechnologies) for bidirectional *in vitro* transcription of a plasmid DNA construct using gene specific primers with flanking T7 transcription initiation sites. For *Hexamerin 2* dsRNA synthesis, we used the following forward and reverse primers fused with the T7 promoter sequence (underlined): 5′-TAATACGACTCACTATAGGGCGACGGCTCTACAAGAGGATAATC-3′ and 5′-TAATACGACTCACTATAGGGCGAGAAATATTTGAAGTTCTCTTC-3′. The resulting dsRNA product was 360 bp long. As an RNAi treatment control, *gfp* dsRNA, was synthesized using primers that cover 520 bases in the open reading frame of the AF097553 template as previously described for gene targeting in honey bees (Wang et al 2010). The dsRNA was purified using phenol:choloform extraction and verified for size and purity by running on 1% agarose gels. The final concentration of dsRNA was 4–5 μg/μl. The sequence used to generate *hexamerin 2* dsRNA is underlined in Figure 2.

Larval sex determination

We determined larval sex by genotyping 9 microsatellite loci that were previously developed from expressed sequence tag sequences [52]. We used 2 μg of total RNA for cDNA synthesis using SuperScript VILO cDNA Synthesis Kit (Invitrogen) and then used the cDNA for microsatellite genotyping. PCR reactions and fragment visualization were performed as previously described [31].

Verification of hexamerin 2 knockdown

Fifth-instar larvae received 2 feedings per day for 2 days of either *hexamerin 2* dsRNA or *gfp* dsRNA. Two days later, in the field, we processed single specimens as each was removed from its nest cell. Larval gut contents, called meconium, are contained within a chitinous peritrophic sac. The meconium volume is particularly large in fifth-instar larvae. To remove the meconium makes subsequent processing of the specimen easier and also obviates any possible RNA degradation. Using a pointed scalpel, we pierced the body and midgut walls and removed the meconium by grasping the peritrophic sac with fine-tipped forceps and quickly extracting it. The larva was then flash frozen in liquid N_2. Specimens were further processed in the lab by homogenization and storage in Trizol reagent. We randomly selected 18 female larvae per treatment group for knockdown verification via real-time quantitative reverse-transcriptase PCR (qRT-PCR) using a one-step qRT-PCR kit (Ambion). Total RNA was isolated and treated with turbo DNase I (Ambion). Briefly, total RNA (25 ng/μl) was used as the template for reverse transcription. All qPCR reactions were run in triplicate using an ABI Prism 7500 (Applied Biosystems), and we analyzed the data with the Delta-Delta CT method [53] with *rps8* as reference gene [17,54]. The *hexamerin 2* forward and reverse primers were 5′-AAACAGCATTATGGGCTCGT-3′ and 5′-GACGTTGAAC-GACAGCATT-3′ respectively. For *rps*8, the forward and reverse primers were 5′-GAAGCGAAAGCCTCTCAGAA-3′ and 5′-AGGCGCAGTGCTCTGTATTT-3′. A negative control (without reverse transcriptase enzyme) was run for every sample to monitor DNA contamination. We used melt curve analysis to verify the absence of primer-dimers.

Field rearing and in vivo dsRNA feedings

We reared and monitored colonies of *P. metricus* using established protocols [30,55,56] (Fig. 7A). We placed three hundred nest boxes in weedy fields and at the margins of lawns and hay fields, all of which were near woodlands at sites owned and managed by North Carolina State University or the North Carolina Department of Agriculture in Wake Co., NC, within 10 km of the North Carolina State University (NCSU) campus. Forty-seven *P. metricus* nests were founded, all but one by a single foundress each. We worked only with single-foundress colonies. During the course of the field work we monitored colony development daily by removing the nest box top (Fig. 7B) and recording the content of each nest cell (empty, egg, larval instar, or cocoon) on a 'nest map' drawn on hexagon-imprinted paper. We collected colony data in the early mornings, generally before 9 am EDT, which enabled counting the number of adults present before daily foraging had begun.

We conducted all dsRNA treatments in the field using late 5th instar larvae in single-foundress colonies *in situ* in the nest boxes. Hexamerins 1 and 2 are not expressed in 1st through 4th instar larvae [49]. Instead, they are expressed in 5th (final) instar larvae and reach peak abundances at the larva to pupa molt [31,49]. Our RNAi treatments were applied to late 5th instar larvae to coincide with peak Hexamerin expression, thus our results are not a consequence of having treated larvae in a time window that might not yield a response signal.

Only a single dsRNA type was used per nest. For treatments, tubes of dsRNA were placed on ice and transported to the field, where we mixed the dsRNA with 30% sucrose (Fig. 7C) immediately before feeding it to larvae using a standard laboratory pipette (Fig. 7D). At each feeding we gave each larva 15 μl of the dsRNA/sucrose mix, which they readily accepted. Each larva was treated morning and evening on 2 consecutive days. Each feeding contained 20 μg of dsRNA, therefore a total of 80 μg of dsRNA was administered to each larva. A total of 72 larvae from 10 nests were fed *hexamerin 2* dsRNA, and a total of 56 from 8 nests were fed *gfp* dsRNA (Table S2). We identified larvae by location on the nest map. When a fed larva completed growth and spun its cocoon, we marked that cocoon with dots of Testors® enamel (Fig. 7B inset). We continued nest mapping daily until all fed larvae in a nest had spun cocoons.

When all fed larvae in a nest had spun cocoons, we transferred that colony to a rearing cage as described in [57]. Adults taken with the nest were anesthetized with CO_2 and examined for wing wear. From 1 to 5 females with wing wear, which characterizes foundresses and workers [39], were marked with a dot of enamel on the metanotum and caged with the nest. We released any males and females with unworn wings, which were putative gynes [39]. We removed and discarded all unmarked cocoons and the untreated developing wasps in them as well as any subsequently cocooning individuals. Eggs and pre-cocooning larvae were undisturbed. We held caged colonies in a sheltered outdoor location at the NCSU Honey Bee Research Facility, where each colony was provided with *ad libitum* water and sucrose (rock candy) and a fresh mid-3rd instar *Manduca sexta* caterpillar daily. When all treated wasps had emerged from a nest, the remaining colony was returned to its nest box in the field and thereafter not disturbed. The field studies did not involve any endangered or protected species. No specific permits were required for the described field studies.

Treatment bioassay

We inspected caged colonies daily at mid-morning and identified newly-emerged wasps by nest map location prior to transporting them to the laboratory and placing each in an isolation cage. Isolation cages had wooden tops and bottoms ~8.5 cm on each side. Three sides of the cage were wire mesh (~3 mm openings) ~23 cm in height. The fourth side was closed with corrugated cardboard into which a small doorway was cut for daily maintenance. A small, empty, pre-emergence nest (7 to 14 cells) constructed in the preceding year by a *P. metricus* foundress near St. Louis, MO, was affixed to the underside of the top of each cage before the newly-emerged wasp was introduced. Most wasps spent time on those nests.

Caged wasps were placed in growth chamber C25 in the NCSU Phytotron. The chamber contained incandescent lights on a 16L/8D cycle; fluorescent lights came on 2 h following and went off 2 h before the incandescent lights. Light intensity was 21 micromoles/sec/m^2 in the incandescent-only morning and evening periods and 225 micromoles/sec/m^2 during the both-lights midday. Chamber temperature was 20°C at night and 30°C during full light, with a gradual ramp-up in the morning and ramp-down in the evening. We provided each isolation cage with a small plastic Petri dish containing sucrose (rock candy), a water source (a 1.5 ml microfuge tube plugged with cotton), and, daily, a single small *M. sexta* caterpillar: late 2nd-instar or early 3rd-instar. Caterpillar head capsules were crushed immediately before provisioning, which ensured that they did not crawl away. Cage maintenance, conducted daily at midday, consisted of replenishing the water (deionized), removing the remains of any uneaten caterpillars, and providing a fresh caterpillar. Time outside the chamber was no more than 2–3 min. Each day we repositioned cages within the chamber.

Treatment response variables

In a previous laboratory experiment with newly-emerged isolated *P. metricus* females, putative workers showed ovarian development whereas putative gynes did not [11]. Wasps entering ovary development could be expected to show higher rates of feeding from protein sources than wasps not entering ovary development. Because ovary development occurs in putative workers but not putative gynes, and because ovary development could reflect post-emergence feeding, we quantified both feeding rate and ovary development in order to test whether *hexamerin 2* treatment might induce ovaries more worker-like than those of controls. Beginning the day after being introduced to the chamber, we scored each wasp daily for 12 days as to whether the caterpillar had been eaten. On the 13th day we removed each wasp from the chamber, chilled it ~10 min in a refrigerator, and removed it from its cage.

We dissected ovaries (together with the digestive tract) by grasping the terminal segment of the metasoma with fine-point forceps and pulling. Removed tissue was placed in 100% EtOH in a labeled microfuge tube and stored at −20°C. Approximately 2 weeks after the last specimen had been prepared, we transferred the tissue samples to 70% EtOH and returned them to −20°C. Approximately 2 months later we dissected the tissue specimens in 70% EtOH under a dissecting microscope to separate the ovaries from other tissues. We then placed the ovaries in 70% EtOH in number-coded 0.2 ml microfuge tubes and held them at laboratory ambient temperature. Approximately 1 month later we digitally photographed each ovary using an Evolution MP Color camera on an Olympus SZX12 microscope with a 1.0× objective lens and the adjustable magnification set at 25×. All images were captured in a single session with no adjustments to the apparatus.

Four ovary images were selected as representing 4 development classes: 0 = ovaries filamentous with no discernable oocytes; 1 = oocytes discernable but small in size and somewhat translucent; 2 = one or more oocytes opaque (with yolk deposition); 3 = one or more large ova present (Fig. S3). We removed the selected images from the image set, and the remaining images were independently scored 0 to 3 by five non-biologists who compared each image to printed copies of Figure S3. For every ovary, we used the mean of these 5 objective scores for analyses. Scorers were blind to the identities of the specimens.

We clipped the right forewing of each wasp at its base using iris scissors and taped it to a labeled microscope slide with clear tape. We then captured digital images of the wings with the same camera/microscope equipment as for ovaries but with a 0.5× objective lens and with the adjustable magnification set at 10×. All images were captured in a single session with no adjustments to the equipment. We quantified images using ImageJ software. A linear measurement (in pixels) was taken from the proximal closure of the C cell to the distal posterior corner of the 2Rs cell (nomenclature as in [58]). Wing wear by some wasps when in isolation precluded total wing length measurement. Measurements were blind with regard to which dsRNA a wasp had received. We converted pixel values to mm by quantifying a scale photographed at the same magnification.

Statistical analyses

The nests that received *gfp* and *hexamerin 2* dsRNA treatment to verify *hexamerin 2* knockdown were equally represented at two study sites. For RNAi validation, we collected *gfp* control larvae from six nests and putative *hexamerin 2* knockdown larvae from four nests. Only a single dsRNA was administered in a single nest. Gene expression data were log transformed. We then used a main effects ANOVA in a model with mRNA level as the dependent variable and with qPCR plate (technical blocking factor) and treatment as the dependent variables. The dataset conformed to assumptions of ANOVA, as determined by Bartlett's tests and normality plots of residuals.

For analyses of individual development and adult bioassays, the *gfp* controls represented five nests at three locations, and the *hexamerin 2* dsRNA treatment group represented four nests and three locations. The majority of the data (20 out of 22 wasps) came from two overlapping sites. To test whether treatment effects could be confounded by environmental influences or heterogeneity between location and nests, we first tested the effect of these two study sites on phenotypic characteristics, using a general linear model with location as random factor, treatment as a fixed effect, and ovary development, total developmental time, forewing length, and number of caterpillars eaten as dependent variables.

All analyses were performed with Statistica 6.0.

Supporting Information

Figure S1 Amino acid alignment of the putative *P. metricus* hexamerin protein with other known insect hexamerin sequences. *Polistes* sequence was aligned with sequences of from other hymenopteran insects species, *Apis meliferra* (ABR45904), *Nasonia vitripennis* (XP_001607029) and *Camponotus festinatus* (CAB62053), and dipteran insects, *Aedes aegypti* (XP_001663961), *Culex quinquefasciatus* (XP_001843494) and *Anopheles gambiae* (XP_321434) using ClustalW version 1.82. The hemocyanin_N, (residues 29–151), hemocyanin_M domains (residues 158–415), and the hemocyanin_C, (residues 435–519) are all under-lined with dashed, solid and double-dashed lines respectively. Conserved residues among all taxa are highlighted in black bars, conserved substitutions among all taxa are highlighted in gray bars and semi-conserved substitutions are highlighted by light gray bars. Gaps in the alignment are indicated by dashes.

Figure S2 Electrospray mass spectrometric scan of Hexamerin 2 peptides. Peptide corresponding to dsRNA sequence LTTYFDQFDTSLNNGLVVESQK with mass 840.4 Da, indicated with the arrow, is only found in Hex2.

Figure S3 Ovary categories 0 to 4 that were used as comparators for scoring ovary development.

Table S1 Primer sequences used for cloning *P. metricus hexamerin 2*.

Table S2 Data records for all treated specimens. Columns, left to right, are the specimen identifier, treatment, date of first feeding with dsRNA, date of the fourth (final) feeding with dsRNA, total number of feedings received, date on which the pupal cocoon was spun, number of days from the first feeding with dsRNA until the cocoon was spun, number of days from cocoon spinning until adult emergence, sex, and fate of the specimen.

Table S3 Specimen identities, dsRNA treatments, and bioassay response data for the 22 adult wasps used for all analyses.

Acknowledgments

We reared experimental wasp colonies on properties owned by North Carolina State University (NCSU) and the North Carolina Department of Agriculture. We thank Ellen E. Lenz and especially Matthew K. Howe for assistance with field work and data collection, Jennifer Keller and David Tarpy for accommodating our bioassay work at the NCSU Honey Bee Research Facility, Carole Saravitz and Janet Shurtleff for accommodating our work at the NCSU Phytotron, Freddie-Jeanne Richard for the photographs for Figure 7, Ying Wang and Florian Wolschin for advice with the *hexamerin* cloning and primer design, and Hilde Garberg, Frode Berven and Anne Døskeland for expert technical support. The mass-spectroscopy experiments were performed by PROBE Proteomic Unit at University of Bergen.

Author Contributions

Conceived and designed the experiments: JHH NSM HH GVA. Performed the experiments: JHH NSM HH MTH. Analyzed the data: NSM HH MTH GVA. Wrote the paper: JHH NSM HH MTH GVA.

References

1. Resetarits WJ, Jr., Bernardo J, eds. Experimental Ecology: Issues and Perspectives. New York: Oxford University Press. xvii+470 p.
2. Garland T, Jr., Rose MR, eds. Experimental Evolution: Concepts, Methods, and Applications of Selection Experiments. Berkeley: University of California Press. xvii+730 p.
3. Rabb RL (1960) Biological studies of *Polistes* in North Carolina (Hymenoptera: Vespidae). Annals of the Entomological Society of America 53: 111–121.
4. Rau P, Rau N (1918) Wasp studies afield. Princeton: Princeton University Press. xv+372 p.
5. Pardi L (1948) Dominance order in *Polistes* wasps. Physiological Zoology 21: 1–13.
6. West Eberhard MJ (1969) The social biology of polistine wasps. Miscellaneous Publications, Museum of Zoology, University of Michigan 140: 1–101.
7. Evans HE (1958) The evolution of social life in wasps. Proceedings of the 10th International Congress of Entomology (Montreal, 1956) 2: 449–457.
8. Hunt JH (2007) The Evolution of Social Wasps. New York: Oxford University Press.
9. Turillazzi S, West-Eberhard MJ (1996) Natural History and Evolution of Paper-Wasps. Oxford: Oxford University Press. xiv+400 p.
10. Reeve HK (2001) In search of unified theories in sociobiology: Help from social wasps. In: Dugatkin LA, ed. Model Systems in Behavioral Ecology. Princeton: Princeton University Press. pp 57–71.
11. Bohm MK (1972) Effects of environment and juvenile hormone on ovaries of the wasp, *Polistes metricus*. Journal of Insect Physiology 18: 1875–1883.
12. Metcalf RA, Whitt GS (1977) Relative inclusive fitness in the social wasp *Polistes metricus*. Behavioral Ecology and Sociobiology 2: 353–360.
13. Gamboa GJ, Dropkin JA (1979) Comparisons of behaviors in early vs. late foundress associations of the paper wasp, *Polistes metricus* (Hymenoptera: Vespidae). The Canadian Entomologist 111: 919–926.
14. Metcalf RA (1980) Sex ratios, parent-offspring conflict, and local competition for mates in the social wasps *Polistes metricus* and *Polistes variatus*. The American Naturalist 116: 642–654.
15. Toth AL, Varala K, Newman TC, Miguez FE, Hutchison S, et al. (2007) Wasp gene expression supports an evolutionary link between maternal behavior and eusociality. Science 318: 441–444.
16. Consortium HGS (2006) Insights into social insects from the genome of the honeybee *Apis mellifera*. Nature 443: 931–949.
17. Hunt JH, Wolschin F, Henshaw MT, Newman TC, Toth AL, et al. (2010) Differential gene expression and protein abundance evince ontogenetic bias toward castes in a primitively social wasp. PLoS ONE 5: e10674.
18. Fire A, Xu SQ, Montgomery MK, Kostas SA, Driver SE, et al. (1998) Potent and specific genetic interference by double-stranded RNA in *Caenorhabditis elegans*. Nature 391: 806–811.
19. Montgomery MK, Fire A (1998) Double-stranded RNA as a mediator in sequence-specific genetic silencing and co-suppression. Trends in Genetics 14: 255–258.
20. Patel A, Fondrk MK, Kaftanoglu O, Emore C, Hunt G, et al. (2007) The making of a queen: TOR pathway is a key player in diphenic caste development. PLoS ONE 2(6): e509.
21. Wolschin F, Mutti NS, Amdam GV (2010) Insulin receptor substrate influences female caste development in honeybees. Biology Letters 7: 112–115.
22. Zhou X, Wheeler MM, Oi FM, Scharf ME (2008) RNA interference in the termite *Reticulitermes flavipes* through ingestion of double-stranded RNA. Insect Biochemistry and Molecular Biology 38: 805–815.
23. Nunes FMF, Simoes ZLP (2009) A non-invasive method for silencing gene transcription in honeybees maintained under natural conditions. Insect Biochemistry and Molecular Biology 39: 157–160.
24. Amdam GV, Simões ZLP, Guidugli KR, Norberg K, Omholt SW (2003) Disruption of vitellogenin gene function in adult honeybees by intra-abdominal injection of double-stranded RNA. BMC Biotechnology 3: 1.
25. Lu H-L, Vinson SB, Pietrantonio PV (2009) Oocyte membrane localization of vitellogenin receptor coincides with queen flying age, and receptor silencing by RNAi disrupts egg formation in fire ant virgin queens. FEBS Journal 276: 3110–3123.
26. Korb J, Weil T, Hoffmann K, Foster KR, Rehli M (2009) A gene necessary for reproductive suppression in termites. Science 324: 758.
27. Starks PT (1998) A novel 'sit and wait' reproductive strategy in social wasps. Proceedings of the Royal Society of London B 265: 1407–1410.
28. Tibbetts EA, Dale J (2004) A socially enforced signal of quality in a paper wasp. Nature 432: 218–222.
29. Queller DC, Strassmann JE (1988) Predation and the evolution of sociality in the paper wasp *Polistes bellicosus*. Ecology 69: 1497–1505.
30. Hunt JH, Dove MA (2002) Nourishment affects colony demographics in the paper wasp *Polistes metricus*. Ecological Entomology 27: 467–474.
31. Hunt JH, Kensinger BA, Kossuth J, Henshaw MT, Norberg K, et al. (2007) A diapause pathway underlies the gyne phenotype in *Polistes* wasps, revealing an evolutionary route to caste-containing insect societies. Proceedings of the National Academy of Sciences USA 104: 14020–14025.
32. Chapman RF (1998) The insects: Structure and function. Cambridge: Cambridge University Press.
33. Hahn DA, Denlinger DL (2007) Meeting the energetic demands of insect diapause: nutrient storage and utilization. Journal of Insect Physiology 53: 760–773.
34. Jones G, Brown N, Manczak M, Hiremath S, Kafatos FC (1990) Molecular-cloning, regulation, and complete sequence of a hemocyanin-related, juvenile hormone-suppressible protein from insect hemolymph. Journal of Biological Chemistry 265: 8596–8602.
35. Perkins DN, Pappin DJ, Creasy DM, Cottrell JS (1999) Probability-based protein identification by searching sequence databases using mass spectrometry data. Electrophoresis 20: 3551–3567.
36. Rappsilber J, Ishisama Y, Mann M (2003) Stop and go extraction tips for matrix-assisted laser desorption/ionization, nanoelectrospray, and LC/MS sample pretreatment in proteomics. Analytical Chemistry 75: 663–670.
37. Deleurance É-P (1955) Contribution à l'étude biologique des *Polistes* (Hyménoptères - Vespides). II. Le cycle évolutif du couvain. Insectes Sociaux 2: 285–302.
38. Toth AL, Varala K, Henshaw MT, Rodriguez-Zas SL, Hudson ME, et al. (2010) Brain transcriptomic analysis in paper wasps identifies genes associated with behaviour across social insect lineages. Proceedings of the Royal Society of London B 277: 2139–2148.
39. Toth AL, Bilof KBJ, Henshaw MT, Hunt JH, Robinson GE (2009) Lipid stores, ovary development, and brain gene expression in *Polistes* females. Insectes Sociaux 56: 77–84.
40. Martins JR, Nunes FM, Cristino AS, Simões ZLP, Mitondi MMG (2010) The four hexamerin genes in the honey bee: structure, molecular evolution and function deduced from expression patterns in queens, workers and drones. BMC Molecular Biology 11: 23.
41. Cristino AS, Nunes FMF, Barchuk AR, Aguiar-Coelho VM, Simões ZL, et al. (2010) Organization, evolution and transcriptional profile of hexamerin genes of the parasitic wasp *Nasonia vitripennis* (Hymenoptera: Pteromalidae). Insect Molecular Biology 19 SI Suppl. 1: 137–146.
42. Ryu K-S, Lee J-O, Kwon TH, Choi H-H, Park H-S, et al. (2009) The presence of monoglucosylated N196-glycan is important for the structural stability of storage protein, arylphorin. Biochemical Journal 421: 87–96.
43. Graves JD, Krebs EG (1999) Protein phosphorylation and signal transduction. Pharmacology & Therapeutics 82: 111–121.
44. Cohen P (2000) The regulation of protein function by multisite phosphorylation - a 25 year update. Trends in Biochemical Sciences 25: 596–601.
45. Arif A, Manohar D, Gullipalli D, Dutta-Gupta A (2008) Regulation of hexamerin receptor phosphorylation by hemolymph protein HP19 and 20-hydroxyecdysone

directs hexamerin uptake in the rice moth *Corcyra cephalonica*. Insect Biochemistry and Molecular Biology 38: 307–319.
46. Zhou X, Oi FM, Scharf ME (2006) Social exploitation of hexamerin: RNAi reveals a major caste-regulatory factor in termites. Proceedings of the National Academy of Sciences USA 103: 4499–4504.
47. Tomoyasu Y, Denell RE (2004) Larval RNAi in *Tribolium* (Coleoptera) for analyzing adult development. Development Genes and Evolution 214: 575–578.
48. Mutti NS, Wang Y, Kaftanoglu O, Amdam GV (2011) Honey bee PTEN - description, developmental knockdown, and tissue-specific expression of splice-variants correlated with alternative social phenotypes. PLoS ONE 6: e22195.
49. Hunt JH, Buck NA, Wheeler DE (2003) Storage proteins in vespid wasps: characterization, developmental pattern, and occurrence in adults. Journal of Insect Physiology 49: 785–794.
50. Nasrin N, Kaushik VK, Fortier E, Wall D, Pearson KJ, et al. (2009) JNK1 phosphorylates SIRT1 and promotes its enzymatic activity. PLoS One: e8414.
51. Havukainen H, Halskau Ø, Skjarven L, Smedal B, Amdam GV (2011) Deconstructing honeybee vitellogenin: novel 40 kDa fragment assigned to its N-terminus. Journal of Experimental Biology 214: 582–592.
52. Henshaw M, Toth A, Young T (2011) Development of new microsatellite loci for the genus *Polistes* from publicly available expressed sequence tag sequences. Insectes Sociaux, (In press).
53. Livak KJ, Schmittgen TD (2001) Analysis of relative gene expression data using real-time quantitative PCR and the $2^{-\Delta\Delta C}$T method. Methods 25: 402–408.
54. Kucharski R, Maleszka R (2005) Microarray and real-time PCR analyses of gene expression in the honeybee brain following caffein treatment. Journal of Molecular Neuroscience 27: 269–276.
55. Rossi AM, Hunt JH (1988) Honey supplementation and its developmental consequences: evidence for food limitation in a paper wasp, *Polistes metricus*. Ecological Entomology 13: 437–442.
56. Hines HM, Hunt JH, O'Connor TK, Gillespie JJ, Cameron SA (2007) Multigene phylogeny reveals eusociality evolved twice in vespid wasps. Proceedings of the National Academy of Sciences USA 104: 3295–3299.
57. Karsai I, Hunt JH (2002) Food quantity affects traits of offspring in the paper wasp *Polistes metricus* (Hymenoptera: Vespidae). Environmental Entomology 31: 99–106.
58. Naumann ID (1991) Hymenoptera. The Insects of Australia. Ithaca: Cornell University Press. pp 916–1000.

Long-Term Fluctuations in Circalunar Beach Aggregations of the Box Jellyfish *Alatina moseri* in Hawaii, with Links to Environmental Variability

Luciano M. Chiaverano[1]*, Brenden S. Holland[2], Gerald L. Crow[3], Landy Blair[4], Angel A. Yanagihara[5]

1 Kewalo Marine Laboratory, Pacific Biosciences Research Center, University of Hawaii at Mānoa, Honolulu, Hawaii, United States of America, **2** Center for Conservation Research & Training, Pacific Biosciences Research Center, University of Hawaii at Mānoa, Honolulu, Hawaii, United States of America, **3** Waikiki Aquarium, University of Hawaii at Mānoa, Honolulu, Hawaii, United States of America, **4** Ocean Safety and Lifeguard Services, City and County of Honolulu, Honolulu, Hawaii, United States of America, **5** Bekesy Laboratory, Pacific Biosciences Research Center, and Department of Tropical Medicine, John A. Burns School of Medicine, University of Hawaii at Mānoa, Honolulu, Hawaii, United States of America

Abstract

The box jellyfish *Alatina moseri* forms monthly aggregations at Waikiki Beach 8–12 days after each full moon, posing a recurrent hazard to swimmers due to painful stings. We present an analysis of long-term (14 years: Jan 1998– Dec 2011) changes in box jellyfish abundance at Waikiki Beach. We tested the relationship of beach counts to climate and biogeochemical variables over time in the North Pacific Sub-tropical Gyre (NPSG). Generalized Additive Models (*GAM*), Change-Point Analysis (*CPA*), and General Regression Models (*GRM*) were used to characterize patterns in box jellyfish arrival at Waikiki Beach 8–12 days following 173 consecutive full moons. Variation in box jellyfish abundance lacked seasonality, but exhibited dramatic differences among months and among years, and followed an oscillating pattern with significant periods of increase (1998–2001; 2006–2011) and decrease (2001–2006). Of three climatic and 12 biogeochemical variables examined, box jellyfish showed a strong, positive relationship with primary production, >2 mm zooplankton biomass, and the North Pacific Gyre Oscillation (NPGO) index. It is clear that that the moon cycle plays a key role in synchronizing timing of the arrival of *Alatina moseri* medusae to shore. We propose that bottom-up processes, likely initiated by inter-annual regional climatic fluctuations influence primary production, secondary production, and ultimately regulate food availability, and are therefore important in controlling the inter-annual changes in box jellyfish abundance observed at Waikiki Beach.

Editor: Graeme Clive Hays, University of Wales Swansea, United Kingdom

Funding: Field components of this project were partially supported by grants from the National Institutes of Health (U54 NS039406, P20 RR016453, G12 RR003061, P20 RR016453, R21 DA024444) and the Victoria S and Bradley L Geist Foundation of Hawaii Community Foundation (HCF 958935, 991879, 20001741, 20011908, 20061497, 20071368 and 47031) (AY). BSH and LMC were supported by the Oahu Army Natural Resources Program during preparation of this manuscript. The funders had no role in study design, data collection and analysis, decision to publish, or preparation of the manuscript.

Competing Interests: The authors have declared that no competing interests exist.

* E-mail: lchiave@hawaii.edu

Introduction

Jellyfish (cubozoans, hydrozoans and scyphozoans) are conspicuous, ecologically important constituents of coastal and oceanic systems. In the last three decades, jellyfish have received growing attention due to fluctuation in abundance often resulting in population explosions (e.g., blooms) in marine ecosystems worldwide, and frequently interfering directly with human activities [1]. When abundant, jellyfish cause widespread problems by clogging fishing nets [2,3], causing fish mortality in aquaculture pens [4,5], clogging intake screens in power generation and desalination plants [6], and impact tourism by stinging swimmers [7].

Predicting changes in jellyfish aggregations over time has been a difficult task due to a number of factors, including difficulties associated with sampling [8], scarcity of historical records and long-term time series datasets [1,9], and the unusual characteristics of their life cycle [10,11]. Many hydrozoans and scyphozoans, and all cubozoans, have complex life cycles including a benthic sessile polyp phase and a planktonic medusa phase [12]. Under favorable conditions, polyps can asexually produce large numbers of new polyps and ephyrae (i.e., immature medusae) [13], which can quickly grow into adult medusae [14]. Under adverse conditions, polyps can form dormant cysts [15] and medusae can also stop somatic and reproductive growth [16,17] until favorable conditions return [18,19]. This remarkable plasticity makes jellyfish suited to highly variable environments, and can result in large temporal and spatial fluctuations in abundance at various time scales [20].

Although many jellyfish populations appear to respond to anthropogenic processes, such as coastal eutrophication, overfishing, translocation of species, benthic-trawling and increased substrate availability for polyp settlement [1,19,21,22,23,24], there is increasing evidence that jellyfish populations are affected by large-scale climate variation and regional environmental conditions associated with climate fluctuations [25,26,27]. In the Irish Sea, 68% of jellyfish abundance was explained by variation in the North Atlantic Oscillation, sea surface temperature, zooplankton biomass, and precipitation [28]. Within the northern California current the highest catches of medusae correlated with cool temperatures during spring-summer and low winter-summer runoff of the Columbia River, and negative anomalies of the Pacific Decadal Oscillation (PDO) [29].

Understanding the relationships among large-scale climate forces and associated regional environmental variables with fluctuations in jellyfish populations over time is crucial to explaining and predicting trends in jellyfish abundance on a global scale [9,30]. Long-term studies on climate fluctuations and jellyfish have increased in the last decade. However, most of these studies have taken place in highly productive, temperate, coastal environments, and only a few attempts have been made to assess trends in jellyfish abundances in oligotrophic, open-ocean gyres [31,32]. In addition, long-term studies of jellyfish abundance and climate exist for only a few species of scyphozoans, hydrozoans, and ctenophores [1], and no studies exist for cubozoan species. This is somewhat surprising as box jellyfish are among the most venomous animals in the world [33]. Quite often, their habitat overlaps with areas of human recreation resulting in dangerous encounters due to painful, even lethal stings [34,35], causing beach closures at various global localities [33].

We have assessed long-term trends in abundance of box jellyfish and investigated the role of environmental conditions in an oligotrophic environment. The North Pacific Sub-tropical Gyre (NPSG) is the earth's largest contiguous oligotrophic biome, extending from 15°N to 35°N and from 135°E to 135°W, encompassing the entire Hawaiian Archipelago [36]. *Alatina moseri* Gershwin, 2005 (previously *Carybdea alata* Reynaud, 1830) is a tropical box jellyfish (Cubozoa) with a complex metagenic life cycle involving a benthic polyp and pelagic medusae [37]. On the Hawaiian Island of Oahu, reproductive cubomedusae have been observed near shore for a 2–4 day period 8–12 days after each full moon (i.e., during the waning crescent phase) along Waikiki beach since 1994 [38], and sporadically along beaches of other leeward Oahu bays. Box jellyfish aggregations often cause mass stinging events and beach closures [39]. Occasional circalunar sightings of box jellyfish on the south shore of Oahu (Waikiki Beach and surrounding areas) were first noted the late 1980s [38], and the timing of this monthly phenomenon has remained consistent. However, questions regarding whether the number of box jellyfish arriving at the beach has changed over time, or if fluctuations are affected by environmental conditions, have not been previously addressed. We have conducted a 14-year daily beach census of *Alatina moseri* medusae occurrence in leeward Oahu focused on Waikiki in order to: 1) assess seasonal patterns in abundance, 2), determine whether abundance of box jellyfish arriving at the beach has changed over time, and 3) explore potential links in variance of box jellyfish abundance at the beach to fluctuations in environmental conditions within the NPSG. This approach will provide useful information regarding the ecology of the Cubozoa, insights into jellyfish responses to environmental conditions in open-ocean, subtropical gyres, and will ultimately enhance our ability to predict the extent of onshore influxes of venomous jellyfish species in areas where they impose a health hazard.

Methods

Box jellyfish abundance and bell height at Waikiki Beach

Box jellyfish beach surveys were carried out monthly at Waikiki Beach, Oahu, Hawaii (21° 16.2′ N, 157° 49.4′ W; Figure 1) by University of Hawaii, Ocean Safety and Lifeguard Services (OSLS) personnel, and volunteers from January 1998 through December 2011. Waikiki Beach, a ~3 km beach on the leeward end of Mamala Bay along the south shore of Oahu (Figure 1), was chosen due to the consistent majority prevalence of box jellyfish arrivals in this area, and its popularity as a tourist destination; it is also the most visited beach in the state of Hawaii. Waikiki Beach was patrolled and checked for the presence of box jellyfish on a daily basis. When *Alatina moseri* medusae were present on or near the beach (typically 8–12 days after each full moon), surveys were initiated. Box jellyfish arrive at Waikiki Beach exclusively during the early morning ebbing tide. Thus, surveys were typically performed from 0200 to 0700 in order to ensure that observers were present for the duration of the influx. Surveys consisted of walking and wading into the shallow water, along the designated portion of shoreline (see below), in one direction. When one end of the section was reached, the search proceeded in the opposite direction. This protocol was repeated until all jellyfish that arrived during a given influx were collected and counted. Individuals were recently beached or swimming in the shallow waters at the time of collection, and visualized with head-lamps and/or flashlights, then collected by hand. Since arrival of medusae to Waikiki Beach can occur on multiple consecutive nights, searches were carried out daily until no medusae were found. Abundance of box jellyfish at the beach was determined by standardizing: 1) *Area searched*: all box jellyfish included in this study were collected from a 400 m portion of Waikiki, up to 1.5 m deep (Figure 1), via 14 years of monthly surveys. The study area was selected because of regular arrival of box jellyfish each month (hereafter referred to as Waikiki Beach). The geographic distribution of box jellyfish arrival on Oahu is not uniform around the island, such that jellyfish counts from Waikiki Beach cannot be extrapolated to other Hawaiian beaches. During the study period, OSLS personnel monitored box jellyfish influxes in other beaches along leeward Oahu, such as Hanauma Bay, Pokai Bay, and Yokohama Bay (Figure 1), and only Mamala Bay (Waikiki Beach) showed consistent monthly influx events (AAY, unpubl. data). 2) *Number of participants*: All box jellyfish medusae were collected within the designated area by 2–4 people. Bell height was recorded for all individuals collected monthly during 2001. As *Alatina moseri* medusae arriving at the beach are relatively large adult individuals (mean bell height = 64 mm, *SD*: 9; Table S1 & Figure S1 in File S1), they can be easily spotted and collected in the relatively small study area. Adult specimens of *A. moseri* can be easily distinguished from adults of the other two cubozoan species occurring in Hawaii, *Carybdea sivickisi* [40] and *Carybdea arborifera* [41]. Although not all individuals counted during this 14 year study were measured, all were examined and confirmed to be adult *A. moseri*. This research did not require collection permits, since the box jellyfish *A. moseri* is neither an endangered nor a protected species.

Climate data

Climatic phenomena within the subtropical North Pacific were described using bi-monthly indices of the Multivariate ENSO Index (MEI), and monthly indices of the Pacific Decadal Oscillation (PDO) and the North Pacific Gyre Oscillation (NPGO), for 1998–2011. The MEI is used to evaluate the variability in ENSO forcing, and it is obtained as the first unrotated principal component of six observed variables (sea-level pressure, zonal and meridional components of surface winds, sea surface temperature, surface air temperature, and total cloud cover) confined to the tropical Pacific [42]. This index is available online from NOAA Climate Diagnostic Center at http://www.cdc.noaa.gov/people/klaus.wolter/MEI/table.html. MEI values refer to El Niño and La Niña conditions, respectively. The PDO forcing emerges as the first mode of sea surface temperature (SST) and sea surface height (SSH) anomalies in the North Pacific Ocean poleward of 20° N [43], and is strongly correlated with variability in atmospheric circulation around the Aleutian low pressure system. Positive and negative PDO values refer to warm and cool

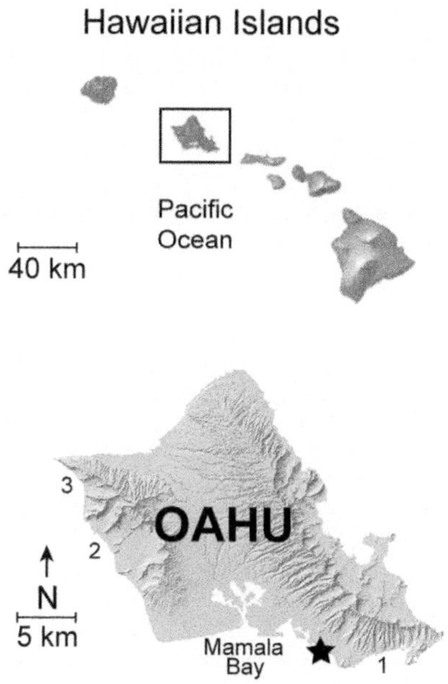

Figure 1. Map of the island of Oahu showing study area along Waikiki Beach. Box jellyfish, *Alatina moseri*, were collected monthly from 1998–2011 at Waikiki Beach, which is represented by the star symbol. Numbers indicate beach areas along leeward Oahu where *A. moseri* medusae arrive sporadically: 1) Hanauma Bay, 2) Pokai Bay, 3) Yokohama Bay.

phases, respectively. The PDO indexes were obtained from the Joint Institute for the Study of the Atmosphere and Ocean website at http://jisao.washington.edu/pdo/PDO.latest. The NPGO index represents the second principal component of sea surface height variability in the Northeast Pacific and, as such, is strongly correlated with fluctuations in the intensity of the geostrophic circulation in this region and its signal is strong south of 38° N [44]. The NPGO indices are available online from Dr. Emanuele Di Lorenzo's website at http://www.o3d.org/npgo/npgo.php. Positive and negative index values refer to more and less intense gyre circulations, respectively.

Biogeochemical data

Thirteen biogeochemical variables collected at station ALOHA (A Long-term Oligotrophic Habitat Assessment), 22°45′ N, 158°W, were downloaded from the HOT Data Organization and Graphical System (HOT-DOGS©) website at http://hahana.soest.hawaii.edu/hot/hot-dogs/interface.html. The location of station ALOHA was carefully chosen to be representative of biogeochemical processes characteristic of the North Pacific Subtropical Gyre (NPSG) [45], which encompasses all main Hawaiian Islands. Variables included: sea surface temperature (SST, °C), surface salinity (S, ppt), particulate nitrogen (PN mmol m^{-2}), and carbon (PC, mmol m^{-2}) concentrations, chlorophyll a concentration (CHa, mg m^{-2}), *Prochloroccocus* (PL, $\# \times 10^{11}$ m^{-2}), eukaryotes (EU, $\# \times 10^{11}$ m^{-2}) and heterotrophic bacteria (HB, $\# \times 10^{11}$ m^{-2}) abundance, primary production (PRI, mgC m^{-2} day^{-1}) and zooplankton biomass (ZO, g DW m^{-2}), which included "day" and "night" zooplankton biomass. Each group was separated into two range sizes: <2 mm and >2 mm (for zooplankton collection and processing methods, see [46]). We integrated values (100 m to the surface, except zooplankton biomass, which was integrated from 160 m to the surface [46]) for all variables except SST and S, for which we obtained surface (0–5 m) monthly means. Variables downloaded spanned the study period from January 1998 to December 2010. Information and description of protocols used for sampling, experimentation and analysis are available from the HOT-DOGS© website at http://hahana.soest.hawaii.edu/hot/hot-dogs/interface.html.

Local weather data

Weather conditions for the Waikiki Beach area were examined using seven local weather parameters, acquired from the National Climatic Data Center (NCDC) at http://www.ncdc.noaa.gov/ including: daily air temperature (°C), rainfall (mm), cloud cover (%), humidity (%), atmospheric pressure (mm Hg), wind speed (km/hr), and wind direction (degrees). Daily values for the period from January 1998 to December 2011 were used to calculate monthly and annual means. For monthly and annual wind direction, we used the method described by Gilhousen (1987) [47] to obtain daily "x" and "y" vector components of wind direction. Average x and y values were then calculated per month and per year, and the resultant wind direction (i.e., vector) was derived from arc-tangent (y/x). Air temperature and rainfall data were obtained from the weather station Waikiki 717.2 (USC00519397: 21° 16′ 17.8" N, 157° 48′ 57.6" W), the remaining parameters were recorded at the weather station at Honolulu International Airport (USW00022521: 21° 19′ 26 N, 157° 55′ 46" W).

Statistical analysis

Box jellyfish seasonality, medusa size, and long-term trends of abundance. Prior to analysis, beach counts were $\log_{10}$-transformed to achieve normal distribution (*Kolmogorov-Smirnov* test, $d = 0.74$, $p = 0.25$). In order to evaluate seasonality of box jellyfish abundance, monthly data from all years were combined into 12 months, and a one-way ANOVA was used to test for significant differences among months. An additional one-way ANOVA was applied to assess seasonality in the bell height of *Alatina moseri* medusae collected throughout 2001. For trend analysis, we used monthly data and mean annual abundances (log-transformed). Potential long-term trend changes in box jellyfish abundance (i.e., net increase or decrease) were assessed using *General Linear Regression Models* (*GRM*). *Generalized Additive Models* (*GAM*) were used to assess potential temporal trends (monthly and annual scales) in the dataset. The *GAM* is a nonlinear regression technique that uses non-parametric smoothing functions (i.e., cubic splines) to model the relationships among response and forcing variables [48,49,50]. An appealing feature of *GAMs* is that they do not assume a functional form for the fitted curve *a priori*. Instead, each predictor variable in the model is separated into sections delimited by "knots", and polynomial functions are fit to each section separately, revealing underlying trends in the observed data [48]. Once periods of decrease and increase in jellyfish abundance were determined by *GAM*, *GRM* models were used to test whether the linear slope deviates significantly from zero within each period. Trend analysis was carried out using the statistical software package Statistica© v7 with the associated distribution set to "normal" and the default smoothing function (i.e., cubic spline). Non-linear $r-$values were tested for significance at $\alpha = 0.05$.

To further evaluate whether significant shifts in monthly abundances (log-transformed) occurred within the periods determined by the *GAM*, we performed a change-point analysis (*CPA*) using the software Change-Point Analyzer © v2.3 [51]. The *CPA* utilizes a combination of time varying cumulative sum charts (*CUSUM*) and bootstrapping to detect significant shifts in the mean of time-ordered datasets, providing number of shifts, direction of each change, and the time of each shift with respective confident levels and confident intervals. This procedure tests the null hypothesis of "no change" across the dataset. Number of bootstrap samples was set to 2,000, without replacement. The confidence level for identifying candidate changes was set at 90%; however, a regime shift was considered to be significant only when confidence levels were higher than 95%, and confidence intervals around a significant change were set at 99%.

Climate indices, biogeochemical variables, and weather parameters versus box jellyfish abundance. Since non-stationary effects are the rule rather than the exception in ecology, box jellyfish abundance, climate, biogeochemical variables, and weather parameters were detrended (except wind direction) by linearly regressing each variable ($\log_{10}$-transformed) against time (monthly scale) to remove temporal trends, and the residuals were saved for subsequent analyses [52]. No significant autocorrelations were detected in the dataset (*Chelton Method*; [53]). Mean annual residual values were then obtained. Potential associations among detrended variables (monthly and annual scales) were evaluated using *GAM* with a univariate approach, with jellyfish abundance as the response variable. When a relationship was detected as non-significant by the *GAM* (non-linear $p>0.05$), a *GRM* was used. The *GAM* and *GRM* analyses were performed using Statistica v7. In order to test for delayed response, or lag of jellyfish to environmental variables, we performed a *Cross-correlation* analysis considering ±1-year (annual scale). All cross-correlation analyses were carried out in Minitab v13. In order to test the potential association between box jellyfish counts and wind direction (monthly and annual means), *Circular-Linear* correlation analyses were performed using the circular statistics software Oriana© v.4.

Results

Seasonality, medusa size, and long-term trends of box jellyfish abundance at Waikiki Beach

Abundance of *Alatina moseri* medusae at Waikiki Beach did not vary significantly among months, indicating no seasonality ($F_{11, 156}=0.31$, $p=0.98$). During this study, *A. moseri* medusae arrived at Waikiki Beach 8–12 days after each full moon (i.e., during waning crescent phase of the moon). Arrivals of box jellyfish to the beach happened on 1–4 consecutive nights, with the exception of February 2002 when medusae arrived on 9 consecutive nights. Circalunar aggregations frequently consisted of 100–1000 box jellyfish (65% of the counts), and arrivals of less than 100 individuals, or more than 1,000, occurred 28% and 7% of the time, respectively. A total of 66,605 medusae were counted between January 1998 and December 2011, with an average of 396.5 individuals (range: 5–2,365; median: 271.5) arriving 8–12 days after each full moon and 4,757 per year (range: 2,155–8,696; median: 4,773) (Table S2 in File S1).

The bell height of individuals ($n = 4228$) measured during beach surveys in 2001 varied significantly among months ($F_{1,\ 11}=70$, $p<0.0001$); however differences in mean bell height throughout the year did not exceed 1.2 cm. The significant differences detected by the ANOVA were due to the consistently small variation in medusae size observed within months (Table S1 & Figure S1 in File S1).

The *GRM* did not reveal a long-term trend in box jellyfish abundance at the beach (i.e., a net increase or decline over time); the linear slope of box jellyfish abundance (log-transformed) did not deviate significantly from zero during the period of study (1998–2011: Slope: 0.0012±0.002; $p=0.51$; $n=168$). However, the *GAM* indicated that box jellyfish abundance co-varied with time in a cyclical, non-linear fashion, explaining 12% of the variation in monthly abundance (non-linear $p=0.003$), and 53% of the variation in mean annual abundance (non-linear $p=0.01$) during the 14-year study, respectively (Figure 2A & 2B). Thus, based on periods determined by the *GAM*, box jellyfish aggregations: 1) significantly increased from 1998 to 2001 (*GRM*, $r=0.46$, $r^2=0.26$; $p<0.01$), 2) significantly decreased from 2001 to 2006 (*GRM*, $r=-0.45$, $r^2=-0.21$, $p<0.001$), and 3) then significantly increased until the end of data collection in 2011 (*GRM*, $r=0.29$, $r^2=0.18$; $p<0.001$). Thus, the 14-year dataset was characterized by four intermittent periods of increase and decrease in box jellyfish abundance, each period lasting approximately four years (Figure 2A).

Change-Point Analysis (*CPA*) detected three significant shifts in average abundance throughout the 14-year period, each one within periods of increase or decrease previously detected by the *GAM*. The first shift was detected in 2000 (October ±6 months, 100% confidence level), with average abundance increasing from 273 to 534. The second shift was detected in 2004 (August ±6 months, 100% confidence level), with average abundance decreasing from 534 to 227, and the third one was detected in

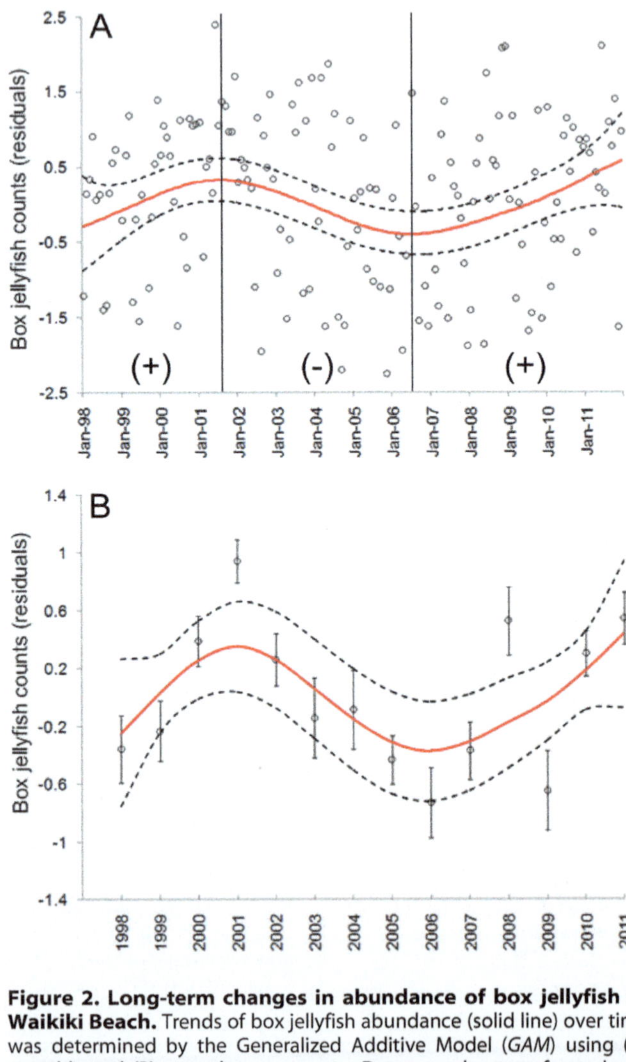

Figure 2. Long-term changes in abundance of box jellyfish at Waikiki Beach. Trends of box jellyfish abundance (solid line) over time was determined by the Generalized Additive Model (*GAM*) using (A) monthly and (B) annual mean counts. Data were log-transformed and detrended before analysis. Dashed lines indicate 95% confidence limits. Vertical solid lines indicate periods of significant (*Linear regression*, $p<0.05$) increase (+) or decrease (−) in box jellyfish counts. Bars show standard errors.

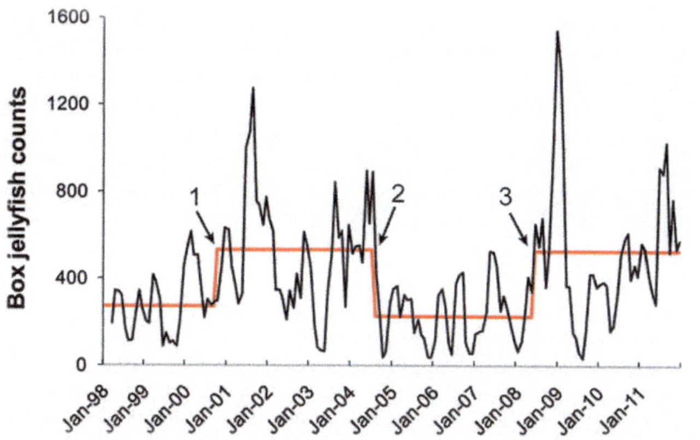

Figure 3. Monthly time-series of box jellyfish counts at Waikiki Beach. A three-month moving average smoothing approach was used to enhance visualization of the trend, depicted by the solid back line. The solid red line indicates mean box jellyfish beach counts during periods of no significant change determined by the Change Point Analysis (CPA). Three significant regime shifts in jellyfish numbers occurred during the 14-year period of study, corresponding to the periods previously identified by the *GAM* analysis: 1) 2000, 2) 2004, and 3) 2008. Dataset had no zeroes.

2008 (June ±8 months, 98% confidence level), with increased average beach abundance from 227 to 526 (Figure 3).

Relationships between box jellyfish counts and climate indices, biogeochemical variables, and weather parameters

Of a total of three climate indices, 13 biogeochemical variables, and seven weather parameters analyzed with a univariate approach using the *GAM* and *GRM*, one climatic and two biogeochemical variables showed a significant relationship with beach counts (Figure 4). No significant relationships were detected between beach counts and any weather parameters examined at both monthly and annual scales (Table S3 in File S1). In addition, the *Circular-Linear* correlation analyses did not detect a significant relationship between jellyfish counts and wind direction at any time scale (Table S3 in File S1). On a monthly scale, the *GRM* indicated a weak positive, significant linear relationship between box jellyfish abundance and both the NPGO (*GRM*, $r^2 = 0.27$, $p < 0.001$; Figure 4A) and >2 mm night zooplankton biomass (*GRM*, $r^2 = 0.19$, $p = 0.036$; Figure 4E). However, on an annual scale, the *GRM* revealed strong, positive linear relationships of beach counts with the NPGO (*GRM*: $F_{1,\ 12} = 13$, $r = 0.72$, $r^2 = 0.52$; $p = 0.003$; Figures 4B & 5A), primary production (*GRM*: $F_{1,\ 11} = 13.5$, $r = 0.74$, $r^2 = 0.54$; $p = 0.004$; Figures 4D & 5B), and >2 mm night zooplankton biomass (*GRM*: $F_{1,11} = r = 0.6$; $r^2 = 0.36$; $p = 0.03$; Figures 4F & 5C). Significant inter-relationships (on an annual scale) were detected among variables that correlated significantly with jellyfish abundance at Waikiki Beach. Primary production (as covariate) showed a strong positive linear relationship with >2 mm night zooplankton biomass (*GRM*, $r = 0.62$; $r^2 = 0.39$; $p = 0.01$; Figure 6A). Interestingly, the *GAM* detected a strong non-linear relationship (i.e., "u" shape) between the NPGO (covariate) and primary production (*GAM*, $r^2 = 0.75$, non-linear $p = 0.001$; Figure 6B). Further analysis indicated that the non-linear relationship between primary production and NPGO was caused by an unusual pairwise observation for 2005. Disregarding this outlier, there was a strong, positive linear relationship between NPGO and primary production (*GRM*: $r = 0.86$; $r^2 = 0.7$; $p = 0.0004$).

Discussion

This study represents the first long-term assessment of trends in abundance of cubozoan medusae, and the first to evaluate potential links between box jellyfish abundance and environmental conditions in a highly oligotrophic environment. Although *Alatina moseri* medusae arrived consistently at Waikiki Beach 8–12 days after full moon throughout the 14-year period, the number of medusae at the beach showed no seasonality and size (bell height) was relatively uniform throughout 2001 (Table S1 & Figure S1 in File S1). Given the lack of seasonality in the NPSG, where sea surface temperature (SST) varies only 3°C throughout the year (HOT data), this may not be surprising. In temperate regions, occurrence of medusae is usually seasonal as onset of strobilation is associated with seasonal variation in SST; however, in tropical and sub-tropical areas, jellyfish polyps typically exhibit continuous or semi-continuous strobilation thus perennial presence of medusae [54]. Continuous or semi-continuous production of medusae from polyps likely explains the lack of seasonality observed in the size of beach aggregations of box jellyfish at Waikiki. In Australia's Great Barrier Reef region, medusae of *Alatina moseri* (= *A. mordens* [41]) also occur year-round, and laboratory observations suggest that polyps are capable of continuous asexual reproduction, and metamorphosis into young medusae (ephyrae) is independent of water temperature variation [55]. In Puerto Rico (Caribbean Sea), polyps of *A. moseri* (= *Carybdea alata*) are capable of year-round asexual reproduction, and year-round occurrence of medusae of this species has been documented in this area [37]. In addition, polyps of scyphozoan jellyfish, such as *Aurelia* spp. [56] and *Mastigias* spp. [57] from tropical and sub-tropical regions, release ephyrae throughout the year in the absence of significant SST variation. Our data support earlier observations that medusae of *A. moseri* arrive to Waikiki Beach 8–12 days after each full moon [38]. Field observations indicate that box jellyfish arrive at Waikiki Beach with fully mature and active gonads, and medusae become senescent a few hours post-reproduction (sperm are broadcast, strands of fertilized eggs are expelled; AAY, unpubl. data), suggesting medusae of this species likely reproduce once in their lifetime (~1 year, [37]). Interestingly, polyps of *A. moseri* go through a metamorphosis (at ~75 days of age, raised at 26–29°C in the laboratory) in which the whole polyp becomes a single young medusa, and no remnant is left behind [37]. Therefore, continuous or semi-continuous metamorphosis from polyps to medusae may be required to maintain the adult medusae cohort over time, while monthly sexual reproductive events may be necessary to sustain an ephemeral polyp population.

Box jellyfish abundance at Waikiki Beach showed neither net increase nor net decrease throughout the 14-year period of study; instead, it showed dramatic monthly and annual variability in a cyclical, non-linear pattern with intermittent periods (pulses) of increase (positive phase) and decrease (negative phase) (Figures 2 & 3). These results are consistent with recent long-term studies suggesting that jellyfish populations, on a global scale, oscillate over different time scales with no apparent long-term trend [20,30]. The arrival of box jellyfish medusae at Waikiki Beach was recurrent with consistent, predictable timing, however, monthly aggregation size varied substantially. Thus the model (*GAM*) showed a poor fit (12%) with monthly beach counts from Jan 1998 to Dec 2011. Month to month comparison of jellyfish beach counts are notorious for their high temporal variability [58,59], and a number of physical factors, including waves, currents,

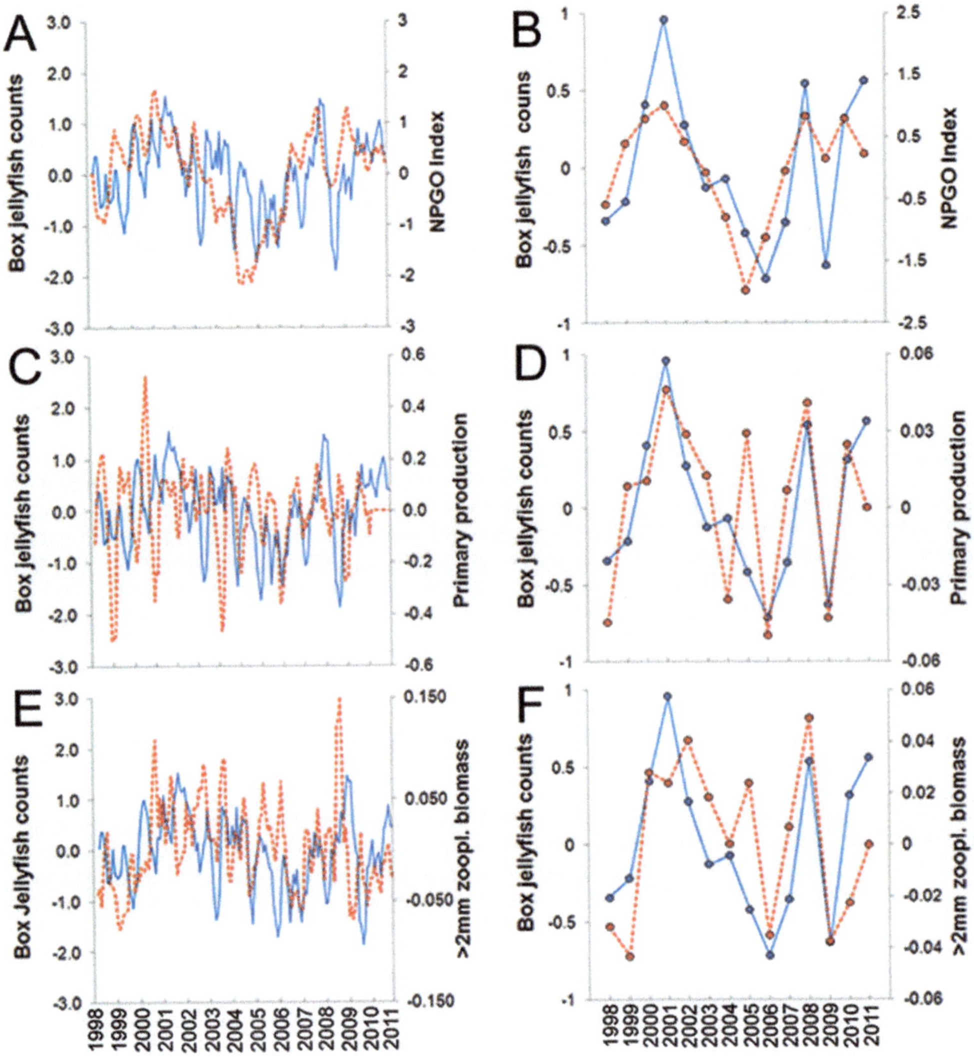

Figure 4. Time-series of climatic and biogeochemical variables with box jellyfish beach counts at Waikiki. Log-transformed, detrended variables are shown, Y-axes are residual values. Graphs A, C, and E represent monthly time-series (three-month average smoothing), while graphs B, D, and F represent annual means. Solid blue lines represent box jellyfish numbers, dashed red lines represent climate and biogeochemical variables, including NPGO (A, B), primary production (C, D), and zooplankton biomass (E, F). Note different scale in the Y-axis of variables between monthly and annual analyses. Error bars were not included in order to enhance visualization. Biogeochemical variables were obtained at Station ALOHA, from the Hawaiian Oceanographic Time-series program (HOT) at http://hahana.soest.hawaii.edu/hot/.

topography, individual behavior and life history, could all play a role in the monthly variation in numbers of medusae at the beach [58,59,60]. Predictability of our model, however, increased (53%) when applied to annual mean beach counts, which are typically considered a more reliable proxy for population abundance for beach counts [61,62].

Our understanding of inter-annual changes in abundance of box jellyfish at Waikiki Beach was enhanced by comparison with long-term local and regional environmental heterogeneity: counts at Waikiki Beach did not correlate significantly with any of the local weather parameters studied (Table S3 in File S1), whereas years of high jellyfish beach counts coincided with years of elevated NPGO (North Pacific Gyre Oscillation) index, primary production, and macro-zooplankton abundance. Thus, fluctuations in abundance of *Alatina moseri* medusae at Waikiki Beach appear to follow decadal climatic oscillations occurring within the sub-tropical North Pacific, independent of local weather conditions (wind speed, wind direction, air temperature, rainfall, cloud cover, humidity, and atmospheric pressure; Table S3 in File S1). Beach counts showed a strong, positive correlation with the North Pacific Gyre Oscillation (NPGO), while no significant relationship was detected with any other large-scale climate force, such as the Pacific decadal Oscillation (PDO) and the Multivariate ENSO Index (MEI). Unlike the PDO and MEI [42,43], the NPGO is associated with variation in wind-driven circulation including upwelling and horizontal advection at both regional and basin scales, and physical processes controlling salinity and nutrient concentrations [44], which in turn drive changes in phytoplankton levels, leading to variability at higher trophic levels. Thus the NPGO is an indicator of fluctuations in the mechanisms controlling planktonic ecosystem dynamics [44]. The NPGO is considered the primary climate index linked to decadal biological variability in the NPSG. Recent studies [63,64] using biogeochemical ocean circulation models, showed NPGO to be the only climate index significantly correlated with primary production at Station ALOHA, in agreement with results of this study. In fact,

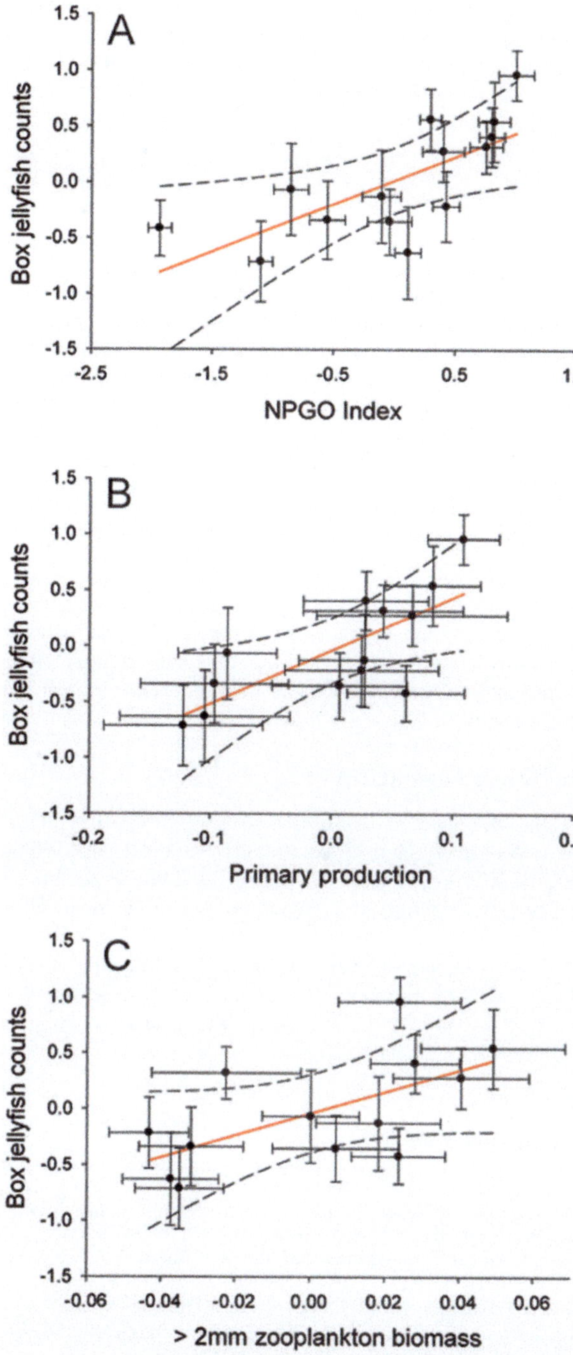

Figure 5. Significant relationships among box jellyfish abundance at Waikiki Beach and environmental variables. A: North Pacific Gyre Oscillation (NPGO), B: Primary production, C: zooplankton biomass within the sub-tropical North Pacific. Significant linear relationships were detected using General Regression Models (*GRM*). Biogeochemical variables were obtained at Station ALOHA, from the Hawaiian Oceanographic Time-series program (HOT) at http://hahana.soest.hawaii.edu/hot/. Solid red line represents the model fit, and dashed lines represent 99% confidence levels. Bars represent standard errors.

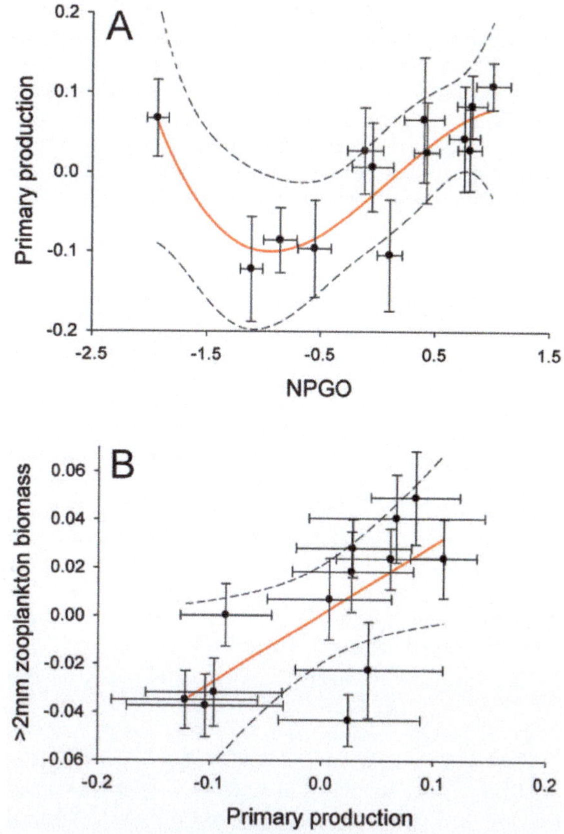

Figure 6. Significant relationships among environmental variables that correlated significantly with box jellyfish abundance at Waikiki Beach. A) Non-linear relationship between primary production and NPGO detected by using Generalized Additive Models (*GAM*). B) Linear relationship between zooplankton biomass and primary production detected by using General Regression Models (*GRM*). Biogeochemical variables were obtained at Station ALOHA, from the Hawaiian Oceanographic Time-series program (HOT) at http://hahana.soest.hawaii.edu/hot/. Solid red line represents the model fit, and dashed lines represent 99% confidence levels. Bars represent standard errors.

neither PDO nor MEI was significantly correlated with beach counts or with any biogeochemical variable that showed a significant relationship with box jellyfish abundance. Although the actual mechanism has not been determined, Dave and Lozier (2010) [64] suggest that the correlation between the NPGO and primary productivity in the NPSG may be driven by changes in diazotrophic activity (N_2-fixation). Approximately 80% of all nitrogen fixation at Station ALOHA occurs within 60 m of the surface, and evidence suggests that diazotrophs account for up to half of vertically integrated new production in the region [65,66,67]. An increase in N-fixation activity translates into introduction of new nitrogen in the system, which can in turn cause an increase in zooplankton abundance via direct trophic transfer [68,69].

The strong positive relationship between annual primary production and zooplankton biomass (secondary production) observed in this study may reflect trophic interactions previously observed at Station ALOHA [69], where copepods comprise the dominant meso-zooplankton taxa (~175 species) as well as the primary consumers. Such periods of high primary production can support elevated secondary productivity [69]. In this study, we detected a strong positive correlation between box jellyfish counts and zooplankton, suggesting that abundance of the box jellyfish *Alatina moseri* at Waikiki Beach could be influenced by regional prey availability. Cubomedusae are predators that feed on a variety meso- and macro-zooplankton taxa [70,71,72,73]. Although detailed prey preference data for *A. moseri* is not currently available, preliminary molecular analysis of gut contents has confirmed the presence of oceanic meso- and macro-zooplankton taxa (Holland et al., in prep). Jellyfish are known to increase growth rates and reach larger adult sizes in response to increased food availability [74], and since body size positively correlates with fecundity in jellyfish, more eggs (and more larvae) are produced when food is readily available [75,76], potentially increasing larval recruitment [54]. At the same time, increased food availability in the water column can also affect the amount of food reaching the benthos [77,78,79], and jellyfish polyps are known to increase budding rates (i.e., asexual production of new polyps) in response to an increase in food supply [80,81,82,83], potentially translating into higher numbers of medusae released via strobilation during periods with high zooplankton biomass compared to years of low zooplankton biomass. In polyps of the box jellyfish *A. moseri* (= *A. mordens*, [41]) increased feeding frequency resulted in increased budding rates [55], with similar results obtained for a related species [84]. As previously mentioned, abundance of *A. moseri* medusae is not seasonal in Hawaii (US, Waikiki Beach), the Caribbean (Puerto Rico) [37], and the Great Barrier Reef (Australia) [55], probably due to continuous production of medusae via polyp metamorphosis, which may account for the lack of a significant lag between jellyfish abundance of jellies arriving at the beach and zooplankton biomass observed in this study.

In spite of the distance between collection localities for jellyfish (Waikiki) and biogeochemical variables (Station ALOHA: 22.45°N, 158°W, ~200 km north of Oahu), correlations are of value for a number of reasons. Station ALOHA was selected because it is representative of oceanographic processes occurring throughout the NPSG [45]. In fact, *Alatina moseri* is one of the few pelagic cubozoan medusae [41], and is seen near shore for only 2–4 days, 8–12 days after each full moon. During non-aggregation periods, medusae have been collected from near the surface offshore, in waters several hundred meters deep [85], and on the day of arrival at the beach, *A. moseri* medusae have been observed ~2 km south of Waikiki Beach actively swimming straight towards the shore against or across currents (AAY, unpubl. data), evidencing the swimming capabilities of large cubozoans [86]. In addition, the Hawaiian Islands were formed via volcanic hotspot located near the center of the NPSG, where oceanic conditions occur close to shore, and depths >400 m are reached within 5 km of the coastline. Thus, regional offshore climatic and biogeochemical processes recorded at Station ALOHA likely have relevance to inter-annual fluctuations in numbers of cubozoan medusae arriving at Waikiki Beach. Given the fact that only mature *A. moseri* medusae arrive at Waikiki Beach from offshore 8–12 days after each full moon, annual fluctuation in beach counts could reflect inter-annual changes in adult off-shore population size. However, the possibility that beach counts do not reflect offshore abundance cannot be ruled out. In the Celtic Sea, for example, medusae of *Chrysaora hysoscella* were shown to be abundant offshore but were rarely observed at the shoreline [58]. However, our observations of *A. moseri* actively swimming towards the beach during the ebbing tide, against as well as across surface currents and tidal flow, contrasts with passive stranding events observed for instance, in scyphozoan medusae [58,59]. Lunar cycles have previously been shown to affect swimming behavior of jellyfish [33] and reproductive cycles of marine Cnidaria [87]. We suggest that the moon cycle may play a role in synchrony of gonad development and timing of *Alatina moseri* arrival at the shore, while environmental variables, such as regional food availability may play a role in fluctuations in Waikiki beach counts.

Long-term, optimally multi-decadal studies are necessary to understand complex ecological factors driving population fluxes in pelagic marine taxa. Box jellyfish aggregations are a major, and poorly understood public safety hazard in populated coastal areas, and present persistent challenges to resource managers and lifeguard services. Only through an enhanced understanding of the relevant ecological drivers will we be able to generate predictive models to minimize public health impacts. Perhaps the most important finding here is that annual Waikiki beach counts appear to be influenced by fluctuation in environmental conditions within the NPSG, including climate (NPGO) and certain biological variables, independent of local weather conditions. These findings suggest that food availability within the NPSG may be an important predictor of inter-annual fluctuations in abundance of *A. moseri* medusae during beach aggregations in Hawaii, likely via bottom-up control. Additional studies will be required to test whether observed fluctuations in beach counts correspond to abundance offshore for Waikiki box jellyfish. Our work will continue to focus on the oscillating aggregation patterns revealed in this study, and on elucidation of the environmental mechanisms governing fluctuations of this important and poorly understood species.

Supporting Information

File S1 Monthly average bell height of Alatina moseri medusae collected at Waikiki Beach throughout 2001 (Table S1), bell height frequency distribution of Alatina moseri medusae collected at Waikiki Beach throughout 2001 (Figure S1), monthly counts (raw data) of Alatina moseri medusae recorded at Waikiki Beach from January 1998 to December 2011 (Table S2), and non-significant correlations between numbers of Alatina moseri medusae at Waikiki Beach and climate indices, biogeochemical variables, and weather parameters (Table S3).

Acknowledgments

This paper is dedicated to the memory of Charles Edmondson, renowned Hawaii invertebrate specialist. We recognize contributions of Kim Bridges and Mahealani Kaneshiro-Pineiro to an early draft of this manuscript, and M. Kaneshiro-Pineiro's presentation at the 3rd International Jellyfish Blooms Symposium in 2010. We are grateful to John Chung, Karen Quinn, Darin Hayakawa, Louise Oliver, Kikiana Hurwitz, Amanda Lee, Cole Nishikawa, Paige Baum, Brian Hew, Karen Tatum and Teresa Carrette for field assistance. The support of Rob Miller, Ralph Goto and OSLS is greatly appreciated. We gratefully acknowledge the valuable and ongoing work of the Hawaiian Ocean Time-series (HOT) program. Thanks also to Lu Eldredge, Daniela del Valle, Shinji Sugiura, and Kelly Robinson for helpful discussions, and Jennifer Crites, Robert Kinzie III, Zachary Williams, Jamie Seymour and Andrew Rossiter for reviewing drafts of this manuscript.

Author Contributions

Conceived and designed the experiments: AAY GLC. Performed the experiments: GLC AAY LB. Analyzed the data: LMC. Wrote the paper: LMC BSH.

References

1. Purcell JE (2012) Jellyfish and ctenophore blooms coincide with human proliferations and environmental perturbations. Annual Review of Marine Science 4: 209–235.
2. Nagata RM, Haddad MA, Nogueira M Jr. (2009) The nuisance of medusae (Cnidaria, Medusozoa) to shrimp trawls in central part of southern Brazilian Bight, from the perspective of artisanal fishermen. Pan-American Journal of Aquatic Science 4: 312–325.
3. Dong Z, Liu D, Keesing JK (2010) Jellyfish blooms in China: dominant species, causes and consequences. Marine Pollution Bulletin 60: 954–963.
4. Doyle TK, De Haas H, Cotton D, Dorschel B, Cummins V, et al (2008) Widespread occurrence of the jellyfish *Pelagia noctiluca* in Irish coastal and shelf waters. Journal of Plankton Research 30: 963–968.
5. Delannoy CMJ, Houghton JDR, Fleming NEC, Ferguson HW (2011) Mauve stingers (*Pelagia noctiluca*) as carriers of the bacterial fish pathogen *Tenacibaculum maritimum*. Aquaculture 311: 255–257.
6. Daryanabard R, Dawson MN (2008) Jellyfish blooms: *Crambionella orsini* (Scyphozoa: Rhizostomeae) in the Gulf of Oman, Iran, 2002–2003. Journal of the Marine Biological Association of United Kingdom 88: 477–483.
7. Fenner PJ, Lippmann J, Gershwin L-A (2010) Fatal and nonfatal severe jellyfish stings in Thai waters. Journal of Travel Medicine 17: 133–138.
8. Pierce J (2009) Prediction, location, collection and transport of jellyfish (Cnidaria) and their polyps. Zoo Biology 28: 163–176.
9. Brotz L, Cheung WWL, Kleisner K, Pakhomov E, Pauly D (2012) Increasing jellyfish populations: trends in large marine ecosystems. Hydrobiologia 690: 3–20.
10. Dawson MN, Hamner WM (2009) A character-based analysis of the evolution of jellyfish blooms: adaptation and exaptation. Hydrobiologia 616: 193–215.
11. Hamner WM Dawson MN (2009) A review and synthesis on the systematics and evolution of jellyfish blooms: advantageous aggregations and adaptive assemblages. Hydrobiologia 616: 161–191.
12. Collins AG (2002) Phylogeny of Medusozoa and the evolution of cnidarians life cycles. Journal of Evolutionary Biology 15: 418–432.
13. Arai MN (1997) A Functional Biology of Scyphozoa. Chapman and Hall, London: 300.
14. Palomares MLD, Pauly D (2009) The growth of jellyfishes. Hydrobiologia 616: 11–21.
15. Arai MN (2009) The potential importance of podocysts to the formation of scyphozoan blooms: a review. Hydrobiologia. 616: 241–246.
16. Hamner WM, Jensen RM (1974) Growth, degrowth, and irreversible cell differentiation in *Aurelia aurita*. America Zoologist 14: 833–849.
17. Chiaverano LM, Mianzan HW, Ramirez F (2004) Gonad development and somatic growth patterns of *Olindias sambaquiensis* (Limnomedusae, Olindiidae). Hydrobiologia 530–31: 373: 381.
18. Boero F, Bouillon J, Gravili C, Miglietta MP, Parsons T, et al (2008) Gelatinous plankton: irregularities rule the world (sometimes) Marine Ecolgy Progress Series 356: 299–310.
19. Richardson AJ, Bakun A, Hays GC, Gibbons MJ (2009) The jellyfish joyride: causes, consequences and management responses to a more gelatinous future. Trends in Ecolgy and Evolution 24: 312–322.
20. Condon RB, Graham WM, Duarte CM, Pitt KA, Lucas CH, et al (2012) Questioning the rise of gelatinous zooplankton in the world's oceans. Bioscience 62: 160–169.
21. Mills C (2001) Jellyfish blooms: are populations increasing globally in response to changing ocean conditions? Hydrobiologia 451: 55–68.
22. Graham WM, Martin DL, Felder DL, Asper VL, Perry HM (2003) Ecological and economic implications of a tropical jellyfish invader in the Gulf of Mexico. Biological Invasions 5: 53–69.
23. Purcell JE, Uye S, Lo W (2007) Anthropogenic causes of jellyfish blooms and their direct consequences for humans: a review. Marine Ecology Progress Series 350: 153–174.
24. Pauly D, Alder J, Booth S, Cheung WWL, Christensen V, et al (2009) Fisheries in Large Marine Ecosystems: Descriptions and Diagnoses, 23–40 in *The UNEP Large Marine Ecosystem Report: a Perspective on Changing Conditions in LMEs of the World's Regional Seas*, Sherman K and Hempel G (ed). UN Environment Programme Regional Seas Report and Studies No. 182, Nairobi, Kenya.
25. Molinero JC, Buecher E, Lučić D, Malej A, Miloslavić M (2008) Climate and Mediterranean jellyfish: assessing the effect of temperature regimes on jellyfish outbreak dynamics. Annales Series Historia Naturalis 19: 11–18.
26. Lynam CP, Attrill MJ, Skogen MD (2010) Climatic and oceanic influences on the abundance of gelatinous zooplankton in the North Sea. Journal of the Marine Biological Association of United Kingdom 90: 1153–1159.
27. Robinson KL, Graham WM (2013) Long-term change in the abundances of northern Gulf of Mexico scyphomedusae *Chrysaora* sp. and *Aurelia* spp. with links to climate variability. Limnology and Oceanography 58: 235–253.
28. Lynam CP, Lilley MKS, Bastian T, Doyle TK, Beggs SE, et al (2011) Have jellyfish in the Irish Sea benefited from climate change and overfishing? Global Change Biology 17: 767–82.
29. Suchman CL, Brodeur RD, Emmett RL, Daly EA (2012) Large medusae in surface waters of the Northern California Current: variability in relation to environmental conditions. Hydrobiologia 690: 113–125.
30. Condon RH, Duarte CM, Pitt KA, Robinson KL, Lucas CH, et al (2013) Recurrent jellyfish blooms are a consequence of global oscillations. Proceedings of the National Academy of Science 110: 1000–1005.
31. Ferraris M, Berline L, Lombard F, Guidi L, Elineau A, et al (2012) Distribution of *Pelagia noctiluca* (Cnidaria, Scyphozoa) in the Ligurian Sea (NW Mediterranean Sea). Journal of Plankton Research 34: 874–885.
32. Lilley MKS, Beggs SE, Doyle TK, Hobson VJ, Stromberg KHP, et al (2011) Global patterns of epipelagic gelatinous zooplankton biomass. Marine Biology 158: 2429–2436.
33. Fenner PJ, Williamson JA (1996) Worldwide deaths and severe envenomation from jellyfish stings. Medical Journal of Australia 165: 658–661.
34. Fenner PJ, Hadok JC (2002) Fatal envenomation by jellyfish causing Irukandji syndrome. Medical Journal of Australia 177: 362–363.
35. Huynh TT, Pereira P, Mulcahy R, Cullen P, Seymour J, et al (2003) Severity of Irukandji syndrome and nematocyst identification from skin scrapings Medical Journal of Australia 178: 38–41.
36. Karl DM (1999) A sea of change: Biogeochemical variability in the North Pacific Subtropical Gyre. Ecosystems 2: 181–214.
37. Arneson AC, Cutress CE (1976) Life history of *Carybdea alata* Reynaud, 1830 (Cubomedusae). In: Mackie GO (ed) Coelentrate Ecology and Behavior. Plenum Press, New York.
38. Thomas CS, Scott SA, Galanis DJ, Goto RS (2001) Box jellyfish (*Carybdea alata*) in Waikiki: Their influx cycle plus the analgesic effect of hot and cold packs on their stings to swimmers at the beach: A randomized, placebo-controlled, clinical trial. Hawaii Medical Journal 60: 100–107.
39. Yoshimoto CM, Yanagihara AA (2002) Cnidarian (coelenterate) envenomations in Hawaii improve following heat application. Transactions of the Royal Society of Tropical Medicine and Hygiene 96: 300–303.
40. Matsumoto GI, Crow GL, Cornelius PFS, Carlson BA (2002) Discovery of the Cubomedusa *Carybdea sivickisi* (Cubozoa:Carybdeidae) in the Hawaiian Islands. Bishop Museum Occassional Papers 69: 44–46.
41. Bentlage B, Cartwright P, Yanagihara AA, Lewis C, Richards GS, et al (2010) Evolution of box jellyfish (Cnidaria: Cubozoa), a group of highly toxic invertebrates. Proceedings of the Royal Society of London 277: 493–501.
42. Wolter K, Timlin MS (1998) Measuring the strength of ENSO: How does 1997/98 rank? Weather 53: 315–324.
43. Mantua N, Hare S, Zhang Y, Wallace J, Francis R (1997) A Pacific interdecadal climate oscillation with impacts on salmon production, Bulletin of the Americal Meteorolgy Society 78: 1069–1079.
44. Di Lorenzo E, Schneider N, Cobb KM, Chhak K, Franks PJS, et al (2008) North Pacific Gyre Oscillation links ocean climate and ecosystem change. Geophysical Research Letters 35 L08607, doi:10.1029/2007GL032838.
45. Karl D, Lukas R (1996) The Hawaii Ocean Time-series (HOT) program: Background, rationale and field implementation. Deep-Sea Research II 43: 129–156.
46. Sheridan C, Landry M (2004) A 9-year increasing trend in mesozooplankton biomass at the Hawaii Ocean Time-series Station ALOHA. ICES Journal of Marine Science 61: 57–463.
47. Gilhousen DB (1987) A field evaluation of NDBC moored buoy winds. Journal of Atmospheric and Oceanic Technology 4: 94–104.
48. Hastie TJ, Tibshirani RJ (1990) Generalized Additive Models, Chapman and Hall, New York.
49. Wood SN (2004) Stable and efficient multiple smoothing parameter estimation for generalized additive models. Journal of America Statistics Association 99: 637–686.
50. Wood SN (2006) Generalized additive models: an introduction with *R*: Chapman and Hall, London.
51. Taylor WA (2003) Change-Point Analyzer, Version 2.3, Taylor Enterprises, Inc., Libertyville, IL (www.variation.com*cpa*).
52. Molinero JC, Ibanez F, Nival P, Buecher E (2005) North Atlantic climate and northwestern Mediterranean plankton variability. Limnology and Oceanography 50: 1213–1220.
53. Pyper BJ, Peterman RM (1998) Comparison of methods to account for autocorrelation in correlation analyses of fish data. Canadian Journal of Fisheries and Aquatic Sciences 55: 2127–2140.
54. Lucas CH (2001) Reproduction and life history strategies of the common jellyfish, *Aurelia aurita*, in relation to its ambient environment. Hydrobiologia 451: 229–246.
55. Courtney R (2012) Seasonality in the tropical *Alatina mordens*. Master thesis 74.
56. Schroth W, Jarms G, Streit B, Schierwater B (2002) Speciation and phylogeography in the cosmopolitan marine moon jelly, *Aurelia* sp. BMC Evolutionary Biology 2: 1–10.
57. Hamner WM, Gilmer RW, Hamner PP (1982) The physical, chemical, and biological characteristics of a stratified, saline, sulfide lake in Palau. Limnology and Oceanography 27: 896–909.
58. Houghton JDR, Doyle TK, Davenport J, Lilley MK, Wilson RP, et al (2007) Stranding events provide indirect insights into the seasonality and persistence of jellyfish medusae (Cnidaria: Scyphozoa). Hydrobiologia 589: 1–13.
59. Doyle TK, Houghton JD, Buckley SM, Hays GC, Davenport J (2007) The broad-scale distribution of five jellyfish species across a temperate coastal environment. Hydrobiologia 579: 29–39.

60. Graham WM, Pages F, Hamner WM (2001) A physical context for gelatinous zooplankton aggregations: a review. Hydrobiologia 451: 199–212.
61. Baker JD, Howell EA, Polovina JJ (2012) Relative influence of climate variability and direct anthropogenic impact on a Sub-tropical Pacific top predator, the Hawaiian monk seal. Marine Ecology Progress Series 469: 175–189.
62. Eberhardt LL, Garrott RA, Becker BL (1999) Using trend indices for endangered species. Marine Mammals Science 15: 766−785.
63. Saba VS, Friedrichs MAM, Carr M-E, Antoine D, Armstrong RA, (2010) The challenges of modeling depth-integrated marine primary productivity over multiple decades: A case study at BATS and HOT. Global Biogeochemical Cycles 24: GB3020.
64. Dave AC, Lozier SM (2010) Local stratification control of marine productivity in the subtropical North Pacific. Journal of Geophysical Research 115: C12032.
65. Karl DM, Letelier R, Tupas L, Dore J, Christian J, et al (1997) The role of nitrogen fixation in biogeochemical cycling in the subtropical North Pacific Ocean. Nature 388: 533–538.
66. Dore JE, Brum JR, Tupas LM, Karl DM (2002) Seasonal and interannual variability in sources of nitrogen supporting export in the oligotrophic subtropical North Pacific Ocean. Limnology and Oceanography 47: 1595–1607.
67. Dore JE, Letelier RM, Church MJ, Lukas R, Karl DM (2008) Summer phytoplankton blooms in the oligotrophic North Pacific Subtropical Gyre: Historical perspective and recent observations, Progress in Oceanography 76: 2–38.
68. Valiela I (1995) Marine Ecological Processes, 2nd edn. New York: Springer-Verlag, 686.
69. Hannides CCS, Popp BN, Landry MR, Graham BS (2009) Quantification of zooplankton trophic position in the North Pacific Subtropical Gyre using stable nitrogen isotopes. Limnology Oceanography 54: 50–61.
70. Stewart S (1996) Field behavior of *Tripedalia cystophora* (Class Cubozoa). Marine and Freshwater Behavior and Physiology 27: 175–188.
71. Carrette T, Alderslade P, Seymour J (2002) Nematocyst and prey in two Australian cubomedusans, *Chironex fleckeri* and *Chiropsalmus* sp. Toxicon 40: 1547–1551.
72. Underwood AH, Seymour JE (2007) Venom ontogeny, diet and morphology in *Carukia barnesi*, a species of Australian box jellyfish that causes Irukandji syndrome. Toxicon 49: 1073–1082.
73. Nogueira M, Haddad A (2008). The diet of cubomedusae (Cnidaria, Cubozoa) in Southern Brazil. Brazilian Journal of Oceanography 56: 157–164.
74. Ishii H, Båmstedt U (1998) Food regulation of growth and maturation in a natural population of *Aurelia aurita* (L.). Journal of Plankton Research 20: 805–816.
75. Lucas CH (1996) Population dynamics of *Aurelia aurita* (Scyphozoa) from an isolated brackish lake, with particular reference to sexual reproduction. Journal of Plankton Research 18: 987–1007.
76. Lucas CH, Lawes S (1998) Sexual reproduction of the scyphomedusa *Aurelia aurita* in relation to temperature and variable food supply. Marine Biology 131: 629–638.
77. Boero F, Belmonte G, Fanelli G, Piraino S, Rubino F (1996) The continuity of living matter and the discontinuities of its constituents: do plankton and benthos really exist? Trends in Ecology and Evolution 11: 177–180.
78. Kahn AS, Ruhl HE, Smith KL (2012) Temporal changes in deep-sea sponge populations are correlated to changes in surface climate and food supply. Deep Sea Research I: Oceanographic Research Papers 70: 36–41.
79. Perea-Blázquez A, Davy SK, Bell JJ (2012) Estimates of particulate organic carbon flowing from the pelagic environment to the benthos through sponge assemblages. PLoS ONE 7 e29569. doi:10.1371/journal.pone.0029569.
80. Ma X, Purcell JE (2005) Temperature, salinity, and prey effects on polyp versus medusa bud production by the invasive hydrozoan *Moerisia lyonsi*. Marine Biology 147: 225–234.
81. Hoover RA, Purcell JE (2009) Substrate preferences of scyphozoan *Aurelia labiata* polyps among common dock-building materials. Hydrobiologia 616: 259–67.
82. Han CH, Uye I (2010) Combined effects of food supply and temperature on asexual reproduction and somatic growth of polyps of the common jellyfish *Aurelia aurita*. Plankton and Benthos Research 5: 98–105.
83. Webster CN, Lucas CH (2012) The effects of food and temperature on settlement of *Aurelia aurita* planula larvae and subsequent somatic growth. Journal of Experimental Marine Biology and Ecology 436–437: 50–55.
84. Fischer AB, Hofmann DK (2004) Budding, bud morphogenesis, and regeneration in *Carybdea marsupialis* Linnaeus, 1758 (Cnidaria: Cubozoa). Hydrobiologia 530–531: 331–337.
85. Garm A, Bielecki J, Petie R, Nilsson D-E (2012). Opposite patterns of diurnal activity in the box jellyfish *Tripedalia cystophora* and *Copula sivickisi*. Biological Bulletin 222: 35–45.
86. Gordon MR, Seymour JE (2009) Quantifying movement of the tropical Australian cubozoan *Chironex fleckeri* using acoustic telemetry. In Jellyfish Blooms: Causes, Consequences, and Recent Advances, Springer Netherlands (87–97).
87. Babcock RC, Bull GD, Harrison PL, Heyward AJ, Oliver JK, et al (1986) Synchronous spawnings of 105 scleractinian coral species on the Great Barrier Reef. Marine Biology 90: 379–394.

Fish Assemblages Associated with Natural and Anthropogenically-Modified Habitats in a Marine Embayment: Comparison of Baited Videos and Opera-House Traps

Corey B. Wakefield[1]*, Paul D. Lewis[1], Teresa B. Coutts[1¤], David V. Fairclough[1], Timothy J. Langlois[1,2]

1 Western Australian Fisheries and Marine Research Laboratories, Department of Fisheries, Government of Western Australia, Perth, Western Australia, Australia, 2 School of Plant Biology, The University of Western Australia, Perth, Western Australia, Australia

Abstract

Marine embayments and estuaries play an important role in the ecology and life history of many fish species. Cockburn Sound is one of a relative paucity of marine embayments on the west coast of Australia. Its sheltered waters and close proximity to a capital city have resulted in anthropogenic intrusion and extensive seascape modification. This study aimed to compare the sampling efficiencies of baited videos and fish traps in determining the relative abundance and diversity of temperate demersal fish species associated with naturally occurring (seagrass, limestone outcrops and soft sediment) and modified (rockwall and dredge channel) habitats in Cockburn Sound. Baited videos sampled a greater range of species in higher total and mean abundances than fish traps. This larger amount of data collected by baited videos allowed for greater discrimination of fish assemblages between habitats. The markedly higher diversity and abundances of fish associated with seagrass and limestone outcrops, and the fact that these habitats are very limited within Cockburn Sound, suggests they play an important role in the fish ecology of this embayment. Fish assemblages associated with modified habitats comprised a subset of species in lower abundances when compared to natural habitats with similar physical characteristics. This suggests modified habitats may not have provided the necessary resource requirements (e.g. shelter and/or diet) for some species, resulting in alterations to the natural trophic structure and interspecific interactions. Baited videos provided a more efficient and non-extractive method for comparing fish assemblages and habitat associations of smaller bodied species and juveniles in a turbid environment.

Editor: Sharyn Jane Goldstien, University of Canterbury, New Zealand

Funding: This research received funding from Fremantle Ports, Oceanica consulting and the Department of Fisheries, Government of Western Australia to assist with the environmental impact assessment for the Kwinana Quays port development in Cockburn Sound. The funders had no role in study design, data collection and analysis, decision to publish, or preparation of the manuscript.

Competing Interests: The authors have the following interests: this study was partly funded by Oceanica Consulting.

* E-mail: corey.wakefield@fish.wa.gov.au

¤ Current address: Department of Environment and Conservation, Government of Western Australia, Broome, Western Australia, Australia

Introduction

Marine embayments and estuaries provide important habitats during the life histories of many fish species [1,2,3]. The evolutionary importance of these areas is evident through their role in facilitating genetic subdivision and divergence in marine and estuarine fish populations [4], for species that exhibit both high or low dispersal capabilities [5,6]. However, on a much shorter time scale, anthropogenic activities have modified their seascape through the addition of infrastructure (e.g. piers and rockwalls), removal of substrate (e.g. dredging and mining), or chemical (e.g. eutrophication) and biological (e.g. introduced species) contamination. Despite extensive examples of anthropogenic induced impacts [7], the ecological processes governing sheltered nearshore areas continue to come under stress from further coastal development. Given the ecological importance of embayments and estuaries it is important to understand how such impacts influence their faunal assemblages.

The capacity of anthropogenically-modified (hereafter referred to as modified habitats) seascapes to resemble natural habitats has been evaluated through comparisons of their faunal assemblages [8]. Modified habitats have typically been shown to support a subset of fish species that occur in adjacent natural habitats with similar physical attributes (e.g. topography), and in relatively higher or lower abundances depending on species-specific requirements for shelter, reproduction and diet [9]. Any significant change in the composition of fishes from that of a natural state should be considered an impact based on alterations to the trophic structure and other interspecific interactions which are likely to occur [9]. Results of studies comparing species richness, abundance and composition between natural and modified habitats (ranging from artificial reefs to piers) are inconsistent and have been shown to be influenced by the materials used in their construction and the sampling method [8]. Large habitat modifications could lead to significant changes in the ecological and/or physical (e.g. hydrodynamics) processes within these embayments and estuaries (e.g. influence spawning and recruitment processes, [10]).

Marine embayments are rare on the west coast of Australia. The temperate embayment of Cockburn Sound (32°12′ S, Fig. 1) is recognised as an important spawning and nursery area for many commercially and recreationally targeted fish and crab species [3,10,11,12,13]. Cockburn Sound is located in close proximity (ca 20 km) to the only capital city along this extensive coastline (ca 3000 km). As such, this embayment supports competing human uses, including heavy industry, shipping, a naval base, aquaculture and commercial and recreational fishing, with further large-scale port developments proposed. Large-scale declines in seagrass cover (ca 80% loss) have been linked to development activities in this embayment [14,15,16]. The few studies of the fish communities of Cockburn Sound have focussed on seagrass beds using beach seine nets, trawling and set (gill) nets [17,18,19,20] and soft sediments using trawling [21,22]. However, there have been no comparisons of the fish faunal composition using the same sampling method among the different natural and modified habitats, as not all the above methods are able to sample these various habitats effectively.

Baited video offers a non-extractive and effective method for describing and comparing fish communities across multiple habitats [23,24]. However, this method has had limited application within marine embayments and estuaries [25]. Technological improvements have led to baited video equipment becoming more affordable and once acquired, the method is relatively inexpensive, can be easily repeated and can sample complex topography and sensitive habitats, such as seagrass [26]. While traps are an extractive method, they have been used to effectively sample the fish communities and populations in various habitats [27,28,29,30,31]. Comparisons between sampling efficiencies of baited video and commercial fish traps from tropical regions have demonstrated that baited video sampled higher numbers of species and abundances, and thus provided greater statistical power for detecting differences in the structure of fish communities [27]. The majority of fish species that occur in Cockburn Sound are either smaller bodied or are typically juveniles [18,22], compared with those sampled in the previous comparison by Harvey et al. [27].

Given Cockburn Sound is an important recruitment area for *Pagrus auratus* and *Sillaginodes punctatus* [10,32], our objective was to investigate whether baited videos or traps were the more effective method for assessing small-bodied fish assemblages to determine their composition, relative abundances and associations with four natural (seagrass, limestone outcrops and soft sediment at< and >10 m depth) and two modified habitats (rockwall and dredged channels). As such, the ability of each method to discriminate fish species compositions, species richness and relative abundances at identical locations were investigated. The capacity of the modified habitats to function as natural habitats was assessed by comparing fish species compositions, species richness and relative abundances of fishes recorded in each habitat. If the two modified habitats were adequate substitutes for natural habitats in Cockburn Sound, it was hypothesised that based on the similarities of their physical characteristics, fish assemblages associated with rockwalls would resemble those of limestone outcrops, whereas assemblages in dredged channels would resemble those of soft sediment habitats.

Materials and Methods

Study Area and Sampling Regime

Cockburn Sound is a semi-enclosed marine embayment ca 16 km long by 9 km wide and has a sea surface area of ca 100 km^2 and maximum depth of 23 m (Fig. 1). This embayment is bounded by the mainland to the east and south, Garden Island to the west and the shallow (<10 m) Parmelia Bank to the north (Fig. 1). The southern entrance of the sound has been partially closed through the construction of a rock-filled causeway in 1971–73. All margins of the sound have shallow banks (<10 m) comprising seagrass [33], small outcrops of limestone and extensive soft sediment (Fig. 1). Shipping channels have been dredged into the eastern and northern banks. A dominant feature of the benthos of Cockburn Sound is the deeper central basin (ca 20 m), which is a relatively uniform expanse of soft sediment (silt, Fig. 1). There is extensive industrial development and associated structures along the eastern margin of the sound and boat harbours at the south-eastern end of Garden Island (naval base), either side of the southern end of the causeway and in the north-east corner of the sound (Fig. 1).

Sampling was undertaken in June and July 2008. Baited videos were deployed over four days at 51 sites, followed one week later by traps at the same sites also over four days. Traps were intentionally used after baited videos, as the extraction of fish using traps could potentially reduce abundances and thus influence the results from the baited videos. Sampling sites were determined from ArcGIS© habitat maps (provided by Oceanica Consulting Pty Ltd) and targeted the major natural habitat types including sand banks (<10 m deep, eight sites), central sand basin (>10 m deep, 23 sites), seagrass (six sites) and limestone outcrops (seven sites, Fig. 1). Modified habitats including rockwalls (three sites) and dredged channels (four sites) were also sampled. Three replicate baited videos and traps were sampled concurrently 100–150 m apart at each site. The bait was refreshed for each baited video and trap set with ca 150 g of diced Australian pilchards (*Sardinops sagax*). Baited videos were left to record for 35 minutes, based on the effective duration determined by Morrison and Carbines [34]. Traps were left to soak for 90 minutes as Ferrell and Sumpton [35], using the same type of traps, found catch rates of teleosts reached an asymptote after this period.

Baited Video and Trap Construction

The baited videos consisted of a single high definition (1920×1080 pixels) video camera (Canon HV20) placed in an underwater housing and fastened to a bar situated 75 cm from the floor and centrally within a galvanised-steel trapezium frame with the video orientated horizontally (Fig. 2). Bait was placed in a circular (13.5 cm diameter, 4 cm height) black plastic meshed container and suspended within the field of view 100 cm in front of the camera (Fig. 2). A rope and float for retrieval of the baited video was attached via a rope bridle at the top of the trapezium frame.

A pilot study was conducted to determine 1) an optimal length for the bait pole for baited videos to allow sufficient monitoring around the bait container in the turbid conditions, and 2) the most effective construction for fish traps. The different configurations of fish traps trialled included green vs black 25 mm stretched mesh over both a rounded opera-house trap (90 cm long, 60 cm wide and 35 cm high) and a rectangular trap (93 cm long, 57 cm wide and 33 cm high). The green meshed opera-house trap sampled a greater range of species at higher relative abundances, and was thus used for the remainder of the study. Each opera-house trap had a 10 cm diameter PVC ring spliced into each of the two openings that were located at either end. Two 25 mm square steel channels were attached to the base of each trap to provide ballast and rigidity, thereby minimising potential motion induced from water surge/currents (Fig. 2). Bait holders, identical to those used for the baited videos, were secured within the traps but offset from the two openings (Fig. 2). The traps were retrieved using a rope attached to a float on the waters surface, with the rope attached to

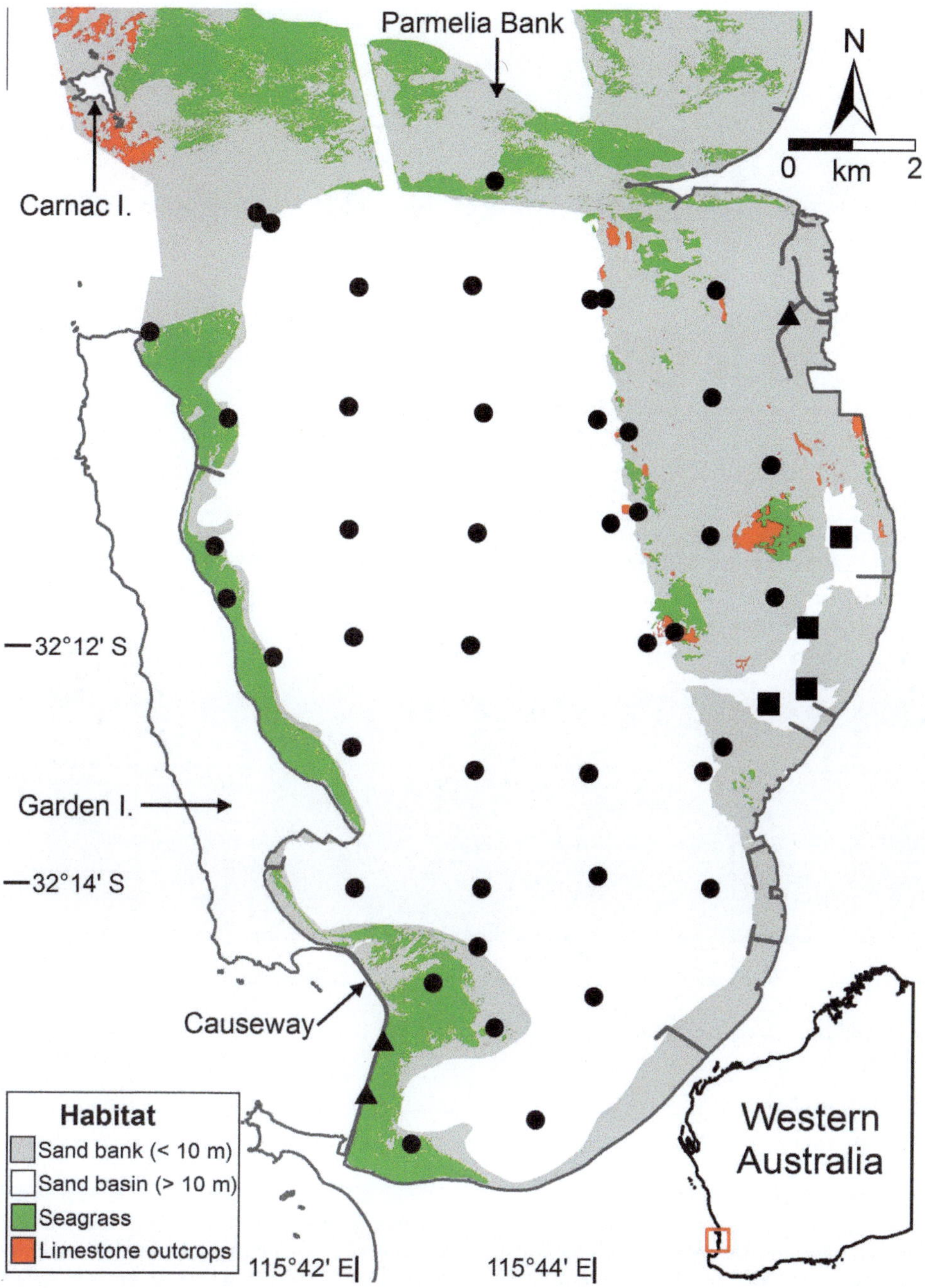

Figure 1. Sites sampled (n = 51) in sand, seagrass and limestone outcrop habitats (black circles) and in dredged channels (black squares) and along rockwalls (black triangles) in Cockburn Sound (habitat map provided by Oceanica Consulting Pty Ltd).

the trap at the base midway along one side and perpendicular to the two openings.

Data Collection and Analysis

Footage recorded from baited video were analysed using a custom interface (BRUVS version 2.1, developed by the Australian Institute of Marine Science) to incorporate data collected from the field, the timing of events, reference images of the seafloor and fish in the field of view. The habitat classifications determined from GIS maps were confirmed for each replicate from video images. Natural habitats were classified into four categories, i.e. shallow sand bank (<10 m, SP), deep sand basin (>10 m, SB), seagrass (GR) and limestone outcrops (LM), and modified habitats into two categories, i.e. rockwall (RW) and dredged channel (DG). Fish were identified to the lowest possible taxa. The relative abundance of each species was determined as the maximum number visible in the field of view at any one time (N_{max}) for baited video and the number caught for traps. The mean relative abundance of all demersal fish and the total number of species recorded were compared for each habitat and method separately.

The multivariate analyses were performed in PRIMER with the PERMANOVA add-on (version 6.1.13, [36,37]). The abundance measure of fish species at each site was calculated as the mean relative abundance (N_{max}, baited videos; numbers, traps) for the

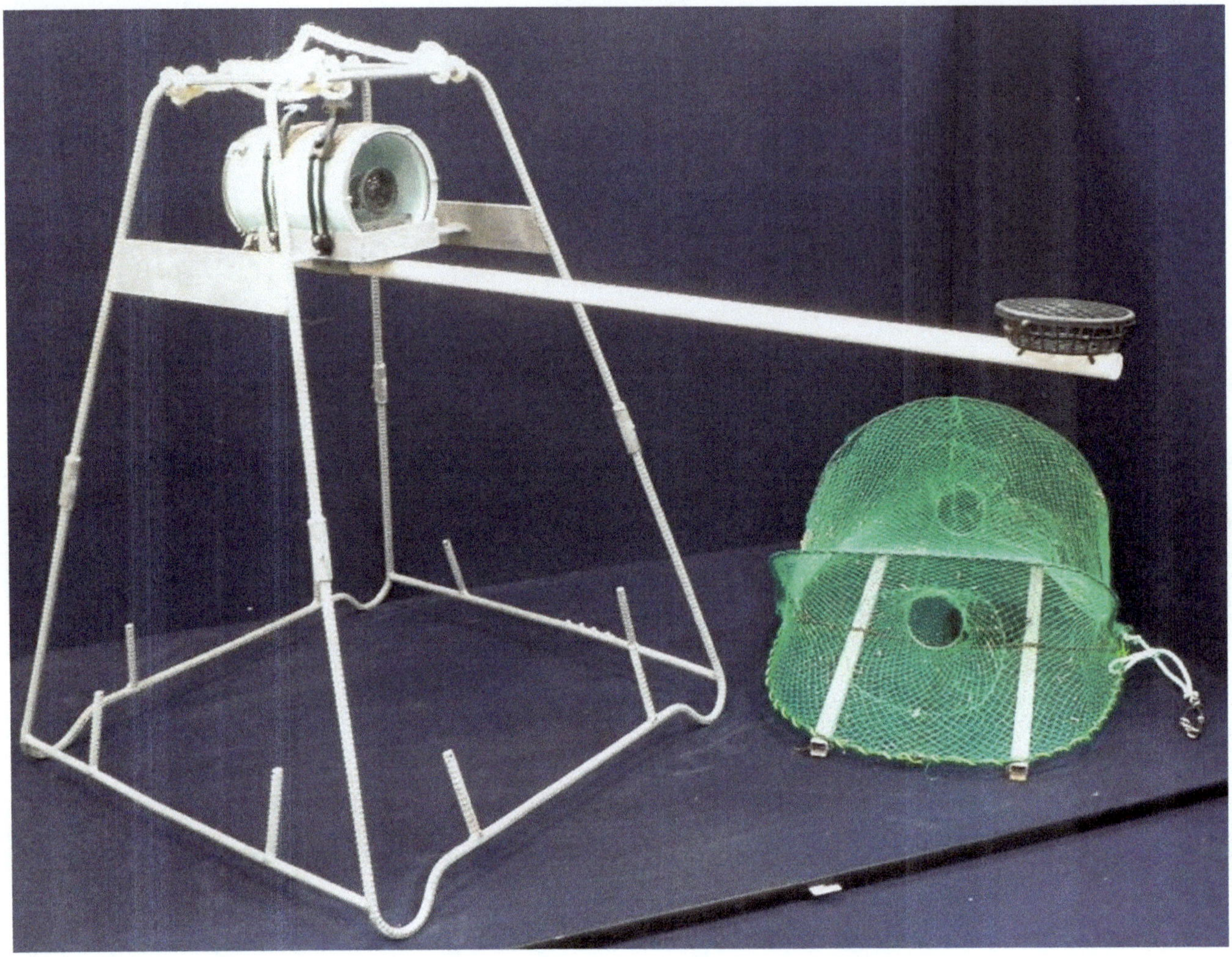

Figure 2. Baited underwater video (left) and opera-house trap (right). Scale reference: camera to bait holder 100 cm, floor to camera 75 cm and trap height 35 cm.

three replicates. Considering the more efficient sampling method was to be used for future monitoring following the completion of this study, analyses of each data set were performed separately and thus relative abundances were not standardised between methods. Relative abundance data for both methods were fourth root transformed prior to analyses based on the gradient of the lineal relationship between the logarithms of standard deviation and mean abundances of species, to down-weight the contribution of the most dominant species [38]. Given their tendency to emphasise species composition and relative abundance within community data, zero-adjusted Bray-Curtis and both modified Gower Log_{10} and Log_2 resemblance measures were considered prior to analyses [39]. The stress performance derived from Shepard Diagrams, that displayed the departure of pairwise distances from the best-fitting increasing regression line produced from non-metric multi-dimensional scaling (nMDS) ordination, indicated that baited video and trap data were better treated using a zero-adjusted Bray-Curtis similarity matrix [40].

Assemblage analysis was performed using permutational multivariate analysis of variance (PERMANOVA, [41]). An unconstrained ordination using principal coordinates analysis (PCO) combined with cluster analysis were used to determine fish assemblage groupings among sites. The significance of these groupings were assessed using a similarity profile test (SIMPROF, [36]). These groupings were compared to a constrained ordination using canonical analysis of principal coordinates (CAP), which maintained *a priori* habitat classifications. An appropriate subset of axis (m) for the CAP analysis was determined by maximising the leave-one-out allocation success ($m = 5$). The first squared canonical correlation (δ^2) and leave-one-out allocation success were used as an indication of how well groups were discriminated within the CAP analysis, as they provide a useful statistical estimate of misclassification error and demonstrate how distinct groups of sites are in multivariate space [42]. A Spearman correlation >0.35 was used as an arbitrary limit to display potential correlations between individual species abundances and habitats relative to the canonical axes.

When significant P values were obtained from pairwise tests using PERMANOVA for baited video data, similarity percentages (SIMPER) were used to identify significant distinguishing fish species. This criteria was based on dissimilarity to standard deviation ratios (Diss/SD) >2 and percentage contributions $>10\%$. The mean relative abundance (± 1 se) of distinguishing species was then compared between each habitat type. Mean relative abundances were also compared between habitats for the targeted species *Pagrus auratus* and *Sillaginodes punctatus*.

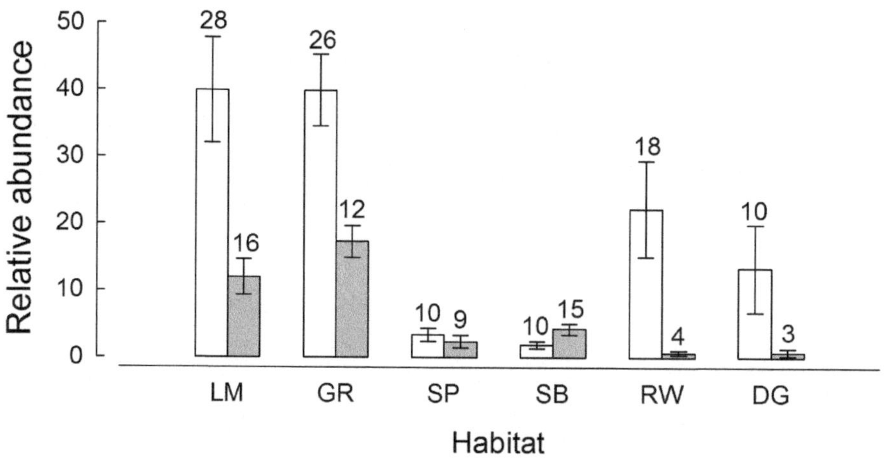

Figure 3. Mean (±1 se) relative abundance (N_{max}, baited video; numbers, trap) of fish sampled per replicate from each habitat using baited videos (white bars) and traps (grey bars). The numbers of fish species sampled from each habitat by each method are shown above bars. LM, limestone outcrops; GR, seagrass; SP, sand bank (<10 m); SB, sand basin (>10 m); RW, rockwall; DG, dredged channel.

Results

Numbers of Fish and Species Recorded by Baited Videos versus Traps

Based on the sum of N_{max} values, baited videos sampled at least 3,944 individual fish from 43 species and 27 families compared with only 1,040 individuals from 27 species and 18 families in traps. There were only two species sampled by the traps that were not sampled by the baited videos (*Platycephalus longispinis* and *Gymnapistes marmoratus*), whereas there were 16 species sampled by baited videos that were not caught in traps. For all species caught by both methods, a greater number of individuals were recorded from baited videos than in the traps, with the exception of *Pentapodus vitta*. Similarly, for each habitat type the mean abundance and species richness recorded by baited videos were greater than those determined from traps, except at sand basin sites (Fig. 3). This was particularly evident for the modified habitats. In rockwall habitats, baited videos recorded an average of 22.2 fish per replicate (±7.2 se) from a total of 18 species, compared with <1 (±0.3 se) fish per replicate from four species caught by traps (Fig. 3). Likewise, in dredged channel habitats baited videos recorded an average of 13.3 fish per replicate (±6.6 se) from 10 species, compared to traps that captured an average of less than one fish per replicate (±0.6 se) from three species (Fig. 3). The lower mean abundances and species richness recorded by traps may be attributed to the fact that 31% of trap sets caught invertebrate piscivorous predators (predominantly *Portunus pelagicus* or *Octopus* spp.), which were likely to deter fish from entering.

Table 1. PERMANOVA results comparing the composition of fish assemblages between methods and across habitats.

Source	df	SS	MS	Pseudo-F	P
Method	1	18872	18872	13.461	0.001
Habitat	5	55459	11092	7.912	0.001
Method*Habitat	5	17957	3591.3	2.5617	0.001
Residual	90	1.26E+05	1401.9		
Total	101	2.21E+05			

Fish Assemblage Comparisons between Sampling Methods and Habitats

The fish assemblages differed significantly between sampling methods and among habitat types and an interaction was detected between these two variables ($P = 0.001$, Table 1). Within the natural habitats, both sampling methods recorded higher relative abundances and numbers of species in limestone outcrop and seagrass habitats (~40 fish per replicate and >25 species) than

Table 2. Results of pairwise PERMANOVA on the composition of fishes recorded with each method between habitats.

Groups			Videos		Traps	
			t	P	t	P
Sand bank	v.	Sand basin	1.614	**0.041**	1.505	0.077
Sand bank	v.	Seagrass	3.122	**0.001**	2.896	**0.001**
Sand bank	v.	Limestone outcrops	2.903	**0.001**	1.966	**0.008**
Sand bank	v.	Rockwall	1.806	**0.024**	1.156	0.290
Sand bank	v.	Dredged channel	1.282	0.180	0.678	0.694
Sand basin	v.	Seagrass	4.068	**0.001**	3.048	**0.001**
Sand basin	v.	Limestone outcrops	3.694	**0.001**	1.757	**0.015**
Sand basin	v.	Rockwall	2.386	**0.002**	1.666	**0.039**
Sand basin	v.	Dredged channel	1.156	0.273	1.253	0.198
Seagrass	v.	Limestone outcrops	2.802	**0.001**	2.294	**0.001**
Seagrass	v.	Rockwall	1.962	**0.016**	2.206	**0.026**
Seagrass	v.	Dredged channel	2.324	**0.006**	2.610	**0.003**
Limestone outcrops	v.	Rockwall	1.231	0.215	1.782	**0.026**
Limestone outcrops	v.	Dredged channel	1.847	**0.029**	1.888	**0.021**
Rockwall	v.	Dredged channel	1.400	0.167	1.336	0.194

Significant differences in bold.

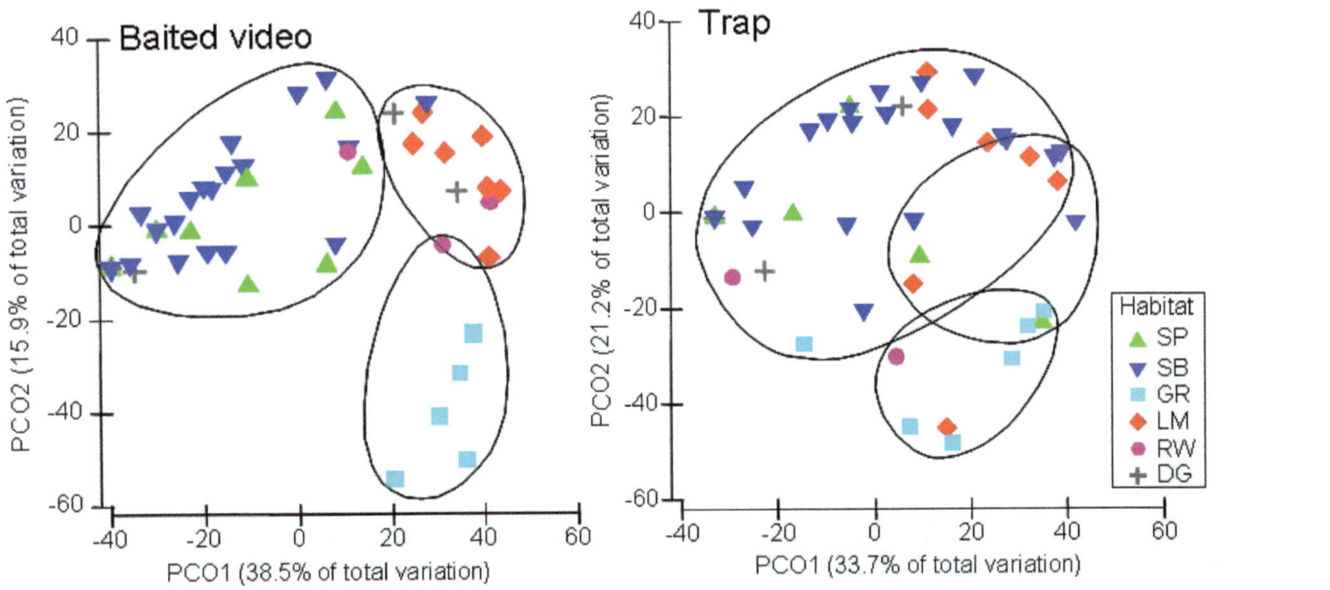

Figure 4. Principal coordinates ordination (PCO) of baited video (left) and trap (right) data overlayed with SIMPROF cluster analysis (35% similarity) defining significant groups. Habitat types include SP, sand bank (<10 m); SB, sand basin (>10 m); GR, seagrass; LM, limestone outcrops; RW, rockwall; DG, dredged channel.

sand bank and sand basin habitats (<4 fish per replicate and 10 species, Fig. 3). This was reflected in the pairwise comparisons, where based on data collected by baited videos, fish assemblages differed significantly between all natural habitats ($P<0.025$, Table 2), with the two soft sediment habitats, i.e. sand banks and sand basin, only marginally different ($P=0.041$). Similar trends in pairwise comparisons were evident between natural habitats using trap data, except that fish assemblages in sand bank and sand basin habitats were not significantly different (Table 2). Fish assemblages associated with the two modified habitats were poorly sampled using traps compared to baited videos (Fig. 3). At the rockwall locations, baited videos recorded approximately half the relative abundance of fish than in limestone outcrop and seagrass habitats (40 vs 22 fish per replicate, Fig. 3). However, baited videos recorded markedly higher relative abundances of fish from dredged channels than sand bank and sand basin habitats (13 vs 3 fish per replicate, Fig. 3). Pairwise tests comparing fish assemblages of natural and modified habitats with similar physical characteristics, as recorded by baited videos, found no significant differences between dredged channels and both sand banks and sand basin habitats ($P=0.273$ and 0.180, respectively) and between rockwall and limestone outcrop habitats ($P=0.215$, Table 2). Fish assemblages recorded by baited videos in seagrass were significantly different from both modified habitats i.e. rockwalls and dredged channels (Table 2). While traps were also unable to detect a significant difference between dredged channel and soft sediment habitats ($P>0.05$), limestone outcrops and rockwall habitats did differ significantly ($P=0.026$, Table 2).

Using an unconstrained PCO ordination and SIMPROF analysis, the natural habitat sites were distinguished into three main groups for both sampling methods at a Spearman correlation >0.35 using Bray-Curtis similarity (Fig. 4). According to the baited video data, these three groupings were clearly discriminated into seagrass, limestone outcrop and soft sediment (combining sand banks and sand basin, Fig. 4). Points for the rockwall habitats were located towards the top right of the plot, with one in each of the limestone outcrop and soft sediment groups and one within an overlap of the limestone outcrop and seagrass groups (Fig. 4). In contrast, the four dredged channel sites were dispersed with two sites in each of the limestone outcrop and soft sediment habitat groups (Fig. 4). According to the trap data, there was greater dispersion of natural and modified habitat sites within the PCO ordination, with many sites included in an overlap between groupings, which was not apparent in the analysis of the composition of fishes recorded by baited video (Fig. 4).

Plots of the principal coordinates from the constrained CAP analysis showed closer clustering among sites from the same habitats for baited videos than traps (Fig. 5). This was confirmed by the higher values of leave-one-out allocation success and canonical correlation (δ^2). Fish assemblages associated with natural habitats had an allocation success of 79.5% ($\delta^2=0.88$) when sampled by baited videos, compared to 54.5% ($\delta^2=0.79$) when sampled by traps. For the baited video data, the seagrass and limestone outcrop categories both had 100% allocation success, revealing the lower allocation success was due to misclassification between the two soft sediment groups (i.e. 62.5% for sand banks and 73.9% for sand basin habitats).

Distributions of Important and Distinguishing Fish Species among Habitats

Using data derived from baited videos, there were distinct groups of fish identified from CAP analysis with relative abundances significantly correlated (Spearman >0.35) with each of the three main natural habitat groups (i.e. seagrass, limestone outcrop and soft sediment, Fig. 5). There were eight fish species with relative abundances significantly correlated with seagrass (*Apogon rueppellii*, *Arripis georgianus*, *Meuschenia freycineti*, *Acanthaluteres spilomelanurus*, *Sillaginodes punctatus*, *Pelates octolineatus*, *Sphyraena novaehollandiae* and *Torquigener pleurogramma*, Fig. 5). Similarly, there were eight fish species with relative abundances significantly correlated with limestone outcrops (*Parequula melbournensis*, *Pseudocaranx sp.*, *Trachurus novaezelandie*, *Pagrus auratus*, *Pentapodus vitta*, *Notolabrus parilus*, *Coris auricularis* and *Upeneichthys vlamingii*, Fig. 5). However, there were only two species of ray, *Trygonorhina fasciata*

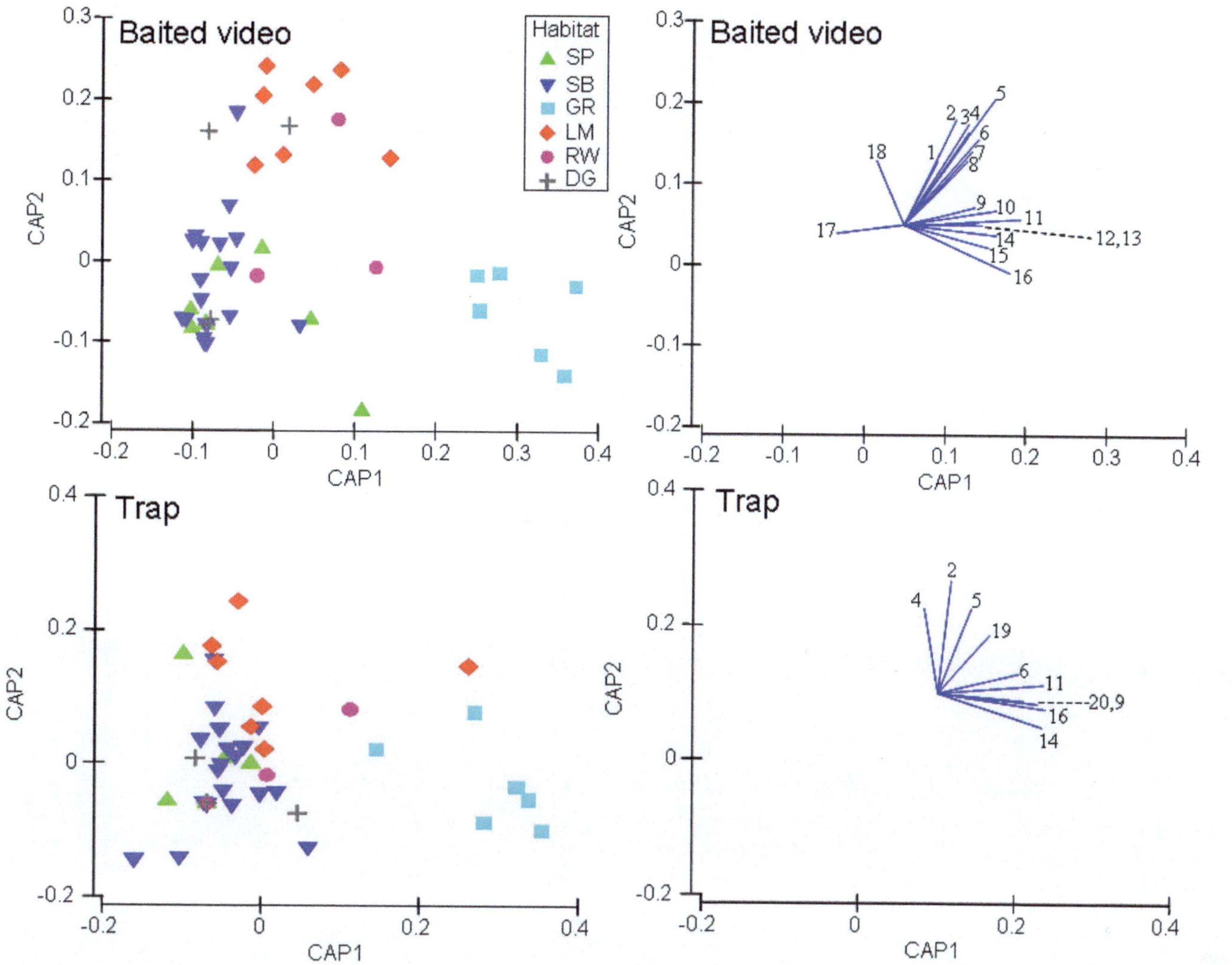

Figure 5. Canonical analysis of principal coordinates (CAP, left) ordination of baited video (above) and trap (below) data with corresponding strength and direction of Spearman correlation >0.35 of fish species shown as line vectors (right). SP, sand bank (<10 m); SB, sand basin (>10 m); GR, seagrass; LM, limestone outcrops; RW, rockwall; DG, dredged channel. Fish species include *Parequula melbournensis*[1], *Pseudocaranx sp.*[2], *Trachurus novaezelandie*[3], *Pagrus auratus*[4], *Pentapodus vitta*[5], *Notolabrus parilus*[6], *Coris auricularis*[7], *Upeneichthys vlamingii*[8], *Apogon rueppellii*[9], *Arripis georgianus*[10], *Meuschenia freycineti*[11], *Acanthaluteres spilomelanurus*[12], *Sillaginodes punctatus*[13], *Pelates octolineatus*[14], *Sphyraena novaehollandiae*[15], *Torquigener pleurogramma*[16], *Trygonorhina fasciata*[17], *Myliobatis australis*[18], *Scobinichthys granulatus*[19], *Haletta semifasciata*[20].

and *Myliobatis australis*, with relative abundances correlated with the soft sediment group, which most likely reflected the markedly lower abundances and species of fish recorded from these habitats (Fig. 5). In comparison, the CAP analysis of data derived from traps determined that only the relative abundances of a subset of fish species could be associated with seagrass from the other habitats (Fig. 5).

Based on SIMPER analysis using data derived from baited videos, there were three fish species, i.e. *Torquigener pleurogramma*, *Pelates octolineatus* and *Meuschenia freycineti*, that distinguished fish assemblages associated with seagrass from soft sediment and dredged habitats (contribution >10%, Diss/SD ratios >2.0). Only *T. pleurogramma* distinguished seagrass from limestone outcrop habitats. Notably, each of these three species occurred almost exclusively in seagrass (Fig. 6). There were three fish species including *Pseudocaranx sp.*, *Trachurus novaezelandie* and *Pentapodus vitta* that distinguished assemblages between limestone outcrop and soft sediment habitats. Although the mean relative abundances of *Pseudocaranx sp.* and *P. vitta* were markedly higher in limestone outcrop and dredged channel habitats, they were also recorded consistently but at lower abundances in all other habitats (Fig. 6). In contrast, *T. novaezelandie* was recorded almost exclusively in limestone outcrop and rockwall habitats (Fig. 6).

The distributions of the recreationally and commercially important *P. auratus* and *S. punctatus* were based on 134 and eight observations, respectively, from the 153 baited video replicates. All of the *P. auratus* except one individual were considered to belong to the 0+ age cohort (ca six months of age), based on their length relative to the size of the bait holder and known age-length relationship [10,43]. The relative abundance of juvenile *P. auratus* was highest in the dredged channel, and to a lesser extent, the limestone outcrop habitats, with occasional occurrences in the remaining habitats (Fig. 6). The ages of the eight *S. punctatus* observed by baited videos were indeterminate based on their relative length, they were however relatively small and most likely from young age classes. The individuals of that species were

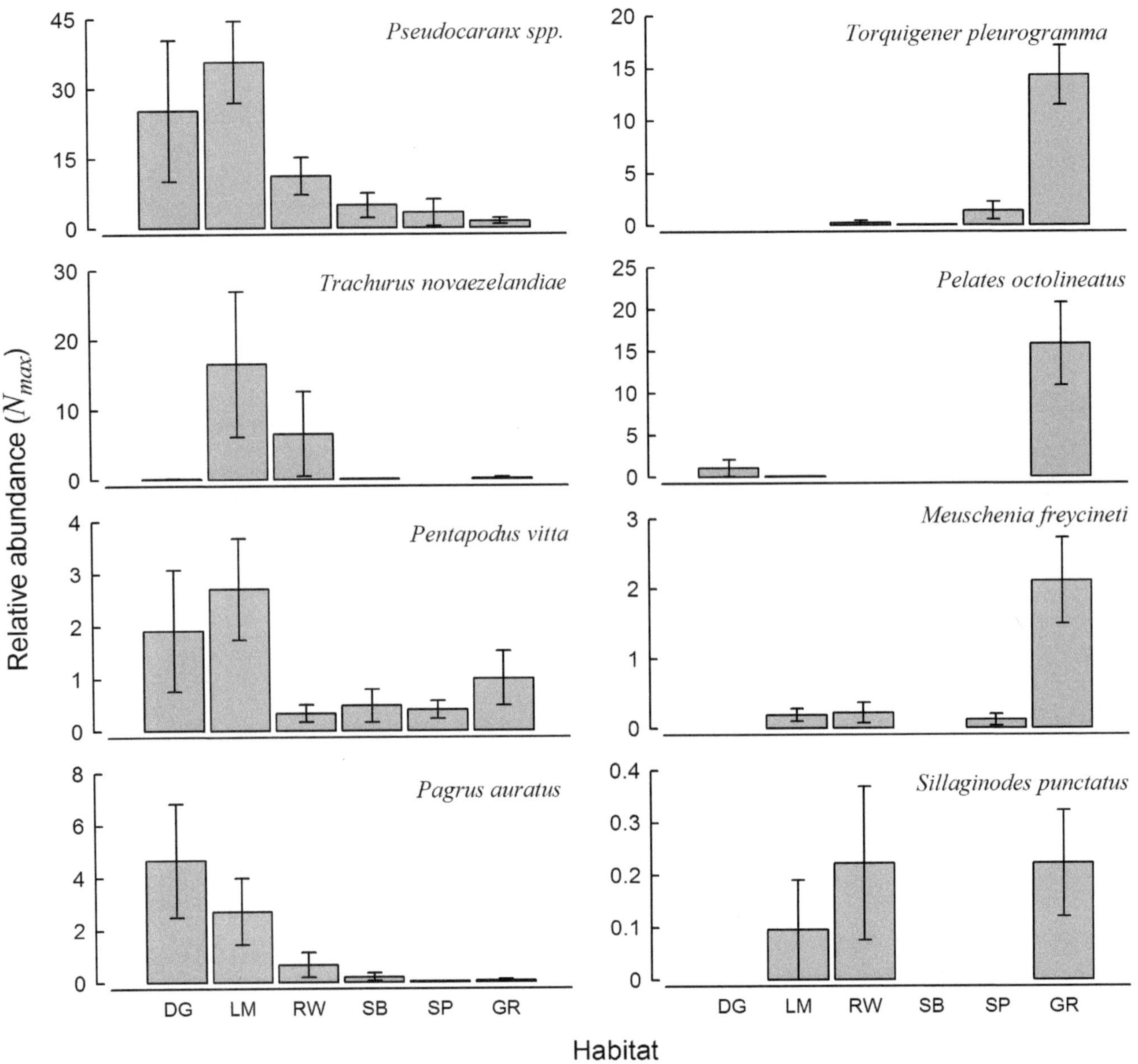

Figure 6. Mean (±1 se) relative abundance per replicate (N_{max}) of distinguishing and important (commercially and recreationally fished) fish species for each habitat type from baited videos. DG, dredged channel; LM, limestone outcrops; RW, rockwall; SB, sand basin (>10 m); SP, sand bank (<10 m); GR, seagrass.

recorded in three of the six habitat types, including rockwall, seagrass and limestone outcrops (Fig. 6).

Discussion

Baited videos recorded much greater relative abundances and numbers of fish species than opera-house traps from identical locations in four natural and two anthropogenically-modified habitats. As such, baited videos allowed for greater discrimination of fish assemblages between habitats and were thus considered to be a more efficient sampling method. It is likely that the capture of invertebrate piscivorous predators (i.e. crabs and octopus) in traps (31% of sets) greatly increased the likelihood of predator-prey interactions and thus reduced their sampling effectiveness [44]. Such predator-prey interactions would have been overcome using baited videos, as it is likely the fish would have still been observed within the wide field of view during interactions. Baited video may have also sampled higher numbers of fish species by recording those that were attracted to the bait (evident through their feeding behaviour) as well as those that swam past, thereby increasing the numbers of predatory or scavenging species while also including herbivorous and omnivorous species [45]. The baited video was also able to collect images of the habitat, which facilitated direct links between fish assemblages and their habitat associations that would have not been achieved from the use of traps alone [30]. However, both baited videos and traps were only able to sample relative rather than absolute abundances of fish, given the complexities associated with estimating sampled area based on bait plume dispersal. In comparison, Morrison and Carbines [34] used a towed video method capable of estimating concentrations of juvenile teleosts by calculating swept area.

Fish assemblages sampled in natural habitats by baited videos were discriminated into three distinct groups, i.e. seagrass, limestone outcrops and soft sediment. Within the soft sediment group, there were only marginal differences between fish assemblages in habitats shallower (sand banks) or deeper (sand basin) than ten metres. In comparison, fish assemblages associated with natural habitats, as determined from traps, displayed weak discriminating power and tended to misclassify sites between the three groups, irrespective of whether analysis was constrained by *a priori* habitat classifications.

The majority of habitat in Cockburn Sound is flat and relatively featureless soft sediment, particularly in the deep basin and eastern sand bank areas. Despite occupying the majority of the sound, fish assemblages recorded by baited video in these habitats comprised markedly fewer species (31% of species recorded) in relatively lower abundances. In comparison, the very limited area occupied by seagrass and limestone outcrop habitats had a 15 fold greater abundance of fish and comprised 60% and 65% of all species sampled, respectively. It thus appears important that to maintain fish diversity in Cockburn Sound, the natural seagrass and limestone outcrop habitats need to be conserved, particularly in light of the large scale and continued loss (>77%) of seagrass in this embayment [46].

The anthropogenically-modified habitats were poorly sampled by traps compared to baited videos. This was not surprising for rockwalls, as traps had to be located adjacent to them and relied on fish to leave the high vertical relief and complex shelter provided by the piles of large limestone blocks used in their construction. Baited video could overcome this by being orientated in the direction of the rockwall, thus identifying fish within this complex structure. Discrepancies in efficiency were however surprising for the low relief dredged channels, where the exposure of fish to baited videos and traps would have been similar. These inconsistencies provided little confidence in the ability of traps to provide useful comparisons between natural and modified habitats.

In regards to the overall fish abundance and species richness, the fish assemblages associated with modified habitats did resemble those of natural habitats with similar physical characteristics. Whereby, the overall abundances recorded using baited videos, were not significantly different between rockwall and limestone outcrops and between dredged channel and soft sediment habitats (sand banks and sand basin). Further investigation using PCO and SIMPROF analysis revealed subtle differences in fish assemblages sampled in rockwall sites, where assemblages were found to consist of a subset of species that were recorded from neighbouring natural habitats. This resulted in the three rockwall sites being grouped with one belonging to each of limestone outcrop and sand groups, and one combined within an overlap of limestone outcrop and seagrass groups. This suggests that fish assemblages sampled from rockwall habitat represented an altered composition from that associated with natural limestone outcrop habitat. Thus, the characteristics of rockwall habitat may not have provided the necessary resource requirements (e.g. shelter and/or diet) for some species, resulting in an alteration to the natural trophic structure and interspecific interactions [8,9].

The fish assemblages sampled by baited video in dredged channel habitat showed mixed results, with two sites included within the soft sediment group, as was hypothesised, and two sites unexpectedly included within the limestone outcrop group. The two dredged channel sites that were grouped within the limestone outcrop group resulted from a high relative abundance of *Pseudocaranx* sp., *P. auratus* and *P. vitta*. The other two dredged channel sites that were included within the soft sediment group consisted of only *T. fasciata* from six replicates. The frequency of dredging of these channels in Cockburn Sound is low (ca every 8–10 years), which suggests a limited number of fish species will recolonise these disturbed areas, but their succession appears variable.

The results of this study support the use of baited videos over traps for broad fish ecology studies [27], and provided a non-extractive application for sampling of predominantly smaller bodied and juvenile fish species in sensitive or turbid environments. It also confirms the advantages of using baited videos for such studies compared to many traditional sampling techniques, such as line, trap and trawl [26,47,48,49]. This method would also be useful for collecting information on the relative abundance of 0+ aged recruits, thus contributing information on recruitment strength toward the stock assessments of exploited teleosts, e.g. *P. auratus* [10]. In addition, this study provides sound quantitative data and repeatable methods for assessing changes in fish communities, which could contribute toward ecological assessments of developments involving anthropogenic modification of marine embayments.

Acknowledgments

We extend our gratitude to staff from the Department of Fisheries for providing logistical support. Thank you also to Bob Clarke for his useful suggestions regarding statistical analysis; and to Mike Travers and Jeff Norriss and two anonymous reviewers whose comments on earlier versions of the manuscript were greatly appreciated.

Author Contributions

Conceived and designed the experiments: CBW PDL. Performed the experiments: CBW PDL TBC. Analyzed the data: CBW PDL DVF TJL. Wrote the paper: CBW PDL TBC DVF TJL.

References

1. Potter IC, Beckley LE, Whitfield AK, Lenanton RCJ (1990) Comparisons between the roles played by estuaries in the life cycles of fishes in temperate Western Australia and Southern Africa. Environ Biol Fishes 28: 143–178.
2. Sadovy Y, Cheung WL (2003) Near extinction of a highly fecund fish: the one that nearly got away. Fish and Fisheries 4: 86–99.
3. Wakefield CB (2010) Annual, lunar and diel reproductive periodicity of a spawning aggregation of snapper *Pagrus auratus* (Sparidae) in a marine embayment on the lower west coast of Australia. J Fish Biol 77: 1359–1378.
4. Watts RJ, Johnson MS (2004) Estuaries, lagoons and enclosed embayments: habitats that enhance population subdivision of inshore fishes. Mar Freshw Res 55: 641–651.
5. Ayvazian SG, Johnson MS, McGlashan DJ (1994) High levels of genetic subdivision of marine and estuarine populations of the estuarine catfish *Cnidoglanis macrocephalus* (Plotosidae) in southwestern Australia. Mar Biol 118: 25–31.
6. Johnson MS, Creagh S, Moran M (1986) Genetic subdivision of stocks of snapper, *Chrysophrys unicolor*, in Shark Bay, Western Australia. Aust J Mar Freshw Res 37: 337–345.
7. Orth RJ, Carruthers TJB, Dennison WC, Duarte CM, Fourqurean JW, et al. (2006) A global crisis for seagrass ecosystems. Bioscience 56: 987–996.
8. Clynick BG, Chapman MG, Underwood AJ (2008) Fish assemblages associated with urban structures and natural reefs in Sydney, Australia. Austral Ecol 33: 140–150.
9. Glasby TM, Connell SD (1999) Urban structures as marine habitats. Ambio 28: 595–598.
10. Wakefield CB, Fairclough DV, Lenanton RCJ, Potter IC (2011) Spawning and nursery habitat partitioning and movement patterns of *Pagrus auratus* (Sparidae) on the lower west coast of Australia. Fish Res 109: 243–251.
11. de Lestang S, Hall NG, Potter IC (2003) Changes in density, age composition, and growth rate of *Portunus pelagicus* in a large embayment in which fishing

pressures and environmental conditions have been altered. J Crustac Biol 23: 908–919.
12. Breheny NB, Beckley LE, Wakefield CB (2012) Ichthyoplankton assemblages asscoiated with pink snapper (*Pagrus auratus*) spawning aggregations in coastal embayments of southwestern Australia. J R Soc West Aust 95: 103–114.
13. Lenanton RCJ (1974) The abundance and size composition of trawled juvenile snapper *Chrysophrys unicolor* (Quoy and Gaimard) from Cockburn Sound, Western Australia. Aust J Mar Freshw Res 25: 281–285.
14. Kendrick GA, Aylwarda MJ, Hegge BJ, Cambridge ML, Hillman K, et al. (2002) Changes in seagrass coverage in Cockburn Sound, Western Australia between 1967 and 1999. Aquat Bot 73: 75–87.
15. Cambridge ML, Chiffings AW, Brittan C, Moore L, McComb AJ (1986) Loss of seagrass in Cockburn Sound, Western Australia: II. Possible causes of seagrass decline. Aquat Bot 24: 269–285.
16. Silberstein K, Chiffings AW, McComb AJ (1986) The loss of seagrass in Cockburn Sound, Western Australia. III. The effect of epiphytes on productivity of *Poidonia australis* hook. F. Aquat Bot 24: 355–371.
17. Vanderklift MA (1996) Influence of adjacent seagrass on the fish assemblages off sandy beaches [Masters Thesis]. Western Australia: Edith Cowan University. 208 p.
18. Vanderklift MA, Jacoby CA (2003) Patterns in fish assemblages 25 years after major seagrass loss. Mar Ecol Prog Ser 247: 225–235.
19. Scott JK, Dybdahl RE, Wood WF (1986) The ecology of *Posidonia* seagrass fish communities in Cockburn Sound, Western Australia. Perth, Western Australia: Department of Conservation and Environment. Technical Series No. 11.
20. Dybdahl RE (1979) Technical report on fish productivity. An assessment of the marine faunal resources of Cockburn Sound. Perth, Western Australia: Department of Conservation and Environment, Report No. 4.
21. Penn JW (1977) Trawl caught fish and crustaceans from Cockburn Sound. Department of Fisheries and Wildlife Western Australia. Report No. 20. 24 p.
22. Johnston DJ, Wakefield CB, Sampey A, Fromont J, Harris D (2008) Developing long-term indicators for the sub-tidal embayment communities of Cockburn Sound. Western Australia: Department of Fisheries. Fisheries Research Report No. 181.
23. Harvey ES, McLean DL, Frusher S, Haywood MDE, Newman SJ, et al. (2012) The use of BRUVs as a tool for assessing marine fisheries and ecosystems: a review of the hurdles and potential. University of Western Australia. FRDC Report Project No. 2010/002. 44 p.
24. Cappo M, Harvey E, Shortis M (2006) Counting and measuring fish with baited video techniques - an overview. Australian Society for Fish Biology Workshop Proceedings: 101–114.
25. Gladstone W, Lindfield S, Coleman M, Kelaher B (2012) Optimisation of baited remote underwater video sampling designs for estuarine fish assemblages. J Exp Mar Biol Ecol 429: 28–35.
26. Cappo M, Speare P, De'ath G (2004) Comparison of baited remote underwater video stations (BRUVS) and prawn (shrimp) trawls for assessments of fish biodiversity in inter-reefal areas of the Great Barrier Reef Marine Park. J Exp Mar Biol Ecol 302: 123–152.
27. Harvey ES, Newman SJ, Mclean DL, Cappo M, Meeuwig JJ, et al. (2012) Comparison of the relative efficiencies of stereo-BRUVs and traps for sampling tropical continental shelf demersal fishes. Fish Res 125–126: 108–120.
28. Travers MJ, Newman SJ, Potter IC (2006) Influence of latitude, water depth, day *v.* night and wet *v.* dry periods on the species composition of reef fish communities in tropical Western Australia. J Fish Biol 69: 987–1017.
29. Ferrell D, Avery R, Blount C, Hayes L, Pratt R (1994) The utility of small, baited traps for surveys of snapper (*Pagrus auratus*) and other demersal fishes. NSW Fisheries Research Institute, Cronulla.
30. Thrush SF, Schultz D, Hewitt JE, Talley D (2002) Habitat structure in soft-sediment environments and abundance of juvenile snapper *Pagrus auratus*. Mar Ecol Prog Ser 245: 273–280.
31. Jackson G, Burton C, Moran M, Radford B (2007) Distribution and abundance of juvenile pink snapper, *Pagrus auratus*, in the gulfs of Shark Bay, Western Australia, from trap surveys. Department of Fisheries, Western Australia. Fisheries Research Report No. 161.
32. Hyndes GA, Platell ME, Potter IC, Lenanton RCJ (1998) Age composition, growth, reproductive biology, and recruitment of king george whiting, *Sillaginodes punctata*, in coastal waters of southwestern Australia. Fish Bull 96: 258–270.
33. Cambridge ML, McComb AJ (1984) The loss of seagrass in Cockburn Sound, Western Australia. I. The time course and magnitude of seagrass decline in relation to industrial development. Aquat Bot 20: 229–243.
34. Morrison M, Carbines G (2006) Estimating the abundance and size structure of an estuarine population of the sparid *Pagrus auratus*, using towed camera during nocturnal periods of inactivity, and comparisons with conventional sampling techniques. Fish Res 82: 150–161.
35. Ferrell D, Sumpton W (1996) Assessment of the fishery for snapper (*Pagrus auratus*) in Queensland and New South Wales. Queensland: QDPI. FRDC 93/074. 143 p.
36. Clarke KR, Gorley RN (2006) PRIMER v6: User Manual/Tutorial. Plymouth, UK: PRIMER-E.
37. Anderson M, Gorley R, Clarke K (2008) PERMANOVA+ for PRIMER: guide to software and statistical methods. Plymouth, UK: Primer-E Ltd.
38. Clarke KR, Warwick RM (2001) Change in marine communities: an approach to statistical analysis and interpretation, 2nd edition. Plymouth, United Kingdom: PRIMER-E.
39. Anderson MJ, Crist TO, Chase JM, Velland M, Inouye BD, et al. (2011) Navigating the multiple meanings of β diversity: a roadmap for the practicing ecologist. Ecol Lett 14: 19–28.
40. Clarke KR, Somerfield PJ, Chapman MG (2006) On resemblance measures for ecological studies, including taxonomic dissimilarities and a zero-adjusted Bray-Curtis coefficient for denuded assemblages. J Exp Mar Biol Ecol 330: 55–80.
41. Anderson MJ (2001) A new method for non-parametric multivariate analysis of variance. Austral Ecol 26: 32–46.
42. Anderson MJ, Willis TJ (2003) Canonical analysis of principal coordinates: a useful method of constrained ordination for ecology. Ecology 84: 511–525.
43. Wakefield CB (2006) Latitudinal and temporal comparisons of the reproductive biology and growth of snapper, *Pagrus auratus* (Sparidae), in Western Australia [Ph.D. Thesis]. Western Australia: Murdoch University. 162 p.
44. High WL, Ellis IE (1973) Underwater observations of fish behavior in traps. Helgol Wiss Meeresunters 24: 341–347.
45. Harvey ES, Cappo M, Butler JJ, Hall NG, Kendrick GA (2007) Bait attraction affects the performance of remote underwater video stations in assessment of demersal fish community structure. Mar Ecol Prog Ser 350: 245–254.
46. Walker DI, Kendrick GA, McComb AJ (2006) Decline and recovery of seagrass ecosystems - the dynamics of change. In: Larkum AWD, Orth RJ, Duarte CM, editors. Seagrasses: biology, ecology and conservation. Dordrecht, Netherlands: Springer. 551–565.
47. Langlois TJ, Fitzpatrick BR, Fairclough DV, Wakefield CB, Hesp SA, et al. (2012) Similarities between Line Fishing and Baited Stereo-Video Estimations of Length-Frequency: Novel Application of Kernel Density Estimates. PLoS ONE 7: e45973.
48. Wakefield CB, Moran MJ, Tapp NE, Jackson G (2007) Catchability and selectivity of juvenile snapper (*Pagrus auratus*, Sparidae) and western butterfish (*Pentapodis vitta*, Nemipteridae) from prawn trawling in a large marine embayment in Western Australia. Fish Res 85: 37–48.
49. Murphy HM, Jenkins GP (2010) Observational methods used in marine spatial monitoring of fishes and associated habitats: a review. Mar Freshw Res 61: 236–252.

9

The Role of Habitat in the Persistence of Fire Ant Populations

Walter R. Tschinkel[1]*, Joshua R. King[2]

1 Department of Biological Science, Florida State University, Tallahassee, Florida, United States of America, 2 Biology Department, University of Central Florida, Orlando, Florida, United States of America

Abstract

The association of the exotic fire ant, *Solenopsis invicta* with man-modified habitats has been amply demonstrated, but the fate of such populations if ecological succession proceeds has rarely been investigated. Resurvey of a fire ant population in a longleaf pine plantation after 25 years showed that the recovery of the site from habitat disturbance was associated with a large fire ant population decline. Most of the persisting colonies were associated with the disturbance caused by vehicle tracks. In a second study, mature monogyne fire ant colonies that had been planted in experimental plots in native groundcover of the north Florida longleaf pine forest had mostly vanished six years later. These observations and experiments show that *S. invicta* colonies rarely persist in the native habitat of these pine forests, probably because they are not replaced when they die. A single site harbored a modest population of polygyne fire ants whose persistence was probably facilitated by reproduction through colony fission.

Editor: Corrie S. Moreau, Field Museum of Natural History, United States of America

Funding: This work was carried out with support from National Science Foundation grant DEB 1020415. The funders had no role in study design, data collection and analysis, decision to publish, or preparation of the manuscript.

Competing Interests: The authors have declared that no competing interests exist.

* E-mail: tschinkel@bio.fsu.edu

Introduction

The monogyne exotic fire ant *Solenopsis invicta* Buren, is one of many species, both plant and animal, that is adapted to exploit ecological disturbance. Communities that develop in such disturbed habitat are unstable and revert to a late stage community whose composition varies with geographic location, soil type and climatic variables. Early succession of such plant communities has been subjected to a great many studies, and general theories of succession have been proposed [1]. Not surprisingly, the communities of insects associated with these plants also change as succession proceeds (e.g.[2–5]). Considering the profound changes of the physical and biotic habitat that are usually associated with plant succession, it is perhaps not surprising that ant communities also undergo changes of species composition as succession proceeds. For example, ant species characteristic of early successional stages of reclaimed bauxite mines disappear with time while those of later stages appear [6]. Fire ants decline in abundance as time since habitat disturbance increases [7]. The profound differences in the composition of ant communities in natural and human-disturbed habitats imply that many of the ant species found in human-disturbed habitat would decline if succession were allowed to proceed (e.g. [8,9].

The fire ant, *S. invicta* is a globally distributed exotic species that has been intensively studied for decades without much attention to how variation in habitat may affect its local and regional distribution and abundance, and thus its impacts [10–14]. We should care about both the persistence and relative abundance of fire ants in different ecosystems and for different levels of anthropogenic disturbance because such knowledge will help us understand the ecology and long-term impacts of this exotic species. Recently, long-term observational studies [15], careful habitat gradient studies [16–17], and large-scale addition and removal experiments [18–21] conducted in the southern USA have provided a better understanding of the local distribution of fire ants in relation to natural and man-made ecosystems and their impacts on co-occurring arthropod populations.

Fire ants are, by far, most abundant in the most altered ecosystems, especially pastures and roadsides [22–23]. Heavily disturbed sites in Texas harbored a much higher abundance of *S. invicta*, and *S. invicta* was one of only two ants species found under complete ecosystem alteration [24–25]. In the deciduous forests of South Carolina, *S. invicta* appeared in clear-cuts (but not uncut controls) within as little as 3 months post-cutting [26]. *S. invicta* was the most abundant species in highly disrupted habitats such as sugar cane fields [10], lawns [27], roadsides [28,29,22], pastures and old fields [30–33] and powerline rights of way [34]. In Florida pine flatwoods, a habitat naturally free of fire ants, experiments combining fire ant additions with habitat alteration (mowing and plowing) of intact habitats revealed that establishment and growth of fire ant populations, occurred naturally (without experimental supplementation) primarily in plowed sites [19]. In short, in situations in which disturbed and undisturbed habitats are juxtaposed, *S. invicta* is sharply limited to the disturbed portion [19,35].

The correlative and experimental evidence reviewed above shows that ecological disturbance is the most likely *cause* of the association of *S. invicta* with early succession habitat. This association is apparent throughout the introduced range, as well as in its native range in South America [36]. However, anthropogenically disturbed ecosystems are unstable, requiring continuous maintenance and, left alone, revert to some late stage condition through plant community succession. The fate of *S. invicta* during the course of such ecological succession has rarely been investigated. Among the few such investigations, Pass [37] found that *S. invicta* was essentially absent once crown closure in pine plantations exceeded 40%. King and Tschinkel [19] showed that mature colonies transplanted to native habitat (in which they do not occur on their own) can persist for considerable time, but their longer-term fate (>3 yr) is unknown. LeBrun et al. [7] report a negative relationship between time since disturbance and fire ant density. Here we report on the fates of *S. invicta* colonies transplanted into native habitats, and of populations self-founded in disturbed habitat after more than 25 years of ecological succession.

Materials and Methods

Ethics statement

This study was carried out in the Apalachicola National Forest under US Forest Service permits APA583 and APA56302. No protected species were involved.

Two previous studies served as the starting point for our study—Tschinkel [35] and King and Tschinkel [19]. In 1985, Tschinkel [35] surveyed two species of fire ants (*Solenopsis invicta, S. geminata*) in the coastal plain longleaf pine forests of the Apalachicola National Forest (ANF) south of Tallahassee, Florida. Of the areas he surveyed, the most relevant to the present study was a young longleaf pine plantation in ANF compartment 228 that hosted a substantial population of *S. invicta* as a result of the soil disturbance that occurred during the establishment of the plantation. Using Tschinkel's published site map and a satellite image for the Hilliardville, FL, DOQQ 5136 (Florida Land Boundary Information System: www.labins.org/) the original transects were run again in March, 2012, more than 25 years later, after the site had largely recovered from the original disturbance. Colonies of *S. invicta* were mapped using a Trimble GeoExplorer (CE series) GPS. The data were differentially corrected using the base station maintained by the Florida Department of Environmental Protection, and displayed on the satellite image. The site was also searched more generally for *S. invicta* colonies. As a control, a 750 m section of the closest unpaved forest road that had been surveyed for *S. invicta* in 1985 was resurveyed in June, 2013.

The study by King and Tschinkel [19] reported a fully-factorial experiment in which the native ground cover of 40×40 m plots in the Apalachicola National Forest was mowed, plowed or left undisturbed, and these treatments were crossed with the addition of mature fire ant colonies planted into shallow holes, the addition of soil (without ants) into shallow holes or no additions. This yielded 9 treatments that were replicated at 5 different sites. This experiment addressed the effect of habitat disturbance and fire ants, separately and together, on the native ant fauna. During the study, plots scheduled for fire ant addition were visually surveyed for live colonies twice a year. If the number was too low for the purposes of the experiment, more colonies were added during the cool season. Colonies to be added were collected from roadsides under conditions that favored capturing the queen, that is, cool, sunny mornings. Only queenright colonies could be expected to persist even under the best of circumstances. Experience has shown that about half of the added colonies were probably queenright. The last additions were made early in 2006, and live colonies in these addition plots were surveyed and mapped in the summer of 2006.

In 2012, more than six years after the last additions, teams of four observers visually searched all of the plots for fire ant colonies, this time mapping all colonies in all plots using the GeoExplorer (CE series) GPS, as above. The object was to determine whether the elevated populations of the addition plots persisted in 2012. These data were added to the satellite images along with the approximate locations (estimated from sketch maps) of the 2006 colonies. For both surveys we focused on counting colonies, rather than using pitfall traps or baits, as mature fire ant colonies are conspicuous, and thus accurately counted within the plots [19,35].

Results

The pine plantation in ANF compartment 228 had been naturally colonized by *S. invicta* in the mid 1970s following the soil disturbance associated with establishing the plantation (Fig. 1a). But had succession in the intervening 27 years changed this fire ant population? Determination of such effects requires a comparison with similar populations whose habitat has remained unchanged over this period. A survey of a nearby section of unpaved, occasionally-graded roadside served this purpose. In 1985, this section of forest road harbored 27 fire ant colonies in 730 m of roadside, or 37 per km. In the 2012, this same section contained 29 colonies, or 40 per km. Similar roadsides within 5–10 km averaged about 40–50 colonies per km in both 1985 and 2012. These roadside surveys therefore suggest that in the absence of succession, no change in the population density of fire ants was to be expected.

At the time of the 1985 survey of the pine plantation [35], soil disturbance was still apparent (Fig. 1a), and 29 colonies of *S. invicta* were found along 1.6 km of transects, a density of about 20 colonies per km of transect (Fig. 1b). By 2012, 27 years after the original survey, the ground cover had mostly reestablished itself and little soil disturbance was apparent. A resurvey of the same transects detected only 10 colonies (6.3 colonies per km; Fig. 1b), a significant decline (Chi-square = 6.49, d.f. = 1, $P<0.01$). Of the 29 colonies present in 1985, 26 were not associated with vehicle tracks, whereas in 2012, 5 of the 10 colonies were not thus associated. The decline of colonies not associated with vehicle tracks was thus even greater (Chi-square = 10.66, d.f. = 1, $P<0.001$).

During the intervening 27 years, the number of colonies associated with vehicle tracks changed little (3 vs. 5). A survey of a longer section of vehicle track not surveyed by Tschinkel (1988) detected another 6 colonies (Fig. 1b, lower left), indicating that *S. invicta* has a strong preference for the disturbance associated with such tracks and populations can persist there for decades.

Experimental Plots

Fig. 2 shows the changes in appearance of a representative of one of the five experimental sites between 2006, the last year of disturbance treatments, and 2011, a year before the final survey. In 2006, the plowed and mowed plots are clearly visible as pale squares (Fig. 2, upper panel), but by 2011 these disturbances were no longer visible (Fig. 2, lower panel). On the ground, some of the effects of plowing were still discernible in some of the plots in 2012, but most of the mowed plots were no longer distinguishable from the general ground cover. Control plots are not visible in either year because ground cover was not disturbed.

From 2003 to 2006, fire ant abundance in the King and Tschinkel study [19] was assessed annually with pitfall traps and twice-annual mound counts of addition plots. The 2006 condition of the plots provided the starting point for the comparisons in this paper. The analysis below takes into account that about half of the added mature colonies were probably queenless and did not survive long. Only queenright colonies have prospects of longer survival. In 2006, only the addition plots (n = 15) were visually

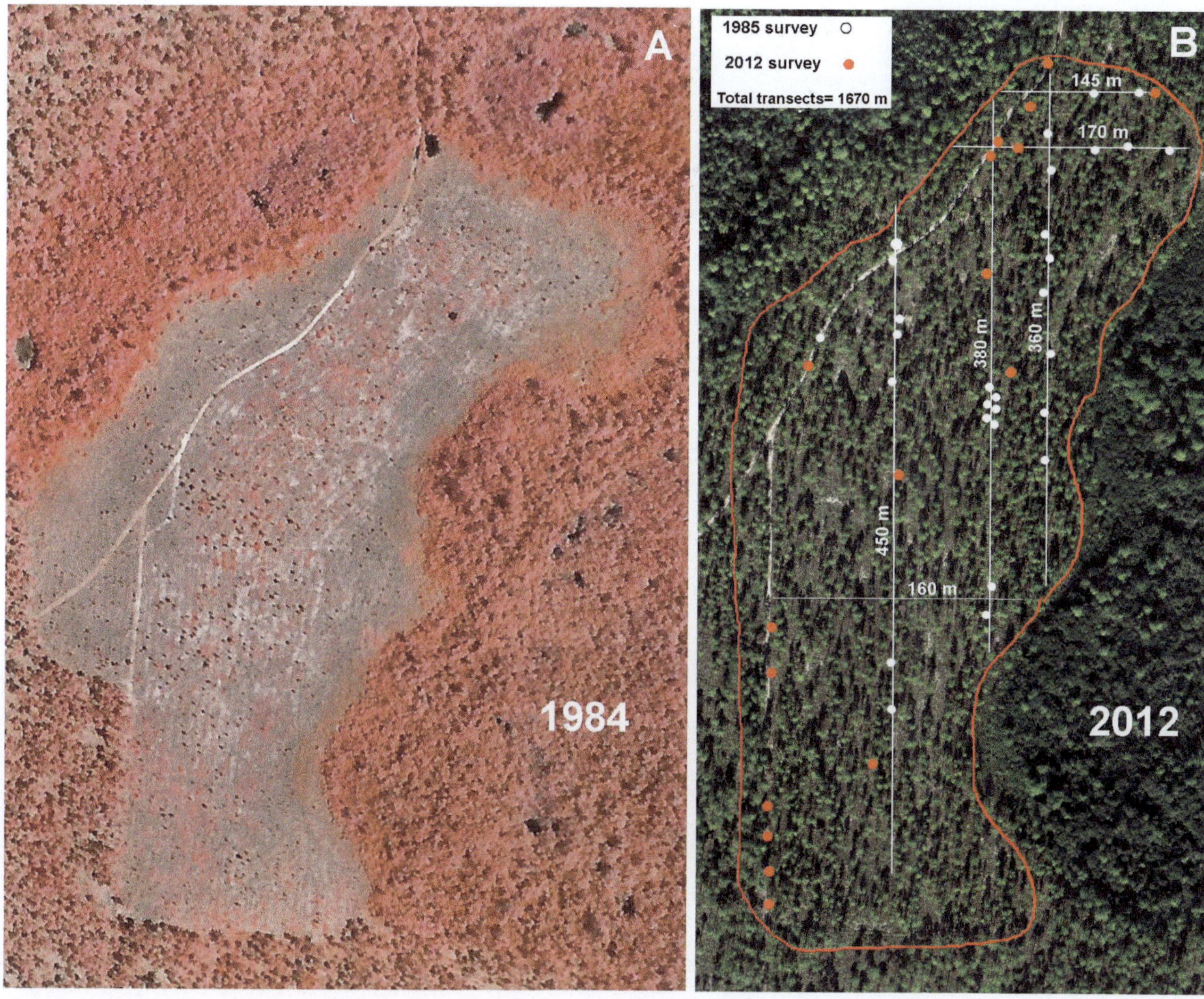

Figure 1. A flatwoods longleaf pine plantation surveyed in 1984 (left panel, false color) and again in 2012 (right panel, true color), showing *S. invicta* colonies detected in 1984 (white symbols) and 2012 (red symbols). Note the furrowed soil and small size of the trees in 1984. White lines indicate transects surveyed both in 1984 and 2012, and the red line shows the boundaries of the12 ha pine plantation established ca. 1975. Transect lengths are indicated. The vehicle track near the western boundary was surveyed only in 2012. Left panel: Aerial photo purchased from USDA Forest Service. Right panel: satellite image from the Land Boundary Information System (www.labins.org/), Hilliardville, FL, True Color DOQQ 5136. Conditions for use are as follows: "Since this data was developed and collected with U.S. Government and the State of Florida funding, no proprietary rights may be attached to it, nor may it be sold to the U.S. Government or the Florida State Government as part of any procurement of products or services."

searched for *S. invicta* mounds, but in 2012, all of the plots were searched (n = 45) to verify the typical density of fire ants throughout the entire area under study.

Adding fire ant colonies to plots greatly increased their abundance. In 2006, the last year of the study, addition plots hosted a mean of 10.6 colonies (95% CI: 8.15–13.1), and had a mean of 24 *S. invicta* per pitfall, whereas the non-addition plots averaged 1.5 *S. invicta* per pitfall (ANOVA, addition treatments: $F_{2,1611} = 630.2$; $P < 0.0000001$; Table 1; for details, see King and Tschinkel [19]). This 16-fold difference in trap catch suggests a parallel difference in mound counts (although these were not counted in 2006).

By 2012, six years later, the mound count in the addition plots had decreased 74% (s.d. = 26%), from 10.6 to only 2.3 colonies (95% CI: 0.92–3.8) (*t*-test by year, *t*-value = 6.27; d.f. = 28; $P < 0.0001$; Table 2). No pitfall trapping was carried out in 2012, but in light of the decrease in mound counts, a large decrease in pitfall catch would have occurred.

Mound counts for 2006 can be approximated indirectly from pitfall captures. In 2012 there was no significant difference in mound numbers among any of the treatments, addition or non-addition (ANOVA), which means that by 2012 the addition plots had declined to the same level as the non-addition plots (mean 2012 non-addition = 1.61 colonies, s.d. = 0.97; one outlier was deleted, more on that below) Non-addition plots consisted of control plots with 0.7 colonies and mow plots with 1.3 colonies. Thus between 1 and 2 colonies per plot can be regarded as the "natural" background density, a density that produced about 1.5 *S. invicta* per pitfall in 2006. Therefore, supplemented fire ant populations returned to background densities, whether measured as mound counts or pitfall catches. For the addition plots in 2012, the background abundance of 1–3 *S. invicta* per pitfall represents a

Figure 2. Satellite views modified from U.S. Geological Survey images of a representative of the five experimental sites in 2006 (upper panel) and in 2011 (lower panel). The plowed and mowed 40×40 m plots are clearly visible as pale squares in 2006, but not in 2011, indicating that most of the effects of disturbance had disappeared. On the ground, some effects of plowing were still visible in some plots in 2012, but most mowed plots were no longer distinguishable from the surrounding ground cover. Undisturbed controls are not visible in either year. For conditions for use, see Fig. 1.

large decrease from about 24 per trap in 2006. These background densities are probably attributable to the pervasive presence of small scale soil disturbance from vehicles, "stump pulls," "borrow pits" and other forest service and recreational activity in this forest.

Table 1. ANOVA for number of *S. invicta* in pitfalls in 2006 in relation to habitat manipulation and addition of fire ant colonies.

Factor	SS	d.f.	MS	F	p
Intercept	128587	1	128587	912	<0.000001
Disturbance	1265	2	632	4.49	<0.02
Fire ants	177687	2	88843	630	<0.000001
Disturbance *Fire ants	2230	4	557	3.96	<0.004
Error	227092	1611	141		

The significant interaction term is the result of higher fire ant numbers in non-addition plowed plots.

The lack of persistence of *S. invicta* at even moderate densities in later successional stage native longleaf pine savanna habitat is emphasized by the fact that colonies had to be added every year in order to maintain the *S. invicta* densities called for in the experiment (Fig. 3). Not surprisingly, more colonies were added in 2004 than the other two years (means = 24, 13, 18 respectively; repeated measures ANOVA, effect of year, $F_{2,24} = 10.1063$, $P < 0.001$). Live mounds were counted each year just before the additions were made, and the number added adjusted accordingly (about half the colonies were probably queenright and had no prospects of surviving long, see Methods). Through the initial addition of 24 colonies per plot in 2004, mean colony counts increased from 4.8 to 10.8 in 2005 (Table 2). Addition of another 13 colonies per plot in 2005 increased the counts to a mean of 11.6 in 2006 (effect of year on counts: repeated measures ANOVA, $F_{2,24} = 23.53$, $P < 0.00001$). However, when habitat treatments are considered, additions increased the counts in mow and plow plots more than in control plots, so that by 2005 and 2006, mow and plow counts were significantly higher (means 12–13) than controls (means 8–10) (year by treatment interaction, $F_{4,24} = 3.112$, $p < 0.04$). This was in spite of 15 to 26 colonies being added to control plots in 2005 and 2006, but only 12 to 14 to the mow and plow plots. In other words, losses of transplanted colonies were

Table 2. Mound counts in the "addition plots" in 2006 and 2012.

compartment	plot	treatment	total count, 2006	total count, 2012	change, 2006–2012	% change, 2006–2012
231	1	control	13	1	−12	−92%
232	1	control	14	7	−7	−50%
245-13	1	control	9	0	−9	−100%
245-50	1	control	3	2	−1	−33%
246	1	control	4	2	−2	−50%
231	3	mow	21	0	−21	−100%
232	3	mow	8	2	−6	−75%
245-13	3	mow	7	1	−6	−86%
245-50	3	mow	14	1	−13	−93%
246	3	mow	10	1	−9	−90%
231	6	plow	10	1	−9	−90%
232	6	plow	10	9	−1	−10%
245-13	6	plow	10	1	−9	−90%
245-50	6	plow	13	4	−9	−69%
246	6	plow	13	3	−10	−77%

Mounds were not counted in "non-addition" plots in 2006.

higher in the control plots than in the disturbed plots, requiring more annual additions, but even then not achieving the same colony counts in controls as in the other two treatments. This suggests that undisturbed native habitat is less congenial to fire ant persistence than disturbed habitat.

Projecting forward to 2012, even the three repetitions of plowing did not alter the habitat enough to allow *S. invicta* colonies to persist for 6 years, so whatever it was about disturbance that improved year-to-year persistence had disappeared within 6 years.

An outlier

Plot 7 in compartment 232 was an interesting exception to the general pattern. It was a control-add soil treatment in which no *S. invicta* were detected in pitfalls in 2006 (Fig. 4, left). However, when surveyed for mounds in 2012, it contained 26 mounds (Fig. 4,

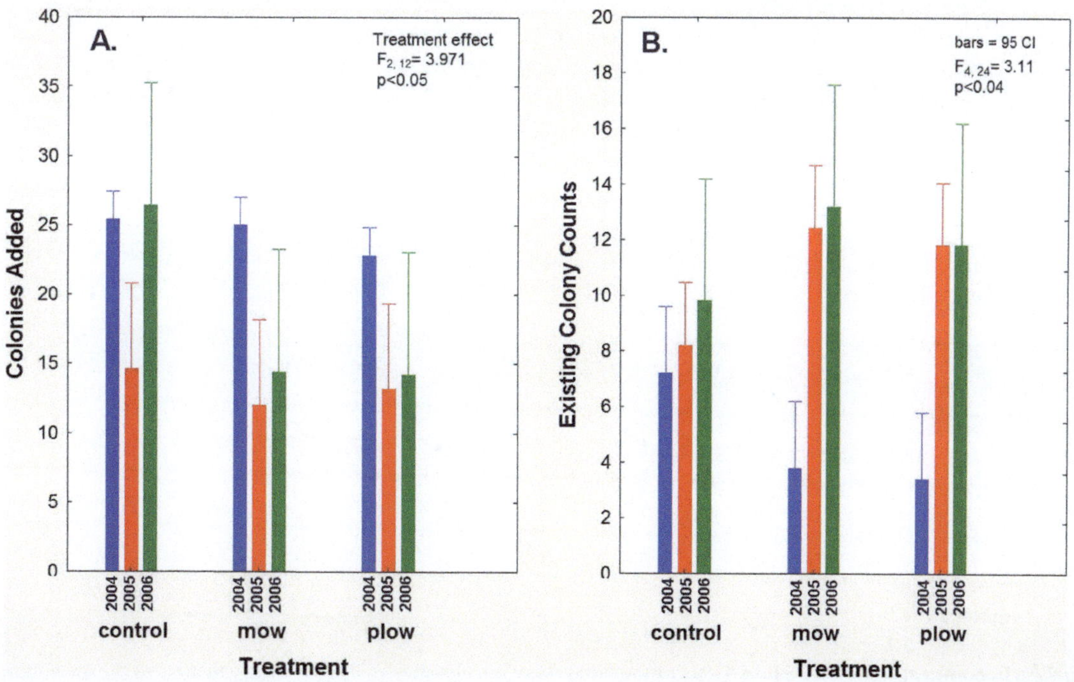

Figure 3. Mature colony transplantations. A. Number of colonies added to treatment plots by year. Initial additions were the same for all plots, but controls required more additions in subsequent years. B. Existing colonies were counted before additions. Persistence of transplanted colonies was higher in the mow and plow plots than in the controls.

right). On a sunny, cool morning in February 2013, 18 mounds in and near plot 7 were checked for queen number by scattering mound soil in large photo trays and inspecting for queens. Of the 18 mounds checked, 11 were confirmed as polygyne by the detection of 2 to 5 queens that were both physogastric and attractive to workers. Of the remaining 7 nests, a single queen was found in 2 and no queen in 5. However, almost all nests, whether queens were seen or not, had the small, pale workers and absence of sexual brood that is characteristic of polygyne colonies, and were probably polygyne. Polygyny was probably wide-spread in this site in general, as several mounds were present in other non-addition plots (Fig. 4, right).

Polygyne colonies were unknown in the Tallahassee region until recently, but since their arrival, they have been spread by the county road department during road maintenance [38]. Polygyne colonies are not known to reproduce successfully through mating flights and independent founding [39]. It is possible that some of the colonies we added were polygyne, although it seems unlikely given the complete absence of fire ants in 2006 pitfalls and colony surveys. Whether they will persist for long is an interesting question.

Discussion

Our data show that monogyne *S. invicta* can persist in this native habitat at only low levels and often in association with small-scale soil disturbance. About half of the non-persistence from year-to-year was probably because half of the transplanted nests in the previous study were queenless [19]. However, even considering that approximately only half the colonies were queenright, persistence was much less than 100%. Thus, over the three years, 65 colonies were added to the control plots, of which perhaps 30 were queenright, but by 2006, only 10 colonies (one-third of queenright) were present. Similarly, about 50 colonies were added to the mow and plow plots. Assuming about 25 were queenright, only 12 to 13 persisted in 2006, a survival rate of about 50%. However this was a better survival rate than resulted in King and Tschinkel's [20] study. They added 234 colonies to experimental plots from 2004-6 in a longleaf pine savanna ecosystem on Brickyard soil (a perched clay layer, as opposed to the pure sandy spodosols of this study) and only 66 colonies were found in the addition plots at the end of the study. In King and Tschinkel's [20] study, fire ants were already present in higher densities (6 colonies, on average, per 40×40 m plot in control plots). This suggests a survival rate of around 15% for added colonies, much lower than the present study, which is likely explained by the poor habitat (for ants) in the King and Tschinkel [20] study.

The non-persistence that is a common theme in all of these studies may in part have been due to the difficulties of establishing a new nest and territory in the new environment, leading directly to colony death. Of colonies that established successfully and persisted for at least a year, longer-term non-persistence was most likely the result of a lack of recruitment— planted colonies probably lived out their life span (average 7–8 years, [40,41], but after they died were not replaced by newly founded colonies. Many of the planted colonies will not have been in the bloom of youth, so that colony mortality was probably continuous during the six year interval. The age of transplanted colonies was unknown, but assuming they were sampled from a stable population, probably averaged 3–4 yr with a right-skewed age distribution [23].

The fact that mature colonies experimentally transplanted to undisturbed habitat in these same forests can persist for more than a year, but mostly vanish within 6 years suggests that the road block to persistence is associated with colony reproduction, i.e. recruitment. Recruitment itself can be dissociated into the habitats

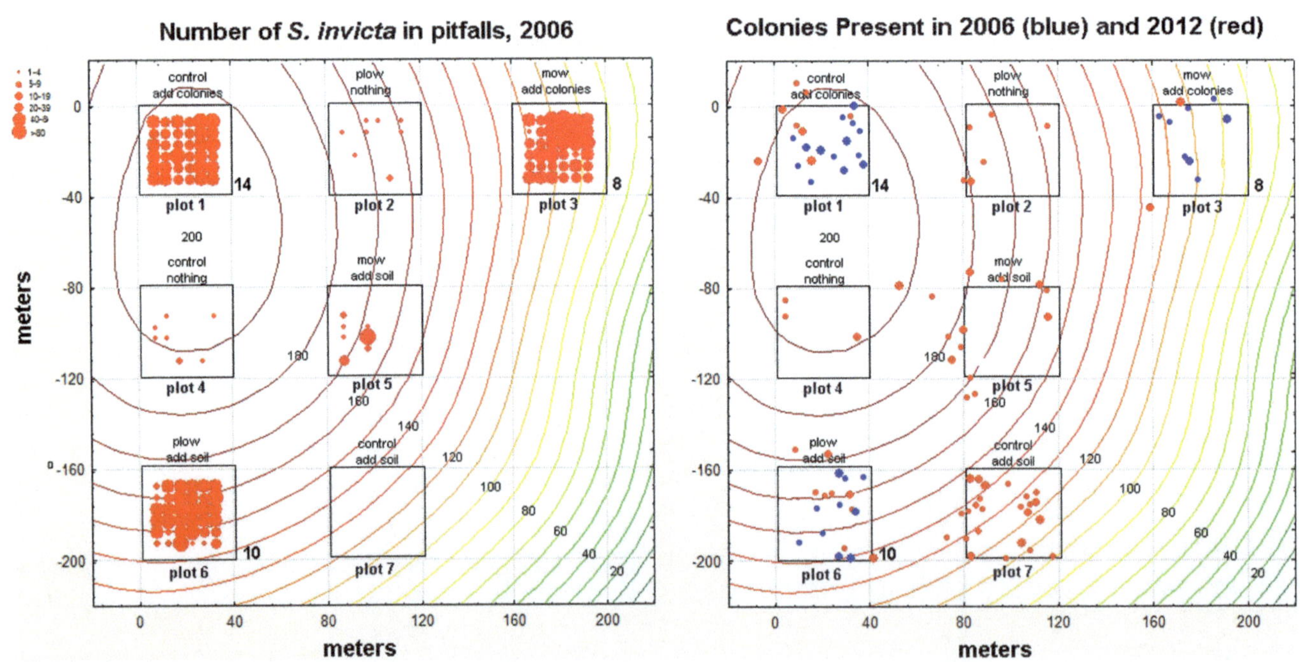

Figure 4. Presence of *S. invicta* in compartment 232, plots 1–7. Contour lines show the elevation (cm) above the wetlands to the lower right and bottom. Boldface numbers next to plots 1, 3 and 6 are the number of fire ant mounds present in 2006. Left panel: *S. invicta* abundance in pitfalls in 2006. The size of the symbol indicates the number captured (legend at upper left). Right panel: mounds detected by survey in 2006 (blue symbols) and 2012 (red symbols). Although only "add colonies" plots were surveyed for mounds in 2006, the low 2006 capture rates in pitfalls of "non-addition" plots suggests that few mounds were present in those plots. Survey of plots 6 and 7 in 2013 found most of the colonies to be polygyne.

that mated queens choose to settle in after mating and their success in founding a colony in the chosen habitat. Newly mated *S. invicta* queens show a strong preference for disturbed habitat and fail to found successfully in native habitat (King and Tschinkel, unpublished data). That is, queens choose to settle in habitat in which they can successfully found new colonies. Just what habitat characteristics produce this difference is unknown, but should be the focus of future studies because the combination of preference and success is a fundamental process in the assembly and evolution of communities. Morrison [15] observed a decline in fire ant densities, possibly related to succession and Morrison and Porter [42] observed a positive association between fire ant densities and generalized ant and arthropod diversity further suggesting that habitat quality is a key feature pf early successional habitats for many insect species.

In a sense, the persistence and possibly novel appearance of polygyne *S. invicta* in compartment 232 supports the hypothesis that the observed failure of colony founding by newly mated monogyne queens in native habitat is a major barrier to persistence. Because polygyne *S. invicta* reproduce by colony fission, they avoid this bottleneck and persist. How long they can persist is an open question. Even the massive polygyne invasion described by Porter and Savingano [43] had largely abated within 10 years [15].

The habitat disturbance caused by mowing and plowing faded over the six elapsed years, although the effects of plowing on the ground cover were still visible. Colony-founding queens of *S. invicta* are attracted to disturbed habitats (unpublished data), but are apparently either unable to establish colonies even in the plowed plots, or the colonies did not survive for 6 years. Whatever the cause, it is clear that even though *S. invicta* colonies are abundant along graded forest roads, they rarely establish and/or persist within the adjacent native habitat. Even those that do persist within the native ground cover are often (but not always) associated with small disturbances that have exposed bare soil (unpublished observations). Human incursions into this habitat are many— hunters, off-road vehicles, old logging scars, old drainage ditches and so on.

Longer-term, successional decline of fire ant populations similar to our pine plantation study has also been reported by LeBrun et al. [7] whose south Texas study showed in a chronosequence that fire ant abundance decreased with the time since disturbance. Similarly, ant species typical of early succession declined in reclaimed bauxite mine sites as they aged and revegetated [6]. The fact that early succession habitats are populated by distinct communities of ants that are different than those of late successional stages [6,8,9] suggests that if succession were allowed to take place, many of the ant species in the early-stage communities would decline or disappear.

Studies of unusual and persistent occurrences of fire ants populations in a few types of intact ecosystems lacking anthropogenic disturbance provide further insight into the attributes of altered habitats that favor fire ants. Specifically, moist habitats with high water tables, little or no canopy, clay or loamy soils, and frequent natural disturbances such as seasonal flooding, and fires [23,35,44] sometimes support fire ant populations averaging approximately 1/6 the densities of pastures and roadsides in the same region [17,20,40]. Wetland margins, prairies, and frequently flooded areas within some longleaf pine forests are the most common examples [17,20]. Such natural habitat disturbances may act as barriers to most native species while simultaneously favoring fire ants [20], and may be somehow quite similar to the habitat features created by anthropogenic disturbances. Additional support for this comes from Stuble et al. [21] who treated plots in native habitat that had substantial densities of both fire ants and native ants with poison baits, reducing all ant densities. Fire ants did not preferentially recolonize these treated plots.

The association of *S. invicta* with mostly human-caused disturbance has been amply demonstrated [23]. The focus now needs to shift to the question of just what life history characteristics are matched to the characteristics of disturbed habitats, and what behavioral and physiological attributes of the ant allow it to seek out, colonize and persist in such habitats. We have shown that populations self-founded in disturbed Florida coastal plain pine forest dwindle as the habitat recovers from the disturbance and reverts to a condition approximating the original undisturbed state. This suggests that certain characteristics of the late-succession habitat are not congenial to monogyne *S. invicta* survival or recruitment, or both. The transition from congenial to not congenial can be as narrow as a few meters between the edge of a maintained forest dirt road and the adjoining native ground cover.

Acknowledgments

We are grateful to Christina Kwapich, David Smiley and Zach Stivers for their uncomplaining help with surveys under trying Florida summer conditions.

Author Contributions

Conceived and designed the experiments: WRT JRK. Performed the experiments: WRT JRK. Analyzed the data: WRT JRK. Contributed reagents/materials/analysis tools: WRT JRK. Wrote the paper: WRT JRK.

References

1. Walker LR, Chapin FS (1987) Interactions among processes undergoing successional change. Oikos 50: 131–135.
2. Brown VK, Hyman PS (1986) Successional Communities of Plants and Phytophagous Coleoptera Journal of Ecology. 74: 963–975.
3. Hendrix SD, Brown VK, Dingle H (1988) Arthropod guild structure during early old field succession in a new and old world Site. Journal of Animal Ecology. 57: 1053–1065.
4. Siemann E, Haarstad J, Tilman D (1999) Dynamics of plant and arthropod diversity during old field succession. Ecography. 22: 406–414.
5. Steffan-Dewenter I, Tscharntke T (2001) Succession of bee communities on fallows. Ecography 24: 83–93.
6. Majer JD, Nichols OG (1998) Long-term recolonization patterns of ants in western Australian rehabilitated bauxite mines with reference to their use as indicators of restoration success. Journal of Applied Ecology, Vol. 35: 161–182.
7. LeBrun EG, Plowes RM, Gilbert LE (2012) Imported fire ants near the edge of their range: disturbance and moisture determine prevalence and impact of an invasive social insect Journal of Animal Ecology 81: 884–895.
8. Wetterer JK, Espadaler X, Wetterer AL, Cabra SGM (2004) Native and exotic ants of the Azores (Hymenoptera: Formicidae). Sociobiology 44:1–19.
9. King JR, Andersen AN, Cutter AD (1998) Ants as bioindicators of habitat disturbance: validation of the functional group model for Australia's humid tropics. Biodiversity and Conservation. 7: 162–1638.
10. Cherry RH, Nuessly GS (1992) Distribution and abundance of imported fire ants (Hymenoptera: Formicidae) in Florida sugarcane fields. Environ. Entomol. 21:767–770.
11. Lofgren CS, Banks WA, Glancey BM (1975) Biology and control of imported fire ants. Ann. Rev. Entomol. 20:1–30.
12. Markin GP, O'Neal J, Collins HL (1974) Effects of mirex on the general ant fauna of a treated area in Louisiana. Environ. Entomol. 3:895–898.
13. Phillips SA, Jr., Rogers WM, Wester DB, Chandler L (1987) Ordination analysis of ant faunae along the range expansion front of the red imported fire ant in south-central Texas. Texas Journal of Agricultural and Natural Resources 1:11–15.
14. Stimac JL, Alves SB (1994) Ecology and biological control of fire ants. p. 353–380 in Pest Management in the Subtropics, Biological Control—A Florida Perspective. D. Rosen, F. D. Bennett and J. C. Capinera, editors. Intercept Ltd., Andover, New Hampshire.

15. Morrison LW (2002) Long-term impacts of an invasion of an arthropod community by the imported fire ant, *Solenopsis invicta*. Ecology 83:2337–2345.
16. Lebrun EG, Plowes RM, Gilbert LE (2012) Imported fire ants near the edge of their range: disturbance and moisture determine prevalence and impact of an invasive social insect. J. Anim. Ecol. doi: 10.1111/j.1365-2656.2012.01954.x
17. Stuble KL, Kirkman LK, Carroll CR (2009) Patterns of abundance of fire ants and native ants in a native ecosystem. Ecological Entomology 34:520–526.
18. King JR, Tschinkel WR (2006) Experimental evidence that the introduced fire ant, *Solenopsis invicta*, does not competitively suppress co-occurring ants in a disturbed habitat, J. Anim. Ecol. 75: 1370–1378
19. King JR, Tschinkel WR (2008) Experimental evidence that human impacts drive fire ant invasions and ecological change. PNAS 105:20339–20343.
20. King JR, Tschinkel WR (2013) Experimental evidence for weak effects of fire ants in a naturally invaded pine-savanna ecosystem in north Florida. Ecol. Entomol. 38: 68–75
21. Stuble KL, Kirkman LK, Carroll CL, Sanders NJ (2011) Relative effects of disturbance on red imported fire ants and native ant species in a longleaf pine ecosystem. Conservation Biology. 25: 618–622
22. Porter SD, Fowler HG, Mackay WP (1992) Fire ant mound densities in the United States and Brazil (Hymenoptera: Formicidae). J. Econ. Entomol. 85, 1154–1161.
23. Tschinkel WR (2006) *TheFire Ants*. Belknap/Harvard Univ. Press
24. Camilo GR, Phillips SA Jr (1990) Evolution of ant communities in response to invasion by the fire ant *Solenopsis invicta*. Pages 190–198 in Applied Myrmecology: A World Perspective. R. K. Vander Meer, K. Jaffe, and A. Cedeno, editors. Westview Press, Boulder, Colorado.
25. Claborn DM, Phillips SA Jr (1986) Temporal foraging activities of *Solenopsis invicta* (Hymenoptera: Formicidae) and other predominant ants of central Texas. Southwestern Nat. 31:555–557.
26. Zettler JA, Taylor MD, Allen CR, Spira TP (2004) Consequences of forest clear-cuts for native and nonindigenous ants (Hymenoptera: Formicidae). Ann. Entomol. Soc. Amer. 97:513–518.
27. Cherry R (2001) Interrelationship of ants (Hymenoptera: Formicidae) and southern chinch bugs (Hemiptera: Lygaeidae) in Florida lawns. J. Entomol. Sci. 36: 411–415.
28. Porter SD (1993) Stability of polygyne and monogyne fire ant populations (Hymenoptera: Formicidae: Solenopsis invicta) in the United States. J. Econ. Entomol. 86: 1344–1347.
29. Wojcik DP, Allen CR, Brenner RJ, Forys EA, Jouvenaz DP (2001) Red imported fire ants: impact on biodiversity. Amer. Entomol. 47, 16–23.
30. Howard FW, Oliver AD (1979) Field observations of ants (Hymenoptera: Formicidae) associated with red imported fire ants, *Solenopsis invicta* Buren, in Louisiana pastures. J. Georgia Entomol. Soc. 14: 259–263.
31. Apperson CS, Powell EE (1984) Foraging activity of ants (Hymenoptera: Formicidae) in a pasture inhabited by the red imported fire ant. Florida Entomol. 67: 383–393.
32. Phillips SA Jr, Jones SR (1986) Temporal foraging patterns of *Solenopsis invicta* and native ants of central Texas. Fire ants and leaf cutting ants: Biology and management. C. S. Lofgren and R. K. Vander Meer. Boulder, CO, Westview Press: 114–122.
33. Fraelich BA (1991) Interspecific competition for food between the red imported fire ant, *Solenopsis invicta*, and native Floridian ant species in pastures, M.S thesis, University of Florida, 158.
34. Stiles JH, Jones RH (1998) Distribution of the red imported fire ant, *Solenopsis invicta*, in road and powerline habitats. Landscape Ecol. 335:335–346.
35. Tschinkel WR (1988) Distribution of the fire ants *Solenopsis invicta* and *S. geminata* in north Florida in relation to habitat and disturbance. Ann. Entomol. Soc. Amer. 81: 76–81.
36. Wojcik DP (1983) Comparison of the Ecology of Red Imported Fire Ants in North and South America. Florida Entomol. 66: 101–111.
37. Pass BC (1960) Bionomics of the imported fire ant, *Solenopsis saevissima richteri* Forel. M.S. Thesis, Auburn Univ., Auburn, Alabama. 54.
38. King JR, Tschinkel WR, Ross KG (2009) A case study of human exacerbation of the invasive species problem: transport and establishment of polygyne fire ants in Tallahassee, Florida, USA. Biol. Invasions 11:373–377.
39. Ross KG, Keller L (1995) Ecology and evolution of social organization: insights from fire ants and other highly eusocial insects. Ann. Rev. Ecol. System. 26:631–656.
40. Tschinkel WR (1987) Fire ant queen longevity and age: estimation by sperm depletion. Ann. Entomol. Soc. Amer. 80: 263–266.
41. Tschinkel WR, Porter SD (1988) The efficiency of sperm use in queens of the fire ant, *Solenopsis invicta* Buren. Ann. Entomol. Soc. Amer. 81: 777–781.
42. Morrison LW, Porter SD (2003) Positive association between densities of the red imported fire ant, Solenopsis invicta (Hymenoptera: Formicidae), and generalized ant and arthropod diversity. Eviron. Entomol. 32: 548–554.
43. Porter SD, Savignano DA (1990) Invasion of polygyne fire ants decimates native ants and disrupts arthropod communities. Ecol. 71:2095–2106.
44. Lubertazzi D, Tschinkel WR (2003) Ant community change across a ground vegetation gradient in north Florida's longleaf pine flatwoods. J. Insect Sci. 3, 21.

10

Global Trophic Position Comparison of Two Dominant Mesopelagic Fish Families (Myctophidae, Stomiidae) Using Amino Acid Nitrogen Isotopic Analyses

C. Anela Choy[1]*, Peter C. Davison[2], Jeffrey C. Drazen[1], Adrian Flynn[3,4], Elizabeth J. Gier[5], Joel C. Hoffman[6], Jennifer P. McClain-Counts[7], Todd W. Miller[8], Brian N. Popp[5], Steve W. Ross[9], Tracey T. Sutton[10]

1 University of Hawaii, Department of Oceanography, Honolulu, Hawaii, United States of America, **2** Scripps Institution of Oceanography at University of California, San Diego, California, United States of America, **3** The University of Queensland, School of Biomedical Sciences, Brisbane, Australia, **4** Commonwealth Scientific and Industrial Research Organization, Marine and Atmospheric Research, Hobart, Australia, **5** University of Hawaii, Department of Geology & Geophysics, Honolulu, Hawaii, United States of America, **6** U.S. Environmental Protection Agency Office of Research and Development, National Health and Environmental Effects Research Laboratory, Mid-Continent Ecology Division, Duluth, Minnesota, United States of America, **7** United States Geological Survey, Southeastern Ecological Science Center, Gainesville, Florida, United States of America, **8** Ehime University, Global Center of Excellence, Center for Marine Environmental Studies, Matsuyama, Japan, **9** University of North Carolina at Wilmington, Center for Marine Sciences, Wilmington, North Carolina, United States of America, **10** College of William & Mary, Virginia Institute of Marine Science, Gloucester Point, Virginia, United States of America

Abstract

The $\delta^{15}N$ values of organisms are commonly used across diverse ecosystems to estimate trophic position and infer trophic connectivity. We undertook a novel cross-basin comparison of trophic position in two ecologically well-characterized and different groups of dominant mid-water fish consumers using amino acid nitrogen isotope compositions. We found that trophic positions estimated from the $\delta^{15}N$ values of individual amino acids are nearly uniform within both families of these fishes across five global regions despite great variability in bulk tissue $\delta^{15}N$ values. Regional differences in the $\delta^{15}N$ values of phenylalanine confirmed that bulk tissue $\delta^{15}N$ values reflect region-specific water mass biogeochemistry controlling $\delta^{15}N$ values at the base of the food web. Trophic positions calculated from amino acid isotopic analyses (AA-TP) for lanternfishes (family Myctophidae) (AA-TP ~2.9) largely align with expectations from stomach content studies (TP ~3.2), while AA-TPs for dragonfishes (family Stomiidae) (AA-TP ~3.2) were lower than TPs derived from stomach content studies (TP~4.1). We demonstrate that amino acid nitrogen isotope analysis can overcome shortcomings of bulk tissue isotope analysis across biogeochemically distinct systems to provide globally comparative information regarding marine food web structure.

Editor: Myron Peck, University of Hamburg, Germany

Funding: This research was funded by National Science Foundation (NSF) grant # OCE-1041329 to B.N.P. and J.C.D., NSF grant # OCE-0623551 to T.T.S., and by the Pelagic Fisheries Research Program via Cooperative Agreement NA17RJ123 between the Joint Institute for Marine and Atmospheric Research and the National Oceanic and Atmospheric Administration to J.C.D. and B.N.P. Field collections and data analyses for GOM samples were supported by grants to S.W.R. by the Department of the Interior U.S.G.S. (Cooperative Agreement 05HQAG0009, subagreements 05099HS004 and 5099HS0013). The funders had no role in study design, data collection and analysis, decision to publish, or preparation of the manuscript.

Competing Interests: The authors have declared that no competing interests exist.

* E-mail: cachoy@hawaii.edu

Introduction

Deep oceanic waters (offshore depths >~200 m) constitute the largest habitat on the planet. Industrialized fishing has substantially reduced the biomass of large predatory fishes (e.g., tunas, billfishes, sharks) within these deep ocean ecosystems [1]. There is growing evidence that overharvesting of these top trophic level animals may ultimately affect the stability and resilience of marine food webs through changes in system structure and function (e.g., [2,3]). Improved understanding of trophic structure and food web interactions at a time of changing climate dynamics is critical for anticipating future changes in exploited marine populations. Particularly important is the need for comparative evaluation of potential fishery impacts on a global scale across biogeochemically and ecologically diverse systems.

Large-scale marine trophodynamics have traditionally been derived from stomach content (SC) analyses and more recently using stable isotope and fatty acid analyses. However, synthesizing multiple SC and/or biochemical datasets to compare ecosystem function between different oceanic regions can be difficult and is infrequently done. For the first time, we utilize a promising and emergent tool, compound-specific nitrogen isotope analysis of individual amino acids (CSIA), to compare the trophic positions (TPs) of widespread pelagic micronekton fishes from five biogeochemically distinct global ecosystems: Tasman Sea, California (CA) Current, Gulf of Mexico (GOM), northern Mid-Atlantic Ridge (MAR), and the North Pacific Subtropical Gyre (NPSG) near Hawaii.

In pelagic ecosystems, micronekton (small fishes, squids, and crustaceans ~2–20 cm in size) are a critical trophic link between primary producers and higher trophic level consumers (e.g., tunas, seabirds, marine mammals). Dragonfishes (family Stomiidae) are considered the most diverse and numerically important higher-trophic level predatory meso- and bathypelagic fish group, while lanternfishes (family Myctophidae) are commonly the dominant micronekton organisms in terms of biomass and abundance in mesopelagic ecosystems (e.g., [4,5]), and are thought to be the primary prey of most dragonfishes (e.g., [6,7]). Widespread distributions and high biomass levels coupled with extensive diel vertical migrations suggest that these fishes are important mediators in the transfer of organic carbon between trophic levels and through a large part of the water column [8], often including benthic communities at continental margins [9].

Carbon (C) and nitrogen (N) stable isotope (SI) techniques have been extensively used in aquatic and terrestrial ecosystems, complimenting SC analyses by delineating TPs and tracing energy/nutrient flows [10,11]. The basic premise underlying these studies is that preferential incorporation of ^{15}N and ^{13}C in consumer tissues results in predictable ~2.0–3.4‰ increases in $\delta^{15}N$ values and ~0.5–0.8‰ increases in $\delta^{13}C$ values relative to their prey at each subsequent trophic level [12,13]. Inferring trophic connectivity from SI data requires sampling across multiple TPs, often a considerable logistic challenge in deep ocean systems. Ecological interpretation of SI data is often complicated by the inability to constrain temporal and spatial variability in the isotopic compositions of primary producers at the food web base [14]. In marine ecosystems like the NPSG for example, primary producers can seasonally switch between N_2-fixation and upwelled nitrate-based production [15]. The $\delta^{15}N$ values for atmospheric N_2 ($\delta^{15}N = 0‰$) and inorganic deep-water nitrate ($\delta^{15}N = \sim 5–7‰$) sources are distinct (e.g., [16]), and these differences are reflected in a consumer's N isotopic composition [17].

Compound-specific isotope analysis of individual amino acids (AAs) is a developing technique that overcomes many of the limitations of bulk SI analysis. Instead of attempting to concurrently sample organisms representing multiple TPs in a food web, the CSIA approach uses the $\delta^{15}N$ values of AAs of a consumer to constrain food web baseline isotopic variability and estimate TPs [18]. Laboratory experiments by McClelland and Montoya [19] demonstrated that certain "source" AAs (after [20]) (e.g., phenylalanine, glycine) fractionate very little with trophic processing and are indicative of the isotopic composition of the food web base. Other "trophic" AAs (e.g., glutamic acid, alanine) involved in transamination and deamination reactions undergo significant enrichment in ^{15}N (~7‰ per trophic level) and are thus indicative of the fractional TP of the consumer [21]. Using this approach, consumer TP can be estimated using a reasonably well-established relationship between trophic and source AAs [18], providing valuable information that can be utilized by ecosystem modelers and managers alike.

Many previous studies have successfully combined bulk SI and CSIA datasets across diverse phyla to demonstrate the advantages of the CSIA approach over traditional SI analysis (e.g., [20] in tuna, [17] in marine copepods, [22] in elasmobranchs). However, no previous studies have applied this approach across global marine ecosystems, and few have provided comparative information from multiple TPs. In this study, we conducted the first cross-system trophic comparison of two dominant marine fish consumer groups with well-characterized and distinct TPs across five unique biogeochemical regions. Although the TPs of the two fish groups appear consistent between regions based on available SC analyses, considerable regional variability in bulk tissue $\delta^{15}N$ ($\delta^{15}N_{bulk}$) values exists. Our CSIA data demonstrate that regional biogeochemistry directly influences fish $\delta^{15}N_{bulk}$ values and suggest that across five global oceanic regions lanternfishes and dragonfishes may not be separated by a whole trophic level, which has implications for the exploited status of large marine ecosystems.

Materials and Methods

Sample Collection and Preparation

Fish specimens were independently collected during 2007–2011 by five research groups (one group per region) using a variety of midwater trawling equipment in five distinct regions (Figure 1; Table 1). For each region, fish species known to represent two distinct TPs (one species each of lanternfish and dragonfish) were carefully selected using existing SC data, were identified to the species level and measured (standard (SL) or total (TL) length), and frozen at sea until analysis (sample sizes in Table 1). Due to limited sample availability two dragonfish species were analyzed for the NPSG region, one of which was also sampled in the Tasman Sea and the GOM (Table 1). All species selected had region-specific SC data supporting the interpretation that the lanternfishes were zooplanktivorous (TP ~3) and the dragonfishes were piscivorous (TP ~4) (Table S1). In the laboratory, scales and skin were removed and white muscle tissue dissected from each specimen. Samples were oven-dried at ~60°C for ~48 hrs, ground and homogenized with a mortar and pestle, and shipped to the University of Hawaii (UH) for analysis. Tissue homogenates were split; splits were weighed and packaged into either tin capsules for bulk tissue SI analysis or combusted glass reaction vials for CSIA. This study was carried out in accordance with the animal use protocols of the University of Hawaii (protocol #10-984) and was approved by the UH Institutional Animal Care & Use Committee.

Bulk Tissue Stable Isotope Analysis

Bulk SI analyses were performed at UH using an isotope ratio mass spectrometer (Delta^{Plus}XP) coupled to an elemental analyzer (ConFlo IV/Costech ECS 4010). Isotopic values are reported in conventional δ-notation relative to the international standards atmospheric N_2 and V-PDB, for N and C respectively. Accuracy and precision were <0.2‰ and were calculated using in-house reference materials analyzed every 10 samples (glycine and a tuna tissue homogenate, extensively characterized using NIST certified reference materials and verified independently in other isotope laboratories). Bulk tissue C isotope ($\delta^{13}C_{bulk}$) values were corrected for lipid contribution using isotope mass balance based on deep-sea fish [23]. Tissue mass was limiting for some samples from two regions so previously measured $\delta^{15}N_{bulk}$ and $\delta^{13}C_{bulk}$ values for all samples from these regions were used (CA Current, $n = 6$, lanternfishes; GOM, $n = 10$, both fish groups). SI analyses for the CA Current followed Nam et al. [24] and analyses for the GOM followed McClain-Counts [25]. Of these 16 samples, enough tissue was available from nine samples to determine good agreement between analyses conducted at UH and those at other laboratories (Figure S1).

TPs from Bulk Stable N Isotope Data

TP is commonly estimated using $\delta^{15}N_{bulk}$ values (TP_{bulk}) of consumers and their prey (e.g., [26]). We estimated TP_{bulk} using the following equations and regional data from the literature (see Table S3):

$$TP_{bulk} = 1 + \frac{(\delta^{15}N_{consumer} - \delta^{15}N_{POM})}{TEF} \quad (1)$$

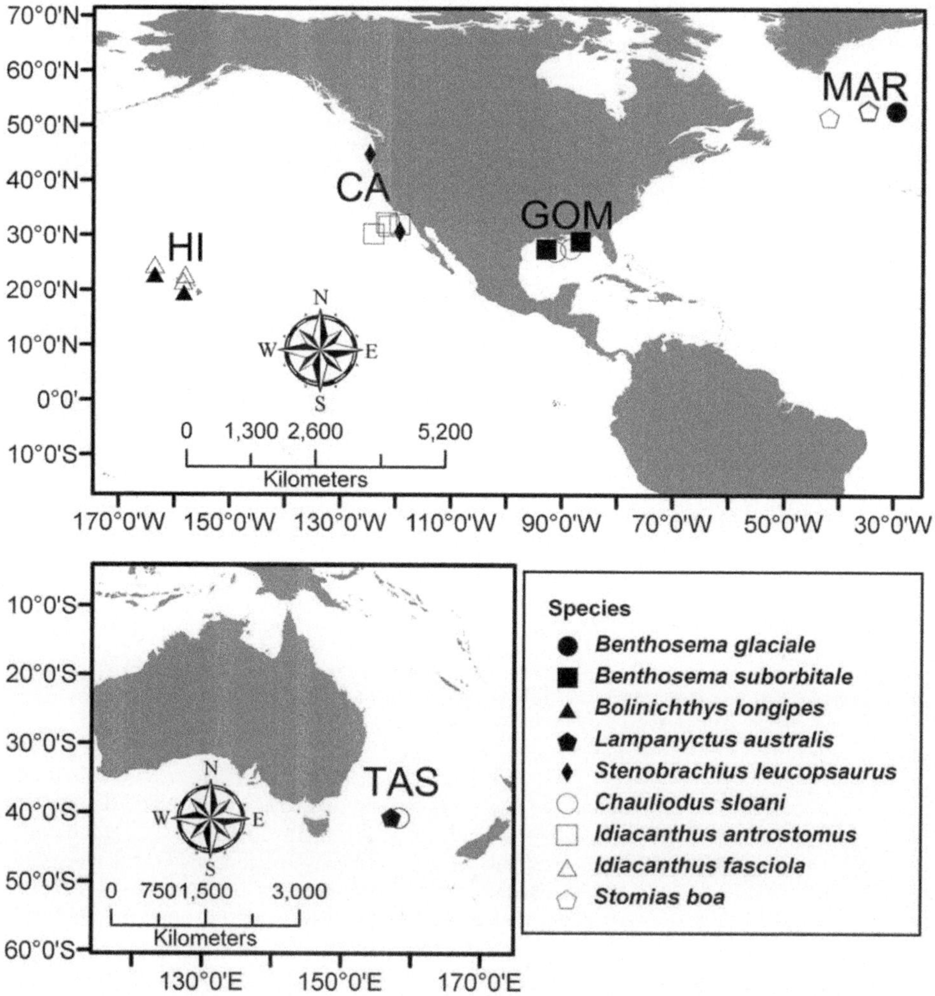

Figure 1. Map of sample collection locations. Approximate capture locations for species of lanternfish (closed symbols) and dragonfish (open symbols) specimens analyzed in this study, from five distinct and globally distributed regions (Tasman Sea (TAS), California Current (CA), Gulf of Mexico (GOM), Hawaii (HI), and the Mid-Atlantic Ridge (MAR)).

$$TP_{bulk} = 2 + \frac{(\delta^{15}N_{consumer} - \delta^{15}N_{zooplankton})}{TEF} \qquad (2)$$

An average trophic enrichment factor (TEF) of 3‰ was used, a value within the range of reported variation amongst diverse organisms [13].

Stable N Isotope Analysis of Individual AAs

A subset of fishes (33 of 66) was selected for CSIA based on broad ranges in size and $\delta^{15}N_{bulk}$ values. Preparation for CSIA followed methods of Hannides et al. [17]. Dried samples were subjected to acid hydrolysis, esterification of the carboxyl terminus, and trifluroacetylation of the amine group [27]. Samples were redissolved in 50–100 µl of ethyl acetate, and the $\delta^{15}N$ values of AAs were measured using an isotope ratio mass spectrometer (either a Delta^{Plus}XP or Delta V Plus) interfaced with a gas chromatograph (Trace GC) through a GC-C III combustion furnace (980°C), reduction furnace (650°C) and liquid-N cold trap. Samples (1–2 µL) were injected (split/splitless injector, using a 10:1 split ratio) onto a capillary column (BP×5 forte, 30 m×0.32 mm×1.0 µm film thickness) at an injector temperature of 180°C with a constant helium flow rate of 1.2 mL min^{-1}. The column oven was held at 50°C for 2 min and then ramped to 190°C at a rate of 8°C min^{-1}. At 190°C, temperature was increased to 300°C (at a rate of 10°C min^{-1}) and held for 7.5 min. Samples were analyzed in triplicate, and the measured N isotopic compositions were normalized to known $\delta^{15}N$ values of two co-injected internal reference compounds (norleucine and aminoadipic acid, $\delta^{15}N$ reference values of 19.06‰ and −6.6‰ respectively). Reproducibility of isotopic analysis of glutamic acid and phenylalanine averaged ±0.5‰ (1 S.D.) and ranged from ±0.1‰ to ±2.4‰. Accuracy of the isotopic analysis was estimated using the known $\delta^{15}N$ norleucine value to determine a measured $\delta^{15}N$ value of aminoadipic acid, treating it as an unknown. Accuracy averaged ±1.3‰ (1 S.D.) and ranged from ±0.0‰ to ±3.5‰.

TPs from CSIA Data

Chikaraishi et al. [18] measured the $\delta^{15}N$ values of AAs in a variety of photoautotrophs and consumers and found that the relationship between glutamic acid (glu) and phenylalanine (phe) accurately described fractional TPs for a diversity of organisms:

Table 1. Collection and size information of lanternfish (L) and dragonfish (D) specimens included in this study.

Region [Collection Year(s)]	Oceanographic Characterization	Species	Size Range Analyzed (mm)	Bulk $\delta^{15}N$ (‰)	Bulk $\delta^{13}C$ (‰)
Hawaii (NPSG) [2010–2011]	oligotrophic, subtropical	*Bolinichthys longipes* (L)	25–46 SL, n = 4 (a, b)	5.8±0.5	−18.6±0.7
		Idiacanthus fasciola (D)	72–275 TL, n = 4 (a, b)	6.9±1.7	−17.5±1.2
		Chauliodus sloani (D)	137 SL, n = 1 (a, b)	7.2	−17.6
Tasman Sea Abyssal Basin [2008]	subtropical convergence, temperate	*Lampanyctus australis* (L)	86–107 SL, n = 6 (a); 86–103 SL, n = 3 (b)	11.3±1.0	−18.0±0.6
		C. sloani (D)	190–280 SL, n = 5 (a); 255–280 SL, n = 2 (b)	11.0±0.9	−18.7±0.3
Gulf of Mexico [2007]	oligotrophic, subtropical	*Benthosema suborbitale* (L)	19–27 SL, n = 4 (a, b)	6.9±0.7	−18.9±0.6
		C. sloani (D)	27–105 TL, n = 5 (a); 27–105 TL, n = 3 (b)	8.0±0.6	−19.0±1.0
Northern Mid-Atlantic Ridge [2009]	high productivity, temperate-subtropical	*Benthosema glaciale* (L)	33–71 TL, n = 12 (a); 33–71, n = 5 (b)	9.9±1.2	−18.8±0.5
		Stomias boa (D)	126–168 TL, n = 9 (a); 142–168 TL, n = 3 (b)	10.4±0.7	−18.2±0.4
California Current [2009–2010]	high productivity, upwelling, temperate	*Stenobrachius leucopsaurus* (L)	41–100 SL, n = 11 (a); 58–100 SL, n = 4 (b)	13.5±0.9	−19.6±1.5
		Idiacanthus antrostomus (D)	153–490 SL, n = 5 (a); 158–318 SL, n = 2 (b)	16.0±0.8	−18.1±0.4

Fish size ranges are reported as standard length (SL) or total length (TL) measurements; sample sizes are also provided for specimens included in bulk tissue isotopic analyses (a) and AA nitrogen isotope analyses (b). Bulk tissue $\delta^{15}N$ and $\delta^{13}C$ values are summarized as mean ± S.D.

$$TP = \frac{(\delta^{15}N_{glu} - \delta^{15}N_{phe}) - 3.4}{7.6} + 1 \qquad (3)$$

In equation (3), 3.4 is the difference between the $\delta^{15}N$ values of glu and phe in marine primary producers (defined as β [18]), and 7.6 is the ^{15}N TEF between glu and phe for each trophic level. Uncertainty resulting from AA-TP was calculated using propagation of errors by combining the uncertainty in β (±0.9‰) and TEF (±1.1‰) as determined by Chikaraishi et al. [18] and the measured analytical reproducibility of glu and phe $\delta^{15}N$ values in each sample. Uncertainty in TP ranged from 0.1 to 0.9 (mean 0.3).

Results

Bulk Tissue Isotopic Analyses

$\delta^{15}N_{bulk}$ values for both lanternfishes and dragonfishes differed significantly by region (ANOVA: $p<0.001$, $F(4,33) = 76.07$ for lanternfishes; $p<0.001$, $F(4,24) = 74.41$ for dragonfishes), and are reported in Table 1. $\delta^{13}C_{bulk}$ values also differed significantly by region for lanternfishes (ANOVA: $p<0.05$, $F(4,33) = 2.98$) and dragonfishes (ANOVA: $p<0.05$, $F(4,24) = 3.90$). Temporal collection parameters were variable but relationships between individual fish $\delta^{15}N_{bulk}$ and $\delta^{13}C_{bulk}$ values and collection year were not significant and weak for $\delta^{15}N_{bulk}$ ($p>0.05$, $r^2 = 0.05$), and $\delta^{13}C_{bulk}$ values ($p>>0.05$, $r^2 = 0.02$). Individuals analyzed spanned a wide size range across the five regions (Table 1). Linear regressions of $\delta^{15}N_{bulk}$ values on fish size per region and species groups were significant with negative slopes ($p<0.05$) for CA Current lanternfishes, and significant with positive slopes for MAR lanternfishes ($p<0.05$), but not significant for any other group and region pair (Figure S2). Comparison of $\delta^{15}N_{bulk}$ and $\delta^{13}C_{bulk}$ values indicated no significant differences between the two fish groups within a region ($\delta^{15}N_{bulk}$: two-tailed paired t-test: $t = 2.78$, $p>0.05$; $\delta^{13}C_{bulk}$: two-tailed paired t-test: $t = 2.78$, $p>0.05$). Estimates of TP_{bulk} were variable across the five regions for both lanternfishes and dragonfishes (ranges for individual specimens were TPs 1.8–4.5 and TPs 1.5–4.8, respectively) and did not align with TPs estimated from SC studies (Table S1, Table 2, Table S3).

Stable N Isotope Analysis of Amino Acids & TP Estimates

Similar to regional differences in $\delta^{15}N_{bulk}$ values, variability in the $\delta^{15}N$ values of the source AA phenylalanine ($\delta^{15}N_{phe}$) was apparent (mean 0.0‰, range −4.9 to 6.6‰). Importantly, $\delta^{15}N_{phe}$ values for both fish groups differed significantly by region (ANOVA, $p<0.05$). Conversely, the $\delta^{15}N_{phe}$ values of lanternfishes and dragonfishes within a region were not significantly different (two-tailed paired t-test: $t = 1.03$, $p>0.05$). The significant correlation between $\delta^{15}N_{phe}$ and $\delta^{15}N_{bulk}$ values (Figure 2) suggested that $\delta^{15}N_{bulk}$ values predominantly reflect the $\delta^{15}N$ values of primary producers in each region (i.e., the regional isotopic baseline).

Variability in the $\delta^{15}N$ values of the trophic AA glutamic acid ($\delta^{15}N_{glu}$) mirrored regional patterns in $\delta^{15}N_{phe}$ values (mean 18.6‰, range 12.1–24.1‰); the highest $\delta^{15}N_{glu}$ values were observed in both fish groups from the productive regions of the CA Current, the Tasman Sea and the MAR (Table S2). Conversely, the lowest $\delta^{15}N_{glu}$ values were observed in the fishes from the tropical oligotrophic waters of the GOM and the NPSG. Similar to regional differences in $\delta^{15}N_{phe}$ values, differences in $\delta^{15}N_{glu}$ values across regions were also significant (ANOVA, $p<0.05$).

TPs calculated from CSIA data (AA-TPs) for lanternfishes and dragonfishes using eq. 1 were very consistent across the five regions despite great variability in $\delta^{15}N_{bulk}$ and $\delta^{15}N_{phe}$ values, as well as fish size (Figure 3). In all five regions the AA-TPs of dragonfishes were much lower than the expected TP of 4.1 based on SC studies. There was no statistical difference in mean dragonfish AA-TPs among regions (ANOVA, $F = 1.62$, $p>0.05$),

Table 2. Regional comparison of lanternfish and dragonfish trophic positions estimated by amino acid and bulk tissue isotopic data.

Region	Mean Lanternfish TP_{bulk} *	Mean Dragonfish TP_{bulk}*	TP_{bulk} Difference	Mean Lanternfish AA-TP#	Mean Dragonfish AA-TP#	AA-TP Difference
North Pacific Subtropical Gyre (Hawaii)	2.0±0.1	2.3±0.5	0.3 TP	2.6±0.2	3.2±0.1	0.6 TP
Tasman Sea	2.7±0.3	2.6±0.3	0.1 TP	2.8±0.0	3.0±0.1	0.2 TP
Gulf of Mexico	2.0±0.2	2.4±0.2	0.4 TP	2.9±0.2	3.0±0.3	0.1 TP
Mid-Atlantic Ridge	4.1±0.4	4.3±0.2	0.2 TP	3.2±0.4	3.4±0.2	0.2 TP
California Current	3.4±0.3	4.2±0.3	1.2 TP	2.8±0.1	3.3±0.1	0.5 TP

#Calculated using Eq. 3.

$$TP = \frac{(\delta^{15}N_{glu} - \delta^{15}N_{phe}) - 3.4}{7.6} + 1 \quad (3)$$

*Calculated using Eq. 1 and Eq. 2 as described in "Methods." Region-specific δ15NPOM and δ15Nzooplankton values from the literature are presented in Table S3.

$$TP_{bulk} = 1 + \frac{(\delta^{15}N_{consumer} - \delta^{15}N_{POM})}{TEF} \quad (1)$$

$$TP_{bulk} = 2 + \frac{(\delta^{15}N_{consumer} - \delta^{15}N_{zooplankton})}{TEF} \quad (2)$$

Summarized values include mean trophic positions (TPs) calculated from bulk tissue $\delta^{15}N$ values (TP_{bulk}) using a trophic enrichment factor (TEF) of 3‰ (mean ± S.D.), and TPs calculated from AA-CSIA data (AA-TP) (mean ± S.D.). Differences in the calculated means between dragonfishes and lanternfishes are shown. Dragonfish values for Hawaii include specimens of both *Chauliodus sloani* and *Idiacanthus fasciola*.

indicating similar TPs across all ecosystems studied. Mean lanternfish AA-TPs were significantly different across the five regions (ANOVA, F = 4.16, $p<0.05$). However, this difference was primarily driven by MAR lanternfishes (mean AA-TP: 3.2), which had elevated AA-TPs relative to fishes from the other regions (mean AA-TPs range: 2.6 to 2.9). Mean lanternfish AA-TPs were significantly lower than mean dragonfish AA-TPs within a region (two-tailed paired t-test, $t = 2.78$, $p<0.05$).

Discussion

Despite regional oceanographic influences on bulk SI and CSIA data, strongly uniform AA-TPs were observed across the five global regions for both fish groups. Significant differences in AA-TPs between the two fish groups were consistent with previous SC analyses, though different in magnitude. The biogeochemical diversity present in the five regions was reflected in both $\delta^{15}N_{bulk}$ and $\delta^{15}N_{phe}$ values ($\delta^{15}N_{phe} = -4.9$ to 6.6‰, Table S2), and is consistent with baseline N values produced by known biogeochemical processes (microbially-mediated N-recycling dynamics in the oligotrophic gyres to nitrate-based upwelling in the CA Current, for example). The strong correlation between $\delta^{15}N_{bulk}$ and $\delta^{15}N_{phe}$ consumer values indicates that regional biogeochemical differences influenced $\delta^{15}N_{bulk}$ values in higher consumers (Figure 2).

The lack of comprehensive regional information characterizing isotopic baselines (i.e., encompassing uniform seasons and years) inhibits a more detailed and accurate calculation of TP from $\delta^{15}N_{bulk}$ values across regions. TPs calculated from $\delta^{15}N_{bulk}$ values were more variable across regions than TPs estimated from either SC studies or CSIA (Table S1, Table 2, Table S3). Absolute AA-TPs for lanternfishes were all within the range of those estimated by SC studies, while absolute AA-TPs for dragonfishes were lower than ranges derived from SC studies. AA-TPs also provided new information suggesting that these two fish groups plausibly have more similar TPs across the global oceans than previously documented by SC data. These results highlight the potential advantages of CSIA-based TPs for food web analysis in remote or highly heterogeneous ecosystems where baseline $\delta^{15}N_{bulk}$ values are deficient or difficult to obtain.

One explanation for the apparent disagreement between dragonfish SC (~TP 4.1) and AA-TP (~TP 3.2) estimates is that the TEF used to establish TP is different for the two fish groups. Stark energetic differences could result in different protein turnover rates and potentially different TEFs. Growth and metabolism data for these fish groups are limited, but two studies found that some species of dragonfish have exceptionally low metabolic rates in the deep ocean, about tenfold lower than the more active, diel-vertically migrating lanternfishes [28,29]. Preferential retention of ^{15}N is dependent upon protein-containing meals, wherein animals assimilate a fraction of the protein (somatic growth) and catabolize and excrete the remainder (metabolism). Tissue turnover information for the two fish groups are not available, however sporadic feeding coupled with low locomotory abilities and low metabolic rates could result in slower protein turnover in dragonfishes relative to lanternfishes.

A second explanation that may reconcile the differences between SC- and AA-derived TP estimates is that available SC data failed to integrate the mean fish diets examined. Ecologists have long recognized that SC analyses represent only a "snapshot" of what an animal has recently eaten [30]. Seasonality and ontogeny, as well as variation in prey abundances, can affect the prey documented in fish stomachs [31]. Calculations of TP from SC data require knowledge of the TPs of animals forming a consumer's prey base, many of which may be poorly known or also estimated from SC data. Additionally, depending on the digestibility of a prey item, organisms with resistant hard parts may be over-represented in SC analysis, while easily digested soft-bodied prey items can be overlooked. As a result, these biases could alter the TP estimated from SC analysis alone.

Results from this study show that amino acid CSIA can be a useful tool for elucidating and comparing trophic structure, which can potentially be broadly transferred to other ecosystems and organisms. More field and laboratory testing are needed before CSIA can be used to accurately estimate TPs for organisms for which diet and trophic information may be limited or missing entirely. Comparison of SC and isotopic data highlights the need for caution when establishing TPs from SC analysis alone, or vice versa. AA-TPs integrate dietary inputs over a longer time scale

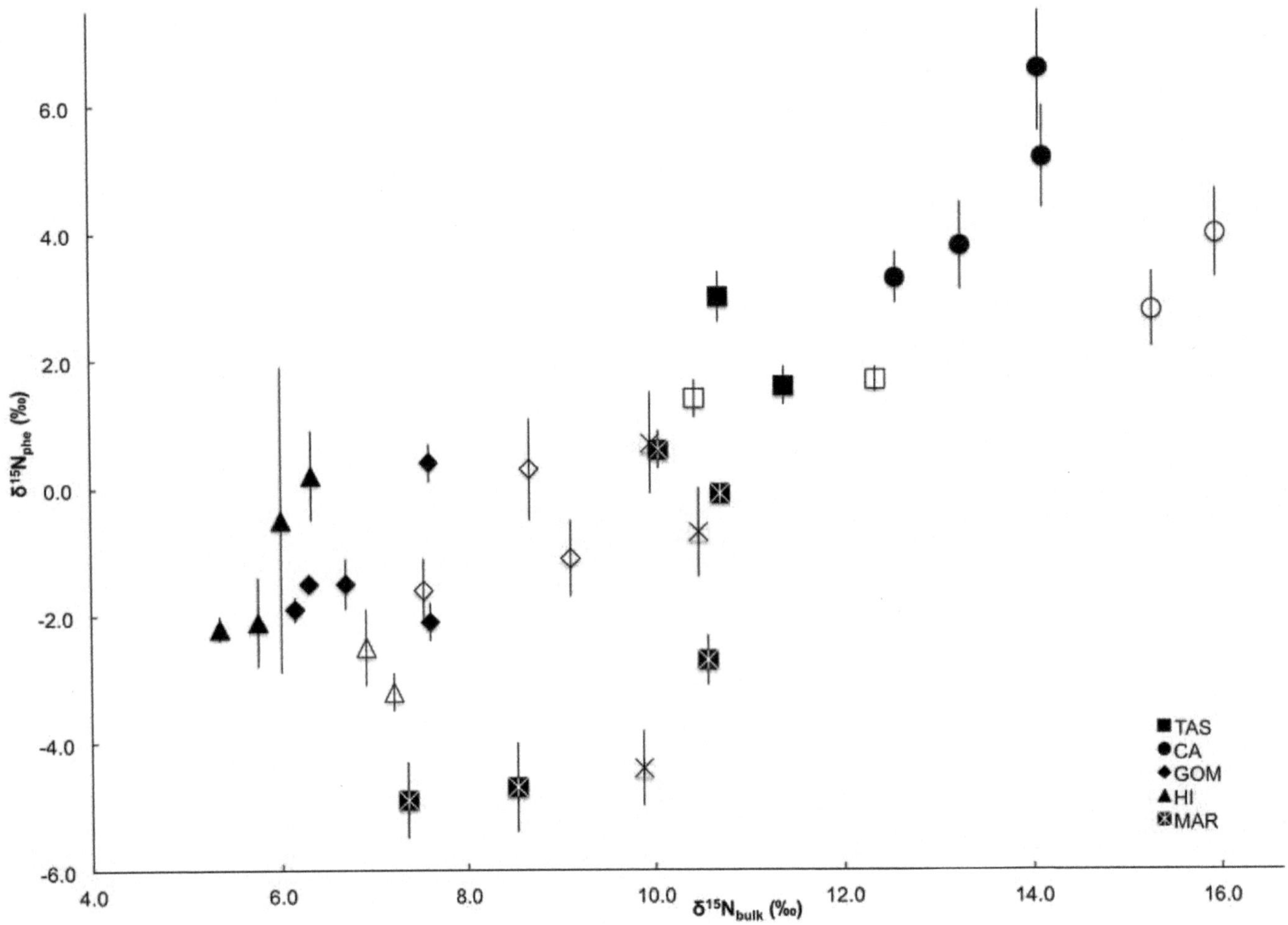

Figure 2. Influence of regional biogeochemistry on consumer isotopic composition. Relationship between $\delta^{15}N$ values of phenylalanine ($\delta^{15}N_{phe}$) (‰) and bulk white muscle tissue ($\delta^{15}N_{bulk}$) (‰) in 33 specimens of mesopelagic lanternfishes (empty symbols) and dragonfishes (filled symbols) from offshore waters of the Tasman Sea (TAS), California Current (CA), Gulf of Mexico (GOM), Hawaii (HI), and the Mid-Atlantic Ridge (MAR). Error bars are standard deviations. $\delta^{15}N_{phe}$ and $\delta^{15}N_{bulk}$ values in fishes across all regions are significantly positively correlated ($p<0.05$, $r^2=0.58$; $y=0.75\times -7.14$).

than SC analysis and yield a quantitative TP across a variety of species and temporal and spatial scales that can be easily contrasted. However, SC studies accomplish what isotopic analyses cannot – taxonomic identification of prey items – and thus cannot and should not be replaced.

Heated debate and pronouncements about the exploited status of large marine ecosystems are often built upon marine fish species mean or fractional TP assignments [2,32]. Thus, TP estimates have widespread implications for describing energy flow, as well as within mathematical models that aim to simulate these ecosystems. Small changes in prey TP estimates can result in substantial changes in the estimates of top predator production. Both lanternfishes and dragonfishes are globally important prey items for many commercially important fishes as well as sharks and marine mammals (e.g., [33,34,35]). Our novel application of CSIA to a global sample set suggests that SC-derived TP estimates (and by extension, food web analyses) will benefit from combining an integrated CSIA approach that overcomes the challenges of using $\delta^{15}N_{bulk}$ values across distinct regions. While SC analysis is a vitally important tool for food web characterization, it may also lead to errors in TP estimation.

The large differences observed in fish $\delta^{15}N_{bulk}$ values represent anywhere from two to three TPs for two fish groups that are less than one TP apart according to available SC data. In contrast to SC studies and despite $\delta^{15}N_{bulk}$ and $\delta^{13}C_{bulk}$ variability, results of CSIA indicate: a) uniform TPs within both fish groups across all five regions, and b) consistent TPs between both fish groups across all five regions (Figure 3). CSIA results indicate that inconsistencies in $\delta^{15}N_{bulk}$ values result from regionally distinct baseline N isotopic compositions. Although we acknowledge that our results raise specific uncertainties regarding this emerging food web tool and interpretation of traditional (SC) TP estimates that merit further investigation, the uniformity in CSIA-based TPs across global regions demonstrates that this is a promising method to compare food webs among ecologically and biogeochemically diverse ecosystems.

Supporting Information

Figure S1 Intra-laboratory comparison of measured bulk tissue $\delta^{15}N$ values. Comparison of bulk tissue $\delta^{15}N$ (a) and $\delta^{13}C$ (b) values measured at the University of Hawaii and two outside laboratories (University of North Carolina Wilmington

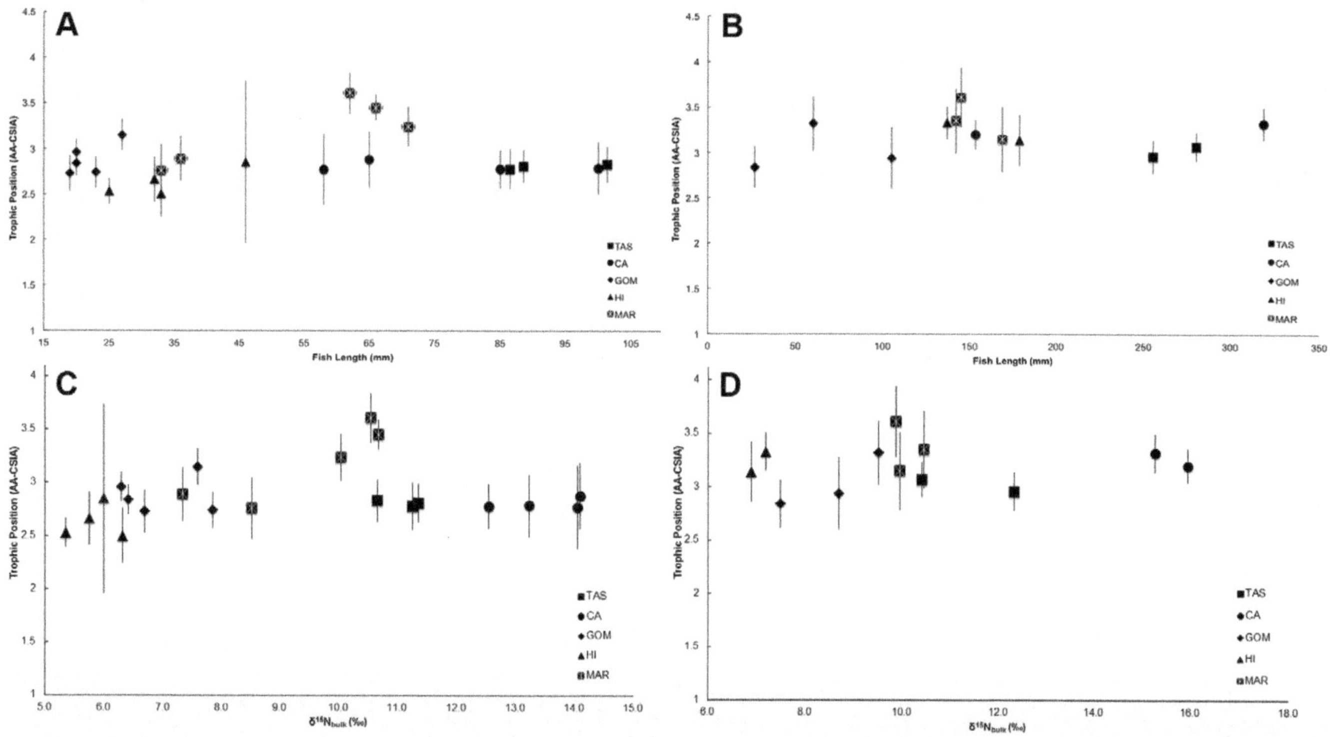

Figure 3. Uniform amino acid based trophic positions for lanternfishes and dragonfishes. Relationship between fish length (mm) and amino acid CSIA estimated trophic positions of a) individual lanternfishes, and b) individual dragonfishes from five regions. Also shown is the relationship between individual fish bulk tissue $\delta^{15}N$ values (‰) and amino acid CSIA estimated trophic positions of c) lanternfishes and d) dragonfishes from five regions. Error bars indicate propagated error from trophic position calculation (see methods).

(n = 3) and Ehime University (n = 6)). Neither the slope nor the intercept is different from 1 and 0, respectively at the 95% confidence interval.

Figure S2 Relationship between fish length and bulk tissue nitrogen isotopic values in fishes. Bulk tissue $\delta^{15}N$ values (‰) versus fish standard length (mm) in a) lanternfishes and b) dragonfishes from five regions (TAS = Tasman Sea, CA = California Current, GOM = Gulf of Mexico, HI = Hawaii, MAR = mid-Atlantic Ridge).

Table S1 Meta-analysis of region-specific published stomach content studies for lanternfish and dragonfish diet. Meta-analysis of region-specific published food items (at the taxonomic level of Order) for lanternfish (L) and dragonfish (D) species and trophic positions (mean ± S.E.) (as published and defined by the *FISHBASE* online database (Froese and Pauly 2012)). Additional primary references listed may not be included in FISHBASE and are specific to fishes analyzed from each region. Food item column headers are as follows: %C is percent copepods, %O is percent ostracods, %E is percent euphausiids, %A is percent amphipods, %F is percent fishes, and %Oth is percent other (includes pteropods, gastropods and other molluscs, debris, salps, unidentified decapoda remains, etc.).

Table S2 Regional values of source and trophic amino acids in lanternfish and dragonfish. Comparison of isotopic compositions of the source amino acid phenylalanine ($\delta^{15}N_{phe}$) and the trophic amino acid glutamic acid ($\delta^{15}N_{glu}$) (mean ± S.D.) in lanternfishes and dragonfishes across all five oceanographic regions. Dragonfish values for Hawaii include specimens of both *Chauliodus sloani* and *Idiacanthus fasciola*.

Table S3 Best available region-specific bulk stable nitrogen isotope data used to estimate fish trophic positions. Summary of best available bulk stable nitrogen isotopic baseline values used to calculate trophic positions of lanternfishes and dragonfishes (TP_{bulk}). Isotopic data characterizing regional food web bases from the same seasons and years was not available, highlighting the need for a more reliable method for calculating TPs from these isotopic data.

Acknowledgments

The views expressed herein are those of the authors and do not necessarily reflect the views of US EPA, NOAA, or any of their subdivisions. Any use of trade, product, or firm names is for descriptive purposes only and does not imply endorsement by the U.S. Government. Samples from the Tasman Sea were generated from a project by CSIRO Marine & Atmospheric Research under the direction of Dr. Rudy Kloser. This study was carried out in accordance with the animal use protocols of the University of Hawai'i (protocol #10-984). This is SOEST contribution number 8502.

Author Contributions

Conceived and designed the experiments: JCD CAC BNP. Performed the experiments: CAC EJG. Analyzed the data: CAC JCD BNP. Contributed reagents/materials/analysis tools: BNP. Wrote the paper: CAC BNP JCD JCH AF JPMC TWM TTS PCD SWR EJG.

References

1. Sibert J, Hampton J, Kleiber P, Maunder M (2006) Biomass, size, and trophic status of top predators in the Pacific Ocean. Science 314: 1773–1776.
2. Pauly D, Christensen V, Dalsgaard J, Froese R, Torres Jr F (1998) Fishing down marine food webs. Science 279: 860–863.
3. Casini M, Hjelm J, Molinero J, Lovgren J, Cardinale M, et al. (2009) Trophic cascades promote threshold-like shifts in pelagic marine ecosystems. Proc Natl Acad Sci U S A 106: 197–202.
4. Brodeur RD, Yamamura O [eds.] (2005) Micronekton of the North Pacific. PICES Science Report No. 30. North Pacific Marine Science Organization, Sidney, BC.
5. De Forest L, Drazen J (2009) The influence of a Hawaiian seamount on mesopelagic micronekton. Deep-Sea Res, Part I 56: 232–250.
6. Sutton TT, Hopkins TL (1996) Species composition, abundance, and vertical distribution of the stomiid (Pisces: Stomiiformes) fish assemblage of the Gulf of Mexcio. Bull Mar Sci 59: 530–542.
7. Sutton TT, Hopkins TL (1996) Trophic ecology of the stomiid (Pisces, Stomiidae) assemblage of the eastern Gulf of Mexico: strategies, selectivity and impact of a mesopelagic top predator group. Mar Biol 127: 179–192.
8. Hidaka K, Kawaguchi K, Murakami M, Takahashi M (2001) Downward transport of organic carbon by diel migratory micronekton in the western equatorial Pacific: its quantitative and qualitative importance. Deep-Sea Res, Part I 48: 1923–1939.
9. Gartner J, Sulak VKJ, Ross SW, Necaise A (2008) Persistent near-bottom aggregations of mesopelagic animals along the North Carolina and Virginia continental slopes. Mar Biol 153: 825–841.
10. Peterson B, Fry B (1987) Stable isotopes in ecosystem studies. Annu Rev Ecol Syst 18: 293–320.
11. Jennings S, Barnes C, Sweeting CJ, Polunin NVC (2008) Application of nitrogen stable isotope analysis in size-based marine food web and macroecological research. Rapid Commun Mass Spectrom 22: 1673–1680.
12. DeNiro M, Epstein S (1981) Influence of diet on the distribution of nitrogen isotopes in animals. Geochim Cosmochim Acta 45: 341–351.
13. Vanderklift M, Ponsard S (2003) Sources of variation in consumer-diet $\delta^{15}N$ enrichment: a meta-analysis. Oecologia 136: 169–182.
14. Post DM (2002) Using stable isotopes to estimate trophic position: models, methods, and assumptions. Ecology 83: 703–718.
15. Dore J, Brum J, Tupas L, Karl D (2002) Seasonal and interannual variability in sources of nitrogen supporting export in the oligotrophic subtropical North Pacific Ocean. Limnol Oceanogr 47: 1595–1607.
16. Wada E, Hattori A (1976) Natural abundance of ^{15}N in particulate organic matter in the North Pacific Ocean. Geochim Cosmochim Acta 40: 249–251.
17. Hannides CCS, Popp BN, Landry M, Graham BS (2009) Quantification of zooplankton trophic position in the North Pacific Subtropical Gyre using stable nitrogen isotopes. Limnol Oceanogr 54: 50–61.
18. Chikaraishi Y, Ogawa N, Kashiyama Y, Takano Y, Suga H, et al. (2009) Determination of aquatic food-web structure based on compound-specific nitrogen isotopic composition of amino acids. Limnol Oceanogr: Methods 7: 740–750.
19. McClelland J, Montoya J (2002) Trophic relationships and the nitrogen isotopic composition of amino acids in phytoplankton. Ecology 83: 2173–2180.
20. Popp BN, Graham BS, Olson R, Hannides CCS, Lott M, et al. (2007) Insight into the trophic ecology of yellowfin tuna, *Thunnus albacares*, from compound-specific nitrogen isotope analysis of proteinaceous amino acids. In Dawson TD, Siegwolf RTW, editors. Stable isotopes as indicators of ecological change. Terrestrial Ecology Series. Elsevier. 173–190.
21. Chikaraishi YY, Kashiyama Y, Ogawa N, Kitazato H, Ohkouchi N (2007) Metabolic control of amino acids in macroalgae and gastropods: implications for aquatic food web studies. Mar Ecol Progr Ser 342: 85–90.
22. Dale JJ, Wallsgrove NJ, Popp BN, Holland KN (2011) Nursery habitat use and foraging ecology of the brown stingray, *Dasyatis lata*, determined from stomach content, bulk and amino acid stable isotope analysis. Mar Ecol Prog Ser 433: 221–236.
23. Hoffman J, Sutton TT (2010) Lipid correction for carbon stable isotope analysis of deep-sea fishes. Deep-Sea Res, Part I 57: 956–964.
24. Nam NT, Miller T, Huan N, Tang V, Omori K (2011) Integrating community structure and stable isotope analysis to assess a heavily exploited coastal marine ecosystem off Central Vietnam. Fish Res 110: 268–276.
25. McClain-Counts J (2010) Trophic structure of midwater fishes over cold seeps in the North-central Gulf of Mexico. M.S. thesis. Univ North Carolina Wilmington.
26. Vander Zanden M, Cabana G, Rasmussen J (1997) Comparing trophic position of freshwater fish calculated using stable nitrogen isotope ratios ($\delta^{15}N$) and literature dietary data. Can J Fish Aquat Sci 54: 1142–1158.
27. Macko S, Uhle M, Engel M, Andrusevich V (1997) Stable nitrogen isotope analysis of amino acid enantiomers by gas chromatography combustion/isotope ratio mass spectrometry. Anal Chem 69: 926–929.
28. Torres J, Belman B, Childress JJ (1979) Oxygen consumption rates of midwater fishes as a function of depth of occurrence. Deep-Sea Res 26: 185–197.
29. Childress JJ, Taylor S, Cailliet G, Price M (1980) Patterns of growth, energy utilization and reproduction in some meso- and bathypelagic fishes off southern California. Mar Biol 61: 27–40.
30. Hyslop E (1980) Stomach contents analysis: a review of methods and their application. J Fish Biol 17: 411–429.
31. Kawaguchi K, Mauchline J (1982) Biology of myctophid fishes (family Myctophidae) in the Rockall Trough, northeastern Atlantic Ocean. Biol Oceanogr 1: 337–373.
32. Branch T, Watson R, Fulton E, Jennings S, McGilliard C, et al. (2010) The trophic fingerprint of marine fisheries. Nature 468: 431–435.
33. Rosecchi E, Tracey DM, Webber WR (1988) Diet of orange roughy, *Hoplostethus atlanticus* (Pisces: Trachichthyidae) on the Challenger Plateau, New Zealand. Mar Biol 99: 293–306.
34. Pilling GM, Purves MG, Daw TM, Agnew DA, Xavier JC (2001) The stomach contents of Patagonian toothfish around South Georgia (South Atlantic). J Fish Biol 59: 1370–1384.
35. Cherel Y, Duhamel G (2004) Antarctic jaws: cephalopod prey of sharks in Kerguelen waters. Deep-Sea Res, Part I 51: 17–31.

Predation Risk Shapes Social Networks in Fission-Fusion Populations

Jennifer L. Kelley[1]*, Lesley J. Morrell[2], Chloe Inskip[3], Jens Krause[4,5], Darren P. Croft[6]

1 School of Animal Biology, University of Western Australia, Perth, Western Australia, Australia, **2** Department of Biological Sciences, University of Hull, Hull, United Kingdom, **3** Durrell Institute of Conservation and Ecology, University of Kent, Kent, United Kingdom, **4** Leibniz-Institute of Biology and Ecology of Fishes, Berlin, Germany, **5** Institute of Integrative and Comparative Biology, University of Leeds, Leeds, United Kingdom, **6** School of Psychology, University of Exeter, Exeter, Devon, United Kingdom

Abstract

Predation risk is often associated with group formation in prey, but recent advances in methods for analysing the social structure of animal societies make it possible to quantify the effects of risk on the complex dynamics of spatial and temporal organisation. In this paper we use social network analysis to investigate the impact of variation in predation risk on the social structure of guppy shoals and the frequency and duration of shoal splitting (fission) and merging (fusion) events. Our analyses revealed that variation in the level of predation risk was associated with divergent social dynamics, with fish in high-risk populations displaying a greater number of associations with overall greater strength and connectedness than those from low-risk sites. Temporal patterns of organisation also differed according to predation risk, with fission events more likely to occur over two short time periods (5 minutes and 20 minutes) in low-predation fish and over longer time scales (>1.5 hours) in high-predation fish. Our findings suggest that predation risk influences the fine-scale social structure of prey populations and that the temporal aspects of organisation play a key role in defining social systems.

Editor: Brock Fenton, University of Western Ontario, Canada

Funding: JLK was supported by the Universities of Western Australia and New South Wales. The Association for the Study of Animal Behaviour and the Fisheries Society of the British Isles funded the project through their small research grants programs. The funders had no role in study design, data collection and analysis, decision to publish, or preparation of the manuscript.

Competing Interests: The authors have declared that no competing interests exist.

* E-mail: jennifer.kelley@uwa.edu.au

Introduction

Linking individual-level behaviours with large-scale social structure is fundamental to ecology and evolution because it allows us to understand patterns of gene, information or disease transfer through populations, predict the impact of invasive species, and reveal the factors that underlie the evolution of sociality [1]. Researchers have often examined the ecological correlates of animal social structure, linking variation in social organisation with factors such as habitat type, kin structure, food availability, sexual dimorphism, body size, technological innovation (in human societies) and predation risk [2,3,4,5,6,7,8]. In particular, predation risk is thought to be a key ecological factor driving the evolution of group formation [9]. Nevertheless, despite a long history of studies demonstrating that predation risk is typically associated with larger prey group sizes [8,10,11,12], little is known about how risk shapes the overall social organisation and dynamics of animal populations.

Fission-fusion societies, which describe populations that are characterised by frequent exchange of individuals among groups [13,14,15,16,17], provide an ideal system in which to examine the ecological correlates of sociality because these systems show rapid responses to changing environmental conditions [18]. While grouping can confer advantages such as increased hunting efficiency, transmission of cultural information and reduced risk of predation, it can also incur potential costs such as increased competition and aggression among group members and greater risk of parasitism [19]. Ecological variability is predicted to drive changes in the relative costs/benefits of group membership, leading to dynamic patterns of fission-fusion [20]. Thus studying the organisational properties of fission-fusion societies not only reveals the factors that influence an individual's grouping decisions, but can also provide insights into how interactions among individuals drive social relationships and ultimately influence population processes [21,22].

Studies linking the social structure of fission-fusion societies with ecological variability have tended to focus on the distribution, abundance or quality of food resources [17,23,24,25,26,27] and its associated effects on competition among individuals within the group [28]. Those studies that have documented a link between predation risk and prey group size have been conducted on populations in the wild [29,30,31], where the role of other interacting factors (e.g. food availability) is difficult to determine. For example, divergent social structures observed in communities of Hawaiian dolphins (fission-fusion societies versus stable associations) could partly be attributed to variation in predation risk, but the availability of prey and suitable habitats was also important [32].

Temporal patterns of social organisation, such as the frequency of fission-fusion events, are an important but often overlooked aspect of sociality [33], and these processes form the basis of large scale effects such as rates of gene flow and speciation. A study with bison at risk of predation from wolves found that the probability of fission-fusion events depended on the habitat, season and the time of day at which the bison were observed [34], but studies relating risk to temporal organisation are generally lacking. Despite recent advances in the analytical techniques used to quantify animal social organisation [33,35], no study has yet adopted an experimental approach for examining the dynamic effects of predation risk on the spatial and temporal organisation of prey populations.

In this paper we use social network analyses to investigate how variation in predation risk affects an individual's grouping decisions and how this contributes to the overall patterns of social organisation. A social network is a graphical representation of the social associations or interactions occurring among individuals within the population(s) [35]. While these analyses originated as a branch of mathematical theory that were applied to fields as diverse as physics [36], systems pharmacology [37], epidemiology [38] and conservation biogeography [39], they have recently attracted the attention of behavioural ecologists [22,35,40,41,42] because calculation of their mathematical parameters can reveal global properties of the population, such as resilience to the removal of particular individuals [43] and the rate at which disease or learned information might be transmitted [44].

We investigated the effect of predation risk on social organisation by comparing the social structure of populations of freshwater fish (guppies) collected from habitats with high and low predation risk and observed in the lab under standard conditions (no risk). Guppies have proved a tractable system for network studies [14,45,46,47,48] and provide a compelling example of the effect of predation risk on a variety of phenotypic and behavioural traits [49,50]. Specifically, classic early work with this species [51] revealed schooling as an evolutionary response to predation, yet we do not know how this group structuring contributes to the overall patterns of social organisation at the population level. For example, different patterns of social organisation may be observed in populations characterised by similar group sizes; between sampling intervals shoal membership may remain stable (low rates of exchange) or be highly dynamic with individuals frequently switching among shoals. These temporal processes are of key importance because they determine how individual movements affect population dynamics [52], such as rates of gene flow, the spread of information or pathogens and the opportunity for social recognition (e.g. familiarity) to develop among individuals.

We anticipated that the larger shoal sizes typically found in high predation guppy populations would produce social networks with higher levels of association among group members than those in low predation habitats. However, we expected temporal patterns of organisation to also play an important part: low rates of fission-fusion are expected to lead to stable, highly associated groups but low connectivity among groups, while high rates will lead to greater movement of individuals among shoals and greater overall connectedness at the level of the community. If movement of individuals (or small groups) among shoals is risky in habitats with high predation risk [53], then within-shoal associations may be high but among-shoal connectedness may be low. In guppies from habitats with low predation risk, within-shoal associations may be low but among-shoal social connectedness may be high if fission/fusion events are frequent. Stable associations can lead to the development of social familiarity among individuals, which can confer foraging and antipredator benefits in fishes [54]. In guppies familiarity takes around 12 days to develop [55]; we therefore used this time frame to determine whether the development of preferred companionships affected the properties of the networks.

Materials and Methods

Ethics statement

This work was approved and performed in accordance with guidelines issued by the Animal Care and Ethics Committee (ACEC) for the University of New South Wales, Australia, under research project approval number 05/34A (issued to J.L.K.). Fish were transported from field sites to the laboratory in Trinidad in sealed plastic bags containing river water, plant material collected from the river (to provide cover/refuge) and Aqua Master Armour Coat^{TM} (for skin protection). Fish were acclimatised in the laboratory in their social groups overnight before being individually tagged (see below for details) the following day. These individual marks were necessary to ensure that each female was individually recognisable (wild female guppies do not have distinguishing colour patterns). We ensured that females were anesthetised briefly (in MS222) for the procedure and allowed to recover in aerated and conditioned water (containing Armour Coat^{TM}). Fish were allowed to recover for a further 24 hours before experiments commenced, during which the health and behaviour of all fish was carefully monitored. We did not observe any ill effects or changes in behaviour as a result of the tagging procedure. At the end of our experiments, the fish were released into large artificial pools at the research station.

Fish collection

Guppies in the Northern Range Mountains of Trinidad, West Indies, inhabit a series of streams that can readily be characterised by predation risk because barrier waterfalls have prevented the upstream migration of most predators. Guppies living in the upper reaches of streams occur with only one fish predator, Hart's Killifish (*Rivulus hartii*), which predominantly feeds on juvenile guppies and invertebrates [56], while guppies inhabiting the lower regions of streams coexist with a variety of fish predators including the pike cichlid (*Crenicichla frenata*) the blue acara (*Aequidens pulcher*) the wolf fish (*Hoplias malabaricus*) and Hart's killifish (*Rivulus hartii*) [57].

We collected guppies from twelve localities in Trinidad comprising habitats with high ($n = 7$) and low ($n = 5$) predation risk. Ten of these sites were upstream (low predation risk) and downstream (high predation risk) populations for the same river (Arima, Aripo, Guanapo, Tacarigua and Turure Rivers). We also collected guppies from a river (the Oropuche) in which upstream and downstream sites are both known to be high predation. Inclusion of these populations allowed us to consider whether any differences observed between upstream/downstream populations might be attributed to ecological parameters other than predation risk. Predator assemblages at all our sample sites have been determined previously and fit the typical upstream (low risk) and downstream (high risk) pattern (see Table S1 for details of collection sites and the predator assemblages).

A seine net was used to collect 8–10 shoals of adult guppies per population along approximately 50 m stretch of each river. The shoals were placed in a single bucket so that fish from each population were derived from mixed shoals and were unlikely to be familiar with one another. We then removed 12 similar-sized females before returning all remaining fish to the river. As the size distribution of females covaries with predation risk [58], we collected females that had a body size that lay between the means for high and low predation risk environments (mean total length of fish used in this study $= 26.9 \pm 0.30$ mm). Females were returned to the laboratory at William Beebe Tropical Research Station (Simla) in Trinidad, where each population was released into a separate circular observation arena (diameter $= 120$ cm, water depth =5 cm) containing rainwater and left to acclimate overnight.

Quantifying social associations

We constructed a total of twelve social networks, each comprising 12 females collected from the same population (seven high predation and five low predation sites, n = 144 individuals). We chose to use only females in this study to avoid confounding general association (shoaling) behaviour with sexual behaviour. Furthermore, previous work on guppy social networks has revealed persistent pairwise associations and cooperative interactions predominantly among females [14,45]. Groups of 12 fish were chosen for each population because of the difficulty of observing the behaviour of a large number of individuals. In addition, laboratory-based network studies utilising small numbers of individuals are a good predictor of association patterns occurring in larger networks constructed in the wild [45]. Small guppy shoals are common in the wild [59] and a sample size of 12 individuals allows for evaluation of associations among several social groups.

As the construction of social networks requires that each individual is recognisable, we anesthetised each female in MS222 and gave her a unique identification tag by injecting VIE (Visible Implant Elastomer, Northwest Marine Technology, Inc.) dye under the surface of the dorsal epidermis [59]. The total length (TL) of each fish was then measured (to the nearest millimetre) using a pair of callipers and all fish were left to acclimate overnight in their respective pools. The following day we observed associations among individuals once per minute over a 30-minute period; previous work has shown that this period is sufficient for revealing non-random social structure in guppies [14]. Females were considered to be associating if they were found within the same shoal, i.e. within four body lengths of one another, a standard method of evaluating shoaling behaviour [60]. Individuals were considered part of the same shoal provided they were within 4 body lengths of another group member (thus a shoal of 3 fish could be 8 body lengths wide). Wild guppy shoals are highly dynamic with shoals exhibiting fission or fusion events (i.e. at least one individual joining or leaving a shoal) on average every 14 s in a population with high predation risk [59]. We chose a considerably longer sampling interval (1 minute) so that sampling periods for high and low predation populations were likely to be independent, as required by social association models [61]. Observations were repeated every 3 days over a 12-day period (= total of 4 observations per population), the time scale over which familiarity has been shown to develop in the laboratory in this species [55]. A total of 41 social networks were therefore constructed and modelled (7 networks were not sampled at the allotted time due to time constraints).

Testing for non-random patterns of social organisation

The social analysis program Socprog 2.3 [33] was used to calculate several measures of association describing different aspects of social structure: association index (AI), gregariousness and population social differentiation (see Suporting Information S1, for further details). AIs provide an estimate of the proportion of time two individuals spend together (0 = not associated, 1 = always associated) [22]. We also calculated gregariousness, which is the mean size of group that an individual experiences [6]. Social differentiation was used as a measure of the variation in associations in our populations: low values (<0.3) suggest little variation and represent homogenous societies while higher values indicate communities that are well differentiated (>0.5) or extremely differentiated (>2.0) [33]. Power analyses were performed to provide the level of confidence at which the true social system has been detected.

We examined whether patterns of social association observed in each of our guppy populations were significantly different from random using the permutation tests in Socprog 2.3. The randomisations were performed for each population because these methods rearrange the empirical data (e.g. row and column sums in the matrices remain the same) so that the overall structure of the data is conserved (in this case, the number of groups each fish is observed in – which may vary among populations). The methods used by Socprog 2.3 have been modified to account for problems that were identified with earlier randomisation techniques [33,62] Non-random social organisation is expected when the coefficient of variation (CV) of association indices is significantly higher or lower in the real data set than in the random data. Permutation tests were also used to examine population differences in gregariousness with the null hypothesis being that all individuals prefer groups of similar size (standard deviations of typical group size for real and random data are similar). These permutations can also be used by Socprog to identify pairs of individuals whose association index is greater than 97.5% or less than 2.5% of their random association indices indicating dyads with significant preferences and avoidances respectively. We performed 5000 permutations for each population (at which point P-values stabilised to within 0.01) with 1000 trials per permutation.

Network analyses

We used five network measures which are particularly useful for the analysis of networks based on association indices [22]: strength (or degree), eigenvector centrality, reach, the clustering coefficient and affinity. Such measures have previously been used to describe social organisation in small experimental populations of animals [63]. The strength of the network (referred to as 'degree' in binary networks) is the sum of association indices with all other individuals minus one and is equivalent to gregariousness [64]. Eigenvector centrality [65] is a measure of how well an individual is associated with others (in terms of the number and strength of connections), and also how well its neighbours are themselves associated. Eigenvector centrality therefore provides a measure of connectedness of individuals within a network. Reach describes how well an individual is indirectly connected to others in the population and reveals the overall strength of an individual's neighbours. The clustering coefficient describes how well an individual's neighbours are connected to one another. Values of zero suggest that none of an individual's neighbours are associated whereas a value of one indicates that all neighbours are linked [22]. Affinity is the weighted mean strength of an individual's neighbours and is weighted by the association index between them. Mean network measures were calculated for each population in Socprog 2.3 and their associated standard errors were calculated using the bootstrap method with 1000 replicates.

MANOVA Models

Prior to testing for an effect of predation risk and time (day) on the association measures and network statistics, we used Principal Component Analysis to look for correlations in our data and reduce the number of variables in our model. We entered population means (averaged over the 12 individuals per population) for association index, gregariousness, social differentiation, strength, eigenvector centrality, reach, clustering coefficient and affinity as factors. Two independent principal components (PC1 & PC2) were extracted from the remaining variables (eigenvalues>1.0), which explained 68.5% and 20.9% of the variance in our data respectively. The first Principal Component (PC1) was loaded by the means of association index, gregariousness, strength, eigenvector centrality, reach, clustering coefficient and affinity (loadings: 0.51–0.99). The second principal component (PC2) was predominantly loaded by mean social differentiation (loading: 0.90); we therefore used the raw data for this response variable.

We used MANOVA with the dependent variable as a repeated measures response (4-levels: days) to examine between-subject (predation risk) and within-subject (day) effects on PC1 and mean population social differentiation. The dependent variables were checked for assumptions of normality, homogeneity of variance and sphericity prior to conducting the statistical tests, which were performed in JMP version 9. We confirmed that mean body size of individuals within each group (population) had no effect on our dependent variables by entering it as a covariate in a MANCOVA (PC1: $F_{1,\ 6} = 0.410$, $P = 0.546$; mean population social differentiation: $F_{1,\ 6} = 1.103$, $P = 0.334$). This term was therefore excluded from subsequent statistical tests.

Temporal patterns of association

Temporal patterns of association were examined using lagged association rates, which describe the probability that two individuals will associate τ time units after a previous association [33,66]. Lagged association rates were plotted continuously (using moving averages) against time lag in Socprog 2.3. We considered changes in lagged and null association rates over two different time scales: time lag in minutes during the 30-minute observation period and time lag in days over the 13-day study period. The precision of lagged association rates was estimated by jacknifing across sampling periods, using a 5-minute sampling period for 'observation' (7 jackknife groups) and a 1-day sampling period for 'day' (4 jacknife groups).

The lagged association graphs were fitted to several mathematical models in order to describe the temporal pattern of organisation in our populations and estimate the parameters of the models. To reduce the number of mathematical models fitted to the data we first inspected the lagged association rate curves to determine the descriptors that were most appropriate. For sampling periods in minutes, we chose four exponential models (Table 1) that included rapid periods of disassociation (breaking up of groups), as these would be expected in fission-fusion societies [22]. We selected the model of best fit for each population according to the quasilikelihood variant of the Akaike Information Criterion (QAIC) where the QAIC value provides an indicator of the level of support for the model (lower values providing the highest level of support) [67]. For time periods of days, inspection of the lagged association rates revealed linear relationships; we therefore fitted a custom model ($g(\tau) = e\tau + f$) describing a constant rate of change in association probability over time (Table 1).

To compare temporal patterns of social organisation among high and low predation risk populations, the overall best-fitting model was applied to all populations and the final model's parameters calculated. We then compared the parameters from these best-fit models (using t-tests) to evaluate the effect of predation risk on patterns of temporal organisation. In order to compare the temporal structure of our populations we had to find the overall model of best-fit; our aim of this part of the study was therefore not to construct accurate mathematical models, but to look for differences in overall patterns of temporal organisation in populations from high and low risk environments.

Results

Testing for non-random patterns of social organisation

Females from both high and low predation risk populations exhibited non-random patterns of social organisation with some individuals preferentially associating with or avoiding others (coefficients of variation for the real association indices were significantly different from those for the random indices; high risk real CV = 0.52, random CV = 0.37, P = 0.01; low risk real CV = 0.77, random CV = 0.56, P = 0.01). Inspection of the dyadic levels of significance revealed that overall, paired associations were characterised by patterns of avoidance (indicated by $P < 0.025$) rather than preference ($P > 0.975$). Patterns of gregariousness did not differ from those expected from random models (i.e. there was no preference by individuals for large or small group sizes); standard deviations of typical group size for real and random data sets were similar for both high (sd real data = 0.70, sd random = 0.52, P = 0.70) and low predation populations (sd real data = 0.77, sd random = 0.55, P = 0.85).

Effect of predation risk on social networks and their development

We found a significant effect of predation risk on PC1 and mean population social differentiation. The variables loaded by PC1 – population means for association index, gregariousness, strength, eigenvector centrality, reach, clustering coefficient and affinity - had higher mean values in high predation populations than low

Table 1. Description of models fitted to lagged association rates (g) in Socprog 2.3. For time lags (τ) of minutes, we chose four models (parameters: *a, b, c, d*) that incorporated rapid periods of disassociation and selected the model of best fit according to the quasiliklihood Akaike Information criterion (QAIC).

Name		Model	Description
RD + CC	Rapid disassociation + constant companions	$g = a$	Some associations decay within 1 sampling period then g is stable. Short-lived, non-random associations
RD + CA	Rapid disassociation + casual acquaintances	$g = ae^{(-b\tau)}$	Some associations decay within one sampling period then g falls to zero.
RD+CC+CA	Rapid disassociation + constant companions + casual acquaintances	$g = a + ce^{(-b\tau)}$	Rapid disassociation within one sampling period and an association rate that falls before levelling off.
RD+2CA	Rapid disassociation + 2 levels of casual acquaintances	$g = ae^{(-b\tau)} + ce^{(-d\tau)}$	Rapid disassociation within one sampling period and levels of disassociation at time intervals of $^1/_b$ and $^1/_d$.
Custom	Gradual change in association/ avoidance over time	$g = e + (-f\tau)$	Linear change in probability that two individuals remain associated following time lag.

For time lags of days we fitted a custom model describing a linear change in association probability over time.

predation ones (Fig 1a, Tables 2 & 3). Overall our populations showed some level of social differentiation (values >0.3) and this effect was stronger for low risk populations than high risk ones (Tables 2 & 3; Fig. 1b). There was no effect of day or the (predation risk x day) interaction term on PC1 or mean population social differentiation in our social association networks (Table 2). Thus overall, the social networks for females from low predation risk environments (Fig. 2a) had fewer connections with lower overall strength than those for fish from high predation risk habitats (Fig. 2b). We accounted for sampling upstream and downstream locations of the same stream by entering both river (6-levels) and day as random effects in a General Linear Mixed Model with predation risk as the fixed factor (it was not possible to have random effects in the MANOVAs). The random effect was non-significant for both PC1 and social differentiation ($F_{5,\ 31}=2.04$, $P=0.10$ and $F_{5,\ 31}=0.86$, $P=0.52$ respectively); the effect of predation risk in this model remained significant for PC1 ($F_{1,\ 31}=8.50$, $P=0.007$) but not social differentiation ($F_{1,\ 31}=0.25$, $P=0.62$).

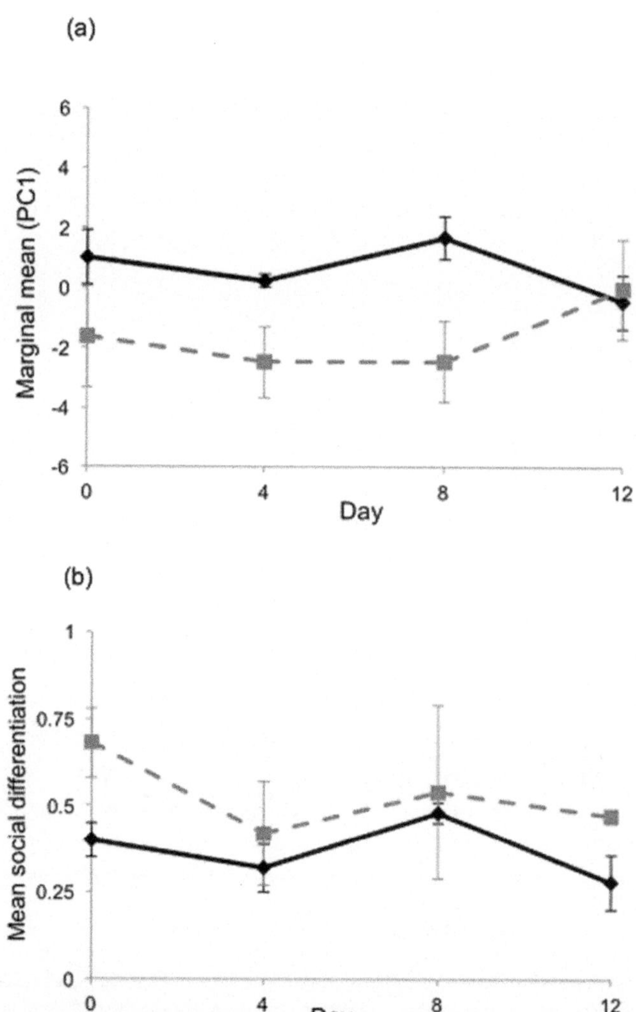

Figure 1. Effect of time (day) and predation risk on guppy social network measures. Figure a represents the marginal means (± SE) for PC1 while figure b shows the mean population social differentiation. Solid lines represent high predation populations; dashed lines are low risk sites.

Table 2. MANOVA models testing the effects of predation risk, day and their interaction on PC1 and mean population social differentiation (both entered as repeated measures responses).

Response variable	*Dependent variable*	*df*	*F*	*P*
PC1	Predation risk	1, 7	5.77	**0.047**
	Day	3, 21	0.26	0.857
	Predation risk *day	3, 21	1.41	0.269
Social differentiation	Predation risk	1, 7	11.39	**0.012**
	Day	3, 21	1.15	0.353
	Predation risk*day	3, 21	0.37	0.779

Significant effects at P<0.05 are shown in bold.

Temporal patterns of association

Lagged association rates plotted for both high and low predation risk populations were best described by an exponential model of rapid disassociation with two levels of casual acquaintances (RD+2CA); QAIC values were lowest for this model in 7 of our 12 populations (the RD + CA model best-fit all the remaining populations except one). In this model ($g(\tau)=ae^{(-b\tau)}+ce^{(-d\tau)}$)parameters a and c are the proportion of associations that break up within a sampling period and $1/_b$ and $1/_d$ represent the time scale of these two periods of disassociation. Thus the first term in the model ($ae^{(-b\tau)}$)represents fission events over short time scales while the latter term ($ce^{(-d\tau)}$) describes longer term changes in social organisation [22]. This type of model may represent fission of short term associations and the slow decay of more permanent relationships or might arise where some individuals form stable groups that move in and out of other groups [22]. The observed association rates were higher than the null rates in each of our populations, confirming that females displayed preferences/avoidances for particular individuals over both time scales (the 30-minute observation period and the 13-day duration of the study). Fitting the RD +2CA model ($g(\tau)=ae^{(-b\tau)}+ce^{(-d\tau)}$)to each of our populations and calculating mean values of the model's parameters gave the following estimates of lagged association rates

Table 3. Network association measures averaged over sampling periods (days) and populations to give overall means and standard errors for high (n = 7) and low risk (n = 5) networks.

Measure	*High predation*	*Low predation*
Association index (AI)	0.265±0.03	0.177±0.03
Gregariousness	0.847±0.08	0.680±0.07
Social differentiation	0.395±0.03	0.414±0.06
Strength	2.911±0.27	1.95±0.29
Eigenvector centrality	0.276±0.001	0.271±0.003
Reach	10.95±1.82	5.69±1.54
Clustering coefficient	0.53±0.03	0.45±0.03
Affinity	3.16±0.29	2.14 ±0.29

Data were obtained from Socprog 2.3.

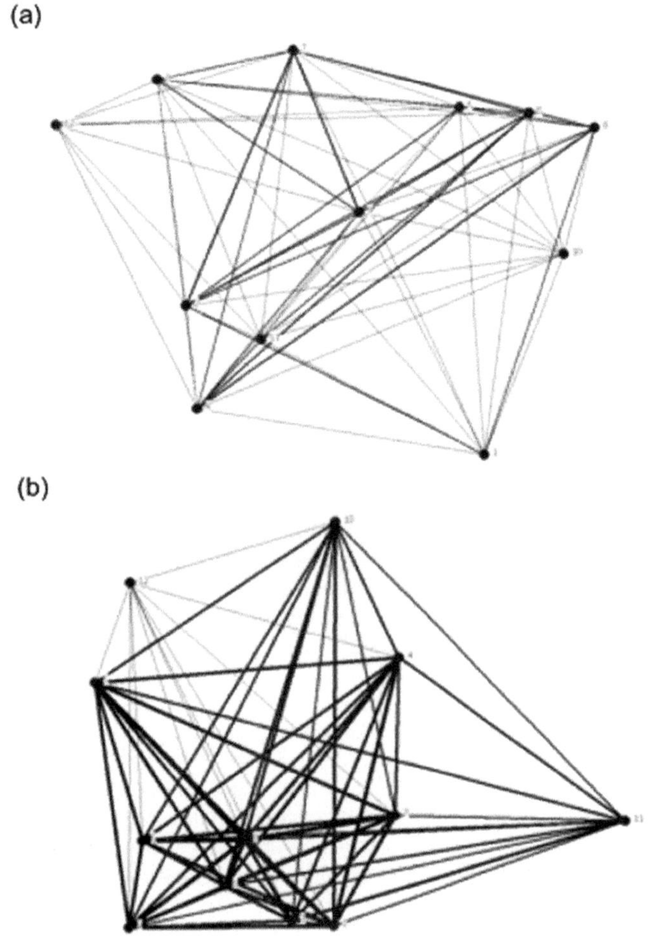

Figure 2. Example of social networks for female guppies collected from habitats differing in predation risk. Fish were from a low predation (a) and high predation (b) population of the Tacarigua River, Trinidad. Associations are represented by lines (edges) between individuals, which are weighted so that stronger associations are shown with darker lines. Drawn in Netdraw [91].

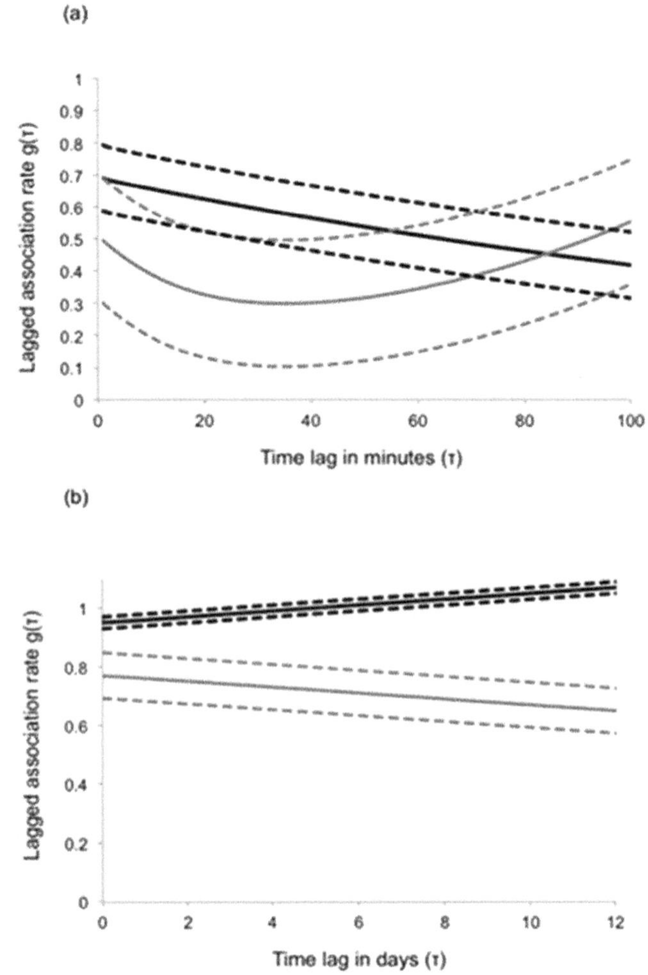

Figure 3. Models of temporal social structure in guppy populations. Lagged association rates are plotted against time lag in minutes (a) and days (b) for networks from high (black lines) and low (grey lines) predation populations. Solid lines indicate mean parameter values; dotted lines indicate the upper and lower boundaries for the standard error of the mean parameter values. The x-axis extends beyond the 30-minute sampling period (in fig. a) to illustrate the longer, 2nd period of fission predicted by the models.

for high and low risk populations (Fig. 3a):

$$\text{High predation: } g(\tau) = 0.01e^{-1.03\tau} + 0.69e^{-0.01\tau}$$

$$\text{Lower predatio: } g(\tau) = 0.32e^{-0.20\tau} + 0.21e^{0.05\tau}$$

Comparison of the estimated parameters of our lagged association models revealed that predation risk contributed to varying patterns of temporal social organisation. For the first term in the model ($ae^{(-b\tau)}$)representing changes in short term associations, parameters a and b differed significantly between populations with high and low predation risk. At short time lags, associations between pairs of low predation females were more likely to break up than those between high predation fish (parameter a, mean ± s.e.: high risk = 0.01±0.05, low risk = 0.32±0.10; $t_{10} = 2.23$, $P = 0.014$). The time scale over which these short-term fission events occurred (given by $1/b$) was also influenced by predation risk; in other words, associations among fish from high predation populations tended to break up over periods of less than one minute while those among fish in low predation populations broke up over periods of around 5 minutes (parameter b, mean ± s.e.: high risk = 1.03±0.09, low risk = 0.20±0.15, $t_{10} = 2.23$, $P < 0.001$).

For the second period of disassociation ($ce^{(-d\tau)}$) the above patterns were reversed. Affiliations between pairs of high predation females were more likely to break up than those between low predation fish (parameter c, mean ± s.e.: high risk = 0.69±0.05, low risk = 0.21±0.10, $t_{11} = 2.20$, $P = 0.001$). Indeed, the model for fish from low predation habitats suggests an increase in the probability of association over longer time periods (e.g. >40 mins; Fig 3a). Here, the models suggest that associations among fish from high predation habitats broke up over periods of 100 minutes while fish from low predation habitats disassociated over a period of about 20 minutes. However, the error around these estimates was high and the differences not significant (parameter d, mean ± s.e.: high risk = 0.01±0.00, low risk = -0.05±0.04, $t_{11} = 2.20$, $P = 0.10$). The effect of predation risk on parameters a, b, c remained significant following correction for multiple comparisons

[False Discovery Rate for dependent comparisons, 68; adjusted critical significance value = 0.024]. When we considered patterns of association over the 13-day familiarity period, predation risk did not influence rates of change in association with time (Fig. 3b; parameter *e*, high risk = 0.01±0.00, low risk = -0.01±0.01, $t_8 = 2.31$, P = 0.19).

Discussion

Our results reveal that predation risk can influence the fine-scale social structure of fission-fusion societies. While a long history of studies has examined the relationship between predation risk and prey group size [5,6,69,70], our findings suggest that more complex elements of social dynamics can be influenced by risk. Specifically, we found that individuals from high predation risk populations were more closely affiliated, and also their neighbours better socially connected, than those from low risk populations that had weaker social ties. Our results also provide one of the first demonstrations of the link between predation risk and temporal patterns of social organisation. Associations among fish from low predation habitats broke up over periods of 5 and 20 minutes, while fish from high predation habitats were more likely to remain associated during these time frames. Thus the strong associations recorded among fish from high predation rivers during our 30-minute observation period likely contributed to the higher levels of overall social connectedness in high predation networks versus low predation ones. Interestingly, the lagged association models suggested that the reverse pattern occurred over longer time periods with associations among fish from low predation rivers more likely to persist. We therefore suggest that the stability of affiliations among females crucially depends on the time frame over which these social preferences are observed.

The dynamic nature of fission-fusion societies provides an ideal framework for testing socioecological theory - the identification of ecological factors that drive variation in social behaviour - because these can provide key insights into large scale evolutionary processes [70,71]. Previous applications of social analysis techniques to wild animal populations have revealed the importance of ecological factors in influencing social organisation; for example, seasonal changes in resource availability can influence the social structure of baboon, bat and elephant societies [17,24,72]. While several studies have considered the role of predation risk in contributing to the observed community social structure [32,34,73], our study provides the first experimental approach to examining the effects of evolutionary/ontogenetic exposure to predation risk on animal sociality.

We predicted that due to the high risk of predation associated with the movement of individuals among groups, shoals collected from high predation environments would exhibit higher stability with reduced frequency of fission-fusion events compared with fish in low predation habitats. Thus we expected that high levels of association would be observed within shoals but lower levels of overall social connectivity would occur among these groups. Our findings do not appear to match these predictions because we found greater overall social connectivity in fish from high predation risk habitats (in our 30-minute observation period). One explanation for this is that our sample populations represented associations among a small number of shoals (n = 12 individuals; median natural shoal size = 5 fish) in a contained area, thus individuals are likely to be more associated than fish in the wild that have greater opportunity for movement among different social groups. Another possible explanation that requires further investigation is that fission/fusion events may involve movements of different sized groups. In fish collected from high predation habitats, small groups of fish may move among larger shoals to reduce the risks associated with travelling among groups. This would likely contribute to high overall levels of social connectivity at the community level. In fish from less risky habitats, individuals may be more likely to move in and out of groups. More research into the temporal dynamics of shoaling behaviour would shed light into mechanisms of movement within and among shoals.

We acknowledge that a limitation of our study (and a common problem in many other social network studies) is the lack of within-population replication. In order to generalise our findings and confidently describe the social structure of a particular population, we would need to construct multiple, independent networks for each sample site. However, this was not the purpose of our study (our networks were constructed in small artificial pools in the laboratory); rather, we aimed to look for overall effects of risk on social behaviour at the level of the population. Whilst we are confident that the divergent patterns of social structure observed in this study correspond to variation in predation risk, we suggest that within-population variation in social organisation is an important area for future research.

It is possible that factors other than predation risk varied between upstream and downstream sites (e.g. population density, sex ratio, water quality, canopy cover) and contributed to our observed differences in social structure. However, we attempted to control for any of these immediate effects by observing same-sex networks comprising similar-sized individuals in the laboratory under identical conditions. Inspection of the association indices for females from upstream and downstream populations of the Oropuche River, which are both high predation environments [74], reveals that they are similar (e.g. mean AI ± se: upstream = 0.18±0.04, downstream = 0.19±0.03), giving us some confidence that differences among populations are a result of predation risk, rather than any other ecological variables. Predation risk is known to influence not only group size and shoaling tendency in guppies, but also a number of other social behaviours such as cooperation, microhabitat use and courtship [50]. Thus the relationship between social organisation and predation risk reflects the combination of the direct and indirect social implications of risk. Like previous work describing the effects of predation risk on guppy shoaling behaviour [11], our observations were also conducted in the absence of predator-related cues, suggesting a fundamental basis to the observed variation in sociality. While our findings are a necessary first step in evaluating the effects of prior predation risk on spatial and temporal social structure, it would be very interesting to investigate the effects of immediate risk (e.g. predator presence) on social structure in fish populations from high and low risk habitats [75].

Relating the structure of social networks to their large-scale properties [76] allows us to make a number of predictions regarding the effect of predation risk on large-scale population processes. Studies linking social organisation with disease transmission have revealed that individuals that play a central role in the network are more likely to become naturally infected with pathogens than those that are less associated [77]. Furthermore, experimental infection of highly socially connected individuals can cause higher rates of disease transmission than if randomly selected individuals are infected [78]. Translating these findings to the present study, we might predict that the transfer of disease or information would be more rapid in high predation populations where individuals are better connected and likely to remain associated, at least over short time periods. Guppies can learn novel foraging tasks, escape routes and antipredator responses through associating with or following other individuals [79,80] and learn more effectively from familiar fish than unfamiliar fish [81]. Specifically, guppies behave differently in the presence of trained/experienced fish and the novel behaviour is retained once they (the trained fish) are removed. Thus, novel information is expected to spread through a population depending on the movements of the knowledgeable/experienced individual(s) and their social connections within the community [82].

The flow of information/disease is likely to be non-random and directed through particular individuals or sub-groups within the community [83]. Previous studies have shown that age [84,85], kinship [73] and behavioural phenotype [48,86] can bias individual-level interactions and affect the centrality of the network [87]. Identifying the role that these genetic and phenotypic factors play in structuring both high and low predation risk networks is an interesting avenue for further research. The level of sub-structuring or 'cliquishness' within the community is important because information or disease may be transmitted rapidly through highly connected sub-groups but less rapidly through the population as a whole [76]. The level of sub-structuring observed in this study (given by the clustering coefficient; population range: 0.4 to 0.74) is higher than that reported for dolphins and sperm whales [88,89] and similar to that reported for bats and other freshwater fishes [47,90]; high levels may act to facilitate the flow of information among members of cliques while increasing their susceptibility to disease. Cliquish networks are also predicted to be less robust as the removal of key individuals or their associations may cause the community sub-structure to fragment [42].

We found no evidence that increasing familiarity among females, which occurs over the course of 12 days in guppies [55], influenced the development of the social networks. This is in contrast to previous work showing that familiarity can develop in small networks comprised of randomly selected females [46]. In this previous study [46], the development of social recognition was examined by testing the preference of females for shoal partners originating from the same social network over females from different networks. One suggestion for the discrepancy in results between our study and that of Darden et al. [46] is that the effects of increasing familiarity among females are not evidenced by changes in association towards one another but rather by their response upon encountering unfamiliar individuals. Thus social recognition among small groups of individuals may develop in the absence of any observed changes in overall network structure over time.

Understanding the impacts of ecological variability on animal social organisation is essential because it allows us to identify factors that may have facilitated the evolution of sociality. The recent advancement of social analysis techniques allows us to go beyond describing the effects of predation risk on prey group size and composition to reveal increasingly complex patterns of spatial and temporal organisation. Our findings suggest that the social structure of a population depends on the temporal scale at which patterns of association are observed. Consideration of the temporal aspects of social organisation is therefore essential, particularly in the case of fission-fusion societies where group splitting/joining events can occur on multiple time scales.

Supporting Information

Table S1 Details of the predator assemblages at each of the twelve guppy populations sampled.

Supporting Information S1 Additional information on the analysis of social structure using Socprog.

Acknowledgments

We are extremely grateful to Hal Whitehead for advice on use of the social analysis program Socprog and Richard James for general discussion of network analysis. Ben Chapman, John Dyer, Christos Ioannou, Chantima Piyapong, Phil Thomas and Amy Wade kindly assisted in the field and Indar Ramnarine provided logistical support. We are also grateful to the referees whose thoughtful comments greatly improved our manuscript.

Author Contributions

Conceived and designed the experiments: JLK JK DPC. Performed the experiments: JLK CI LJM DPC. Analyzed the data: JLK LJM CI. Contributed reagents/materials/analysis tools: JLK LJM DPC JK. Wrote the paper: JLK.

References

1. Couzin ID, Laidre ME (2009) Fission-fusion populations. Current Biology 19: R633–R635.
2. Foley R, Gamble C (2009) The ecology of social transitions in human evolution. Philosophical Transactions of the Royal Society B-Biological Sciences 364: 3267–3279.
3. Rubenstein DI, Wrangham RW (1986) Ecological aspects of social evolution. Princeton: Princton University Press.
4. Clutton-Brock TH, Harvey PH (1977) Primate ecology and social organization. Journal of the Zoological Society of London 183: 1–39.
5. Crook JH, Gartlan JS (1966) Evolution of primate societies. Nature 210: 1200–1203.
6. Jarman PJ (1974) Social organization of antelope in relation to their ecology. Behaviour 48: 215-&.
7. Krob J, Heinze J, eds (2008) Ecology of social evolution. Berlin Heidelberg: Springer.
8. Hill RA, Lee PC (1998) Predation risk as an influence on group size in cercopithecoid primates: implications for social structure. Journal of Zoology 245: 447–456.
9. Hamilton WD (1971) Geometry for the selfish herd. Journal of Theoretical Biology 31: 295–311.
10. Caraco T, Wolf LL (1975) Ecological determinants of group sizes of foraging lions. American Naturalist 109: 343–352.
11. Seghers BH (1974) Schooling behaviour in the guppy (*Poecilia reticulata*): an evolutionary response to predation. Evolution 28: 486–489.
12. Wrona FJ, Jamieson Dixon RW (1991) Group size and predation risk: a field analysis of encounter and dilution effects. American Naturalist 137: 186–201.
13. Connor RC, Wells R, Mann J, Read A (2000) The bottlenose dolphin, *Tursiops* spp.: social relationships in a fission-fusion socieity. In :Mann J, Connor RC, Tyack P, Whitehead H, eds. Cetacean socieities: Field studies of whales and dolphins. Chicago: University of Chicago Press.
14. Croft DP, Krause J, James R (2004) Social networks in the guppy (*Poecilia reticulata*). Proceedings of the Royal Society London B (Biology Letters) 271: S516–S519.
15. Kruuk H (1972) The spotted hyena: a study of predation and social behavior. ChicagoIllinois: University of Chicago Press.
16. Sueur C, Petit O, Deneubourg JL (2010) Short-term group fission processes in macaques: a social networking approach. Journal of Experimental Biology 213: 1338–1346.
17. Wittemyer G, Douglas-Hamilton I, Getz WM (2005) The socioecology of elephants: analysis of the processes creating multitiered social structures. Animal Behaviour 69: 1357–1371.
18. Lehmann J, Boesch C (2004) To fission or to fusion: effects of community size on wild chimpanzee (Pan troglodytes verus) social organisation. Behavioral Ecology and Sociobiology 56: 207–216.
19. Krause J, Ruxton GD (2002) Living in Groups. Oxford: Oxford University Press.
20. Aureli F, Schaffner CM, Boesch C, Bearder SK, Call J, et al. (2008) Fission-Fusion Dynamics New Research Frameworks. Current Anthropology 49: 627–654.
21. Hinde RA (1976) Interactions, relationships and social structure. Man 11: 1–17.
22. Whitehead H (2008) Analyzing animal societies. Chicago: University of Chicago Press. 336 p.

23. Smith JE, Kolowski JM, Graham KE, Dawes SE, Holekamp KE (2008) Social and ecological determinants of fission-fusion dynamics in the spotted hyaena. Animal Behaviour 76: 619–636.
24. Henzi SP, Lusseau D, Weingrill T, van Schaik CP, Barrett L (2009) Cyclicity in the structure of female baboon social networks. Behavioral Ecology and Sociobiology 63: 1015–1021.
25. Chapman CA, Wrangham RW, Chapman LJ (1995) Ecological constraints on group size - an analysis of spider monkey and chimpanzee subgroups. Behavioral Ecology and Sociobiology 36: 59–70.
26. Ramos-Fernandez G, Boyer D, Gomez VP (2006) A complex social structure with fission-fusion properties can emerge from a simple foraging model. Behavioral Ecology and Sociobiology 60: 536–549.
27. Kummer H (1971) Primate Societies: group techniques of ecological adaptation. Chicago: University of Chicago Press.
28. Smythington MM (1988) Food competition and foraging party size in the black spider monkey (*Ateles paniscus chamek*). Behaviour 105: 117–134.
29. Boesch C (1991) The effects of leopard predation on grouping patterns in forest chimpanzees. Behaviour 117: 221–242.
30. Lehmann J, Dunbar R (2009) Implications of body mass and predation for ape social system and biogeographical distribution. Oikos 118: 379–390.
31. Sakura O (1994) Factors affecting party size and composition of chimpanzees (*Pan troglodytes verus*) at Bossou, Guinea. International Journal of Primatology 15: 167–183.
32. Karczmarski L, Wursig B, Gailey G, Larson KW, Vanderlip C (2005) Spinner dolphins in a remote Hawaiian atoll: social grouping and population structure. Behavioral Ecology 16: 675–685.
33. Whitehead H (2009) SOCPROG programs: analysing animal social structures. Behavioural Ecology and Sociobiology 63: 765–778.
34. Fortin D, Fortin ME, Beyer HL, Duchesne T, Courant S, et al. (2009) Group-size-mediated habitat selection and group fusion-fission dynamics of bison under predation risk. Ecology 90: 2480–2490.
35. Croft DP, James R, Krause J (2008) Exploring Animal Social Networks. Princeton, N. J. Princeton University Press. 192 p.
36. Callaway DS, Newman MEJ, Strogatz SH, Watts DJ (2000) Network robustness and fragilitiy: percolation on random graphs. Physical Review Letters 85: 5468–5471.
37. Berger SI, Iyengar R (2009) Network analyses in systems pharmacology. Bioinformatics 25: 2466–2472.
38. Romano CM, Guedes de Carvalho-Mello IMV, Jamal LF, Lucas de Melo F, Iamarino A, et al. (2010) Social networks shape the transmission dynamics of Hepatitus C virus. Plos One 5: e11170.
39. Cummings GS, Bodin O, Ernston H, Elmqvist T (2010) Network analysis in conservation biogeography: challenges and opportunities. Diversity and Distributions 16: 414–425.
40. Sih A, Hanser SF, McHugh KA (2009) Social network theory: new insights and issues for behavioral ecologists. Behavioral Ecology and Sociobiology 63: 975–988.
41. Krause J, Croft DP, James R (2007) Social network theory in the behavioral sciences: potential applications. Behavioral Ecology and Sociobiology 62: 15–27.
42. Wey T, Blumstein DT, Shen W, Jordan F (2008) Social network analysis of animal behaviour: a promising tool for the study of sociality. Animal Behaviour 75: 333–344.
43. Flack JC, Girvan M, de Waal FBM, Krakauer DC (2006) Policing stabilizes construction of social niches in primates. Nature 439: 426–429.
44. Drewe JA (2010) Who infects whom? Social networks and tuberculosis transmission in wild meerkats. Proceedings of the Royal Society B-Biological Sciences 277: 633–642.
45. Croft DP, James R, Thomas POR, Hathaway C, Mawdsley D, et al. (2006) Social structure and co-operative interactions in a wild population of guppies (Poecilia reticulata). Behavioral Ecology and Sociobiology 59: 644–650.
46. Darden SK, James R, Ramnarine IW, Croft DP (2009) Social implications of the battle of the sexes: sexual harassment disrupts female sociality and social recognition. Proceedings of the Royal Society B-Biological Sciences 276: 2651–2656.
47. Croft DP, James R, Ward AJW, Botham MS, Mawdsley D, et al. (2005) Assortative interactions and social networks in fish. Oecologia 143: 211–219.
48. Croft DP, Krause J, Darden SK, Ramnarine IW, Faria JJ, et al. (2009) Behavioural trait assortment in a social network: patterns and implications. Behavioral Ecology and Sociobiology 63: 1495–1503.
49. Endler JA (1995) Multiple-trait coevolution and environmental gradients in guppies. Trends in Ecology and Evolution 10: 22–29.
50. Magurran AE (2005) Evolutionary Ecology: the Trinidadian guppy. Oxford: Oxford University Press. 206 p.
51. Seghers BH (1974) Schooling behaviour in the guppy (*Poecilia reticulata*) - an evolutionary response to predation. Evolution 28: 486–489.
52. Morales JM, Moorcroft PR, Matthiopoulos J, Frair JL, Kie JG, et al. (2010) Building the bridge between animal movement and population dynamics. Philosophical Transactions of the Royal Society B-Biological Sciences 365: 2289–2301.
53. Rodgers GM, Ward JR, Askwith B, Morrell LJ (2011) Balancing the Dilution and Oddity Effects: Decisions Depend on Body Size. Public Library of Science One 6.
54. Ward AJW, Hart PJB (2003) The effects of kin and familiarity on interactions between fish. Fish and Fisheries 4: 348–358.
55. Griffiths SW, Magurran AE (1997) Familiarity in schooling fish: how long does it take to acquire? Animal Behaviour 53: 945–949.
56. Liley NR, Seghers BH (1975) Factors affecting the morphology and behaviour of guppies in Trinidad. In :Baerends GP, Beer C, Manning A, eds. Function and Evolution in Behaviour. Oxford: Clarendon Press. pp 92–118.
57. Magurran AE, Seghers BH, Shaw PW, Carvalho GR (1995) The behavioural diversity and evolution of guppy, *Poecilia reticulata* populations in Trinidad. Advances in the Study of Animal Behavior 24: 155–202.
58. Rodd FH, Reznick DN (1997) Variation in the demography of guppy populations: The importance of predation and life histories. Ecology 78: 405–418.
59. Croft DP, Arrowsmith BJ, Bielby J, Skinner K, White E, et al. (2003) Mechanisms underlying shoal composition in the Trinidadian guppy, *Poecilia reticulata*. Oikos 100: 429–438.
60. Pitcher TJ, Magurran AE, Allan JR (1983) Shifts of behaviour with shoal size in cyprinids. Proceedings of the British Freshwater Fisheries Conference 3: 220–228.
61. Whitehead H, Dufault S (1999) Techniques for analysing vertebrate social structure. Advances in the Study of Behavior 28: 33–74.
62. Krause S, Mattner L, James R, Guttridge T, Corcoran MJ, et al. (2009) Social network analysis and valid Markov chain Monte Carlo tests of null models. Behavioural Ecology and Sociobiology 63: 1089–1096.
63. Jacoby DMP, Busawon DS, Sims DW (2010) Sex and social networking: the influence of male presence on social structure of female shark groups. Behavioral Ecology 21: 808–818.
64. Barthelemy MA, Barrat R, Pastor-Sarorras R, Vespignani A (2005) Characterization and modeling of weighted networks. Physica A 346: 34–43.
65. Newman MEJ (2004) Analysis of weighted networks. Physical Review E 70: 056131.
66. Whitehead H (1995) Investigating structure and temporal scale in social organizations using identified individuals. Behavioral Ecology 6: 199–208.
67. Whitehead H (2007) Selection of models of lagged identification rates and lagged association rates using AIC and QAIC. Communications in Statistics-Simulation and Computation 36: 1233–1246.
68. Benjamini Y, Yekutieli D (2001) The control of the false discovery rate in multiple testing under dependency. Annals of Statistics 29: 1165–1188.
69. Van Schaik CP (1983) Why are diurnal primates living in groups? Behaviour 87: 120–144.
70. Cluttonbrock TH, Harvey PH (1977) Primate ecology and social organization. Journal of Zoology 183: 1–39.
71. Emlen ST, Oring LW (1977) Ecology, sexual selection and the evolution of mating systems. Science 197: 215–223.
72. Vonhof MJ, Whitehead H, Fenton MB (2004) Analysis of Spix's disc-winged bat association patterns and roosting home ranges reveal a novel social structure among bats. Animal Behaviour 68: 507–521.
73. Wittemyer G, Okello JBA, Rasmussen HB, Arctander P, Nyakaana S, et al. (2009) Where sociality and relatedness diverge: the genetic basis for hierarchical social organization in African elephants. Proceedings of the Royal Society B-Biological Sciences 276: 3513–3521.
74. Endler JA, Houde AE (1995) Geographic variation in female preferences for male traits in *Poecilia reticulata*. Evolution 49: 456–468.
75. Botham MS, Hayward RK, Morrell LJ, Croft DP, Ward JR, et al. (2008) Risk-sensitive antipredator behavior in the Trinidadian guppy, *Poecilia reticulata*. Ecology 89: 3174–3185.
76. Newman MEJ (2003) The structure and function of complex networks. Society of Industrial and Applied Mathematics 45: 167–256.
77. Corner LAL, Pfeiffer DU, Morris RS (2003) Social-network analysis of Mycobacterium bovis transmission among captive brushtail possums (Trichosurus vulpecula). Preventive Veterinary Medicine 59: 147–167.
78. Corner LAL, Pfeiffer DU, de Lisle GW, Morris RS, Buddle BM (2002) Natural transmission of Mycobacterium bovis infection in captive brushtail possums (Trichosurus vulpecula). New Zealand Veterinary Journal 50: 154–162.
79. Kelley JL, Evans JP, Ramnarine IW, Magurran AE (2003) Back to school: can antipredator behaviour in guppies be enhanced through social learning? Animal Behaviour 65: 655–662.
80. Reader SM, Kendal JR, Laland KN (2003) Social learning of foraging sites and escape routes in wild Trinidadian guppies. Animal Behaviour 66: 729–739.
81. Swaney W, Kendal J, Capon H, Brown C, Laland KN (2001) Familiarity facilitates social learning of forgaging behaviour in the guppy. Animal Behaviour 62: 591–598.
82. Hoppitt W, Laland KN (2011) Detecting Social Learning Using Networks: A Users Guide. American Journal of Primatology 73: 834–844.
83. Coussi-Korbel S, Fragaszy DM (1995) On the relation between social dynamics and social learning. Animal Behaviour 50: 1441–1453.
84. Patriquin KJ, Leonard ML, Broders HG, Garroway CJ (2010) Do social networks of female long-eared bats vary with reproductive period and age? Behavioral Ecology and Sociobiology 64: 899–913.
85. Wey TW, Blumstein DT (2010) Social cohesion in yellow-bellied marmots is established through age and kin structuring. Animal Behaviour 79: 1343–1352.
86. Pike TW, Samanta M, Lindstrom J, Royle NJ (2008) Behavioural phenotype affects social interactions in an animal network. Proceedings of the Royal Society B-Biological Sciences 275: 2515–2520.
87. Schurch R, Rothenberger S, Heg D (2010) The building-up of social relationships: behavioural types, social networks and cooperative breeding in a

cichlid. Philosophical Transactions of the Royal Society B-Biological Sciences 365: 4089–4098.
88. Lusseau D, Wilson B, Hammond PS, Grellier K, Durban JW, et al. (2006) Quantifying the influence of sociality on population structure in bottlenose dolphins. Journal of Animal Ecology 75: 14–24.
89. Wiszniewski J, Allen SJ, Moller LM (2009) Social cohesion in a hierarchically structured embayment population of Indo-Pacific bottlenose dolphins. Animal Behaviour 77: 1449–1457.
90. Chaverri G (2010) Comparative social network analysis in a leaf-roosting bat. Behavioral Ecology and Sociobiology 64: 1619–1630.
91. Borgatti S (2002) NetDraw: Graph Visualization Software. Harvard: Analytic Technologies.

Soil Food Web Changes during Spontaneous Succession at Post Mining Sites: A Possible Ecosystem Engineering Effect on Food Web Organization?

Jan Frouz[1,2]*, Elisa Thébault[3], Václav Pižl[1], Sina Adl[4], Tomáš Cajthaml[5], Petr Baldrián[5], Ladislav Háněl[1], Josef Starý[1], Karel Tajovský[1], Jan Materna[6], Alena Nováková[1], Peter C. de Ruiter[7]

1 Institute of Soil Biology, Biology Centre Academy of Sciences of the Czech Cepublic, České Budějovice, Czech Republic, **2** Institute for Environmental Studies, Charles University, Prague, Czech Republic, **3** Laboratoire Biogéochimie et Ecologie des Milieux Continentaux Unité mixte de recherche 7618, Centre National de la Recherche Scientifique, Paris, France, **4** Department of Soil Science, University of Saskatchewan, Saskatoon, Canada, **5** Institute of Microbiology Academy of Sciences of the Czech Cepublic, Prague , Czech Republic, **6** Krkonoše National Park, Museum of Krkonoše in Vrchlabí, Vrchlabí, Czech Republic, **7** Biometris, Wageningen University, Wageningen, The Netherlands

Abstract

Parameters characterizing the structure of the decomposer food web, biomass of the soil microflora (bacteria and fungi) and soil micro-, meso- and macrofauna were studied at 14 non-reclaimed 1– 41-year-old post-mining sites near the town of Sokolov (Czech Republic). These observations on the decomposer food webs were compared with knowledge of vegetation and soil microstructure development from previous studies. The amount of carbon entering the food web increased with succession age in a similar way as the total amount of C in food web biomass and the number of functional groups in the food web. Connectance did not show any significant changes with succession age, however. In early stages of the succession, the bacterial channel dominated the food web. Later on, in shrub-dominated stands, the fungal channel took over. Even later, in the forest stage, the bacterial channel prevailed again. The best predictor of fungal bacterial ratio is thickness of fermentation layer. We argue that these changes correspond with changes in topsoil microstructure driven by a combination of plant organic matter input and engineering effects of earthworms. In early stages, soil is alkaline, and a discontinuous litter layer on the soil surface promotes bacterial biomass growth, so the bacterial food web channel can dominate. Litter accumulation on the soil surface supports the development of the fungal channel. In older stages, earthworms arrive, mix litter into the mineral soil and form an organo-mineral topsoil, which is beneficial for bacteria and enhances the bacterial food web channel.

Editor: Luís A. Nunes Amaral, Northwestern University, United States of America

Funding: Czech Science foundation (http://www.gacr.cz/en/) grant no P504/12/1288. The funders had no role in study design, data collection and analysis, decision to publish, or preparation of the manuscript.

Competing Interests: The authors have declared that no competing interests exist.

* E-mail: frouz@natur.cuni.cz

Introduction

Mechanisms behind ecological succession have been a lasting topic in ecosystem studies [1,2]. It has frequently been emphasized that interactions among plants and between plants and their environments drive succession [2–4]. Most studies on this issue, however, have considered only vegetation patterns, ignoring community assembly processes at other trophic levels than primary producers [5]. Here we focus on the development of the decomposer food web during succession. The decomposer food web substantially affects the nutrient status of soil [6,7], and changes in the decomposer food web may have an effect on plant succession [8,9].

The decomposer food web processes a majority of organic matter produced in terrestrial ecosystems [10]. Fungi and bacteria play a principal role in primary decomposition of dead organic matter, while the direct contribution of detritivorous invertebrate consumers in terms their digestion and assimilation is assumed to be low, although they may alter conditions for soil microorganisms and affect the decomposition rate indirectly [10]. Moreover, soil invertebrates feeding on dead organic matter also ingest micro-organisms associated with and embedded in the organic matter. Since the nutritional value of micro-organisms is much higher than that of dead plant material, they become crucial food source for other organisms in decomposer food webs [10,11]. In this context, the bacterial and fungal pathways of the decomposer food web require particular attention in soil food web studies [12]. Proportion of bacterial and fungal pathway in the food web has significant consequences for ecosystem functioning. The bacterial pathway is generally associated with processing more easy decomposable litter and promotes faster mineralization and nutrient release. On the other hand, prevalence of the fungal pathway is associated with slower mineralization, lower nutrient availability and carbon sequestration [6,7,10,13]. During terrestrial succession in temperate zones, the soil food web is assumed to change from a bacterial to a fungal dominated system [6,14]. The soil food web is also hypothesized to be bottom-up regulated [15], while succession changes are driven by changes in the plant community with consequent changes in litter quality [6,16]. However, the soil biota also substantially modifies its environment by bioturbation, aggregate formation, aeration and soil mixing [7,10]. Some soil invertebrates, such as earthworms, may have strong engineering effects on soil and may completely alter humus forms, causing dramatic shifts in soil properties [7,10]. The consequences of these engineering effects on soil food web structure are still largely unknown.

In order to better understand primary succession of soils, we studied the soil food webs in a mine spoil chronosequence. We hypothesized that (1) because the amount of organic matter increases during succession, the soil supports more trophic levels and more complex food webs (higher connectance and diversity) following Odum's theory of ecosystem succession [4,5]; (2) as succession proceeds, the food web changes from a bacterial channel dominated food web towards a more fungal channel dominated one [14]; (3) thickness of fermentation (Oe) layer would be a good predictor of fungal to bacterial channel. Thickness of fermentation layer is driven by several processes, including litter input and bioturbation. Input of litter, namely litter with high CN ratio such as birch, alder or willow litter increase Oe layer thickness on the other hand bioturbation by soil fauna namely earthworms may decrease it [9].

We sampled fourteen decomposer food webs along a chronosequence of post mining sites ranging in age from 1 to 41 years [9,17] and compared these food web structures with previous studies of soil organic layer development. Post-mining sites are very suitable for this type of study because soil develops here anew from the parent material. Moreover, previous studies show that the presence of earthworms at these sites causes massive changes in soil morphology and plant community composition [9,17,18].

Materials and Methods

Ethics Statement

Land, where study was performed, has been owned by mining company "Sokolovská Uhelná a.s. právní nástupce", which permitted access to the sites as well as all filed works done in this study and also provided data about plot age. As concern other rules and regulations no permits were required for the described study, which complied with all relevant regulations.

Study site

The study was carried near the town of Sokolov in the western part of the Czech Republic, Central Europe on four large colliery spoil heaps (from several dozen to several hundred km^2 in area) resulting from open-cast coal mining. The altitude of the study area was 500–600 m a.s.l. with mean annual precipitations of 650 mm and mean annual temperature of 6.8°C. A chronosequence of 14 non-reclaimed sites, created 1–41 years ago using the same substrate, was available for the study [9,19]. At all of the sites selected for the study, no subsequent manipulation of the heap material occurred, and the sites have developed on their own since the heaping. At each site, a sampling plot (10×10 m) was established at least 5 m from the site margin. Plot age (time since heaping) was determined based on historical data supplied by the coal mining company. The substrate of all the sites was composed of alkaline tertiary clay shales, substrate do not show any sign of contamination, site pH gradually decreased with plot age, as plant succession progressed (Table 1). As a result of the heaping process, parallel rows of depressions and elevations were formed at each site. Because of remarkable changes in vegetation soil development, we called these succession stages the initial, the shrub and the forest stage; the initial stage covered four sites 1–14 years old, the shrub stage six sites 15–22 years old and the forest stage four sites 23–41 years old [9,19]. In the initial stage (i.e., 1–14 years before the establishment of the shrub layer), the vegetation was dominated by a sparse herb cover, and the topsoil formed mainly from the dumped spoil material. At the 14–22 year-old sites dominated by shrubs (namely *Salix caprea*), the litter reserve on the soil surface was higher than annual litter production, and a thick fermentation layer developed. In older plots (24 years old or older), dominated by a young birch and aspen (*Betula pendula* and *Populus tremula*) forest, the humus layer was formed, and the thickness of the fermentation layer decreased (Table 1).

Sampling and analysis

Soil microflora sampling and processing is described in Frouz and Nováková [19]. Composite soil samples consisting of 5 individual subsamples (each consisting of about 100 g of soil) were sampled in March 2002 from top 5 cm below the litter layer (see Frouz and Nováková [19] for more details). Both microhabitat depressions and wave tops were sampled separately, but pooled data are presented here. Microbial biomass was measured by the chloroform fumigation and extraction method [20]. Biomass was assumed to be twice the mass of C. The amount of fungal biomass was estimated by measuring ergosterol content (see Baldrian et al. [21] for more details).

The soil fauna was sampled in March 2002 as described in Frouz et al. [9]. To study the soil microfauna and mesofauna, two composite samples (one from depressions and the other from elevations) were taken from each plot, each consisting of five individual samples (area of each 10 cm^2, depth 5 cm). Two sets of samples were collected, one for the extraction of microfauna and enchytraeids and the other for the rest of the mesofauna. Samples of the first set were homogenized; 1 g of the mixed soil was diluted and used to directly count protists following standard methods [22]. Additional 10 g of the soil were exposed in Baermann funnels modified according to Háněl [23] to isolate the metazoan microfauna (nematodes, rotifers and tardigrades) and enchytraeids. A Tullgren apparatus was used for the extraction of other mesofauna groups.

To study the soil macrofauna, two composite samples (one from depressions and the other from elevations), each consisting of five particular samples (area of each 125 cm^2, depth 5 cm), were taken in each plot. The macrofauna were extracted by a Kempson apparatus.

In total 30 guilds of soil biota was distinguished, two guilds of soil microflora (bacteria and fungi), four guilds of Protozoa (naked. amoebae, flagellates, ciliates and testacea) and 24 guilds of soil fauna. The soil fauna were sorted into 24 guilds according to the following ad hoc guild system. Taxonomic groups of the soil fauna with rather uniform ecology were considered as guilds. Other taxonomic groups were divided into several guilds based on their feeding habits. Nematodes (Nematoda) were divided into bacterivores (Nem-Bact), fungivores (Nem-F), root and fungal feeders (Nem-RFF), plant parasites (Nem-PP), predators (Nem-P), and omnivors (Nem-O) [24]. Plant-associated nematodes are herein referred to either as root fungal feeders - consisting from members of family Tylenchidae that feed on plant roots but also on fungi, and plant parasites that feed on various parts of plants and cause plans diseases [25]. Diptera larvae were divided into microsaprophags, saprophags and predators according to Frouz [26]. Earthworms (Lumbricidae) were divided into two groups: litter dwelling epigeic earthworms and mineral soil dwellers. The first group includes species such as *Dendrodrilus rubidus* and *Dendrobaena octaedra*, which live preferentially in the litter layer. The latter includes endogeic species and the epigeic species *Lumbricus rubellus*,

Table 1. Characteristics of soil development and plant production in individual succession stages.

	pH		Oe		A		herb mass		litter input		fine roots	
			cm		cm		g m^{-2}		g m^{-2} year^{-1}		g m^{-2}	
initial	7.9	±0.4	0.1	±0.1	0.0	±0.0	42	±34	8	±14	63	±78
shrub	7.2	±0.3	2.4	±0.1	0.0	±0.0	22	±35	152	±87	594	±3
forest	6.6	±0.3	0.6	±0.2	3.4	±1.7	111	±76	268	±97	463	±194

Values represent mean ± SD for individual phases of succession obtained by pooling data from 4 initial, 6 shrub and 4 forest sites, the initial, shrub and forest stages were 1–13, 14–22 and 23–41 years old, respectively. Oe refers to fermentation layer thickness and A to the thicknes of the darker organo-mineral layer. Based on data in Frouz et al. [9].

which lives in litter as well as at the top of the mineral layer [27,28]. This ad hoc system for earthworms was used because anectic species were completely absent from the investigated plots and endogeic species were rare. The biomass of each group was estimated based on allometric equations. Carbon was estimated based on the assumption that 50% of biomass is carbon. Characteristics of individual feeding guilds were used to produce a topographic stucture of food web (Figure 1).

Estimation of feeding rates in the soil food web

In order to estimate annual feeding rates from the observed trophic group biomasses, we followed Hunt et al. [29] by assuming that species production balances out losses through natural death and predation (assumption of steady state). Trophic group feeding rates can thus be expressed as:

$$F_{ij} = \frac{\omega_{ij} B_i}{\sum_{k=0}^{n} \omega_{kj} B_k} F_j \text{ with } F_j = \frac{d_j B_j + \sum_{k=0}^{n} F_{jl}}{a_j p_j} \quad (4)$$

where F_{ij} is group j feeding rate on group i (g m^{-2} yr^{-1}) which depends on F_j, the group j overall feeding rate (g m^{-2} yr^{-1}) and on the relative biomass B_i of group i and feeding preference ω_{ij} of j on i. The feeding preferences, death rates, assimilation and production efficiencies of the different trophic groups were taken from Hunt et al. [29] and further updated by a review of recent literature on the subject (see Table 2).

The estimation of the feeding rates requires solving a system of equations (4) with all trophic groups. In a few instances (4 out of 14 food webs), it was not possible to solve this system because of intraguild predation between two top predators (Aranea and Formicidae). In this case, we consider that only the most abundant predator consumes the other, and thus we ignore one trophic link when estimating feeding rates.

Analyses of soil food web structure

We analysed several parameters of the food web structure. First, food web richness (S) corresponds to the number of groups in the food web, while food web diversity ($DivS$) is calculated as $DivS = 2^D$ with $D = -\sum_{i=1}^{S} \frac{B_i}{B_{tot}} \log_2\left(\frac{B_i}{B_{tot}}\right)$ and B_i is the biomass of trophic group i and B_{tot} the total food web biomass (Shannon diversity index). Food web connectance is expressed as the number of realized links in the food web over the number of possible links.

Secondly, we also consider measures based on trophic position and measures based on energy channels. The trophic position of a species is defined here by the average of the trophic position of the species it consumes weighted by the diet fraction these species represent:

$TL_i = 1 + \sum_{j=1}^{S} g_{ij} TL_j$ Where TL_i is the trophic level of species i and gij the fraction of the consumer i's diet derived from the prey j. These "flow-based" trophic levels were computed following the method of Levine (1980) [30]. The average trophic level for each consumer is the sum of all entries in each column of $A = [I - G]^{-1}$ with I being the identity matrix and $G = (gij)$. The food web structure is characterized here by the average trophic level (TLm), calculated as average of all values of group trophic levels in the food web and by the maximum trophic level (TLM), corresponding to the maximum group trophic level in the food web.

Fungal and bacterial energy channels are measured by the biomass of all the groups belonging to that channel weighted by their contribution to this channel. The contribution of a group to a channel is defined by $C_i = \sum_{j=1}^{S} g_{ij} C_j$ and thus the contribution of each group is equal to the product of A by a vector V, with $V_i = 1$ for the source of the energy channel (either fungi or bacteria) and 0 otherwise. We measured two different indices to quantify the fungal and bacterial energy channel. First we summed the biomass of all the groups belonging to a given channel weighted by their contribution C to this channel. Then, because the order of magnitude of biomasses differs strongly between the trophic groups, we also calculated the energy channels with standardized biomasses of each group by dividing the biomass of one group by the overall mean of that group over all food webs [31].

We estimated the stability of the soil food webs by following the method of de Ruiter et al. [32] and Neutel et al. [5,33]. Stability is evaluated on the Jacobian community matrix, following May's approach [34]. We estimate the Jacobian matrix corresponding to each food web as a function of the trophic group biomasses and feeding rates calculated above and by assuming that species dynamics can be modelled by using the generalized Lotka-Volterra system near the equilibrium [33]. Food web instability is defined as the level of intraspecific interaction (diagonal strength) needed for all eigenvalues of the Jacobian matrix to have negative real parts. Food webs that need high levels of diagonal strength are more instable.

A one way ANOVA was used to compare individual food web parameters among individual succession stages. Trends between individual parameters and plot age were fitted by linear, logarithmic, exponential and or second order polynom functions. We present function that was significant and explain the largest proportion of data variability.

Results

During succession, the biomass of the whole food web, as well as the amount of C that flows through it increased (Figure 2, Table 2).

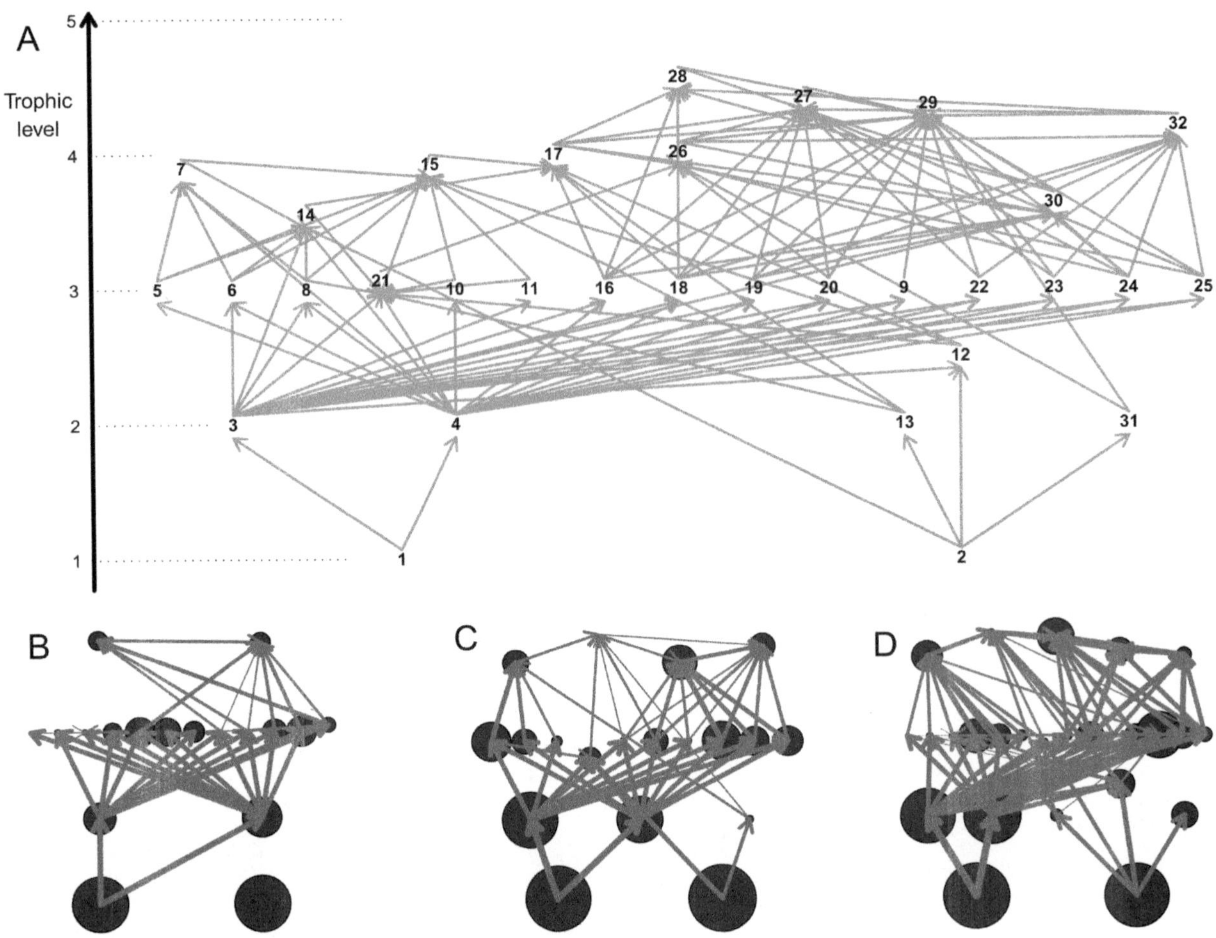

Figure 1. Food web diagrams in the post mining chronosequence. A. General food web diagram as defined in this study. **B**, **C** and **D**. Examples of soil food webs observed in the initial (**B**), shrub (**C**) and forest (**D**) stages. In panel **A**, the numbers refer to the trophic groups: 1, detritus; 2, roots; 3, fungi; 4, bacteria; 5, naked amoebae; 6, flagellates; 7, ciliates; 8, testacea; 9, enchytraeidae; 10, bacteriophagous nematodes; 11, fungivorous nematodes; 12, root and fungal feeder nematodes; 13, plant parasite nematodes; 14, omnivorous nematodes; 15, predatory nematodes; 16, oribatida; 17, predaceous acari; 18, collembolan; 19, other microarthropods; 20, litter dwelling lumbricidae; 21, mineral dwelling lumbricidae; 22, isopoda; 23, diplopoda; 24, microsaprophagous diptera; 25, macrosaprophagous diptera; 26, predaceous diptera; 27, chilopoda; 28, aranea; 29, carabidae; 30 staphylinidae; 31, root feeding colleoptera and 32, formicidae. The arrow on the left indicates the trophic level of the groups, with primary producers and detritus at the bottom of the food web (trophic level = 1). In panels **B**, **C** and **D**, the size of the circles correspond to the relative biomass of the different trophic groups in the food web, while the width of the arrows corresponds to the relative intensity of C flow between trophic groups estimated by the model.

The number of functional groups was significantly positively correlated with plot age (Figure 3). Also the maximum trophic level of the groups in the food web, which indicates the length of the food chain in the food web, increased with succession age. However, this increase was the most pronounced during the first years of succession before shrub vegetation developed. Later on, the maximum trophic level of the groups in the food web did not differ significantly between the shrub and the initial stage (Figure 3). The mean trophic level of the groups did not differ significantly between succession stages (Figure 3). We did not detect any significant changes in food web connectance during succession (Figure 4). Food web instability in the shrub stage was significantly higher than in the other stages (Figure 4).

Initial food webs were dominated by the bacterial channel, but in the shrub stage and intermediate stages of succession, there was a strong and significant increase in the ratio of the fungal over the bacterial channel, and the intermediate stages of succession were strongly dominated by fungi (Figure 5, Table 2). Later on in forest stages, the proportion of the fungal over the bacterial channel decreased significantly. In both the initial and the forest stage, the food webs were therefore dominated by the bacterial channel (Figure 5). Considering the environmental variables in Table 1, thickness of fermentation layer (Oe horizon) was the best predictor of the fungal over the bacterial channel (Figure 6). Other parameters in Table 1 did not show any significant correlation with the fungal/bacterial channel ratio.

Discussion

Remarkable shifts from the bacterial to the fungal channel and back were observed during the succession. The increase of the fungal channel compared to the initial bacteria-dominated channel in the shrub stage is consistent with the general assumption that the importance of the fungal channel should grow during succession [6]. On the contrary, the decrease in the importance of the fungal channel and the prevalence of the bacterial channel observed in the next forest stage seems to be against this expectation. As shown above, the thickness of the

Table 2. Trophic groups.

Trophic group	Initial		Shrub		Forest		A	P	D
Fungi	4106	±4659a	43853	±28161b	31354	±12629ab	1	0.44	3.7
Bacteria	6404	±6791a	6787	±3603a	47031	±18944b	1	0.51	9
Nak. Amoebae[1]	+		+		+		0.55	0.58	7.3
Flagellates	+		+		+		0.52	0.6	7.3
Ciliates	+		+		+		0.55	0.58	7.3
Testacea	34	±29a	715	±219b	223	±51ab	0.55	0.58	7.3
Enchytraeidae	22	±39a	8	±18a	267	±164b	0.28	0.29	1.95
Nem-Bact[2]	89	±112	72	±12	14	±7	0.54	0.49	14.1
Nem-F[3]	4	±5	1	±1	1	±0	0.42	0.31	6
Nem-RFF[4]	1	±1	+		3	±3	0.42	0.31	6
Nem-PP[5]	+		1	±6	5	±6	0.42	0.31	2.3
Nem-O[6]	146	±211	164	±120	77	±49	0.55	0.28	5.8
Nem-P[7]	13	±17	44	±57	54	±49	0.55	0.28	5.8
Oribatida	+		+		+		0.5	0.4	1.42
Pred. acari[8]	+		+		+		0.75	0.3	3.44
Collembola	+	a	24	±25b	47	±25b	0.34	0.37	1.96
Other microart[9]			+		+		0.34	0.37	1.96
Lumb liter[10]			+		65	±101	0.22	0.32	0.14
Lumb min[11]		a	++	1a	3040	±2357b	0.22	0.32	0.14
Isopoda			1	±390	9	±16	0.25	0.18	1
Diplopoda	155	±252	5	±119	254	±407	0.25	0.18	0.5
Dip. micro[12]	46	±40			27	±20	0.50	0.17	3.33
Dip. sap[13]	20	±35a	+	a	400	±265b	0.35	0.125	1
Dip. pred[14]	84	±145	40	±20	12	±21	0.60	0.17	1
Chilopoda		a	284	±195b	546	±125c	0.34	0.20	1
Aranea	15	±13ab	2	±32a	33	±18b	0.72	0.18	1
Carabidae	10	±13	41	±195	13	±13	0.80	0.17	1
Staphylinidae	4	±3	9	±*	1	±1	0.50	0.17	1
Root f. Col[15]	7	±12a		a	34	±26b	0.50	0.17	0.5
Formicidae			19	±19	5	±4	0.30	0.10	1
Sum	11161	±12024a	52584	±29479a	83517	±33404b			

Mean and SD of biomass dry weight (mg m^{-2}) in individual sucession stages and values of physiological parameters of the different trophic groups in the food web. A – assimilation efficiency, P – production efficiency, D – death rate (yr^{-1}), data from Hunt et al. [29]. Biomass values lower than 0.5 mg are marked as +; empty cells indicate the absence of a group. Statistically homogeneous groups of the same guild in various succession stages are marked by the same letter (ANOVA, LSD test $p<0.05$); no letter means that no significant difference was found.

1 – Naked Amoebae, 2 – Nematodes, bacteriophagous, 3 – Nematodes, fungivorous, 4 – Nematodes, root and fungal feeders, 5 – Nematodes, plant parasites, 6 – Nematodes, omnivores, 7 – Nematodes, predators, 8 – Predaceous acari, 9 – Other microarthropods, 10 –Lumbricidae, litter dwelling, 11 – Lumbricidae, mineral dwelling, 12 – Diptera, microsaprophagous, 13 – Diptera, macrosaprophagous, 14 –Diptera, predaceous, 15 – Colleoptera, root feeding.

fermentation (Oe) layer is the best predictor of the fungal/bacterial channel ratio. The decrease in the fungal bacterial channel ratio at the transition from the shrub to the forest stage corresponds with a decrease in the thickness of the fermentation layer. Previous micromorphological observations [9,17,19] on the same chronosequence show that this decrease is associated with colonization of these sites by earthworms, which take part in bioturbation and mixing of litter and the fermentation layer in mineral soil. This mixing is also associated with changes in the soil microbial community, namely an increase in populations of bacteria and a decrease in the fungal/bacterial ratio [18,19,21] and may explain the observed changes in the fungal to bacterial channel ratio. The observed decrease in the importance of the fungal over the bacterial channel and the dominance of the bacterial channel in forest stages of succession might have therefore been caused by engineering effects of earthworms, which remove the fermentation layer that is favourable for fungi by bioturbation. In addition, consumption, digestion and excretion of the fermentation layer also promotes the growth of bacteria. This is consistent with previous studies describing the effects of earthworm invasion on soil and the microbial community in particular [19,35]. In the present study, the removal of the litter accumulation and fermentation (Oe) layer from the soil surface by earthworm bioturbation is assumed to be the major factor promoting bacteria over fungi [36]. Changes of humus forms from mor-moder to moder-mull associated with earthworm colonization were ob-

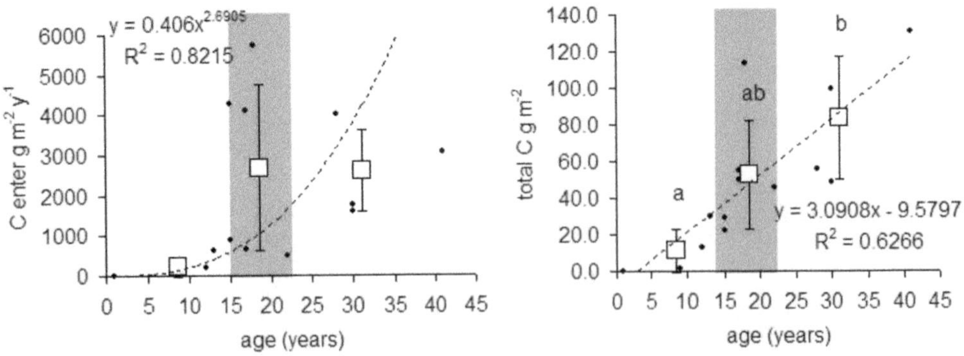

Figure 2. Food web C flow and total biomass of individual food webs related to plot age. Amount of C entering food web ner m^2 and year, and biomas per m^2. Dots indicate the parameter value for individual ages; squares correspond to the mean value (bars mean SD) for initial, shrub and forest stages. Grey areas highlight the shrub stage and delimitate the initial stage on the left and the forest stage on the right. A trend line between a given parameter and time is present only if significant $p<0.05$; statistically homogeneous groups of stages are indicated by the same letter (ANOVA, LSD post hoc test $p<0.05$); if no letter is present in the panel, no significant difference was found.

served not only in this chronosequence but also on other chronosequences of primary succession [37–40]. Consequent changes in the microbial community are in agreement with the concept of humus form where moder-like forms of humus with little bioturbation are dominated by fungi whereas mull-like forms with intensive bioturbation are dominated by bacteria [7,10]. The effect of earthworms bioturbation on the soil community is complex and consists of effects of worm on vegetation and soil [9,41]. Similar mechanisms may apply also in situations when earthworm free ecosystems are invaded by earthworms, such as north temperate forests in USA [35].

In our opinion, this observation cannot be interpreted that this particular chronosequence develops into a bacteria-driven food web at the climax stage. Instead, we presume that the proportion

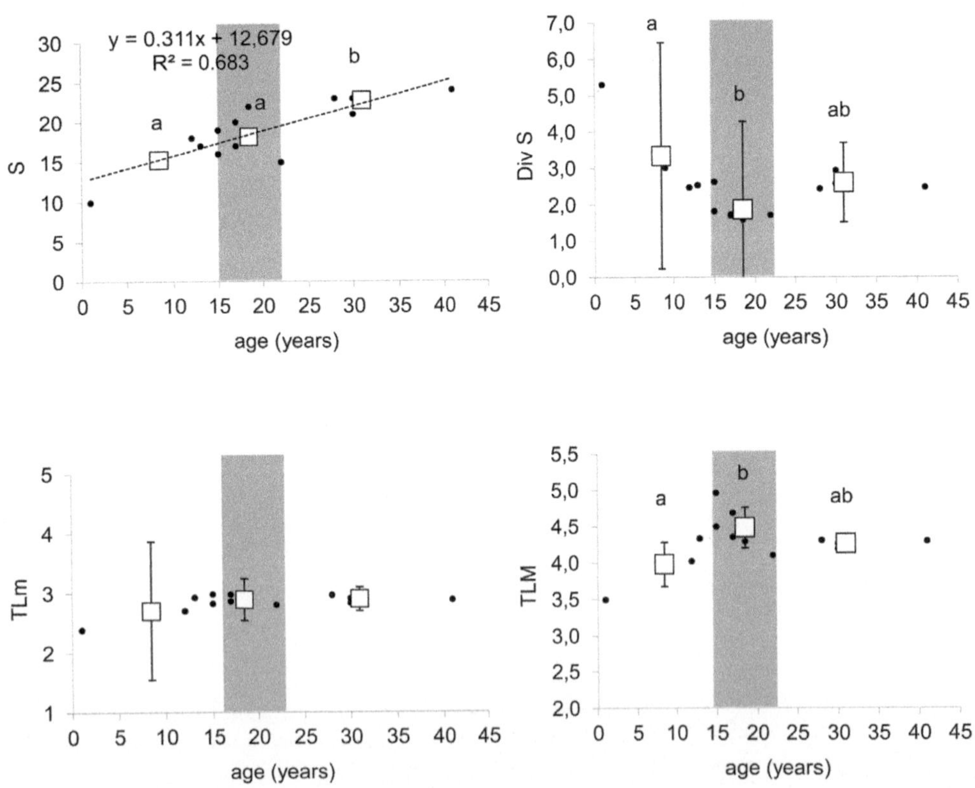

Figure 3. Food web diversity related to plot age. Number of functional groups in the food web (S), exponent of Shannon-Weiner index, of diversity for functional groups present in the food web (Div S), mean (TLm) and maximum (TLM) trophic level of the groups in the food web. Dots indicate the parameter value for individual ages; squares correspond to the mean value (bars mean SD) for initial, shrub and forest stages. Grey areas highlight the shrub stage and delimitate the initial stage on the left and the forest stage on the right. A trend line between a given parameter and time is present only if significant $p<0.05$; statistically homogeneous groups of stages are indicated by the same letter (ANOVA, LSD post hoc test $p<0.05$); if no letter is present in the panel, no significant difference was found.

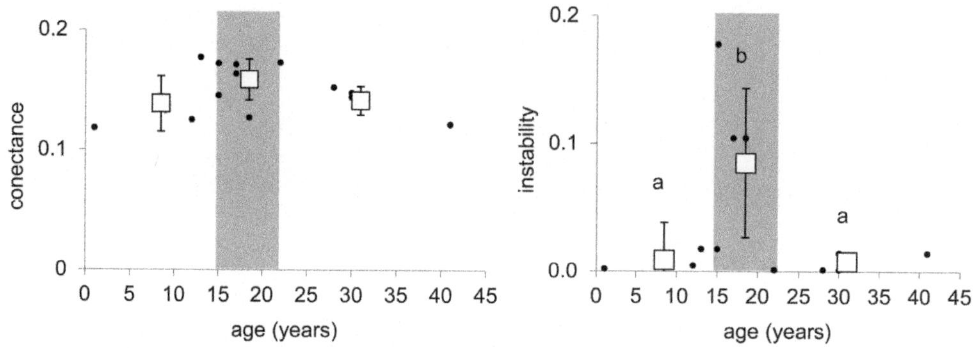

Figure 4. Conectance and instability of food webs related to plot age. Dots indicate the parameter values for individual ages; squares correspond to the mark mean value (bars mean SD) for initial, shrub amd forest stages. Grey areas highlight the shrub stage and delimitate the herb stage on the left and the forest stage on the right. There is no significant trend line between the given parameters and time. Statistically homogeneous groups of stages are indicated by the same letter (ANOVA, LSD post hoc testp<0.05).

between the fungal and the bacterial channel may change several times over the course of succession as soil organic matter accumulates and the soil community matures from a functional ecology perspective as well as from a biodiversity perspective. The oldest observed stages in the present study are now dominated by an aspen and birch forest, but there are already numerous beech and oak seedlings in the understory [9]. A beech and oak forest represents the hypothetical (historical) climax stage in this region. Further development toward a beech and oak dominated climax forests may bring about further changes in litter quality and cause another shift to a fungal food web. We rather tend to interpret the observed shifts in food web channels as intermediate succession changes.

One interesting question is how much can this pattern of successional development be related to soil development in general and how much is the observed successional pattern a mere exception in early soil genesis and primary succession situations. Similar waves in plant diversity are often observed during plant community succession [42–44]. Adl [45] also pointed out that changes in the soil food web and community structure occur after about 20 years of succession, which is likely related to soil structure and horizon development. Kaufmann [46] observed based on comparison of long-term data and chronosequence data that community development at individual sites often does not follow idealized succession trajectories that can be derived from chronosequences, which, again, suggests that successional trajectories are likely to consist of one or several oscillating waves rather than a gradual transition from one stage to another.

The effect of ecosystem engineers on food web structure is frequently discussed [47,48] but seldom observed. One of the reasons for this is that the effect is indirectly related to environmental changes mediated by particular ecosystem engineers. These changes may be slow, so it may take some time after engineering species occur before changes in the ecosystem can be detected and, vice versa, existing structures may persist for some time after engineering species are removed.

The observed increase in overall food web biomass and diversity with increasing succession age is in good agreement with at least one previous study [14]. As concern development of individual taxa oribatids, which are bioindicators of site disturbance, it is likely that the sites are still too immature to provide an adequate habitat. Others such as predatory consumers (nematodes, acarids) appear when there is sufficient prey in the food web, that is, in later succession stages. Similarly, collembola and root feeding nematodes appear later when there is sufficient root biomass. The biomass of protists is always low in soil; their contribution to biomass turnover (bacteriophagy) is nonetheless significant [49–51].

We did not detect changes in food web connectance even though other studies found positive results [5]. Food web

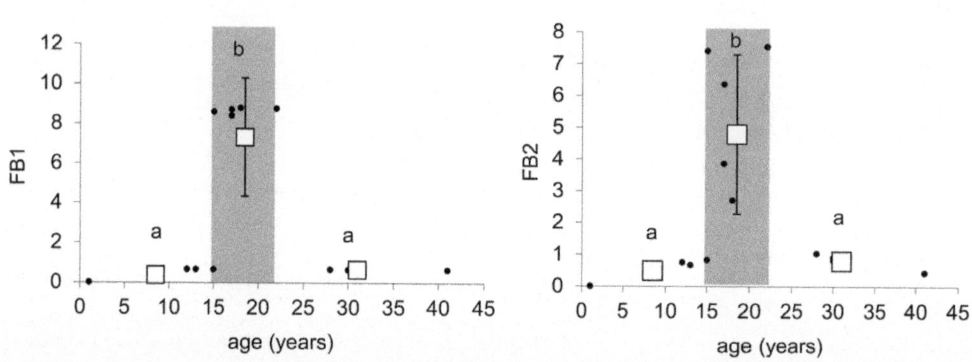

Figure 5. Proportion of the fungal to the bacterial channel. FB1 -based on the biomass of individual guilds, FB2 based on the biomass of individual guilds standardized over the biomass of the same guild in the whole chronosequence. Dots indicate the parameter values for individual ages; squares correspond to the mark mean value (bars mean SD) for initial, shrub and forest stages. Grey areas highlight the shrub stage and delimitate the herb stage on the left and the forest stage on the right. There is no significant trend line between the given parameters and time. Statistically homogeneous groups of stages are indicated by the same letter (ANOVA, LSD post hoc test p<0.05).

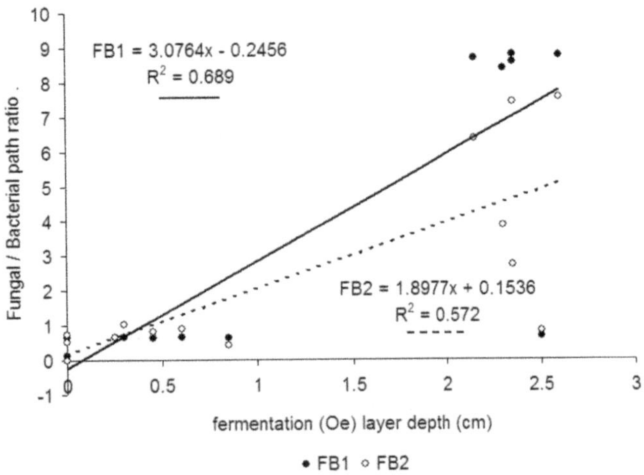

Figure 6. Relationship between fungal bacterial ration and Oe layer. Relationships between thiskness of fermentation Oe layer and two forms of fungal bacterial pathway ratio, FB1 -based on the biomass of individual guilds, FB2 based on the biomass of individual guilds standardized over the biomass of the same guild in the whole chronosequence. Both regression lanes are significant $p<0.05$.

instability in shrub dominated intermediate stages of succession increased, so stability decreased. This is consistent with previous observations of Neutel et al. [5] who suggest an alternating pattern of decreasing and increasing stability in the course of ecosystem succession. According to Neutel et al. [5], this stability fluctuation is connected with biomass accumulation at the top trophic level, and a decrease in stability at the point of creation of a new food web structure is caused by the arrival of a new top predator. In our case, the changes in food web structure included not only additional new trophic levels but also shifts in flows between the fungal and the bacterial pathway (as described above), which may contribute to the alteration of food web stability. A recent theoretical study indeed suggests that coupling of two different energy pathways in food webs can strongly determine ecosystem stability [52]. It would be difficult to generalize the outcome of our result based on one chronosequence, but exploring the role of pathway proportion in food web stability clearly represents a promising field for future research.

Acknowledgments

Coal mining company Sokolovská uhelná právní nástupce a.s. is thanked for research permit and data about plot age. Bruce Jafee (JafeeEdit) is thanked for proofreading of the ms. This study is part of the project "The role of leaf functional traits in soil organic matter accumulation during primary sucession".

Author Contributions

Conceived and designed the experiments: JF. Performed the experiments: JF VP SA TC PB LH JS KT JM AN. Analyzed the data: JF ET PR. Contributed reagents/materials/analysis tools: JF VP SA TC PB LH JS KT JM AN. Wrote the paper: JF ET PR SA.

References

1. Clements EF (1936) Nature and structure of the climax. J Ecol 24: 552–584.
2. Glenn-Lewin DC, Peet RK, Veblen TT (1992) Plant succession, theory and prediction. London: Chapman and Hall. 352 p.
3. Luken JO (1990) Directing ecological succession. London: Chapman and Hall. 251 p.
4. Odum EP (1969) The strategy of ecosystem development. Science 164: 262–270.
5. Neutel AM, Heesterbeek JAP, van de Koppel J, Hoenderboom G, Vos A, et al. (2007) Reconciling complexity with stability in naturally assembling food webs. Nature 449: 599–603.
6. Chapman SK, Langley JA, Hart SC, Koch GW (2006) Plants actively control nitrogen cycling: uncorking the microbial bottleneck. New Phytol 169: 27–34.
7. Ponge JF (2003) Humus form in terrestrial ecosystem: a framework to biodiversity. Soil Biol Biochem 35: 935–945.
8. De Deyn GB, Raaijmakers CE, Zoomer HR, Berg MP, de Ruiter PC, et al. (2003) Soil invertebrate fauna enhance grassland succession and diversity. Nature 422: 711–713.
9. Frouz J, Prach K, Pižl V, Háněl L, Starý J, et al. (2008) Interactions between soil development, vegetation and soil fauna during spontaneous succession in post mining sites. Eur J Soil Biol 44: 109–121.
10. Lavelle P, Bignell D, Lepage M, Wolters V, Roger P, et al. (1997) Soil function in a changing world: the role of invertebrate ecosystem engineers. Eur J Soil Biol 33: 159–193.
11. Wardle DA, Yeates GW (1993) The dual importance of competition and predation as regulatory forces in terrestrial ecosystems: evidence from decomposer food-webs. Oecologia 93: 303–306.
12. Moore JC, de Ruiter PC (2000) Invertebrates in detrital food webs along gradients of productivity. In: Coleman DC, Hendrix PF, editors. Invertebrates as webmasters in ecosystems. CABI Wallingford, UK. pp. 161–184.
13. Hedlund K, Griffiths B, Christensen S, Scheu S, Setälä H, et al. (2004) Trophic interactions in changing landscapes: responses of soil food webs. Basic Appl Ecol 5: 495–503.
14. Wardle DA, Yeates GW, Watson RN, Nicholson KS (1995) Development of the decomposer food-web, trophic relationships, and ecosystem properties during a three-year primary succession in sawdust. Oikos 73: 155–166.
15. Scheu S, Schaefer M (1998) Bottom-up control of the soil macrofauna community in a beechwood on limestone: manipulation of food resources. Ecology 79: 1573–1585.
16. Hobbie SE (1992) Effects of plant species on nutrient cycling. Trends Ecol Evol 7: 336–339.
17. Frouz J, Pižl V, Tajovský K (2007) The effect of earthworms and other saprophagous macrofauna on soil microstructure in reclaimed and un-reclaimed post-mining sites in Central Europe. Eur J Soil Biol 43: 184–189.
18. Frouz J, Elhottová D, Pižl V, Tajovský K, Šourková M, et al. (2007) The effect of liter quality and soil fauna composition on organic matter dynamic in post mining soil: a laboratory study. Appl Soil Ecol 37: 72–80.
19. Frouz J, Nováková A (2005) Development of soil microbial properties in topsoil layer during spontaneous succession in heaps after brown coal mining in relation to soil microstructure development. Geoderma 129: 54–64.
20. Vance ED, Brookes PC, Jenkinson DS (1987) An extraction method for measuring soil microbial biomass C. Soil Biol Biochem 19: 703–707.
21. Baldrian P, Trögl J, Frouz J, Šnajdr J, Valášková V, et al. (2008) Enzyme activities and microbial biomass in topsoil layer during spontaneous succession in spoil heaps after brown coal mining. Soil Biol Biochem 40: 2107–2115.
22. Adl SM, Acosta-Mercado D, Lynn DH (2007) Protozoa. In: Carter MR, Gregorich EG, editors. Soil sampling and methods of analysis. CRC Press. pp. 455–469.
23. Háněl L (1995) Secondary successional stages of soil nematodes in cambisols of South Bohemia. Nematologica 41: 197–218.
24. Yeates CW (1998) Feeding in free-living soil nematodes: a functional approach. In: Perry RN, Wricht DJ, editors. The physiology and biochemistry of free-living and parasitic nematodes. CABI publishing, Wallingford. pp. 245–269.
25. Háněl L (2003) Soil nematodes in cambisoil agroecosystem of the Czech Republic. Biologia 52: 205–216.
26. Frouz J (1999) Use of soil dwelling Diptera (Insecta, Diptera) as bioindicators: a review of ecological requirements and response to disturbance. Agr Ecosyst Environ 74: 167–186.
27. Makulec G (2002) The role of *Lumbricus rubellus* Hoffm. in determining biotic and abiotic properties of peat soils. Pol J Ecol 50: 301–339.
28. Pižl V (1992) Succession of earthworm population in abandoned fields. Soil Biol Biochem 24: 1623–1628.
29. Hunt HW, Coleman DC, Ingham ER, Ingham RE, Elliott ET, et al. (1987) The detrital food web in a shortgrass prairie. Biol Fertil Soils 3: 57–68.
30. Williams RJ, Martinez ND (2004) Stabilization of chaotic and non-permanent food-web dynamics. Eur Physical J B 38: 297–303.
31. Holtkamp R, Kardol P, van der Wal A, Dekker SC, van der Putten WH (2008) Soil food web structure during ecosystem development after land abandonment. Appl Soil Ecol 39: 23–34.
32. de Ruiter PC, Neutel AM, Moore JC (1995) Modelling food webs and nutrient cycling in agro-ecosystems. Trends Ecol Evol 9: 378–383.
33. Neutel AM, Heesterbeek JAP, de Ruiter PC (2002) Stability in real food webs: weak links in long loops. Science 296: 1120–1123.
34. May RM (1973) Qualitative stability in model ecosystems. Ecology 54: 638–641.
35. Bohlen PJ, Groffman PM, Fahey TJ, Fisk MC (2004) Ecosystem consequences of exotic earthworm invasion of north temperate forests. Ecosystems 7: 1–12.

36. Dempsey MA, Fisk MC, Fahey TJ (2011) Earthworms increase the ratio of bacteria to fungi in northern hardwood forest soils, primarily by eliminating the organic horizon. Soil Biol Biochem 43: 2135–2141.
37. Dunger W (1968) Die Entwicklung der Bodenfauna auf rekultivieren Kippen und Halden des Braunkohlentegebaues. Abh Ber Naturkundesmuseums Görlitz 43: 1–256.
38. Dunger W (1991) Zur Primärsukzession humiphager Tiergruppen auf Bergbauflächen. Zool Jahrb Syst 118: 423–447.
39. Frouz J, Keplin B, Pižl V, Tajovský K, Starý J, et al. (2001) Soil biota and upper soil layers development in two contrasting post-mining chronosequences. Ecol Eng 17: 275–284.
40. Rusek J (1971) Pedozootische Sukzessionen warend der Entwicklung von Ökosystemen. Pedobiologia 18: 426–433.
41. Aira M, McNamara N, Trevor P, Domínguez J (2009) Microbial communities of *Lumbricus terrestris* L. middens: structure, activity, and changes through time in relation to earthworm presence. J Soils Sediments 9: 54–61.
42. Prach K, Pyšek P, Šmilauer P (1999) Prediction of vegetation succession in human-disturbed habitats using an expert system. Restor Ecol 7: 15–23.
43. Prach K, Bartha S, Joyce CB, Pyšek P, van Diggelen R, et al. (2001) The role of spontaneous vegetation succession in ecosystem restoration: a perspective. Appl Veg Sci 4: 111–114.
44. Prach K, Pyšek P, Bastl M (2001) Spontaneous vegetation succession in human-disturbed habitats: a pattern across seres. Appl Veg Sci 4: 83–88.
45. Adl SM (2008) Setting the tempo in land remediation: short term and long term pattern in biodiversity recovery. Microbes Environ 23: 13–19.
46. Kaufmann R (2001) Invertebrate succession on an alpine glacier foreland. Ecology 82: 2261–2278.
47. Brussaard L (1998) Soil fauna, guilds, functional groups and ecosystem processes. Appl Soil Ecol 9: 123–135.
48. Worm B, Duffy JE (2003) Biodiversity, productivity and stability in real food webs. Trends Ecol Evol 18: 628–632.
49. Griffiths BS, Bonkowski M, Dobson G, Caul S (1999) Changes in soil microbial community structure in the presence of microbial-feeding nematodes and protozoa. Pedobiologia 25: 383–401.
50. Paustian K, Andrén O, Clarholm M, Hansson AC, Johansson G (1990) Carbon and nitrogen budgets of four agroecosystems with annual and perennial crops, with and without nitrogen fertilization. J Appl Ecol 27: 60–84.
51. Sohlenius B (1990) Influence of cropping system and nitrogen input on soil fauna and microorganisms in a Swedish arable soil. Biol Fertil Soils 9: 168–173.
52. Rooney N, McCann K, Gellner G, Moore JC (2006) Structural asymmetry and the stability of diverse food webs. Nature 442: 265–269.

13

Invasive Fishes Generate Biogeochemical Hotspots in a Nutrient-Limited System

Krista A. Capps[1,2]*, Alexander S. Flecker[1]

1 Department of Ecology and Evolutionary Biology, Cornell University, Ithaca, New York, United States of America, **2** Sustainability Solutions Initiative, University of Maine, Orono, Maine, United States of America

Abstract

Fishes can play important functional roles in the nutrient dynamics of freshwater systems. Aggregating fishes have the potential to generate areas of increased biogeochemical activity, or hotspots, in streams and rivers. Many of the studies documenting the functional role of fishes in nutrient dynamics have focused on native fish species; however, introduced fishes may restructure nutrient storage and cycling freshwater systems as they can attain high population densities in novel environments. The purpose of this study was to examine the impact of a non-native catfish (Loricariidae: *Pterygoplichthys*) on nitrogen and phosphorus remineralization and estimate whether large aggregations of these fish generate measurable biogeochemical hotspots within nutrient-limited ecosystems. Loricariids formed large aggregations during daylight hours and dispersed throughout the stream during evening hours to graze benthic habitats. Excretion rates of phosphorus were twice as great during nighttime hours when fishes were actively feeding; however, there was no diel pattern in nitrogen excretion rates. Our results indicate that spatially heterogeneous aggregations of loricariids can significantly elevate dissolved nutrient concentrations via excretion relative to ambient nitrogen and phosphorus concentrations during daylight hours, creating biogeochemical hotspots and potentially altering nutrient dynamics in invaded systems.

Editor: Diego Fontaneto, Consiglio Nazionale delle Ricerche (CNR), Italy

Funding: This work was funded by the National Science Foundation (Doctoral Dissertation Enhancement Program Grant (183-8371); Integrated Graduate Education and Research in Biogeochemistry and Environmental Biocomplexity (0221658) Small Grant Program), the Fulbright-Hays Doctoral Dissertation Research Abroad Program, Sigma Xi, the Tinker Field Research Grant, the Andrew and Margaret Paul Graduate Fellowship in the Life Sciences, and the American Cichlid Association Paul V. Loiselle Conservation Fund. The funders had no role in study design, data collection and analysis, decision to publish, or preparation of the manuscript.

Competing Interests: The authors have declared that no competing interests exist.

* E-mail: krista.capps@maine.edu

Introduction

Mobile organisms can generate areas of enhanced nutrient recycling rates, or biogeochemical hotspots, that may influence primary productivity in both terrestrial and aquatic ecosystems [1–3]. McClain et al. [2] defined biogeochemical hotspots as small areas within a landscape matrix that show comparably high reaction rates relative to the surrounding areas. For organisms to generate biogeochemical hotspots within an ecosystem, their population densities must vary through space and/or time and the contribution of the species to nutrient remineralization rates must be significant relative to ecosystem demand [3]. Therefore, spatially or temporally heterogeneous aggregations of organisms can potentially generate hotspots of biogeochemical activity that influence patterns of nutrient remineralization and alter ecosystem nutrient dynamics.

In aquatic ecosystems, consumer-driven nutrient remineralization can be significant relative to ecosystem nutrient demand and can enhance periphyton biomass and productivity [4–6]. Fishes can play important roles in nutrient dynamics in freshwater ecosystems through nutrient sequestration and nutrient remineralization [3,7,8] and this may influence primary producer biomass and productivity [9]. For example in a study examining the influence of a native fish assemblage on nitrogen (N) and phosphorus (P) cycling in a tropical river, McIntyre et al. [3] found that the aggregate excretion of fishes was sufficient to turn over the entire ambient pool of N in the water column, the nutrient limiting primary productivity, in less than 0.3km. Though previous investigations have linked ecosystem processes such as biogeochemical cycling with native fish assemblages [3,10–12], few studies have demonstrated how the effects of fish invasion alter these processes [13,14,15].

Armored catfishes (Loricariidae) are native to Central and South America [16,17], and have been introduced to tropical and subtropical freshwater ecosystems throughout the globe [16,18–26]. In invaded ecosystems, loricariids attain high population densities and they are thought to compete with native organisms for food resources and space [27,28]. Natural resource managers have noted that populations of non-native loricariids create large aggregations [29]; however, the potential influence for this behavior to alter nutrient concentrations in space and time has not been explored. The purpose of this study was to document the diel changes in behavior of non-native populations of armored catfish and investigate the potential of these fishes to generate biogeochemical hotspots, or localized areas of increased nutrient concentrations, in invaded river systems. We predicted that nutrient remineralization by loricariids would represent a substantial flux of N and P. Moreover, we posited daytime aggregations of loricariids would generate biogeochemical hotspots, resulting in discontinuities of stream nutrient availability in space and time.

Methods

Study Site

The field work for this study was conducted in the Chacamax River (N17°29′047″ W91°58′430″) in Chiapas, Mexico during the dry season months of March-May 2008–2010. The benthos in the study reach was characterized by large cobble and gravel substrate. During the study, stream discharge averaged ~1,600 L s^{-1} and water temperature in the river ranged from 21 to 28°C. Ambient nutrient concentrations, collected from sites without aggregations of armored catfish, in the study reaches were low to moderate (average values: NH_4^+-N, 10 μg L^{-1}; NO_3^--N, 353 μg L^{-1}; total dissolved nitrogen, 387 μg L^{-1}; soluble reactive phosphorus, <2 μg L^{-1}; total dissolved phosphorus, 3 μg L^{-1}). Nutrient diffusing substrates, constructed after methods outlined in Capps et al. [30], indicated primary producers in the stream were limited by P (Figure S1). Loricariid density was 2.3±3.4 m^{-2} (mean ± SD) and areal biomass was 225±45 g m^{-2} (mean ± SD), two orders of magnitude greater than the native fish biomass in the study reach in 2010 [31].

Non-native loricariids were first documented in the Chacamax River in 2004, and include *Pterygoplichthys pardalis* (Castelnau, 1855), *Pterygoplichthys disjunctivis* (Weber, 1991) and *Pterygoplichthys* that do not adhere to type specimens (Figure S2). The wide variations in pattern suggest the *Pterygoplichthys* population in the Chacamax may be comprised of hybrids of the two species; hence, we refer to the fish as *Pterygoplichthys*.

Diel Patterns in Fish Behavior and Ambient Nutrient Concentrations

To describe diel patterns in *Pterygoplichthys* behavior, we counted the number of fish found in five 1 m×1 m quadrats along the edge of a 100 m reach of stream during daytime and nighttime hours. These measurements were collected to document loricariids were spreading out from their daytime aggregations to graze the entire stream bed. We counted *Pterygoplichthys* every four hours over a period of three days in March 2010. To document diel fluctuations in water chemistry, we collected duplicate water samples from the thalweg of the stream in single a run habitat for NH_4^+ and PO_4^{-3} analysis every four hours on three dates in 2010.

To estimate the influence of time of day (day/night) on loricariid feeding behavior, 77 fish were harvested and weighed during daytime (22 fish; 1200–1730 hrs) or nighttime (55 fish; 1900–0400 hrs) hours, euthanized using an overdose of MS-222, and their gut contents were collected, dried, and weighed according to methods outlined in German and Bittong (2009) under IACUC protocol number: 2006–0169, Cornell University. All values were expressed as the ratio of gut content dry mass (g) to fish wet mass (g) to account for size variation in the fishes we sampled. Again, these measurements were taken to document pronounced, diel changes in loricariid behavior and their nocturnal foraging activities.

Nutrient Remineralization and Hotspot Sampling

To determine nutrient remineralization rates of loricariids and estimate the effects of their remineralization on ambient water chemistry, we conducted fish excretion incubations and sampled water within and outside of aggregations of loricariids in the Chacamax River. We predicted that aggregations of loricariids would generate biogeochemical hotspots, evidenced by increased ambient nutrient concentrations from samples collected in the aggregations relative to samples collected outside of the aggregations. Twenty *Pterygoplichthys* (Standard length 23.5 cm ±5 cm(mean ± SD)) were collected using hand nets and immediately, individually incubated in 15L plastic tubs in 10L of filtered stream water for approximately 1 h. Ten incubations were conducted between 1200 and 1500 hrs (daytime) and 10 incubations occurred between 1900 and 2100 hours (nighttime). Two additional fish-free tubs were maintained as controls during each incubation period. Tubs were filled with 10L of filtered stream water and placed in the shade for the duration of the incubation. At the end of the incubation, we collected filtered water samples for NH_4^+ and total dissolved phosphorus (TDP) analysis. Fish nutrient recycling rates were estimated based on the difference in dissolved N and P concentrations between plastic tubs incubated with and without *Pterygoplichthys* [3,8]. At the end of the incubation period, water samples were filtered through glass-fiber filters (Gelman A/E) and were either acidified and shipped to the USA for P analysis, or were analyzed in the field for NH_4^+. We used standard colorimetric methods to analyze TDP and soluble reactive phosphorus (SRP) samples (APHA 1998) using a Lachat QuickChem 8000 (Lachat Instruments, Loveland, Colorado). All NH_4^+samples were refrigerated and analyzed in the field using the flurometric methods outlined by Taylor et al. [32].

To ascertain if *Pterygoplichthys* generated hotspots of nutrient remineralization (within aggregations) relative to ambient water chemistry (outside of aggregations), we collected paired stream water samples within (i.e. within the white boundary in Figure 1A) and outside of aggregations (i.e. outside of the white boundary in Figure 1A) of loricariids in the Chacamax River in 2008 and 2010. We collected water samples from aggregations of *Pterygoplichthys* with minimum areas of 3 m^2 with at least 50 *Pterygoplichthys* m^{-2} (Figure 1A). Paired sites were located parallel to one another along a transect extending between river banks. They were matched for similar depth (min = 0.5 m, max = 1.5 m, mean = 0.9 m) and water velocity (min = 0.04 m s^{-1}, max = 0.09 m s^{-1}, mean = 0.07 m s^{-1}). Water samples were collected and analyzed for NH_4^+ and SRP using the aforementioned methods.

Statistical Analysis

Nutrient recycling rates and gut content time comparisons (daytime/nighttime) were made using one-way ANOVAs, where time was the fixed factor. We used a mixed model to estimate the effects of fish size on nutrient recycling rates, where wet mass was considered the fixed factor and time and wet mass×time were considered random factors. Aggregation/ambient comparisons were made using a two-way ANOVA where sample site, year, and the interaction term were considered fixed factors and sample pair (aggregation/ambient) was considered a random factor. All data were $\log_{10}$ transformed to meet the assumptions of the models and analyzed using JMP 9 statistical software (SAS Institute, 2010).

Results

Diel Changes in Fish Behavior and Nutrient Concentrations

Loricariids formed large aggregations in the main channel of the Chacamax River during the day, but spread out to graze the entire riverbed at night (Figures 1, 2). The aggregations occurred in same relative area each day unless there was a large discharge event that moved large, woody debris and larger cobbles. Concurrent increases in ambient NH_4^+ occurred at night when loricariids were broadly dispersed and actively feeding; however, there was no similar increase in ambient PO_4^{3-} (Figure 2). Importantly, all

Figure 1. ***Pterygoplichthys* in the Chacamax River (N17°29′047″ W91°58′430″).** (A) Daytime aggregation of loricariids. The white line outlines the aggregation boundary. (B) Underwater photo of loricariid aggregation. Individual fish are marked with white numbers (1–35). (C) Loricariids spreading out from aggregation to begin evening feeding. Each dark spot (C) is at least one *Pterygoplichthys*. A small group of fishes (1–10) have been marked with individual numbers to demonstrate fish abundance. Photo credits: K. A. Capps.

of the PO_4^{3-} samples were at or near detection; thus, any change in ambient levels would have been difficult to detect.

Nutrient Recycling and Hotspot Observations

Average N excretion was approximately 0.6 μmol NH_4^+-N g wet mass $^{-1}$ hr^{-1} and did not differ between the afternoon and evening sample periods ($F_{(1,\ 18)} = 0.209$, $p = 0.6528$; Figure 3A). In contrast, average P excretion was twice as high in samples collected during nighttime sampling periods (approximately 0.077 μmol TDP-P g wet mass $^{-1}$ hr^{-1}) than those collected in the daytime (approximately 0.031 μmol TDP-P g wet mass $^{-1}$ hr^{-1}; $F_{(1,\ 18)} = 5.61$, $p = 0.0292$; Figure 3C). This resulted in a significant decrease in the N:P ratio of excretion from an average of 23 in the daytime to approximately 12 in the nighttime ($F_{(1,\ 75)} = 9.57$, $p = 0.006$; Figure 3C). This pattern may have been driven by nocturnal loricariid feeding behavior, evidenced by more amorphous detritus found in loricariid guts during nighttime sampling hours ($F_{(1,\ 75)} = 12.09$, $p = 0.0008$; Figure 3D). Loricariid size also influenced excretion rates, as larger fishes tended to excrete less N and P per gram of fish than smaller loricariids ($F_{(1,18)} = 7.52$, $p = 0.013$ and $F_{(1,\ 18)} = 6.67$, $p = 0.019$, respectively). However, loricariid size did not influence excretion stoichiometry ($F_{(1,\ 18)} = 0.633$, $p = 0.6966$). By multiplying loricariid excretion rates by their average areal biomass, we estimate that loricariids remineralize approximately 7 μmol P m $^{-2}$ hr^{-1} during daylight hours, 18 μmol P m $^{-2}$ hr^{-1} during nighttime hours, and 135 μmol N m $^{-2}$ hr^{-1} during both time periods.

Aggregations of loricariids (Figures 1, 4) generated hotspots of nutrient recycling relative to ambient water chemistry in paired river sites. Water samples collected within aggregations of loricariids had 41% higher concentrations of NH_4^+-N ($p<0.0001$, $F_{(3,\ 60)} = 9.63$, Figure 4) and 66% higher concentrations of SRP ($p = 0.0005$, $F_{(3,\ 60)} = 11.7$, Figure 4) relative to paired ambient chemistry sites.

Discussion

The results from this study support findings from other studies documenting the important role consumer nutrient recycling can play in stream ecosystems [3,11,33]. For example, fish excretion of the limiting nutrient N exceeded N demand in a Venezuelan stream [3]. Similarly, freshwater shrimp excretion was equivalent to approximately 20% of the N uptake and 5% of the P uptake in Puerto Rican streams [33]. In our study, excretion of N and P by high densities of loricariids appears to be a large and potentially

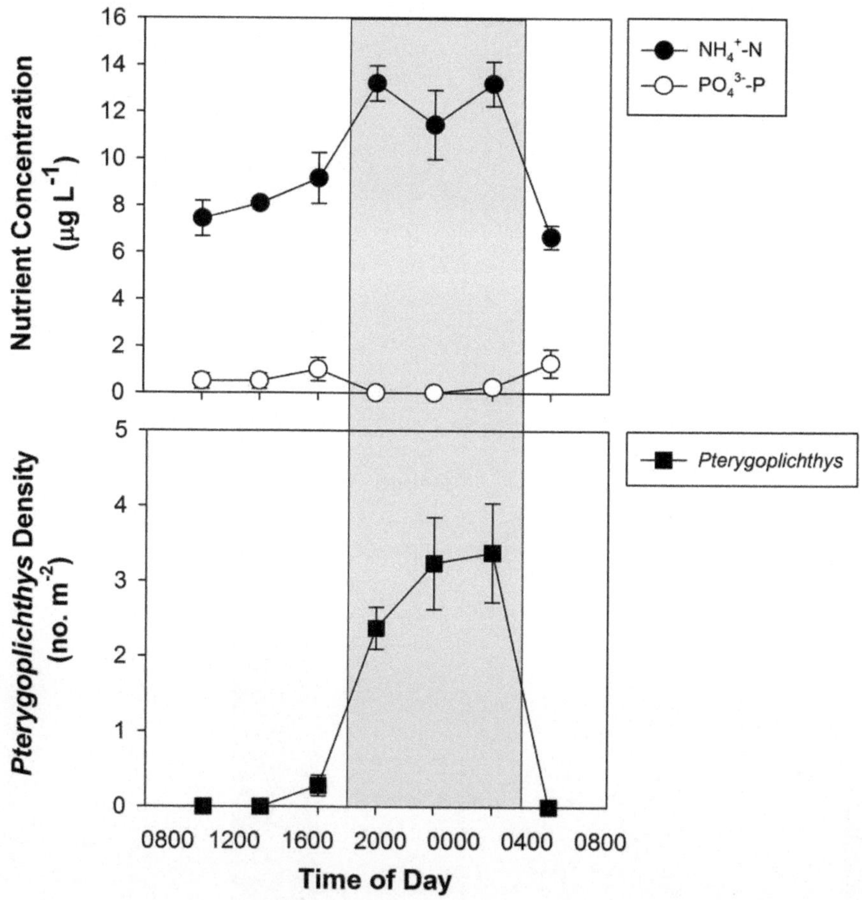

Figure 2. Diel changes in ambient water chemistry and *Pterygoplichthys* behavior. Data were collected in 2008 and 2010 (±1 SE). (A) NH4+-N and PO4-3-P concentrations over time; (B) number of *Pterygoplichthys* counted in 1 m^2 quadrats near the stream bank (within 24 cm) over time. The shaded areas represent nighttime sampling hours.

important flux of nutrients in the Chacamax River that is both spatially and temporally heterogeneous throughout the stream reach. This is one of the first studies to measure novel biogeochemical hotspots generated by an aquarium invader.

Similar to observations reported for loricariids in their native ranges [34] and introduced populations of *Pterygoplichthys* in Florida [35], *Pterygoplichthys* in the Chacamax River were nocturnally active (Figure 2). Gut content mass was greater during evening hours, indicating loricariids are primarily active and feeding at night in the study site (Figure 3D). Loricariid behavior may have evolved to minimize predation from day-active predators, as diel changes in behavior has been attributed to predator avoidance in native populations of loricariids [36]. Unfortunately, due to the limited number of potential predators remaining in the Chacamax River and the potential large body size of *Pterygoplichthys* (up to at least 70cm SL [36]), it is unlikely that predation will control the loricariid population in the region.

For organisms to generate areas of enriched nutrient recycling rates in space, or hotspots, across a landscape, their distribution must change through space and/or time [2,3]. In this study, we observed diurnal aggregating behavior of loricariids (Figure 1) and temporal variation in fish excretion rates (Figure 3), which created a mechanism for armored catfish to generate biogeochemical hotspots in the Chacamax River. Water samples collected within loricariid aggregations had roughly double the concentration of N and P than water collected from outside the aggregations (Figure 4), suggesting that loricariids generate localized pulses of nutrients downstream from aggregations during daylight hours when primary producers are photosynthesizing. Primary producers in the study site were nutrient limited (Figure S1) [31]. Hence, elevated ambient nutrient availability may locally alleviate some nutrient limitation and enhance algal growth and primary productivity immediately downstream of loricariid aggregations. However, loricariid aggregations disperse daily; therefore, any increase in algal biomass driven by proximity to an aggregation of loricariids is likely removed quickly by nocturnal loricariid grazing and would be difficult to detect. Future work should attempt to separate the grazing and nutrient remineralization effects of loricariids to estimate the net effects of these fishes on algal biomass and gross primary production in invaded systems.

Species-specific characteristics and areal biomass are important factors to consider when predicting if a consumer could be a significant driver of nutrient recycling or if they have the potential to increase or relax nutrient limitation [3,12]. For example, the elemental composition, or stoichiometry, of an invader may influence the impact of changing nutrient dynamics in an ecosystem [37]. Phosphorus excretion rates are highly variable among fish species [8,12]; thus, P-cycling in P-limited systems, such as the Chacamax River, may be strongly influenced by changes in fish communities [12]. In such a system, the introduction of a P-rich invader such as loricariids [38] may actually intensify P-limitation of primary producers and microbial

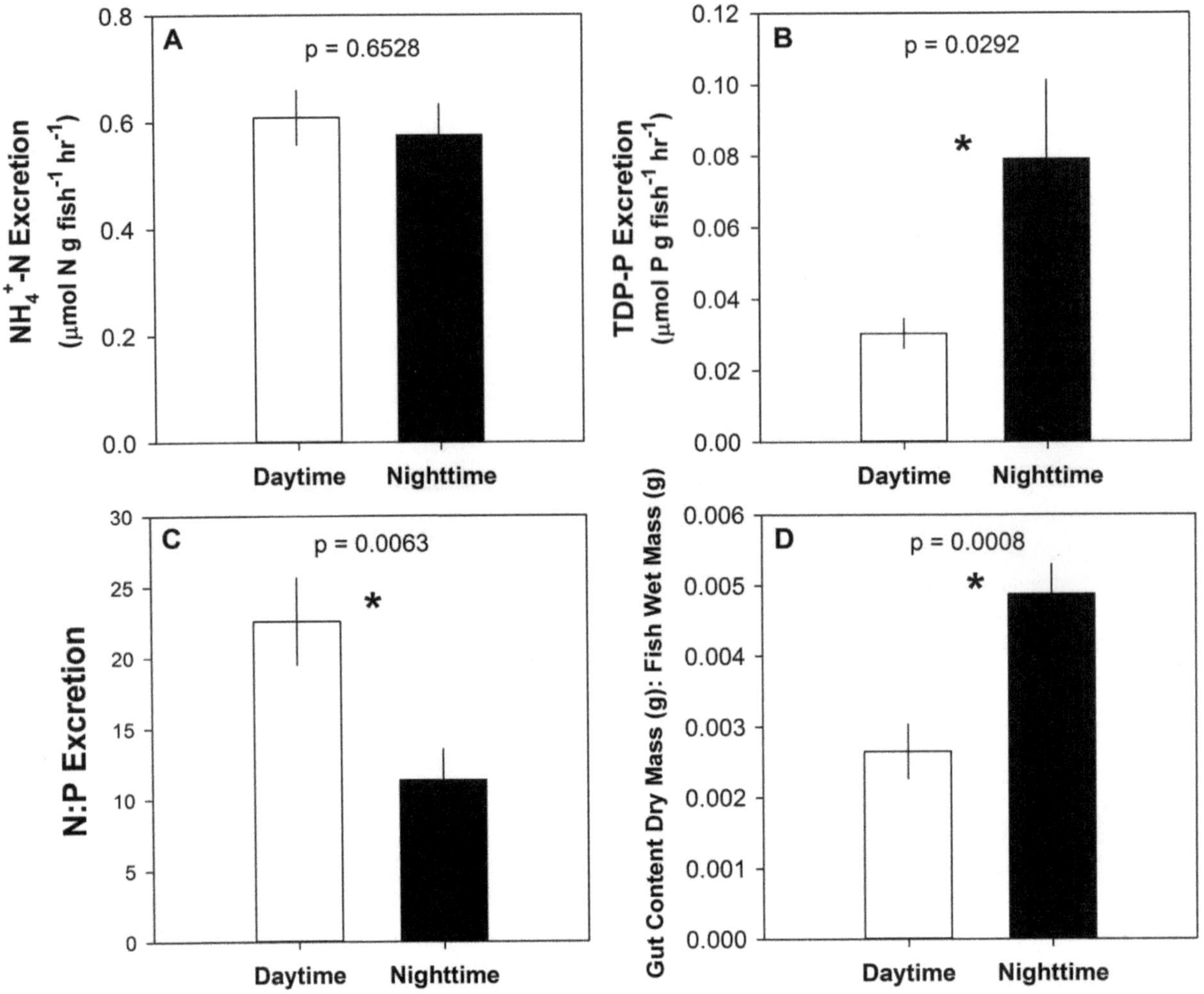

Figure 3. Diel changes in *Pterygoplichthys* excretion and gut content mass. Average *Pterygoplichthys* excretion and gut content mass during daytime (1000–1500 h) and nighttime hours (1900–0400 h). (A) *Pterygoplichthys* NH_4^+-N excretion rates; (B) *Pterygoplichthys* total dissolved phosphorus excretion rates; (C) N:P of *Pterygoplichthys* excretion; (D) gut content dry mass per wet mass of *Pterygoplichthys*. Error bars represent ±1 SE.

heterotrophs, and influence nutrient cycling rates in streams. For example, in our site, nocturnal feeding activity (Figure 3D) may have generated increased nocturnal P excretion by loricariids (Figure 3B) and significantly altered the N:P ratio of excretion between daytime and nighttime hours (Figure 3C). Notably, the N:P ratio of excretion during daylight hours (Figure 3C) was greater than the Redfield ratio [39]; therefore, diurnal loricariid excretion may have actually enhanced P-limitation.

It is important to mention, increased loricariid activity occurred simultaneously with increases in stream water NH_4-N concentrations, but there was no measureable change in P concentrations in the water column (Figure 2). Though we documented diel changes in P excretion by loricariids that we attributed to feeding behavior, we did not see the same pattern in N excretion (Figure 3). Most likely, increased nighttime NH_4-N concentrations in the stream water were due to diel patterns in autotrophic ammonium demand [40] rather than differences in ammonium excretion by loricariids. Though a similar pattern may have been evident with autotrophic uptake of P, all of the P concentrations measured outside of loricariid aggregations were at or very near the level of detection. Thus, stoichiometric changes at the ecosystem level were most likely being driven by large changes in ambient ammonium concentrations. This also suggests pulses of soluble P generated by loricariid aggregations may be locally significant.

Well-mixed rivers like the Chacamax add complexity to studying the formation of biogeochemical hotspots, for unlike terrestrial environments, hotspots may not have discrete boundaries. Hence, it is important to note that loricariid remineralization contributed to the N and P concentrations of samples we collected within and outside of fish aggregations. Moreover, there was likely an array of influences loricariid invasion had on nutrient dynamics in addition to remineralization, such as bioturbation. However, we focused this study on the direct influence of invaders on nutrient cycling via remineralization. Importantly, ambient water chemistry data on the Chacamax were not collected prior to loricariid invasion, so we cannot determine the potential additive effects of loricariid nutrient recycling to total ambient solute concentrations. Moreover, P concentrations in the Chacamax were near the level of detection; thus changes in ambient P after invasion may be difficult to detect.

To be important drivers of nutrient cycling, the contribution of nutrient remineralization by organisms must be significant at the ecosystem-level [2,3]. Our data suggest that loricariids strongly influence nutrient remineralization rates in the Chacamax River.

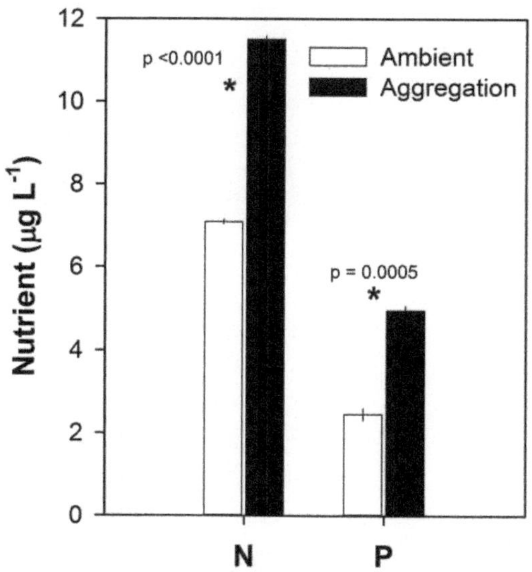

Figure 4. Biogeochemical hotspots created by aggregations of *Pterygoplichthys.* Means of NH_4^+-N and PO_4^{-3}-P (±1 SE) samples taken from paired sites within and outside of loricariid aggregations in the Chacamax River in 2008 (n = 16 ambient sites, n = 16 aggregation sites) and 2010 (n = 16 ambient sites, n = 16 aggregation sites). Aggregations were defined as water samples taken within groups of *Pterygoplichthys* that had an area of at least 5 m^2 with at least 40 *Pterygoplichthys* per m^2. Ambient samples were collected from sites parallel to the aggregations without immediate upstream aggregations of loricariids.

The volumetric excretion estimates for loricariids were greater than ambient water chemistry values indicating that remineralization by loricariids may be an important flux of nutrients within the river. Moreover, loricariid biomass was two orders of magnitude greater than native fish biomass [31]. Consequently, loricariid invasion may have shifted the Chacamax River from a system where fishes were not central drivers of biogeochemical processes to a system where remineralization of nutrients by fishes is a large and important flux of nutrients.

Recent work has suggested invasive species management and eradication efforts should target species that present serious environmental risks and alter the function of ecosystems [41]. When coupled with work demonstrating non-native loricariids are: competing with native species for food resources [42,43], altering the structure of aquatic and riparian ecosystems [44,45], and negatively affecting populations of threatened and endangered species [35,46]; the results from this study indicate high-densities of invading loricariids threaten the structure and function of ecosystems. Moreover, the results from this investigation provide additional evidence [47] that aggregations of non-native organisms have the potential to alter nutrient dynamics and generate biogeochemical hotspots. This work highlights the potential threat invasive, aquarium species present to ecosystem function in freshwater ecosystems.

Supporting Information

Figure S1 Nutrient limitation of periphyton in the Chacamax River. Mean (±1 SE) algal biomass collected from nutrient diffusing substrates (NDS) for each of four nutrient treatments (control (CON), nitrogen (N), phosphorus (P), nitrogen and phosphorus (N+P)). Bars with different letters have significantly different ($p<0.005$) algal biomasses according to Tukey's Honestly Significant Difference test. Nutrient diffusing substrates were constructed using methods outlined in Capps et al. [1]. They were deployed for a total of 14 days. Nutrient diffusion rate estimates were made for each nutrient treatment on day 0 and day14 by subtracting the diffusion rate of nutrient amended NDS from the rate of control NDS [1]. Nutrient diffusion rate estimates were made for each nutrient treatment by subtracting the diffusion rate of nutrient amended NDS from the rate of control NDS. On day 14 (N: $7.2\times10^{-5}\pm3.8\times10^{-6}$; P: $7.1\times10^{-4}\pm3.1\times10^{-5}$; N+P: (N) $9.5\times10^{-6}\pm2.1\times10^{-6}$, (P) $8.1\times10^{-5}\pm1.5\times10^{-6}$, mean ± SE (mol m^{-2} hr^{-1})), all treatments were diffusing less than on day 0 (N: $2.1\times10^{-2}\pm3.1\times10^{-4}$; P: $4.9\times10^{-3}\pm4.3\times10^{-5}$; N+P: (N) $3.9\times10^{-2}\pm3.1\times10^{-4}$, (P) $9.1\times10^{-3}\pm5.8\times$10-5, mean ± SE (mol m^{-2} hr^{-1})). The results from NDS indicated that primary producers in the Chacamax River were P-limited ($p<0.0001$, $F_{(3, 43)} = 13.6$). References Cited: 1. Capps KA, Booth MT, Collins SM, Davison MA, Moslemi JM, et al. (2011) Nutrient diffusing substrata: a field comparison of commonly used methods to assess nutrient limitation. Journal of the North American Benthological Society 30:522–532.

Figure S2 Range of ventral patterns of *Pterygoplichthys* collected in the Chacamax River (N17°29′047″ W91°58′430″). *Pterygoplichthys pardalis* is characterized by a ventral pattern of dark spots (A). *Pterygoplichthys disjunctivus* is characterized by dark, vermiculated lines (G) (Armbruster & Page 2006). The wide variations in pattern suggest the *Pterygoplichthys* population in the Chacamax may be comprised of hybrids of the two species. Photo credit: K. A. Capps.

Acknowledgments

We thank Christy Goodale, Nelson Hairston Jr., and Stuart Findlay for their comments on earlier versions of this manuscript. We also thank Rocío Rodiles-Hernández, Jessica Strickland, and Daniel Rodríguez for help with fieldwork in Mexico. Finally, we would like to thank the personnel of the Hotel Nututun for their support in Mexico. Organisms were harvested in Mexico using Mexican collection permit number DGOPA.07525.25706.3233 and fishes were handled using methods outlined in the IACUC protocol 2006–0169 from Cornell University.

Author Contributions

KAC. Conceived and designed the experiments: KAC ASF. Performed the experiments: KAC. Analyzed the data: KAC ASF. Contributed reagents/materials/analysis tools: KAC ASF. Wrote the paper: KAC ASF.

References

1. Meyer JL, Schultz ET, Helfman GS (1983) Fish schools-an asset to corals. Science 220: 1047–1049.
2. McClain ME, Boyer EW, Dent CL, Gergel SE, Grimm NB, et al. (2003) Biogeochemical hot spots and hot moments at the interface of terrestrial and aquatic ecosystems. Ecosystems 6: 301–312.
3. McIntyre PB, Flecker AS, Vanni MJ, Hood JM, Taylor BW, et al. (2008) Fish distributions and nutrient cycling in streams: Can fish create biogeochemical hotspots? Ecology 89: 2335–2346.
4. Liess A, Hillebrand H (2006) Role of nutrient supply in grazer-periphyton interactions: reciprocal influences of periphyton and grazer nutrient stoichiometry. Journal of the North American Benthological Society 25: 632–642.
5. Evans-White MA, Lamberti GA (2006) Stoichiometry of consumer-driven nutrient recycling across nutrient regimes in streams. Ecology Letters 9: 1186–1197.

6. Knoll LB, McIntyre PB, Vanni MJ, Flecker AS (2009) Feedbacks of consumer nutrient recycling on producer biomass and stoichiometry: separating direct and indirect effects. Oikos 118: 1732–1742.
7. Vander Zanden MJ, Vadeboncoeur Y (2002) Fishes as integrators of benthic and pelagic food webs in lakes. Ecology 83: 2152–2161.
8. Vanni MJ, Flecker AS, Hood JM, Headworth JL (2002) Stoichiometry of nutrient recycling by vertebrates in a tropical stream: linking species identity and ecosystem processes. Ecology Letters 5: 285–293.
9. McIntyre PB, Michel E, Olsgard M (2006) Top-down and bottom-up controls on periphyton biomass and productivity in Lake Tanganyika. Limnology and Oceanography 51: 1514–1523.
10. McIntyre PB, Jones LE, Flecker AS, Vanni MJ (2007) Fish extinctions alter nutrient recycling in tropical freshwaters. Proceedings of the National Academy of Sciences of the United States of America 104: 4461–4466.
11. Reisinger AJ, Presuma DL, Gido KB, Dodds WK (2011) Direct and indirect effects of central stoneroller (*Campostoma anomalum*) on mesocosm recovery following a flood: can macroconsumers affect denitrification? Journal of the North American Benthological Society 30: 840–852.
12. Small GE, Pringle CM, Pyron M, Duff JH (2011) Role of the fish *Astyanax aeneus* (Characidae) as a keystone nutrient recycler in low-nutrient Neotropical streams. Ecology 92: 386–397.
13. Kitchell JF, Koonce JF, Tennis PS (1975) Phosphorus flux through fishes. Verhandlungen International Verein Limnologie 19: 2478–2484.
14. Kraft CE (1993) Phosphorus regeneration by Lake Michigan alewives in the mid-1970s. Transactions of the American Fisheries Society 122: 749–755.
15. Bunnell DB, Johnson TB, Knight CT (2005) The impact of introduced round gobies (*Neogobius melanostomus*) on phosphorus cycling in central Lake Erie. Canadian Journal of Fisheries and Aquatic Sciences 62: 15–29.
16. Nico LG, Martin RT (2001) The South American suckermouth armored catfish, *Pterygoplichthys anisitsi* (Pisces : Loricariidae), in Texas, with comments on foreign fish introductions in the American southwest. Southwestern Naturalist 46: 98–104.
17. Weber C (1991) New taxa in *Pterygoplichthys* S I (Pisces, Siluriformes, Loricariidae). Revue Suisse De Zoologie 98: 637–643.
18. Capps KA, Nico LG, Mendoza-Carranza M, Arevalo-Frias W, Ropicki AJ, et al. (2011) Salinity tolerance of non-native suckermouth armoured catfish (Loricariidae: *Pterygoplichthys*) in south-eastern Mexico: implications for invasion and dispersal. Aquatic Conservation-Marine and Freshwater Ecosystems 21: 528–540.
19. Courtenay WR, Hensley DA, Taylor JN, McCann JA (1986) Distribution of exotic fishes in North America. In: Hocutt CH, Wiley EO, editors. The zoogeography of North American freshwater fishes. New York: John Wiley and Sons. 675–698.
20. Arthington AH, Kailola PJ, Woodland DJ, Zalucki JM (1999) Baseline environmental data relevant to an evaluation of quarantine risk potentially associated with the importation to Australia of ornamental finfish. Canberra: Department of Agriculture, Fisheries, and Forestry.
21. Bomford M, Glover J (2004) Risk assessment model for the import and keeping of exotic freshwater and estuarine finfish. In: Sciences BoR, editor. Canberra: Australian Government.
22. Liang SH, Wu HP, Shieh BS (2005) Size structure, reproductive phenology, and sex ratio of an exotic armored catfish (*Liposarcus multiradiatus*) in the Kaoping River of southern Taiwan. Zoological Studies 44: 252–259.
23. Chavez JM, De La Paz RM, Manohar SK, Pagulayan RC, Vi JRC (2006) New Philippine record of South American sailfin catfishes (Pisces : Loricariidae). Zootaxa: 57–68.
24. Kailola PJ (2007) Risk assessment of ten species of ornamental fish under the Environment Protection and Biodiversity Conservation Act 1999. Sydney: Australian Government.
25. Ozdilek SY (2007) Possible threat for Middle East inland water: an exotic adn invasive species, *Pterygoplichthys disjuntivis* (Weber, 1991) in Asi river, Turkey (Pices: Loricariidae). Journal of Fisheries and Aquatic Sciences 24: 303–306.
26. Sinha RK, Sinha RK, Sarkar UK, Lakra WS (2010) First record of the southern sailfin catfish, *Pterygoplichthys anisitsi*, Eigenmann and Kennedy, 1903 (Teleostei: Loricariidae), in India. Journal of Applied Ichthyology 26: 606–608.
27. Devick WS (1989) Disturbances and fluctuations in the Wahiawa Reservoir ecosystem. Honolulu: Hawaii Department of Land and Natural Resources, Division of Aquatic Resources. 30 p.
28. Hoover JH, Kilgore KJ, Cofrancesco AF (2004) Suckermouth catfishes: Threats to aquatic ecosystems of the United States. Vicksburg: U.S. Army Corps of Engineers.
29. Mendoza RE, B C, Orr R, Balderas SC, Courtenay WR, et al. (2009) Trinational risk assessment guidelines for aquatic alien invasive species. Montréal: Commission for Environmental Cooperation.
30. Capps KA, Booth MT, Collins SM, Davison MA, Moslemi JM, et al. (2011) Nutrient diffusing substrata: a field comparison of commonly used methods to assess nutrient limitation. Journal of the North American Benthological Society 30: 522–532.
31. Capps KA (2012) Changes in community structure and ecosystem processes in response to armored catfish (Silurformes: Loricariidae) invasion. Ithaca: Cornell University. 142 p.
32. Taylor BW, Keep CF, Hall RO, Koch BJ, Tronstad LM, et al. (2007) Improving the fluorometric ammonium method: matrix effects, background fluorescence, and standard additions. Journal of the North American Benthological Society 26: 167–177.
33. Benstead JP, Cross WF, March JG, McDowell WH, Ramirez A, et al. (2010) Biotic and abiotic controls on the ecosystem significance of consumer excretion in two contrasting tropical streams. Freshwater Biology 55: 2047–2061.
34. Casatti L, Castro RMC (2006) Testing the ecomorphological hypothesis in a headwater riffles fish assemblage of the rio Sao Francisco, southeastern Brazil. Neotropical Ichthyology 4: 203–214.
35. Nico LG (2010) Nocturnal and diurnal activity of armored suckermouth catfish (Loricariidae: *Pterygoplichthys*) associated with wintering Florida manatees (*Trichechus manatus latirostris*). Neotropical Ichthyology 8: 893–898.
36. Froese R, Pauly D (2011) FishBase. 10/2011 ed: World Wide Web.
37. Gonzalez AL, Kominoski JS, Danger M, Ishida S, Iwai N, et al. (2010) Can ecological stoichiometry help explain patterns of biological invasions? Oikos 119: 779–790.
38. Hood JM, Vanni MJ, Flecker AS (2005) Nutrient recycling by two phosphorus-rich grazing catfish: the potential for phosphorus-limitation of fish growth. Oecologia 146: 247–257.
39. Redfield AC (1958) The biological control of chemical factors in the environment. American Scientist 64: 205–221.
40. Johnson LT, Tank JL (2009) Diurnal variations in dissolved organic matter and ammonium uptake in six open-canopy streams. Journal of the North American Benthological Society 28: 694–708.
41. Davis MA, Chew MK, Hobbs RJ, Lugo AE, Ewel JJ, et al. (2011) Don't judge species on their origins. Nature 474: 153–154.
42. Pound KL, Nowlin WH, Huffman DG, Bonner TH (2011) Trophic ecology of a nonnative population of suckermouth catfish (Hypostomus plecostomus) in a central Texas spring-fed stream. Environmental Biology of Fishes 90: 277–285.
43. Mendoza-Carranza M, Hoeinghaus DJ, Garcia AM, Romero-Rodriguez A (2010) Aquatic food webs in mangrove and seagrass habitats of Centla Wetland, a Biosphere Reserve in Southeastern Mexico. Neotropical Ichthyology 8: 171–178.
44. Lienart GDH, Rodiles-Hernandez R, Capps KA (2012) Nest burrows and nesting behavior of non-native catfishes (Silurformes: Loricariidae) in the Usumacinta watershed, Mexico. Southwestern Naturalist In press.
45. Nico LG, Jelks HL, Tuten T (2009) Non-native suckermouth armored catfishes in Florida: description of nest burrows and burrow colonies with assessment of shoreline conditions. Aquatic Nuisance Species Research Program 09: 1–30.
46. Bunkley-Williams L, Williams EH, Lilystrom CG, Corujo-Flores I, Zerbi AJ, et al. (1994) The South American sailfin armored catfish *Liposarcus multiradiatus* (Hancock), a new exotic established in Puerto Rican fresh waters. Caribbean Journal of Science 30: 90–94.
47. Bouletreau S, Cucherousset J, Villeger S, Masson R, Santoul F (2011) Colossal Aggregations of Giant Alien Freshwater Fish as a Potential Biogeochemical Hotspot. Plos One 6: 10.1371/journal.pone.0025732.

14

Growth, Development and Temporal Variation in the Onset of Six *Chironex fleckeri* Medusae Seasons: A Contribution to Understanding Jellyfish Ecology

Matthew Gordon*, Jamie Seymour

School of Marine and Tropical Biology, James Cook University, Cairns, Queensland, Australia

Abstract

Despite the worldwide distribution, toxicity and commercial, industrial and medical impacts jellyfish present, many aspects of their ecology remain poorly understood. Quantified here are important ecological parameters of *Chironex fleckeri* medusae, contributing not only to the understanding of an understudied taxon, the cubozoa, but also to the broader understanding of jellyfish ecology. *C. fleckeri* medusae were collected across seven seasons (1999, 2000, 2003, 2005–07 and 2010), with growth rates, temporal variation in the medusae season onset and differences in population structure between estuarine and coastal habitats quantified. With a mean of 2 September ±2 d (mean ±95% confidence limits), the earliest date of metamorphosis was temporally constrained between seasons, varying by only 7 d (30 August to 5 September). Juvenile medusae appeared to be added over an extended period, suggesting polyp metamorphosis was an ongoing process once it commenced. At a maximum of 3±0.2 mm d^{-1} IPD, medusae growth to an asymptotic size of ~190 mm IPD was rapid, yet, with the oldest medusae estimated to be ~78 d in age, medusae did not appear to accumulate along the coastline. Furthermore, a greater proportion of juveniles were observed along the coastline, with estuarine populations typified by larger medusae. With key aspects of *C. fleckeri*'s ecology now quantified, medusae season management protocols can be further developed.

Editor: Jack Anthony Gilbert, Argonne National Laboratory, United States of America

Funding: Mission Beach Tourism Commission (http://www.missionbeachtourism.com/), Lions Foundation of Australia (www.lionsclubs.org.au), National Geographic (www.nationalgeographic.com.au), Australian Geographic (www.australiangeographic.com.au), Cairns City Council (www.cairns.qld.gov.au), Cardwell Shire Council (www.csc.qld.gov.au), Growing the Smart State Queensland PhD grants (www.smartstate.qld.gov.au) and Comalco (http://www.riotintoalcan.com/). The funders had no role in study design, data collection and analysis, decision to publish, or preparation of the manuscript.

Competing Interests: The authors have read the journal's policy and have the following conflicts. Commercial funders: Mission Beach Tourism Commission (http://www.missionbeachtourism.com/), Lions Foundation of Australia (www.lionsclubs.org.au), National Geographic (www.nationalgeographic.com.au), Australian Geographic (www.australiangeographic.com.au), Cairns City Council (www.cairns.qld.gov.au), Cardwell Shire Council (www.csc.qld.gov.au), Growing the Smart State Queensland PhD grants (www.smartstate.qld.gov.au) and Comalco (http://www.riotintoalcan.com/).

* E-mail: matthew.gordon@jcu.edu.au

Introduction

The occurrence of jellyfish, particularly in blooms, negatively affects a range of recreational, industrial and commercial activities. For instance, while some commercially important fisheries have been unable to function when jellyfish have clogged fishing nets [1,2], in other fisheries, jellyfish have become predators of and competitors to those species being targeted [2,3]. Serious industrial issues have also been attributed to increased jellyfish abundances, with power station shut down necessary when water intake pipes have become clogged with medusae [4,5,6]. For other regions, it is the medical liability that jellyfish represent that continues to adversely affect the tourism industry which is often integral to local and regional economies [7]. Despite these significant issues, quantitative data documenting key aspects of jellyfish ecology are, in general, lacking. As a result, determining whether claims of increased season length or intensity and frequency of blooms are difficult to validate. For the Australian tropics, it is the seasonal occurrence of *Chironex fleckeri* Southcott that significantly impacts the way in which coastal areas are utilised, yet many of the currently favoured theories relating to the temporal variation in medusae occurrence, medusae growth and development as well as population structure are based on sting records or qualitative data.

The generalisation that *C. fleckeri's* life cycle is seasonal is based on numerous reports of medusae and stings from the warmer months of the year [8,9,10,11,12,13,14,15,16,17,18] contrasted with a lack of such reports from the winter months. Although the life history of cubozoans is complex [19] in which an asexually reproducing polyp phase alternates with a sexually reproducing medusae phase [15,16,20,21,22,23,24], the timing of the shift from the polyp to medusae phase appears to vary both spatially and temporally [10,12,15,16,18]. For instance, while *C. fleckeri* medusae typically appear along the far north Queensland coastline in December, they have been reported as early as October in some seasons [12,15,18]. There is also a suggestion that the season commences earliest on the west coast, shortly after the first rains of the wet season, with the onset of the stinger season delayed if the wet season is late [10,25]. Other authors discount the relevance of the wet season [8,26], however, suggesting that medusae arrival is associated with rising water temperature [10,16,26,27]. With the polyp habitat thought to be located some distance from the coastline within estuary systems [16], the onset of the stinger season is unlikely to accurately reflect the timing of polyp metamorphosis given that the colloquial term 'stinger season' typically refers to the arrival of medusae along the coastline rather than the timing of polyp metamorphosis.

Few extrapolations can be drawn from other cubozoan species either, with studies identifying a species cue for metamorphosis limited to a handful of species. Here, increasing water temperature [28], an interaction between increasing water temperature and food [29,30] or increasing water temperature and photoperiod [31] have been identified as cues. While the development to a specific number of tentacles [32] and the presence of photosymbiotic algae [33] have also been suggested as potential cues for metamorphosis, these links have not been validated quantitatively. Similarly, while the correlation between rainfall events and successive pulses of juvenile cubomedusae [34,35] may implicate salinity as a cue for metamorphosis, quantitative data demonstrating this link is lacking. Given this paucity in data, a need therefore exists for research distinguishing between those mechanisms driving polyp metamorphosis and those merely correlated with this process.

At metamorphosis, juvenile *C. fleckeri* medusae are approximately 1.2 to 1.4 mm in size [24], but increase in size rapidly [15,16,17], reaching sexual maturity late in the season [10,15]. While such growth patterns are reported for a number of scyphozoan species [36], for cubomedusae, it is only for *Chiropsella bronzei* Gershwin that growth parameters have been quantified [34,35]. This paucity in data is largely due to the lack of a reliable method by which cubomedusae can be aged. Size of cubomedusae, for instance, is an unreliable indicator of age, given that degrowth of the bell can occur in cases where feeding regimes are inadequate (for example, an underfed captive *C. fleckeri* medusa [37]). While some authors have used the number of tentacles per pedalium as an indicator of development [25], tentacles are added across a size range, and hence, may still be somewhat dependent upon feeding regime. More recently, the statoliths contained within the statocysts of cubomedusae rhopalia (eye bearing sensory structures) have been shown to contain fine growth rings that are added on a daily basis [35,38,39,40,41].

Insights into the ecology of a species can also be gained from population structure data. For example, several cohorts of *C. bronzei* medusae occurring within a single season, as well as a correlation between cohort appearance and significant rainfall events, were elucidated from population structure data [34,35]. Comparable data for *C. fleckeri* populations is, however, currently limited to the generalisation that an abundance of small individuals early in the season progresses to fewer but larger medusae late in the season [15]. Given that the shift from estuarine to coastal habitats is thought to accompany the shift from polyp to medusae phases of the life cycle [16] whereby medusae are washed from within estuaries at the onset of the season [15,42], coastal populations are likely to be dominated by larger and older medusa, while smaller and younger medusae would typify estuarine populations. Yet, it would appear that large medusae can be present even in early season samples [10,14] and "no specific rule can be given to the range of sizes encountered in a given area at a specific time" [10]. Whether the presence of only large medusae at a given location and small medusae in another indicates the presence of a nursery location [14], or perhaps the polyp habitat itself, is yet to be demonstrated. A need therefore exists to quantify aspects of population structure if the overall ecology of *C. fleckeri* is to be better understood.

The current paper aims to quantify some of the long held theories relating to the ecology of *C. fleckeri* medusae. Key aspects to be investigated include the:

- relationship between statolith size and growth rings
- temporal variability in the onset of medusae production (metamorphosis)
- growth and development of *C. fleckeri* medusae
- population structure of coastal and estuarine medusae populations.

Methods

Sample Sites

A total of 484 medusae were collected from seven sites at Weipa during the 1999, 2000, 2003, 2005, 2006, 2007 and 2010 stinger seasons. While not all sites were visited on each occasion or within each season, Landfall Point (12°34′53″S, 141°39′50″E), Andoomajetti Point (12°36′20″S, 141°49′21″E), Rocky Point (12°37′10″S, 141°52′38″E), Jessica Point (12°40′05″S, 141°51′42″E), Hey Point(12°44′23″S, 141°53′35″E), Wooldrum Point Beach (12°42′19″S, 141°48′03″E) and Westminster South (12°50′11″S 141°44′56″E) represented Weipa (Figure 1). A further 46 medusae were collected during the 2005, 2006 and 2007 seasons from 12 east coast sites between Cairns and Townsville, which included Mission Beach, Gin Camp, Yorkeys Knob, Port Douglas, Cardwell, Palm Cove, Buchan Point, Lugger Bay, Townsville and the Tully, Murray and Hull Rivers. Given the low numbers collected on each occasion, east coast sites were excluded from most analyses.

Weipa sites were classified as coastal or estuarine sites. Coastal sites were those that occurred along a beachfront or the embayment of Albatross Bay, such as Wooldrum Point Beach. Estuarine sites were those that occurred wholly within an estuary system or at the intersection of an estuary system and the coastal embayment of Albatross Bay, such as Andoomajetti Point.

Statolith Technique

Medusae were collected by hand and their inter pedalia distance (IPD), the distance between the mid line of alternate pedalia along the line passing through the rhopalia, was measured to the nearest mm. Each of the four rhopalial niches and a gonad sample (if gonads were evident) were removed and preserved in 98% ethanol.

For 71 medusae, the statoliths of two rhopalia were dissected from the base of their eye set. Undamaged and unshattered statoliths were cleaned of any cellular material before being embedded in resin in the profile plane (kidney shape was evident). The top 50% of each statolith was ground using 1200 gauge wet and dry sand paper, polished with Brasso, rinsed and then polished with tooth paste. The number of rings present in each statolith was counted under oil immersion using ×400 magnification on a light microscope. Each dark band with a light band either side was considered to be one growth ring. The length of each statolith was measured using a calibrated stereo dissector microscope, with length taken to be the longest distance between curved apical ends of a statolith. Both the average number of rings and average length of each statolith pair were then calculated, with the relationship between the number of growth rings and statolith length investigated using regression analysis.

Two statoliths from a further 437 medusae were dissected and their length measured under a calibrated stereo dissector microscope. The average length of each statolith pair was calculated, from which, the number of rings was estimated using the relationship between statolith length and number of rings.

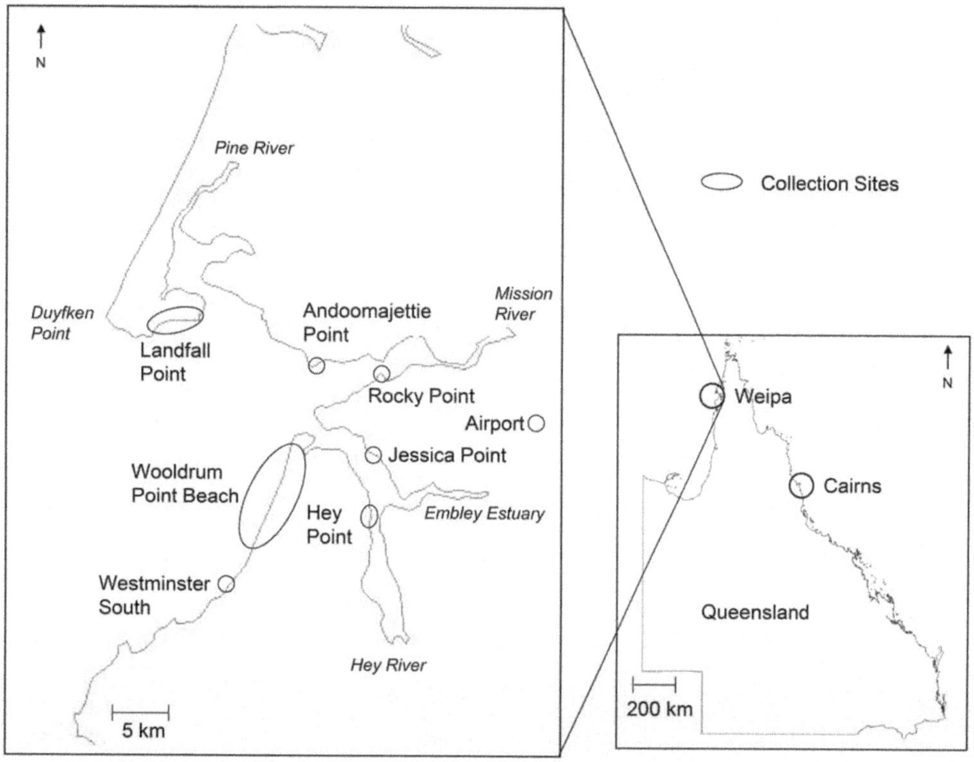

Figure 1. Geographic location of Weipa medusae collection sites, western Cape York, north Queensland, Australia.

Quantifying the Time Frame of Statolith Growth Rings

Establishing a significant relationship between medusae size (mm IPD) and age (number of rings in statolith) was a two stage process. Firstly, regression analysis was used to determine whether medusae size (IPD mm) could be predicted from tentacle number, and secondly, whether tentacle number was correlated with ring number. Given that medusae can undergo both growth and degrowth, tentacle number was considered a more reliable indicator of medusae development as tentacles are not lost once added.

To be able to age medusae, however, it was necessary to quantify the unit of time represented by successive growth rings. While tetracycline is widely used for growth ring time frame validation in fish otoliths, several attempts at applying this technique to *C. fleckeri* medusae were unsuccessful, despite various concentrations of tetracycline being trialled and medusae being housed in large, custom made cylindrical tanks. While this has limited the methods by which the interval between successive rings can be quantified, several pieces of evidence suggest that successive rings are added on a daily basis. Firstly, if growth rings were added at hourly or weekly intervals, not only are these arbitrary units of time that medusae would be unable to measure, but would also make medusae within this study less than four days (hourly) or 1.5 years (weekly) old. Monthly or annual units of time are also unrealistic given that the oldest medusae would have been 7 or 80 yrs of age respectively. Rather, if successive rings were added on a daily basis, medusae collected within this study would have ranged between one and three months of age. Not only are these realistic age estimates, given that medusae are unlikely to survive between seasons, but growth rings are added on a daily basis in three other species of cubozoa [35,38,39,40,41]. In this scenario, alternating dark and light bands would reflect the diurnal behaviours of medusae [25,43]. That is, throughout the day, medusae expend considerable amounts of energy swimming and feeding [15], with less energy available for growth. In contrast, reduced activity levels at night [43] would allow relatively more energy to be devoted to growth. Under this scenario, alternating light and dark bands would arise from the variation in statolith density associated with differential growth between day and night. Collectively then, the most plausible unit of time between consecutive growth rings is daily, as has been shown in other cubozoans [35,38,39,40,41].

Calculating Date of Metamorphosis

Only medusae collected from Weipa sites were included in these analyses, as small sample sizes from east coast sites made analysis unreliable. The age of 461 medusae was taken to be either (a) the number of rings observed within their statoliths (64 medusae) or (b) the number of rings estimated from the relationship between statolith length and ring number (397 medusae). The metamorphosis date of each individual was calculated by subtracting an individual's age from its date of capture. The percentage of each sample that underwent metamorphosis on a given day was calculated and plotted against season number. Season number was used in preference to year number as it allowed successive samples within a season to be plotted together. That is, a sample collected in January 2007 was denoted a season number of 2006 as was a sample collected in November 2006.

Temporal variation in the onset of the 2000, 2003, 2005, 2006, 2007 and 2010 seasons was quantified by calculating the average earliest metamorphosis date of these six seasons and the associated 95% confidence limits.

A number of environmental parameters are potentially relevant to the shift from the polyp to medusae phase of the life cycle, with the following parameters quantified for the five weeks prior to the onset of metamorphosis in each season:

- water temperature was taken to be the daily sea surface temperature at midday at 16°35′23″S, 141°33′36″E (Albatross Bay) and was obtained from the Integrated Marine Observing System at www.marine.csiro.au/remotesensing/imos
- daily rainfall totals were obtained from the Bureau of Meteorology for Weipa Eastern Avenue (location 27042).
- photoperiod, or the total number of hours of daylight per day, was calculated from sunrise and sunset times for Weipa obtained from Geoscience Australia at http://www.ga.gov.au/geodesy/astro/sunrise.jsp
- tidal amplitude was calculated from hourly tide height (m) data for Humbug Wharf, obtained from Maritime Safety Qld at http://www.msq.qld.gov.au/Home/Tides.

Data were assigned to a week category (1 to 5) in which week category represented the number of weeks prior to the onset of metamorphosis for that season. The weekly variation between years in each of these parameters was investigated using a two way Analysis of Variance in which both week category and year were fixed factors. The date of the full moon within the five weeks prior to the onset of metamorphosis in each season was obtained from Geoscience Australia at http://www.ga.gov.au/earth-monitoring/geodesy.

Growth Curve Calculations

Gordon *et al.* [35] established that a Gompertz growth equation most accurately described the growth parameters of a closely related cubozoan, *C. bronzei*, using the criteria described by Kauffman [44]. Parameters for a four criteria Gompertz growth curve were estimated in Sigma Plot 10 using size at age data for 461 medusae. Size was taken to be IPD at time of capture and age was either (a) the number of rings present in an individual's statolith (64 medusae) or (b) an estimate using the relationship between statolith size and ring number (397 medusae). The maximum daily growth rate (mm d^{-1}) and 95% confidence limits were estimated by regression analysis of the linear component of the growth curve, which occurred between 40 and 70 d. The time to the onset of sexual maturity was also estimated using the resulting growth equation.

Investigation of Population Structure

Since each site was not visited on each occasion, data for medusae collected in Weipa were pooled for all seasons and sites within habitat type. East coast sites were not analysed here due to small sample sizes. Those medusae in which gonads had not developed were denoted as immature juveniles. For mature individuals, sex was determined by examining gonad samples under a stereo dissecting microscope. Individuals in which ova were visible were denoted as females, while those in which sheets of convoluted filamentous tissue were evident were denoted as males. In those cases where it was unclear whether tissue was that of a small male or an immature specimen, individuals were classified as undistinguishable (~60 individuals) and grouped with immature specimens. A chi squared homogeneity test was used to determine whether any significant difference between the proportion of males, females and juveniles existed between the estuarine and coastal habitats.

Results

A significant and positive relationship between statolith length (mm) and number of rings was established ($F = 151.243$, $df = 1 \times 69$, $n = 71$, $P < 0.001$, $R^2 = 0.687$), whereby the number of rings present within a statolith increased as did statolith length. Within the size range of statoliths sampled, a linear relationship (Figure 2) provided a better fit than curvilinear equations, with this linear relationship best described by the equation:

$$R = (78.213 \times SL) - 0.088$$

where R is the number of rings within a statolith and SL is statolith length in mm.

A significant and positive relationship was established between the number of tentacles per pedalium and inter pedalia distance (mm) ($F = 1201.176$, $df = 1 \times 420$, $n = 422$, $P < 0.001$, $R^2 = 0.740$) whereby medusae size (IPD in mm) increased as did the number of tentacles per pedalium (Figure 3). A power curve described this relationship most appropriately, with inter pedalia distance increasing at a faster rate as more tentacles were added to each pedalium. This curve provided a minimum medusae size at the one tentacle stage (ie following metamorphosis) of ~1.8 mm and is best described by the equation:

$$S = 1.846 \times T^{1.628}$$

where S is medusa size (IPD) in mm and T is the number of tentacles per pedalium.

Medusae appeared to add tentacles in pairs, with only ~8% of the 422 medusae for which tentacle number was collected possessing an even number of tentacles per pedalium. A significant and positive relationship was identified between the number of rings within a statolith and the number of tentacles suspended from each pedalium ($F = 639.733$, $df = 1 \times 410$, $n = 412$, $P < 0.001$, $R^2 = 0.609$). Within the age range of medusae sampled, a linear relationship in which tentacle number increased as did ring number (Figure 4) provided a better fit than did curvilinear equations and is best described by the equation:

$$T = (0.189 \times R) - 0.899$$

where T is the number of tentacles per pedalium and R is the number of rings per statolith.

The earliest date of metamorphosis was 30 August and occurred in the 2007 season. Despite the earliest date of specimen collection varying by 33 d, the earliest date of metamorphosis did not vary by more than 7 d. That is, while medusae were first collected on the 18th October in the 2000 season in which the earliest date of metamorphosis was 31 August, in the 2010 season for which the earliest metamorphosis date was 5 September, medusae were not collected until the 20th November. The mean earliest date of metamorphosis was 2 September ±2 d (mean ±95% confidence limits).

Within the 2005, 2006 and 2007 seasons in which sample sizes were large, medusae were added to the population on an almost daily basis, with each date of metamorphosis represented by ~2%, with no one date accounting for more than 10% of a sample (Figure 5). Metamorphosis also appeared to be an ongoing process once it commenced. That is, in those seasons where the interval between successive samples (between grey and black arrowhead lines) was approximately one month, as for the 2006 and 2007 seasons, metamorphosis dates were either continuous (2006 season) or overlapped by a small amount (2007 season). In the 2005 season, however, successive samples were collected ~60 d apart (November 2005 and January 2006), with the gap in metamorphosis corresponding to the length of time between sampling occasions.

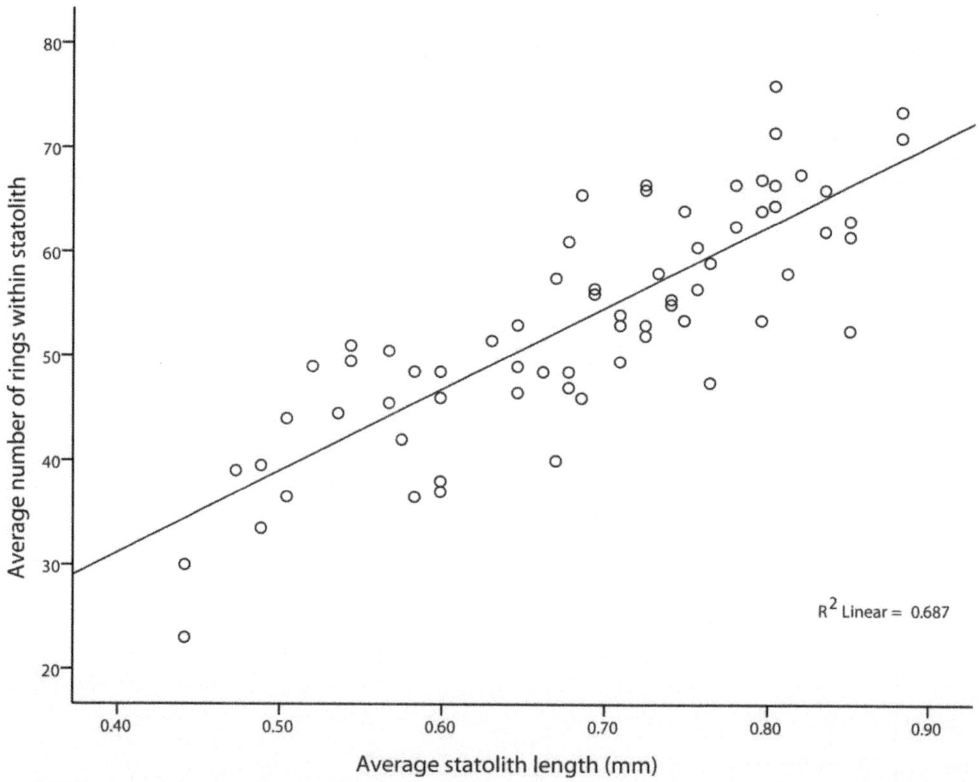

Figure 2. Positive linear relationship between statolith length and number of rings within a *Chironex fleckeri* statolith. Statolith length is in mm and number of rings is averaged for statolith pairs.

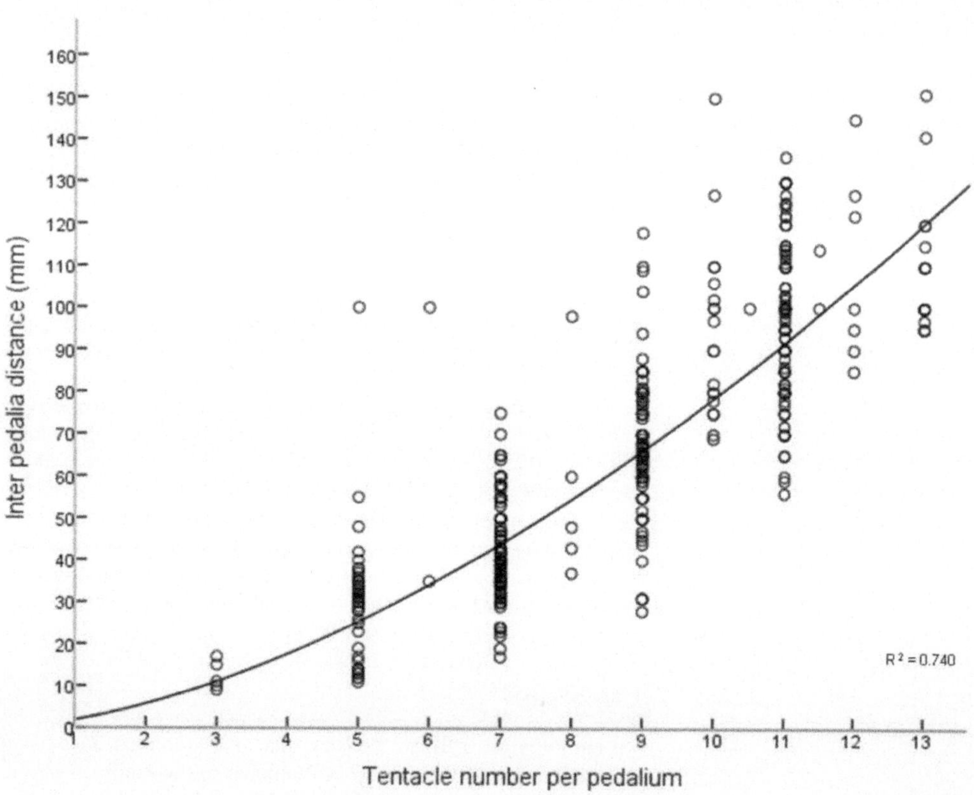

Figure 3. Positive curvelinear relationship between number of tentacles and Inter Pedalia Distance for *Chironex fleckeri* medusae. Inter pedalia distance is in mm and tentacle number is per pedalium.

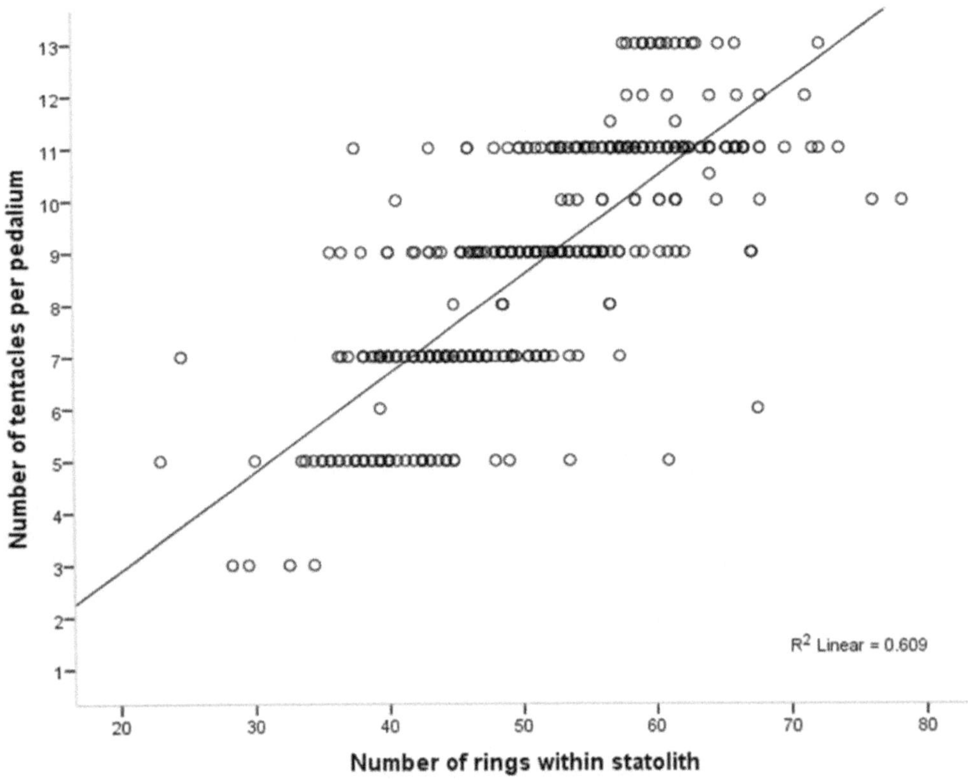

Figure 4. Positive linear relationship between number of rings and number of tentacles for *Chironex fleckeri* medusae. Number of rings is the average per statolith and tentacle number is per pedalium.

Water temperature (°C), daily rainfall (mm) and tidal amplitude (m) each showed significant weekly variation between years in the five weeks prior to the earliest date of metamorphosis within each season (Table 1). Photoperiod (h daylight d^{-1}), however, did not vary significantly by week between years (Table 1). The dates of the full moon in the five weeks preceding the onset of metamorphosis differed by 19 d across the six seasons studied here, ranging from 9 August in 2006 to 28 August in 2008.

The degree to which each environmental parameter varied in the five weeks prior to the onset of metamorphosis was parameter specific. That is, daily rainfall total (DRFT in mm) showed the greatest degree of variation, having a $\log_{10}$ transformed CV value of ~1, while photoperiod showed the least amount of variation with a $\log_{10}$ transformed CV value of ~−3 (Figure 6). The variation in both water temperature (°C) and tidal amplitude (m) was also considerably greater than photoperiod, ranging from ~0 (tidal amplitude in m) to ~−1.5 for water temperature (°C) (Figure 6).

A significant and positive relationship between IPD (mm) and age (days) of medusae ($F = 423.3479$, $df = 3 \times 457$, $n = 461$, $P < 0.0001$, $R^2 = 0.735$) whereby medusae increased in size with age towards an asymptotic size of ~190 mm IPD (Figure 7). The Gompertz growth equation had the following format:

$$S = 1.5 + 186.5617 \times e^{-e^{-\left(\frac{t-51.8408}{21.9887}\right)}}$$

where S is medusa size (IPD) in mm and t is medusa age in days.

Regression analysis of the linear component of this relationship (from 40 to 70 d) revealed a maximum growth rate of ~3 mm d^{-1} (±0.2 mm d^{-1}). The minimum size (IPD in mm) at which males and females could be reliably distinguished was 46 mm IPD and 50 mm IPD respectively. According to the above growth equation, it would take ~45 d to reach the average size (IPD mm) of sexual differentiation. At an average 96 mm IPD for male medusae and 97 mm IPD for females, the mean size of male and female medusae did not vary significantly between the sexes ($F = 0.25$, $df = 1 \times 240$, $n = 242$, $P = 0.621$).

The age structure of the overall population differed significantly between the estuarine and coastal habitats ($\chi^2 = 49.477$, $df = 2$, $n = 281$, $P < 0.001$). Younger medusae were not as well represented within the estuarine habitat, with a predominance of larger and older individuals evident instead. Along the coastline, however, a greater spread of males, females and immature medusae was observed, ranging in age from 30–70 d and from five to 13 tentacles per pedalium (Figure 8). Furthermore, the oldest medusae within the estuarine habitat were older than those along the coastline (Figure 8). Although medusae added new tentacles over an age range, it was typically possible to determine the sex of an individual providing it possessed more than nine tentacles. This was consistent between the estuarine and coastal populations for all but three medusae within the coastal habitat who were classed as immature at the 11 tentacle stage.

Discussion

One area of jellyfish ecology that has received increasing attention of late is the shift from the polyp to the medusae phase, and the factors associated with, or acting as cues for this shift. Not only was the onset of medusae production temporally constrained between seasons, but it also commenced earlier than expected. That is, the earliest metamorphosis date was 30 August (2007

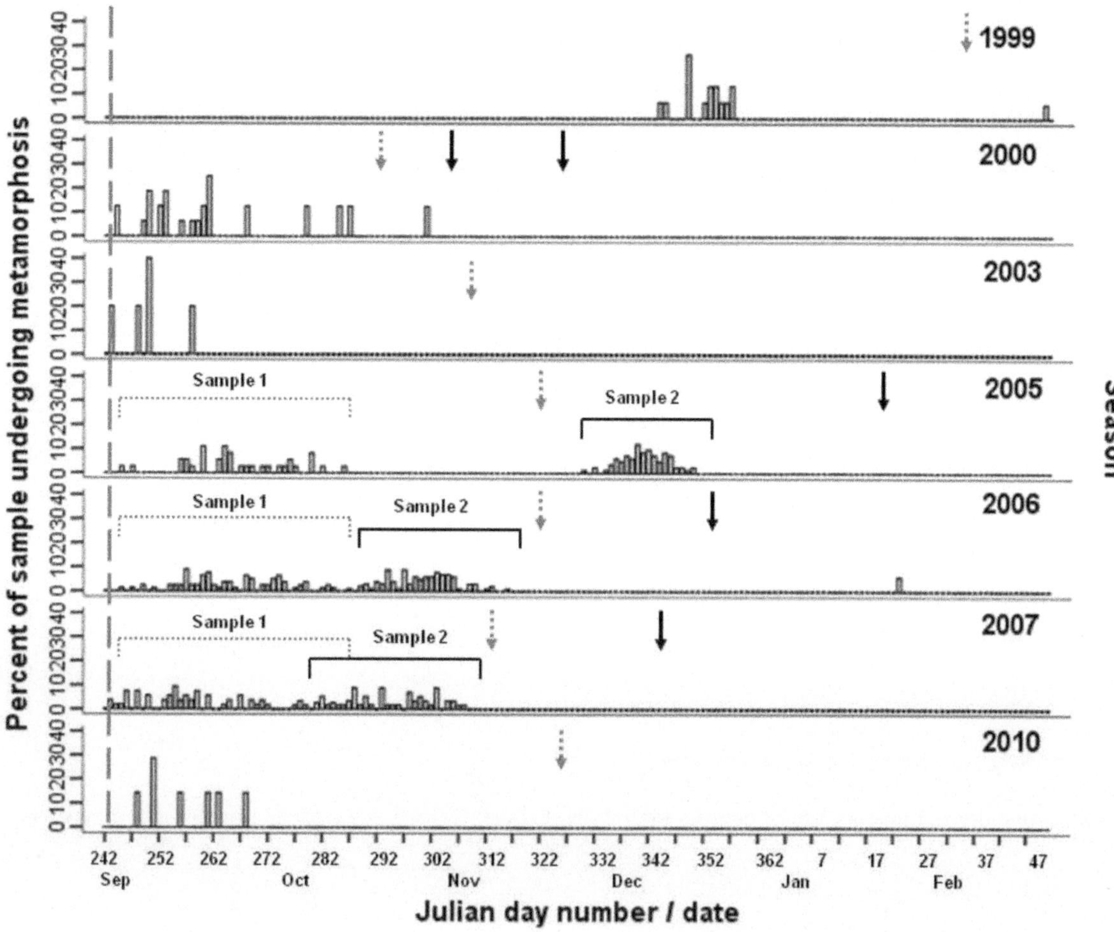

Figure 5. Percent frequency of *Chironex fleckeri* medusae that metamorphosed each day. Frequency is percentage within a sample and date is Julian day number or month, with 30 August referenced by a grey dash line. The first sampling occasion within a season is denoated by a grey arrowhead line, while subsequent sampling occasion(s) are denoted by a black arrowhead line.

season) and varied by only ~7 d across six seasons. Not only is September also the earliest month in which juvenile medusae are reported in plankton tows from east coast studies [15], but medusae of 120 mm collected in December on Magnetic Island [14] would be ~70 d of age (from the above growth curve), also giving them a metamorphosis date in September. This result is particularly significant in terms of modelling the overall medusae season in that the onset of each season can now be defined with greater accuracy.

Quantifying when polyp metamorphosis commenced also allows the factors associated with the shift from the polyp to medusae phase to be identified. It is these factors that polyp based studies should include when quantitatively indentifying the cue for metamorphosis. For instance, the significant between year variation in weekly mean water temperature suggests that water temperature is unlikely to have provided the temporal periodicity observed in the onset of metamorphosis within this study. A similar case exists for both tidal amplitude and rainfall. That is, while unusually high amplitude tides could result in salinity changes at the polyp habitat by pushing higher salinity waters further into estuary systems or allowing fresh waters to drain further down estuaries, the timing of any tidally driven salinity fluctuations would have varied between years. Likewise, significant between year variation in total daily rainfall (mm) suggests that rainfall (or rainfall driven salinity changes) was unlikely to have acted as a cue for the onset of polyp metamorphosis in the seasons investigated within this study. Indeed, the climate of Weipa is dominated by strong seasonal patterns [45,46,47], with low rainfall and elevated, stable salinity typical for late August/early September [46]. Salinity did not appear to be related to the metamorphosis of *C. fleckeri* polyps in laboratory based trials [15,24] either, although dilution rates may have induced encystment rather than

Table 1. ANOVA results for the variation in parameters potentially associated with metamorphosis.

Parameter	Interaction	F	df	*P*
Water Temperature (°C)	week category×year	2.031	20×192	0.008
Total daily rainfall (mm)	week category×year	1.814	20×210	0.022
Tidal amplitude (m)	week category×year	1.696	10×697	0.030
Photoperiod (h daylight d^{-1})	week category×year	0.088	20×210	1.000

Water temperature is in °C, total daily rainfall is in mm, tidal amplitude is in m and photoperiod is the number of h of daylight d^{-1} for each week in the five weeks prior to the onset of metamorphosis in each season.

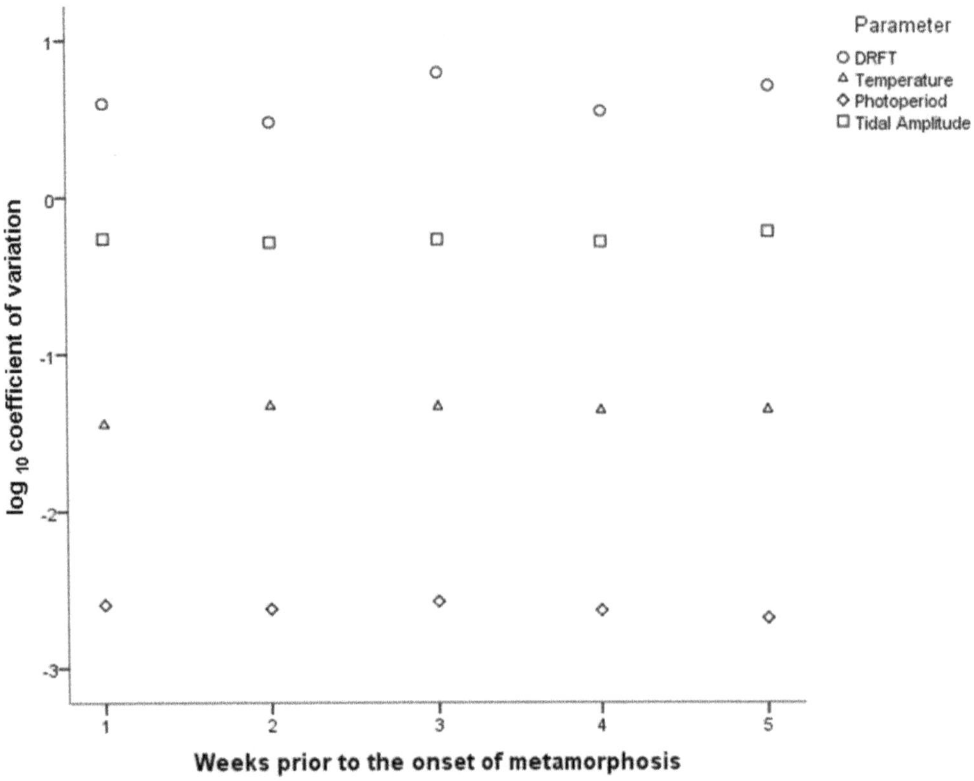

Figure 6. Coefficient of variation ($\log_{10}$ transformed) of environmental parameters. CV for daily rainfall total (mm), tidal amplitude (m), water temperature (°C) and photoperiod (h daylight d^{-1}) for the five weeks preceding the earliest date of metamorphosis in each season.

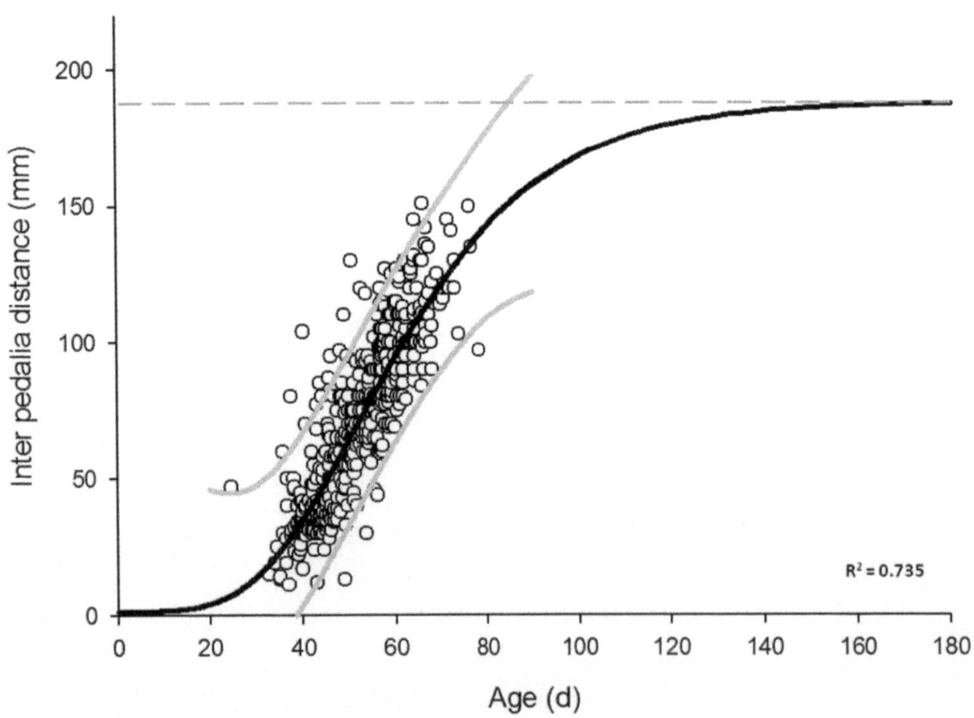

Figure 7. Gompertz growth curve for significant, positive relationship between age and IPD of *Chironex fleckeri* medusae. Growth curve shows medusae size (IPD) in mm at age in d with 95% prediction limits (grey solid line).

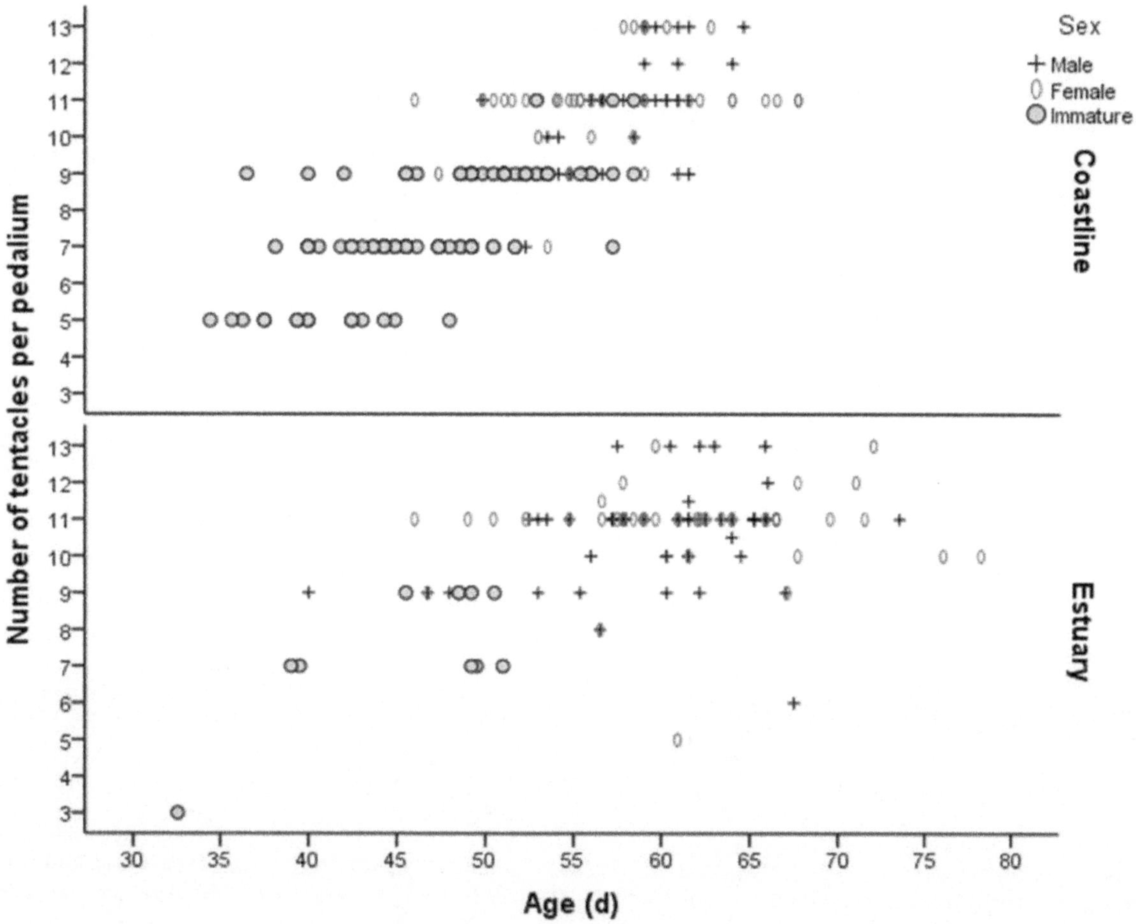

Figure 8. Age of *Chironex fleckeri* medusae with respect to tentacle number. Distinction is made between males, females and immature medusae within the coastal and estuarine habitats of Weipa, where age is in d and tentacle number is per pedalium.

metamorphosis. The influence of temperature is also unclear from polyp based studies, with all trials conducted at 28°C [15,24]. While the results of this study suggest that salinity, temperature, tidal amplitude or moon phase are unlikely to provide the temporal consistency in the onset of metamorphosis as was observed within this study, their role as interacting variables cannot, as yet, be disregarded.

One parameter that could provide a higher degree of temporal consistency in the onset of metamorphosis is photoperiod, with the average hours of daylight consistent between seasons in the five weeks prior to the earliest date of metamorphosis within each season. The influence of photoperiod on polyp metamorphosis remains largely untested, however, with *Carybdea morandinii* the only cubozoan for which a link between photoperiod and metamorphosis has been established [31]. Not only has light been positively correlated with asexual reproduction for some Scyphozoans [48,49], but Purcell [48] suggests that melatonin, a light sensitive hormone, may also play an important role in coordinating the strobilation of the Scyphozoan *Aurelia labiata*. Photoperiod, which has been shown to coordinate breeding cycles in some marine invertebrates [50], may play a similar role in coordinating polyp metamorphosis in *C. fleckeri*, and as such future polyp based research should quantify the significance of photoperiod on polyp metamorphosis.

Metamorphosis did not appear to be a single or pulse event for *C. fleckeri*, rather, an ongoing process whereby low numbers of medusae were produced on an almost daily basis (between 2 and 10% d^{-1}). The collection of a medusa in March 2000 (metamorphosis date of February 19) and another in March 2007 (metamorphosis date of January 23) further suggests that medusae production continued over an extended time frame. Reports of 0.6 to 1.8 mm juvenile *C. fleckeri* medusae in estuarine plankton samples between September and January [15], the presence of both adult and small medusae in the first arrivals along the coastline [10,14], as well as the collection of 6 mm medusae in January and February when 120 mm medusae were collected in December further suggest that metamorphosis occurs over an extended timeframe. Reports of juvenile medusae occurring in successive waves in the only laboratory based study conducted on *C. fleckeri* polyps [24] initially appear contradictory to the results of this study, however, without the timeframe between successive waves quantified, the potential for pulses to have occurred on a daily basis, as observed within this study, cannot be disregarded.

Once within the sexual phase of the life cycle, medusae growth was rapid at up to 3 ± 0.2 mm d^{-1}, which is up to three times that established for a closely related and often co occurring species, *C. bronzei* [34,35]. Although growth rates are likely to vary between individuals due to prey availability [8,15], medusae would typically reach their estimated asymptotic size of ~190 mm (IPD in mm) after ~140 days. That *C. bronzei* has an estimated asymptotic size of 71 mm IPD [34,35] and *Chiropsalmus quadrumanus* Agassiz is reported to reach 110 mm [51] suggests that *C. fleckeri* medusae attain a larger size than do other Chirodropid species. On an applied level, Hartwick [16] has previously suggested that medusae reach a size dangerous to humans within approximately two to three months. Indeed, *C. fleckeri* medusae undergo an ontogenetic shift in their cnidome (and diet) that potentially explains the lethality of larger medusae to humans [52] at ~60–100 mm IPD [52], a size they would reach after ~50–65 d. This is an important parameter to consider in the further development of management protocols in that the time at which medusae are likely to become lethal to humans can now be defined with greater reliability.

That the onset of sexual maturity occurred at approximately ~50 mm IPD compares favourably to estimates provided by Barnes [10] who noted that the development of a very large area of gonad material commenced at the eight tentacle stage (~60 mm IPD based on the regression equation developed here) [25]. At the rapid rate of growth quantified here, medusae would become sexually mature after ~45–50 d, which is considerably less time than the typical length of a season (~180 d). Given that the oldest medusa was ~78 d, it does not appear that medusae accumulate as the season progressed [15]. Potentially, medusae relocated from within the estuarine and coastal areas sampled here, with the infrequent collection of medusae up to several km from shore [8,37,53] suggesting that some form of emigration could take place. Alternatively, medusae that underwent metamorphosis in early September would have had several months of stable conditions and an abundant food supply prior to the onset of the wet season in which to grow and mature. That is, September to December falls within Weipa's dry/pre wet season [47,54] when salinity regimes are typically stable and elevated [47], water temperatures are typically increasing [45,46,47] and an abundance of post larval prawns occurs within the Embley Estuary [46,47]. Whether several generations of medusae occur within a single season is an aspect of *C. fleckeri*'s ecology that future research should examine.

Medusae development can also be considered with respect to tentacle number, with some authors using tentacle number rather than size when discussing the development of medusae (e.g. [25]). The maximum number of tentacles per pedalium observed within this study was 13, which compares favourably to maximums of 12 [10,11] and 13 [25], but suggests that 15 tentacles may be limited to those individuals of ~300 mm IPD which are rarely observed [8,12,55]. Given that newly metamorphosed medusae possess one tentacle per pedalium [24] and only 8.3% of medusae within this study possessed an even number of tentacles, *C. fleckeri* would appear to add tentacles in pairs. With 12 medusae possessing an odd number of tentacles and 10 possessing a even number of tentacles, samples in Kinsey [25] suggest that tentacles are added singularly, however this may be an artefact of a small sample size. *C. bronzei* also appears to add tentacles in pairs, with 92% of the 1652 medusae collected possessing an odd number of tentacles [34]. On a more applied level, tentacle number may provide a more standardised method by which groups such as Surf Life Saving can provide consistent estimates of medusae size and age, given the significant relationships that exist between these variables.

A difference in the population structure of coastal and estuarine habitats would be expected if a seasonal alternation in generations and habitats [16] is occurring for *C. fleckeri*. That is, juvenile medusae would be representative of estuarine populations while coastal populations would be typified by both a greater range in medusae size as well as an accumulation of larger individuals as the season progressed. Not only were medusae from coastal sites typically smaller (fewer tentacles) and younger than those found within the estuarine habitat, but the oldest medusae (~78 d) was also collected from within the estuarine habitat. While these results appear contradictory to expected patterns, juvenile medusae reported by Hartwick [15] were as small as 0.6 mm and collected in plankton tows [15,16], with the visually based collection techniques used within this study possibly possessing an inherent bias against such small individuals. Differences in population structure may also reflect the suitability and availability of prey within the estuarine and coastal habitats. For instance, not only have mangrove areas of the Embley Estuary been shown to be important nursery areas for many species [56], but a greater abundance of fish [54] and prawn species [57] have been reported for intertidal areas adjacent to mangrove stands. In contrast, both species diversity and overall abundance of fish was lower along the coastline [58]. Further data quantifying medusae abundance, gastrovascular cavity content and prey abundance is required to further validate these claims.

Collectively, the results of this study are relevant in both an applied and an ecological context. By quantifying growth and development rates as well as the temporal variation in the onset of polyp metamorphosis between seasons for *C. fleckeri* medusae, this paper has contributed to the understanding of the ecology of an understudied taxon, the cubozoa, as well as to the broader understanding of jellyfish ecology. On an applied level, this study has presented quantitative data upon which models predicting the seasonal occurrence of this species can be developed. For instance, estimates of when medusae are likely to present a considerable risk to humans can now be based on medusae growth and development rates, allowing significant events, such as the ontogenetic shift in cnidome, to be modelled with greater accuracy. It is in this way that the negative effects of the stinger season can be managed more effectively. Such models are of particular relevance given the way in which the seasonal occurrence of *C. fleckeri* impacts the way in which the tropical Australian coastline is utilised throughout the warmer months of the year. However, it is only when a complete understanding of the medusae phases ecological relationships are developed that the occurrence and distribution of *C. fleckeri* can be modelled with accuracy and reliability. This study represents the first attempt at quantifying such parameters, however, further long term studies are required if management practices are to be optimised and broader ecological questions regarding season length or the intensity and frequency of jellyfish blooms are to be validated.

Acknowledgments

Sincere thanks to Teresa Carrette, Robert & Julieanne Gordon, Jennie Gilbert, Glenda, Amelia & Benjamin Seymour and Ben Kelly for their assistance in field work.

Author Contributions

Conceived and designed the experiments: MG JS. Performed the experiments: MG JS. Analyzed the data: MG JS. Contributed reagents/materials/analysis tools: MG JS. Wrote the paper: MG JS.

References

1. Purcell JE, Uye S, Lo W (2007) Anthropogenic causes of jellyfish blooms and their direct consequences for humans: a review. Marine Ecology Progress Series 350: 153–174.
2. Graham WM, Martin DL, Felder DL, Asper VL, Perry HM (2003) Ecological and economic implications of a tropical jellyfish invader in the Gulf of Mexico. Biological Invasions 5: 53–69.

3. Purcell JE, Arai MN (2001) Interactions of pelagic cnidarians and ctenophores with fish: a review. Hydrobiologia 451: 27–44.
4. Rajagopal S, Nair KVK, Asariah J (1989) Some observations on the problem of jellyfish ingress in a power station cooling system at Kalpakkam, east coast, India. Mahasagar Quarterly Bulletin, National Institute of Oceanography (Goa) 22: 151–158.
5. Matsumura K, Kamiya K, Yamashita KHF, Watanabe I, Murao Y, et al. (2005) Genetic polymorphosm of the adult medusae invading an electric power station and wild polyps of *Aurelia aurita* in Wakasa Bay, Japan. Journal of the Marine Biological Association of the United Kingdom 85: 563–568.
6. Masilamoni JG, Jesudoss KS, Nandakumar K, Satpathy KK, Nair KVK, et al. (2000) Jellyfish ingress: A threat to the smooth operation of coastal power plants. Current Science 79: 567–569.
7. Burnett JW (2001) Medical aspects of jellyfish envenomation: pathogenesis, case reporting and therapy. Hydrobiologia 451: 1–9.
8. Burnett JW, Currie B, Fenner PJ, Rifkin J, Williamson JA (1996) Cubozoans (Box Jellyfish). In: Williamson JA, Fenner PJ, Burnett JW, Rifkin J, eds. Venomous and Poisonous Marine Animals - a Medical and Biological Handbook. Sydney: University of New South Wales. pp 98–118.
9. Barnes J (1965) *Chironex fleckeri* and *Chiropsalmus quadrigatus* - morphological distinctions. North Queensland Naturalist. pp 13–22.
10. Barnes J (1966) Studies on three venomous cubomedusae. The Cnidaria and their Evolution: Symposium of the Zoological Society of London. London: Academic Press. pp 307–332.
11. Southcott R (1956) Studies on Australian Cubomedusae, including a new genus and species apparently harmful to man. Australian Journal of Marine and Freshwater Ecology 7: 254–280.
12. Southcott R (1971) The Box-Jellies or Sea-Wasps. Australian Natural History. pp 123–128.
13. Keen TEB (1971) Comparison of tentacle extracts from *Chiropsalmus quadrigatus* and *Chironex fleckeri*. Toxicon 9: 249–254.
14. Brown T (1973) *Chironex fleckeri* - Distribution and movements around Magnetic Island, North Queensland. Townville: James Cook University. 30 p.
15. Hartwick R (1991a) Distributional ecology and behaviour of the early life stages of the box-jellyfish *Chironex fleckeri*. Hydrobiologia 216/217: 181–188.
16. Hartwick R (1987) The Box Jellyfish. In: Covachevich J, Davie P, Pearn J, eds. Toxic Plants and Animals. Brisbane: Queensland Museum Press. pp 99–105.
17. Hamner W, Doubilet D (1994) Australia's Box Jellyfish - A Killer Down Under. National Geographic 186: 116–130.
18. Fenner PJ, Williamson JA (1996) Worldwide deaths and severe envenomation from jellyfish stings. Medical Journal of Australia 165: 658–661.
19. Leonard JL (1980) Cubomedusae belong to the Cubozoa, not Scyphozoa. Nature 284: 377.
20. Werner B, Chapman D, Cutress C (1976) Muscular and nervous systems of the cubopolyp (Cnidaria). Experimentia 32: 1047–1049.
21. Werner B, Cutress C, Studebaker J (1971) Life cycle of *Tripedalia cystophora* Conant (Cubomedusae). Nature 232: 582–583.
22. Werner B (1973) Spermatozeugmen und Paarungsverhalten bei *Tripedalia cystophora* (Cubomedusae). Marine Biology 18: 212–217.
23. Werner B (1975) Bau und Lebensgeschichte des Polypen von *Tripedalia cystohpora* (Cubozoa, class. nov., Carybdeidae) und seine Bedeutund fur die Evolution der Cnidaria. Helgorlander wiss Meeresunters 27: 461–504.
24. Yamaguchi M, Hartwick R (1980) Early life history of the sea wasp, *Chironex fleckeri* (Class Cubozoa). In: Tardent R, Tardent R, eds. Development and Cellular Biology of Coelenterates: Elsevier/North-Holland Biomedical Press. pp 11–16.
25. Kinsey BE (1986) Barnes on box jellyfish. Unpublished folio manuscripts held in the Archives of James Cook University.
26. Jacups S (2010a) Global warming - rising sea surface temperatures - a longer box jellyfish (*Chironex fleckeri*) stinger season for the Northern Territory? The Northern Territory Disease Control Bulletin 17: 25–28.
27. Jacups S (2010b) Warmers waters in the Northern Territory-Herald an earlier onset to the annual *Chironex fleckeri* stinger season. EcoHealth 7: 14–17.
28. Laska-Mehnert G (1985) Cytologische Veranderungen wahrend der metamorphoses des cubopolypen *Tripedalia cystophora* (Cubozoa, Carybdeidae) in die medusae. Helgorlander Meeresuntersuchungen 39: 129–164.
29. Stangl K, Salvini-Plawen L, Holstein TW (2002) Staging and induction of medusae metamorphosis in *Carybdea marsupialis* (Cnidaria, Cubozoa). Life and Environment 52: 131–140.
30. Straehler-Pohl I, Jarms G (2005) Life cycle of *Carybdea marsupialis* Linnaeus, 1758 (Cubozoa, Carybdeidae) reveals metamorphosis to be a modified strobilation. Marine Biology 147: 1271.
31. Straehler-Pohl I, Jarms G (2011) Morphology and life cycle of *Carybdea morandinii*, sp. nov. (Cnidaria), a cubozoan with zooxanthellae and peculiar polyp anatomy. Zootaxa 2755: 36–56.
32. Arneson A, Cutress C (1976) Life history or *Carybdea alata* Reynaud, 1830 (Cubomedusae). In: Mackie G, ed. Coelenterate Ecology and Behaviour: Plenum Press. pp 227–236.
33. Hofmann DK, Kremer BP (1981) Carbon metabolism and strobilation in *Cassiopea andromeda* (Cnidaria: Scyphozoa): Significance of endosymbiotic dinoflagellates. Marine Biology 65: 25–33.
34. Gordon M (1998) Ecophysiology of the Tropical Australian Chirodropid *Chiropsalmus sp.* (Haeckel) [Honours]. Cairns: James Cook University. 114 p.
35. Gordon M, Seymour J, Hatcher C (2004) Growth and age determination of the tropical Australian cubozoan *Chiropsalmus* sp. Hydrobiologia 530/531: 339–345.
36. Arai M (1997) A Functional Biology of Scyphozoa. London: Chapman Hall.
37. Hamner WH, Jones MS, Hamner PP (1995) Swimming, feeding, circulation and vision in the Australian box jellyfish, *Chironex fleckeri* (Cnidaria: Cubozoa). Marine and Freshwater Research 46: 985–990.
38. Kawamura M, Ueno S, Iwanaga S, Oshiro N, Kubota S (2003) The relationship between fine growth rings in the statolith and growth of the cubomedusae *Chiropsalmus quadrigatus* (Cnidaria: Cubozoa) from Okinawa Island, Japan. Plankton Biology and Ecology 50: 37–42.
39. Ueno S, Arimoto Y, Tezuka J (2003) Growth and life history of the cubomedusae, *Carbydea rastoni*, in the southern Japan inshore waters. 7th International Conference on Coelenterate Biology.
40. Ueno S, Imai C, Mitsutani A (1995) Fine growth rings found in statolith of a cubomedusae *Carybdea rastoni*. Journal of Plankton Research 17: 1381–1384.
41. Ueno S, Imai C, Mitsutani A (1997) Statolith formation and increment in *Carybdea rastoni* Haacke, 1886 (Scyphozoa: Cubomedusae): evidence of synchronization with semilunar rhythms. In: den Hartog JC, ed. Proceedings of the sixth International Conference on Coelenterate Biology 1995. Leiden: Nationaal Natuurhistorisch Museum. pp 491–493.
42. Cropp B, Cropp L (1984) The Deadliest Creature on Earth. The Scuba Diver 3: 42–48.
43. Seymour J, Carrette T, Sutherland P (2004) Do box jellyfish sleep at night? Medical Journal of Australia 181: 706.
44. Kaufmann KW (1981) Fitting and using growth curves. Oecologia 49: 293–299.
45. Cyrus DP, Blaber SJM (1992) Turbidity and salinity in a tropical northern Australian estuary and their influence on fish distribution. Estuarine, Coastal and Shelf Science 35: 545–563.
46. Vance DJ, Haywood MDE, Heales DS, Kenyon RA, Loneragan NR (1998) Seasonal and annual variation in abundance of postlarval and juvenile banana prawns *Penaeus merguiensis* and environmental variation in two estuaries in tropical northeastern Australia: a six year study. Marine Ecology Progress Series 163: 21–36.
47. Vance DJ, Haywood MDE, Heales DS, Staples DJ (1996b) Seasonal and annual variation in abundance of postlarval and juvenile grooved tiger prawns *Penaeus semisulcatus* and environmental variation in the Embley River, Australia: a six year study. Marine Ecology Progress Series 135: 43–55.
48. Purcell JE (2007) Environmental effects on asexual reproduction rates of the scyphozoan *Aurelia labiata*. Marine Ecology Progress Series 348: 183–196.
49. Lieu WC, Lo WC, Purcell JE, Chang HH (2009) Effects of temperature and light intensity on asexual reproduction of the scyphozoan, Aurelia aurita (L.) in Taiwan. Hydrobiologia 616: 247–258.
50. Sachlikidis NG, Jones CM, Seymour JE (2005) Reproductive cues in *Panulirus ornatus*. New Zealand Journal of Marine and Freshwater Science 39: 305–310.
51. Guest W (1959) The occurrence of the jellyfish *Chiropsalmus quadrumanus* in Matagorda Bay, Texas. Bulletin of Marine Science of the Gulf and Caribbean 9: 79–83.
52. Carrette T, Alderslade P, Seymour J (2002) Nematocyst ratio and prey in two Australian cubomedusans, *Chironex fleckeri* and *Chiropsalmus* sp. Toxicon 40: 1547–1551.
53. Kinsey BE (1988) More Barnes on box jellyfish. Unpublished folio manuscripts held in the Archives of James Cook University.
54. Blaber SJM, Brewer DT, Salini JP (1989) Species composition and biomasses of fishes in different habitats of a tropical Northern Australian estuary: Their occurrence in the adjoining sea and estuarine dependence. Estuarine, Coastal and Shelf Science 29: 509–531.
55. Barnes J (1960) Observations on jellyfish stings in North Queensland. The Medical Journal of Australia 11: 993–999.
56. Vance DJ, Haywood MDE, Heales DS, Kenyon RA, Loneragan NR, et al. (1996a) How far do prawns and fish move into mangroves? Distribution of juvenile banana prawns *Penaeus merguiensis* and fish in a tropical mangrove forest in northern Australia. Marine Ecology Progress Series 131: 115–124.
57. Vance DJ, Haywood MDE, Staples DJ (1990) Use of a mangrove estuary as a nursery area by postlarval and juvenile banana prawns,*Penaeus merguiensis* de Man, in Northern Australia. Estuarine, Coastal and Shelf Science 31: 689–701.
58. Blaber SJM, Brewer DT, Salini JP (1995) Fish communities and the nursery role of the shallow inshore waters of a tropical bay in the gulf of Carpentaria, Australia. Estuarine, Coastal and Shelf Science 40: 177–193.

Non-Linear Interactions Determine the Impact of Sea-Level Rise on Estuarine Benthic Biodiversity and Ecosystem Processes

Tsuyuko Yamanaka[1,2], David Raffaelli[1], Piran C. L. White[1]*

1 Environment Department, University of York, York, United Kingdom, 2 School of Environmental Sciences, University of Liverpool, Liverpool, United Kingdom

Abstract

Sea-level rise induced by climate change may have significant impacts on the ecosystem functions and ecosystem services provided by intertidal sediment ecosystems. Accelerated sea-level rise is expected to lead to steeper beach slopes, coarser particle sizes and increased wave exposure, with consequent impacts on intertidal ecosystems. We examined the relationships between abundance, biomass, and community metabolism of benthic fauna with beach slope, particle size and exposure, using samples across a range of conditions from three different locations in the UK, to determine the significance of sediment particle size beach slope and wave exposure in affecting benthic fauna and ecosystem function in different ecological contexts. Our results show that abundance, biomass and oxygen consumption of intertidal macrofauna and meiofauna are affected significantly by interactions among sediment particle size, beach slope and wave exposure. For macrofauna on less sloping beaches, the effect of these physical constraints is mediated by the local context, although for meiofauna and for macrofauna on intermediate and steeper beaches, the effects of physical constraints dominate. Steeper beach slopes, coarser particle sizes and increased wave exposure generally result in decreases in abundance, biomass and oxygen consumption, but these relationships are complex and non-linear. Sea-level rise is likely to lead to changes in ecosystem structure with generally negative impacts on ecosystem functions and ecosystem services. However, the impacts of sea-level rise will also be affected by local ecological context, especially for less sloping beaches.

Editor: Tasman Crowe, University College Dublin, Ireland

Funding: The work was supported by the University of York. The funders had no role in study design, data collection and analysis, decision to publish, or preparation of the manuscript

Competing Interests: The authors have declared that no competing interests exist.

* E-mail: piran.white@york.ac.uk

Introduction

Climate change is expected to have significant effects on the marine environment through temperature change, ocean acidification, and accelerated sea-level rise, due to thermal expansion and melting of ice sheets [1]. Warmer sea temperatures are likely to cause changes in abundance, diversity and size composition of zooplankton [2,3], increases in abundance of southern invertebrate species and decreases of northern species in the northern hemisphere [4]. Acidification may have both direct and indirect impacts, via changes in the phytoplankton community, on bacteria and zooplankton [5,6,7]. Some of the greatest impacts of sea-level rise are likely to occur in intertidal sediment habitats. Intertidal habitats will rarely be allowed to transgress inland due to the high value of real estate and land behind the existing high tide zone, so they will be squeezed between rising sea-level and hard coastal defences. Coastal squeeze leads to the loss of intertidal area, sedimentary shifts towards coarser particles and more reflective morphodynamic states, greater tidal velocity, changes in water depth and (for estuaries) salinity, and increased storm surges [1,8,9,10].

Estuarine soft sediment habitats and their associated biodiversity are of great functional importance for the entire marine ecosystem and for human wellbeing in terms of the ecosystem services they produce, including high primary productivity, high secondary productivity, nutrient cycling, climate regulation, pollution control, decomposition, biodegradation and recreation [8,11,12,13,14,15,16,17]. They also provide nursery grounds for marine fish, and feeding and breeding areas for migratory birds and other species [18]. Many of these services are underpinned by ecosystem processes occurring within the benthos [19,20,21]. Understanding the likely impacts of the physical constraints imposed by sea level rise on benthic assemblages is therefore important for evaluating and managing threats to future ecosystem service provision.

Coarser sediment size, steeper beach slopes and higher exposure to wave action are predicted for intertidal areas of the typical V-shaped estuaries that characterise much of the European coastline [9,22]. Changes in these physical beach characteristics are likely to have an impact on benthic invertebrate biomass and body size distribution [22,23,24], and consequently on ecosystem processes. However, the nature of these impacts is likely to be determined by the interactions among the physical constraints of particle size, beach slope and exposure, collectively known as beach morphodynamic state [23,25], rather than any of these factors acting alone. At one end of the morphodynamic range, dissipative beaches with finer particles and gentle slopes have been shown to support a higher number of species and a more abundant macrofauna [26,27,28,29], whilst the opposite is true for reflective beaches at the other end of the range. Meiofauna are likely to be less affected than macrofauna by changes in exposure and sediment particle size [30].

Here we extend previous considerations of the potential impacts on estuarine benthic organisms of physical changes associated with sea-level rise to the ecosystem processes with which they are involved. We examine the relationships between abundance, biomass, and community metabolism of benthic organisms with beach slope, particle size and exposure, using samples that span a range of conditions from three very different locations in the UK. We test the following specific hypotheses [26,27,28]: (1) Total invertebrate abundance, biomass, and species richness will decline in response to increases in average sediment grain, steeper beach profiles and greater exposure to wave action; and (2) these changes will also be reflected in changes in community metabolism, a measure of ecosystem functioning. We also specifically consider the interactions between the different physical constraints and local ecological context in determining the overall impacts on beach fauna.

Methods

Ethics Statement

No specific permits were required for the field studies. All the sites sampled are public beaches, except for the Ythan which is a National Nature Reserve. Samples were taken from the Ythan under the permit issued by Scottish Natural Heritage to OceanLab, University of Aberdeen (Dr Martin Solan). None of the field studies involved endangered or protected species.

Study Sites and Sampling Method

We chose three estuaries within the UK: the Humber (from 53°33′20″ to 53°34′50″ N, and from 0°00′40″ to 0°03′20″ W), the Ythan (from 57°18′50″ to 57°20′20″ N, and from 1°59′20″ to 2°01′ 10″ W) and the Firth of Forth/the Forth estuary (from 55°57′30″ to 56°00′30″ N, and from 3°06′30″ to 3°31′00″ W) on the basis of their wide geographical spread within the UK and previous history of research. We selected five or six sampling stations on each of these estuaries to provide a range of particle sizes, slopes and exposures (Fig. 1). The sampling stations were restricted to a defined section of the outer (marine) salinity gradient (>30, spring high tide) to minimise any potentially confounding effects of salinity [31].

At the mid-tide level at each sampling station, to correspond with maximum invertebrate abundance, a cylindrical core (10.4 cm diameter) was pushed into the sediment to the depth of 10 cm on a randomly chosen surface to sample macrofauna, meiofauna and sediment, thus ensuring that all variables were measured at a similar scale (grain). Sampling stations were established using a map and previous surveys [31,32,33] Four widely-spaced replicates were taken for macrofauna, one sample for meiofauna and one sample for sediment. No replicates were taken for meiofauna as the sampling unit (86 cm^2) is far greater than the scale of patchiness for this group (c. 3–10 cm^2) [34], such that variance between samples of this size is extremely low.

Biological Measurements

Macrofauna were separated from sediment using a 500 µm mesh, preserved in 70% ethanol, identified to species level wherever possible, and counted, using a low-power microscope. Meiofauna were separated from sediment using a 64 µm mesh, preserved in 70% ethanol and stained with Rose Bengal (5 mg l^{-1} 70% ethanol). Because of the large core size used for sampling meiofauna and the large number of individuals collected, meiofauna from each site were sub-sampled after extraction and homogenisation. Meiofauna in these sub-samples (= $1/200^{th}$ of core) for each site were identified to the lowest possible taxon, counted and measured to provide body dimensions for the calculation of body mass (Table S1).

Calculation of body mass. Body dimensions of individual animals were measured under a high-power microscope, and converted to body mass (dry weight) using established relationships ([35]; Table S1). Where densities were high, body dimensions were measured and body mass calculated for a sub-sample, and then scaled up to the number in the full sample.

Community metabolism. Community metabolism was calculated as oxygen consumption by all individual invertebrates using data on mass-specific oxygen consumption factors derived by Banse [36], as used by Gerlach, Hahn, et al. [37]. Abundances of macrofauna and meiofauna were expressed as numbers $core^{-1}$, biomass as Ash Free Dry Weight (AFDW) as mg $core^{-1}$, and oxygen consumption as $\mu lO_2\ hr^{-1}\ core^{-1}$. Thus,

Log_{10} (Oxygen Consumption) = $-0.2599*Log_{10}$ (AFDW) + 0.51 for macrofauna;

Log_{10} (Oxygen Consumption) = $-0.24*Log_{10}$ (AFDW) + 0.0096 for meiofauna; and

Log_{10} (Oxygen Consumption) = $-0.2605*Log_{10}$ (AFDW) − 0.4424 for foraminifera [34,35].

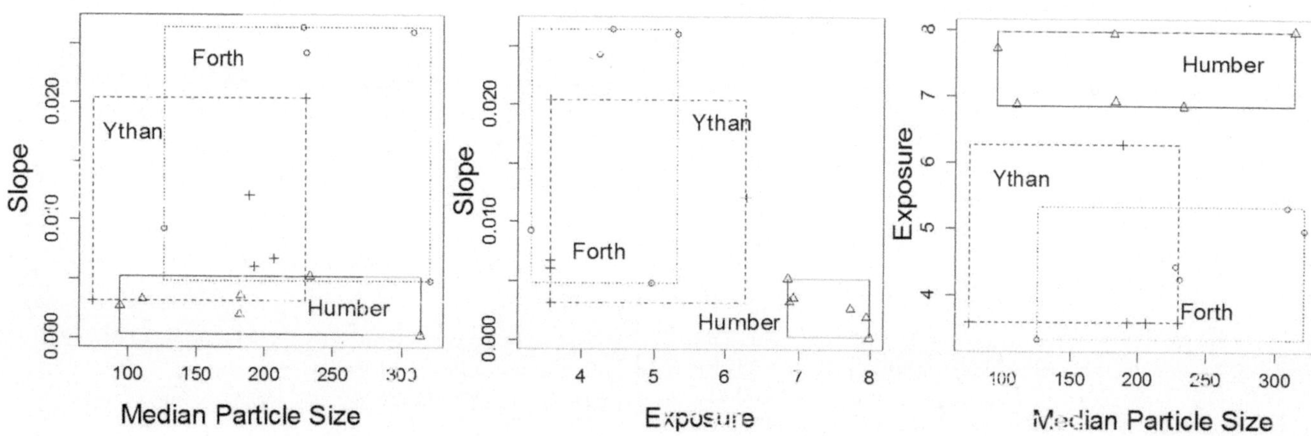

Figure 1. The range of median particle size, slope and exposure for each estuary.

Beach Physical Characteristics

Sediment particle size. Sediment cores were oven-dried at 80°C for at least 24 hours, after which the sample was thoroughly homogenised. Particle size composition was determined by dry sieving the material through a tower of sieves of mesh 4.75 mm, 2.8 mm, 1 mm, 500 μm, 250 μm, 125 μm and 64 μm. Median particle size (μm), 25% and 75% quartiles (Q_{25}, Q_{75} in μm) and the sorting coefficient $QD\varphi = \frac{Q_{25}-Q_{75}}{2}$ were determined by standard probit analysis. Silt content was taken as the weight of material passing through a 64 μm mesh by wet sieving.

Beach slope. The slope at each sampling station was calculated as:

$$Slope = \frac{H}{\sqrt{D^2 - H^2}}.$$

where H = vertical height (m) difference between the low and high shore, and D = distance (m) down the beach between the low and high points. Height difference was measured using a dumpy level and a staff, and distance using a surveyor's tape. Because the values for slope on these beaches are very small, they are presented in the text as slope $\times 10^3$, to facilitate inter-site comparisons.

Exposure. Exposure was calculated using a modified form of the Thomas Exposure Index (TEI) [38], derived from wind velocity, direction, duration and the effective fetch:

$$Exposure = \sum \log W \times \log\left[\frac{1+F}{CS + 0.1\ DS}\right].$$

where W = (percentage of time wind blows in a sector (22.5 degree sectors of the compass rose)/100) × (mean wind speed (kn)^2).

F = Fetch in nautical miles (100 nautical miles maximum).

CS = Extent in nautical miles of water <6 m deep adjoining the shore.

DS = Extent in nautical miles of water within the fetch <6 m deep, but not adjoining the shore.

W was calculated using the wind data from windfinder.com, and F, CS, and DS were derived from UK Admiralty charts.

Statistical Analysis

Statistical models (mixed model for macrofauna and linear regression with GLS extension for meiofauna) were developed [39,40] for abundance, biomass and oxygen consumption of macrofauna and meiofauna, in relation to the physical beach variables. Physical beach variables (median particle size, beach slope and wave exposure) were measured for each station, and four replicates were included in the model for macrofaunal abundance, biomass and oxygen consumption and no replicates were included for meiofaunal abundance, biomass and oxygen consumption. Silt content was not included in the models as it was strongly correlated with median particle size, and the sorting coefficient QDφ was not significant in any models. We included the three estuaries initially as a random local ecological effect in all models; this term was highly significant in the model for macrofauna, although not in that for meiofauna. For the macrofauna model, we therefore used a mixed modelling approach, including median particle size, beach slope and exposure as physical fixed effects and estuary as a random local effect. For the meiofauna model, we included median particle size, beach slope and exposure as effects, and used a linear regression model with a GLS extension to allow for unequal variance associated with median particle size [41,42]. All the possible two and three way interactions were included in both macrofaunal and meiofaunal models.

Results

Beach Physical and Biological Characteristics

The Humber and the Forth estuaries had a similar range of particle size (with the Ythan range slightly lower), and the Humber had much higher exposure than the Ythan and the Forth (no overlap in range). The Ythan and the Forth also had a similar range of beach slopes, whereas those for the Humber were much shallower (Table 1, Fig. 1).

The Ythan and the Humber had similar numerically dominant species and species richness of macrofauna, but differed markedly in the numbers of individuals recorded (Table 2). The Forth had intermediate abundances, but a greater range of species, and was dominated numerically by polychaetes. The Forth had the most meiofaunal taxa represented, including mites (Acarina) and archiannelids. Nematodes and foraminiferans (live as opposed to dead shells) occurred in large numbers, especially on the Humber and the Ythan, and Copepods and Turbellarians were abundant on the Ythan.

Relationships of Macro- and Meiofauna with Beach Physical Characteristics

Macrofaunal abundance, biomass and oxygen consumption. Median particle size and beach slope had a similarly strong influence on the abundance of macrofauna (Table 3), whilst exposure had relatively less influence on abundance. The three-way interaction term particle size × beach slope × exposure was also significant. Sediment particle size, beach slope and wave exposure are usually associated to each other (the beach with high wave exposure tend to have steeper beach profile and coarser sediment). Because the local ecological effect (estuary) was highly significant, the graphs to show the predictions of the mixed model for macrofaunal abundance were made for minimum, mean and maximum exposure and beach slope for each estuary separately (Fig. 2).

For shallow-sloping beaches, there were no consistent relationships between abundance and exposure or median particle size across the estuaries, and the local ecological effect dominated the patterns observed (Fig. 2). Thus, for shallow-sloping beaches, there was a weak negative relationship between abundance and particle size for the Humber stations, a slightly stronger, positive one for all the Ythan stations and no relationship for the Forth stations. However, for intermediate and steep beaches, there were clear and consistent non-linear relationships between abundance and exposure and particle size, which overrode the local ecological effect (Fig. 2). For the most sheltered beaches, there were strong

Table 1. Summary of physical characteristics of each estuary showing the median values (and ranges) encountered within each estuary.

	Estuary		
	Humber	Ythan	Forth
Median particle size (μm)	183 (95–314)	192 (74–230)	230 (126–321)
Silt content (%)	3.8 (2–54.8)	20.4 (12.4–67.6)	1.0 (0.6–22.6)
Sorting coefficient QDφ	44 (29–154)	81 (40–133)	53.5 (49–73)
Slope ($\times 10^3$)	2.9 (0.2–5.2)	6.8 (3.1–20.4)	16.8 (4.8–26.5)
Exposure	7.33 (6.84–7.98)	3.57 (3.57–6.27)	4.34 (3.32–5.32)

Exposure was calculated as a modified Thomas Exposure Index (see text).

Table 2. Summary of biological characteristics of each estuary showing the median values (and ranges) encountered within each estuary.

	Estuary		
	Humber	Ythan	Forth
Macrofauna			
Hydrobia ulvae	8.75 (0–87.5)	48 (0.25–380.5)	0.75 (0–36)
Macoma balthica	2.50 (0.5–42.25)	0.75 (0.25–7.5)	0 (0–4.25)
Pygospio elegans	1.00 (0.25–6.25)	42.5 (1.5–92.25)	1.75 (0.75–170.75)
Cerastoderma edule	0.75 (0.25–1.25)	0 (0–0.75)	0.25 (0–3.5)
Etone longa	0 (0–0.75)	1 (0–0.75)	0.5 (0–3)
Nereis diversicolor	0 (0–0.25)	2.25 (0.25–8.25)	0.25 (0–2)
Corophium volutator	0 (0–0.25)	167.25 (8.25–468.25)	0 (0–0.25)
Midge fly larvae	0 (0–0.25)	0.25 (0–0.75)	0 (0–0)
Collembola	0 (0–1)	0 (0–1.75)	0 (0–0)
Capitellidae	0 (0–1.25)	0 (0–0)	1.75 (0.5–403)
Syllidae	0.375 (0–1.75)	0 (0–0)	0 (0–0)
Scoloplos armiger	0.125 (0–0.75)	0 (0–0)	0 (0–0)
Urothoe brevicornis	0 (0–0.25)	0 (0–0)	0.25 (0–0.5)
Arenicola marina	0 (0–0.25)	0 (0–0)	0 (0–0.25)
Cumacea	0 (0–0.25)	0 (0–0)	0 (0–0.75)
Other Oligochaetes	1.375 (0–17.5)	0 (0–0)	1.5 (0–30)
Nephtys caeca	1.25 (0–2.25)	0 (0–0)	0 (0–0)
Mytilus edulis	0 (0–0)	0 (0–0.5)	2 (0–7.25)
Retusa abtusa	0 (0–0)	0 (0–0.5)	0 (0–5)
Tubifex costatus	0 (0–0)	30.25 (9.75–132.5)	0 (0–0)
Tubificoides benedeni	0 (0–0)	47.75 (1.75–64)	4.25 (0.25–104)
Manayunkia aestuarina	0 (0–0)	11 (05–84.5)	0 (0–1)
Carcinus menas	0 (0–0)	0 (0–0.25)	0 (0–0)
Tetrastemma melanocephalum	0 (0–0)	0 (0–0.25)	0 (0–0)
Streblospio benedicti	0 (0–0)	0 (0–0)	9.75 (1.25–50)
Littorina littorea	0 (0–0)	0 (0–0)	0 (0–0.25)
Meiofauna			
Forminiferans	25369 (3103–111498)	48223 (24896–187471)	4001 (308–59000)
Nematodes	16800 (4673–34669)	19974 (14410–58924)	4212 (2916–13235)
Oligochaetes	571 (178–2140)	571 (143–1855)	778 (454–3256)
Copepods	375 (36–892)	17121 (6335–47253)	243 (178–8181)
Turbellarians	0 (0–0)	25681 (12983–112140)	1231 (405–5054)
Ostracods	0 (0–0)	0 (0–0)	259 (0–4666)
Archiannerids	0 (0–0)	0 (0–0)	194 (0–3094)
Acarina	0 (0–0)	0 (0–0)	0 (0–97)

Values for macrofauna and meiofauna are averages per core.

negative relationships between abundance and particle size. However, as exposure increased, this relationship became more weakly negative and then positive at intermediate exposures, and then became negligible or very weakly positive at the highest exposures (Fig. 2).

Macrofaunal biomass and oxygen consumption exhibited the same non-linear relationships with physical factors as macrofaunal abundance (Figs. S1 and S2). The most significant single factor affecting both biomass and oxygen consumption was beach slope (Table 3), followed by median particle size and exposure. The three-way interaction term particle size × beach slope × exposure was also significant for both macrofaunal biomass and oxygen consumption (Table 3). Biomass and oxygen consumption show contrasting trends because weight-specific oxygen consumption increases as organisms become smaller [36,37].

Meiofaunal abundance, biomass and oxygen consumption. Beach slope had the greatest effect on meiofaunal abundance (Table 3), followed by median particle size and exposure. The three-way interaction term particle size × beach slope × exposure also had a significant effect on meiofaunal abundance; meiofaunal abundance decreased with increasing exposure and increasing particle size on the shallow- and

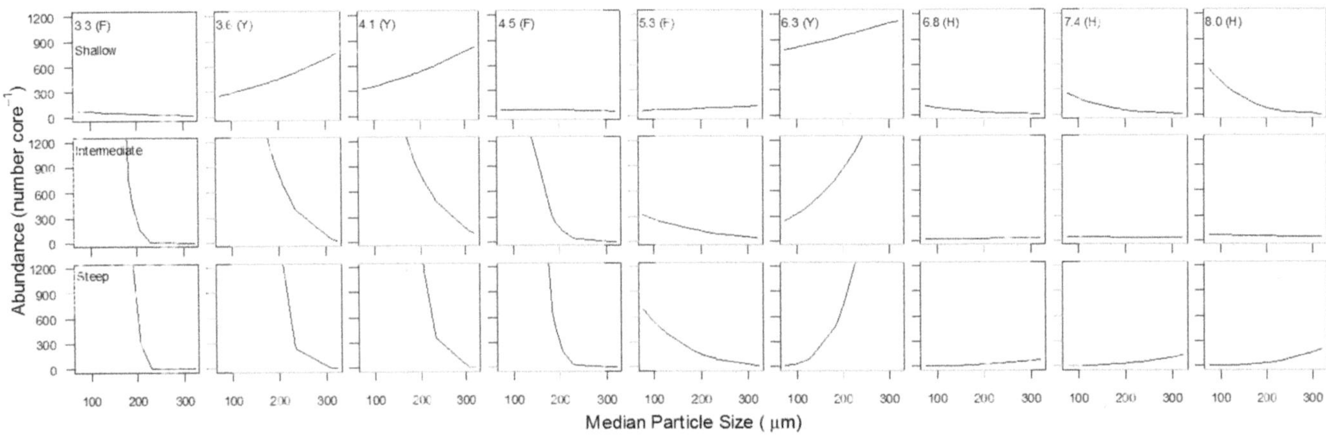

Figure 2. Predicted macrofaunal abundance based on the minimal adequate regression model for each estuary. The values on graphs show the predictions for minimum, mean and maximum exposure values of each estuary. H, Y, and F represent the Humber, the Ythan, and the Forth, respectively. Shallow, Intermediate, and Steep represent the minimum (top row), mean (middle row) and maximum (bottom row) slope of each estuary. Minimal adequate model is the model that contains the minimum number of predictors which are chosen here by stepwise multiple regression.

intermediate-slope beaches (Fig. 3). However, on the steep beaches, there was evidence of a non-linear response to the interaction between exposure and particle size, similar to that observed for macrofauna.

Meiofaunal biomass and oxygen consumption exhibited the same non-linear relationships with physical factors as meiofaunal abundance (Figs. S3 and S4). Meiofaunal biomass and oxygen consumption were affected most significantly by exposure, followed by beach slope and median particle size (Table 3). The three-way interaction term particle size × beach slope × exposure was a significant factor determining both meiofaunal biomass and oxygen consumption. The two-way interactions were also significant for all the models, but these lower term interactions are not discussed here.

Discussion

Our analyses revealed strong relationships between macrofaunal and meiofaunal abundance, biomass and oxygen consumption and beach slope, particle size and exposure. That relationships exist is not unexpected, but here we have been able to tease out the relative importance of the different variables and, more importantly, their linear or non-linear nature. The most significant factor affecting macrofauna was slope, followed by particle size and exposure. For meiofauna, the most significant factor was exposure, followed by slope and particle size. For the shallowest beaches, the local context (which estuary it was) overrode these other relationships, but for intermediate and steeper beaches, the physical factors were the dominant influences. The overall pattern for the relationship for intermediate and steep beaches was negative, with invertebrate abundance, biomass and oxygen consumption declining with increasing slope, particle size and exposure, but the effects were complex and non-linear, with the relationship switching at intermediate levels of exposure.

Previous studies have shown that dissipative beaches with fine particles and gentle slopes support higher numbers of species, in greater abundance and biomass [15,23,26,27,28]. The present study confirms these trends for the meiofauna, showing declines in abundance, biomass, and oxygen consumption with increasing particle size, beach slope, and exposure. However, our results go much further than previous analyses in demonstrating that under certain conditions these relationships are likely to be more

Table 3. L-ratios for regression models of abundance, biomass and oxygen consumption for macrofauna and meiofauna with respect to the single factors median particle size, slope and exposure (each $df = 4$) and the three-way interaction between them ($df = 1$).

	Median particle size	Slope	Exposure	Particle size×slope×exposure
Macrofauna				
Abundance	42.97***	42.91***	34.97***	27.76***
Biomass	24.38**	36.61***	22.82**	8.7*
O_2 consumption	28.01***	37.31***	25.1***	13.85**
Meiofauna				
Abundance	15.36*	15.8*	14.59*	6.74*
Biomass	15.38*	27.63***	30.51***	8.27*
O_2 consumption	20.45**	24.91**	25.4***	11.78**

***$P<0.0001$, **$P<0.001$ and *$P<0.01$.

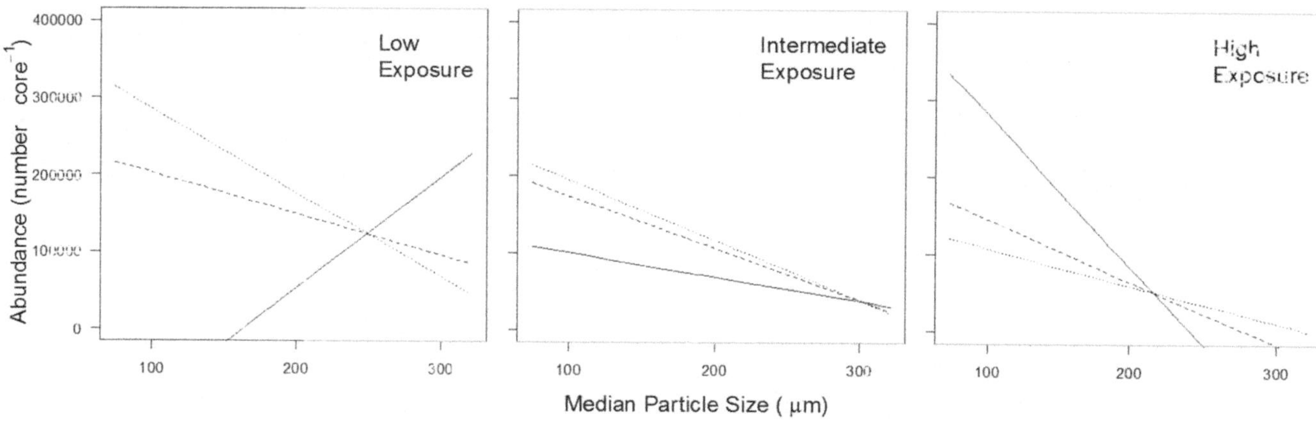

Figure 3. Predicted meiofaunal abundance for low, intermediate and high exposure, based on minimal adequate regression model. The lines in the panels represent steep slope (solid line), intermediate slope (dashed line) and shallow slope (dotted line).

complex and non-linear. This has important implications for the impact of sea-level rise on ecosystem service provision such as reduced food resources, reduced water quality, reduced primary productivity and change in release of nutrient to the overlying water column, since the effects of steeper slopes, increasing particle size and increased exposure on ecosystem structure and function will be moderated by local conditions, especially for less sloping beaches.

McLachlan et al. [30] reported greater numbers of meiofaunal harpacticoid copepods and fewer nematodes in relation to larger median particle sizes. Those authors found that nematodes tended to dominate in finer sands (<330 μm), with overall biomass higher, similar to the present study, where all sites had particle sizes <330 μm and were also dominated by nematodes. In contrast, Rodriguez et al. [43] found increased meiofaunal biomass with increasing wave exposure, but their sites were much more exposed. Taken together, these results emphasise the complexity of the relationships between biodiversity elements and physical constraints on beaches, which probably contribute to the non-linear trends observed by others for ecosystem service provision in coastal ecosystems [14].

Interactions between Beach Physical Factors and Local Ecological Context

McLachlan [25] and Raffaelli and Hawkins [23] argued that the interaction between particle size, beach slope and exposure is probably as important, if not more so, for predicting changes in beach fauna than the effect of any single physical factor. This is formally confirmed in the present study for the first time: the simple bivariate relationships depicted in the literature such as graphs of abundance or species richness against sediment particle size (e.g. [23], p59, Fig. 2.10) are clearly inadequate descriptions of the true, more complex relationships, and would be misleading if used to predict impacts of sea-level rise.

The meiofaunal models revealed linear relationships between variables, but all the macrofaunal models indicate non-linear relationships. Since maximum faunal abundance and diversity may change its exact location along the intertidal gradient depending on exposure, and the sampling stations were fixed at a geographical mid-tide level in this study, it could be argued that the observed non-linear responses of macrofauna with exposure-slope could be due to sampling at this fixed point. However, we think this is unlikely given that the meiofaunal relations might then have been expected to be similarly non-linear. The underlying causes of the non-linear macrofaunal relationships thus remain unclear.

The location of the three estuaries was highly significant for models of abundance, biomass and oxygen consumption of macrofauna, but not for the meiofauna. This suggests that the importance of physical factors outweighs that of local context for meiofauna, although for macrofauna, this is only the case under the limited conditions of less sloping beaches. The exact nature of these local effects is unknown, but they must reflect other local variables which we did not measure, such as differences in predation by birds and fish, or differences in water quality.

Impact of Sea-level Rise on Ecosystem Services from Estuarine Systems

Accelerated sea-level rise is expected to make beaches coarser and steeper, and more reflective in morphodynamic state [10,29]. The predicted increase in the frequency of storm surges [1] will add to this effect. Our results suggest that if beach physical factors change as predicted, then sea-level rise will make estuarine intertidal areas less diverse and less productive, through declines in abundance, biomass, and community metabolism as well as through the loss of area due to coastal squeeze. Mechanistic understanding of the relationships between biodiversity, ecosystem functions and ecosystem services remains uncertain for estuaries [42], although experimental work on benthic biodiversity and ecosystem functioning suggests that reductions in macro- and meiofaunal biomass and metabolism are likely to have significant effects in reducing nutrient cycling within sediments and release of nutrients to the overlying water column [19,39,44,45,46]. Our results have highlighted the complex and non-linear nature of these relationships, and more precise prediction of the impact of sea-level rise on ecosystem service provision will depend on a much improved understanding of the interactive effects of local ecosystem contexts and globally-operating physical constraints on ecosystem structure and function.

Supporting Information

Figure S1 Predicted macrofaunal biomass based on the minimal adequate regression model for each estuary. The values on graphs show minimum, mean, maximum exposure values of each estuary. H, Y, and F stand for the Humber, the Ythan, and the Forth, respectively. Shallow, Intermediate, and

Steep represent the minimum (top row), mean (middle row) and maximum (bottom row) slope of each estuary.

Figure S2 Predicted macrofaunal oxygen consumption based on the minimal adequate regression model for each estuary. The values on graphs show minimum, mean, maximum exposure values of each estuary. H, Y, and F stand for the Humber, the Ythan, and the Forth, respectively. Shallow, Intermediate, and Steep represent the minimum (top row), mean (middle row) and maximum (bottom row) slope of each estuary.

Figure S3 Predicted meiofaunal biomass for low, intermediate and high exposure, based on minimal adequate regression model. The lines in the panels represent steep slope (solid line), intermediate slope (dashed line) and shallow slope (dotted line).

Figure S4 Predicted meiofaunal oxygen consumption for low, intermediate and high exposure, based on minimal adequate regression model. The lines in the panels represent steep slope (solid line), intermediate slope (dashed line) and shallow slope (dotted line).

Table S1 Relationships between morphological dimensions and body weight.

Acknowledgments

We would like to thank M. Bulling for the help in statistical analysis, T. Fujii for providing beach slope data for the Humber Estuary, and M. Bulling, T. Fujii, M. Solan, J. Godbold, J. Finch, and E. Curmi for field assistance.

Author Contributions

Conceived and designed the experiments: TY DR PCLW. Performed the experiments: TY. Analyzed the data: TY. Contributed reagents/materials/ analysis tools: TY. Wrote the paper: TY DR PCLW.

References

1. IPCC Working Group 1 (2007) Observations: Oceanic Climate Change and Sea Level. In: Solomon S, Qin D, Manning M, Marquis M, Averyt K et al., editors. Climate Change 2007 The Physical Science Basis. Cambridge: Cambridge University Press.
2. Richardson AJ (2008) In hot water: zooplankton and climate change. ICES Journal of Marine Science: Journal du Conseil 65: 279–295.
3. Roemmich D, McGowan J (1995) Climatic warming and the decline of zooplankton in the California current. Science 267: 1324–1326.
4. Barry JP, Baxter CH, Sagarin RD, Gilman SE (1995) Climate-related, long-term faunal changes in a California rocky intertidal community. Science 267: 672–675.
5. Hofmann GE, Barry JP, Edmunds PJ, Gates RD, Hutchins DA, et al. (2010) The Effect of Ocean Acidification on Calcifying Organisms in Marine Ecosystems: An Organism-to-Ecosystem Perspective. Annual Review of Ecology, Evolution, and Systematics 41: 127–147.
6. Kurihara H, Shimode S, Shirayama Y (2004) Sub-lethal effects of elevated concentration of CO2 on planktonic copepods and sea urchins. Journal of Oceanography 60: 743–750.
7. Lohbeck KT, Riebesell U, Reusch TBH (2012) Adaptive evolution of a key phytoplankton species to ocean acidification. Nature Geosci 5: 346–351.
8. Fujii T, Raffaelli D (2008) Sea-level rise, expected environmental changes, and responses of intertidal benthic macrofauna in the Humber estuary, UK. Marine Ecology Progress Series 371: 23–35.
9. Goss-Custard JD, McGrorty S, Kirby R (1990) Inshore birds of the soft coasts and sea-level rise. In: Beukema JJ, Wolff WJ, Brouns JJWM, editors. Expected Effects of Climate Change on Marine Coastal Ecosystem. Dordrecht: Kluwer Academic Publishers. 189–193.
10. Pethick J (1993) Shoreline adjustments and coastal management: physical and biological processes under accelerated sea-level rise The Geographical Journal 159: 162–168.
11. Balvanera P, Pfisterer AB, Buchmann N, He J-S, Nakashizuka T, et al. (2006) Quantifying the evidence for biodiversity effects on ecosystem functioning and services. Ecology Letters 9: 1146–1156.
12. Hooper DU, Adair EC, Cardinale BJ, Byrnes JEK, Hungate BA, et al. (2012) A global synthesis reveals biodiversity loss as a major driver of ecosystem change. Nature 486: 105–108.
13. Kennish MJ (2002) Environmental threats and environmental future of estuaries. Environmental Conservetion 29: 78–107.
14. Koch EW, Barbier EB, Silliman BR, Reed DJ, Perillo GM, et al. (2009) Non-linearity in ecosystem services: temporal and spatial variability in coastal protection. Frontiers in Ecology and the Environment 7: 29–37.
15. McLusky DS, Elliot M (2004) The Estuarine Ecosystem Oxford: Oxford University Press. 214 p.
16. Palumbi SR, Sandifer PA, Allan JD, Beck MW, Fautin DG, et al. (2009) Managing for ocean biodiversity to sustain marine ecosystem services. Frontiers in Ecology and the Environment 7: 204–211.
17. White PCL, Godbold JA, Solan M, Wiegand J, Holt AR (2010) Ecosystem services and policy: a review of coastal wetland ecosystem services and an efficiency-based framework for implementing the ecosystem approach. Issues in Environmental Science and Technology In press.
18. Galbraith H, Jones R, Park R, Clough J, Herrod-Julius S, et al. (2002) Global Climate Change and Sea Level Rise: Potential Losses of Intertidal Habitat for Shorebirds. Waterbirds: The International Journal of Waterbird Biology 25: 173–183.
19. Bulling MT, White PCL, Raffaelli D, Pierce GJ (2006) Using model systems to address the biodiversity - ecosystem functioning process. Marine Ecology Progress Series 311: 295–309.
20. Ieno EN, Solan M, Batty P, Pierce GJ (2006) How biodiversity affects ecosystem functioning: roles of infaunal species richness, identity and density in the marine benthos. Marine Ecology Progress Series 311: 263–271.
21. Raffaelli D (2006) Biodiversity and ecosystem functioning: issues of scale and trophic complexity. Marine Ecology Progress Series 311: 285–294.
22. McLachlan A, Brown A (2006) The Ecology of Sandy Shores. London: Elsevier.
23. Raffaelli D, Hawkins S (1996) Intertidal Ecology: Chapman&Hall.
24. Yamanaka T, White PCL, Spencer M, Raffaelli D (2011) Patterns and processes in abundance–body size relationships for marine benthic invertebrates. Journal of Animal Ecology 81: 463–471.
25. McLachlan A (1990) Dissipative Beaches and Macrofauna Communitied on Exposed Intertidal Sands. Journal of Coastal Research 6: 57–71.
26. Lercari D, Defeo O (2006) Large-scale diversity and abundance trends in sandy beach macrofauna along full gradients of salinity and morpodynamics. Estuarine Coastal and Shelf Science 68: 27–35.
27. McLachlan A (1996) Physical factors in benthic ecology: effects of changing sand particle size on beach fauna. Marine Ecology Progress Series 131: 205–217.
28. McLachlan A, Jaramillo E, Donn TE, Wessels F (1993) Sandy beach macrofauna communities and their control by the physical environment: a geographical comparison. Journal of Coastal Research 15: 27–38.
29. Yamanaka T, Raffaelli D, White PCL (2010) Physical determinants of intertidal communities on dissipative beaches: Implications of sea-level rise. Estuarine, Coastal and Shelf Science 88: 267–278.
30. McLachlan A, Wooldridge T, Dye AH (1981) The ecology of sandy beaches in Southern Africa. South African Journal of Zoology 16: 219–231.
31. Fujii T (2007) Spatial patterns of benthic macrofauna in relation to environmental variables in an intertidal habitat in the Humber estuary, UK: Developing a tool for estuarine shoreline management. Estuarine Coastal and Shelf Science 75: 101–119.
32. Bolam SG, Fernandes TF, Huxham M (2002) Diversity, Biomass, and Ecosystem Processes in the Marine Benthos. Ecological Monographs 72(4): 599–615.
33. Leach JH (1970) Epibenthic Algal Production in an Intertidal Mudflat. Limnology and Oceanography 15: 514–521.
34. Hall SJ, Raffaelli D, Thrush SF (1994) Patchiness and disturbance in shallow water benthic assemblages. *British Ecological Society Symposium, April 1992*, Aquatic Ecology: scale, pattern and process. Oxford: Blackwells. 333–375.
35. Feller RJ, Warwick RM (1988) Enegetics. In: Higgins RP, Thiel H, editors. Introduction to the Study of Meiofauna. London: Smithsonian Institution Press. 181–196.
36. Banse K (1982) Mass-Scaled Rates of Respiration and Intrinsic Growth in Very Small Invertebrates. Marine Ecology 9: 281–297.
37. Gerlach SA, Hahn AE, Schrage M (1985) Size spectra of benthic biomass and metabolism. Marine Ecology 26: 161–173.
38. Thomas MLH (1986) A physically derived exposure index for marine shorelines. Ophelia 25: 1–13.
39. Bulling MT, Solan M, Dyson KE, Hernandez-Milian G, Luque P, et al. (2008) Species effects on ecosystem processes are modified by faunal responses to habitat composition. Oecologia 158: 511–520.
40. Zuur AF, Ieno EN, Smith GM (2007) Analysing ecological data. New York: Springer.

41. Pinheiro J, Bates D, DebRoy S, Sarker DS (2006) nlme(3.1–90): an R package for fitting and comparing Gaussian linear and nonlinear mixed-effects models.
42. Pinheiro JC, Bates DM (2000) Mixed-effects models in S and S-plus. New York: Springer.
43. Rodriguez GJ, Mariano L, Lopez J (2002) Meiofauna distribution along a gradient of sandy beaches in northern Spain. Estuarine Coastal and Shelf Science 58: 63–69.
44. Solan M, Raffaelli DG, Paterson DM, White PCL, Pierce GJ (2006) Theme Section. Marine biodiversity and ecosystem function: empirical approaches and future research needs. Introduction.. Marine Ecology Progress Series 311: 175–178.
45. Dyson KE, Bulling M, Solan M, Raffaelli D, White PCL, et al. (2007) Influence of macrofaunal assemblages and environmental heterogeneity on microphytobenthic production in experimental systems. Proceedings of the Royal Society of London B 274: 2547–2554.
46. Bulling MT, Hicks N, Murray L, Paterson DM, Raffaelli D, et al. (2010) Marine biodiversity-ecosystem functions under uncertain environmental futures. Philosophical Transactions of the Royal Society of London B in press.

16

Exploring Macroinvertebrate Species Distributions at Regional and Local Scales across a Sandy Beach Geographic Continuum

Iván F. Rodil[1,2]*, Tanya J. Compton[2], Mariano Lastra[1]

1 Departamento de Ecología y Biología Animal, Universidad de Vigo, Vigo, Spain, **2** National Institute of Water and Atmospheric Research, Hamilton, New Zealand

Abstract

Exposed sandy beaches are highly dynamic ecosystems where macroinvertebrate species cope with extremely variable environmental conditions. The majority of the beach ecology studies present exposed beaches as physically dominated ecosystems where abiotic factors largely determine the structure and distribution of macrobenthic communities. However, beach species patterns at different scales can be modified by the interaction between different environmental variables, including biotic interactions. In this study, we examined the role of different environmental variables for describing the regional and local scale distributions of common macrobenthic species across 39 beaches along the North coast of Spain. The analyses were carried out using boosted regression trees, a relatively new technique from the field of machine learning. Our study showed that the macroinvertebrate community on exposed beaches is not structured by a single physical factor, but instead by a complex set of drivers including the biotic compound. Thus, at a regional scale the macrobenthic community, in terms of number of species and abundance, was mainly explained by surrogates of food availability, such as chlorophyll a. The results also revealed that the local scale is a feasible way to construct general predictive species-environmental models, since relationships derived from different beaches showed similar responses for most of the species. However, additional information on aspects of beach species distribution can be obtained with large scale models. This study showed that species-environmental models should be validated against changes in spatial extent, and also illustrates the utility of BRTs as a powerful analysis tool for ecology data insight.

Editor: Brian R. MacKenzie, Technical University of Denmark, Denmark

Funding: Funding to IFR was provided by a post-doctoral grant (Angeles Alvariño, Xunta Galicia) during the writing of this manuscript. The funders had no role in study design, data collection and analysis, decision to publish, or preparation of the manuscript.

Competing Interests: The authors have declared that no competing interests exist.

* E-mail: irodil@uvigo.es

Introduction

Exposed sandy beaches dominate the world's open coastlines shaping one of the most valuable coastal ecosystems. Increasing recreational and financial demands are placing escalating pressures on sandy beaches at unprecedented scales [1,2]. Although exposed beaches have been considered marine deserts, they are teeming with life that provides numerous goods and services [3]. Specifically, macrobenthos are an integral component of beaches that process large quantities of organic material, recycle nutrients, and are also a fundamental food resource for higher vertebrates of both, commercial importance and conservation significance [3,4].

Macroinvertebrates in exposed beaches show specific adaptations to the dynamic environment provided by the interactions of sand, waves, and tides, also called beach morphodynamics [3]. An understanding of the relationships between beach morphodynamics and the macroinvertebrate community has been critical in beach ecology studies [4–8]. Studies on beach macroinvertebrate have mainly focused on the elucidation of spatial and temporal patterns along a gradient of contrasting morphodynamic types from harsh reflective (coarse sands, steep slopes) to benign dissipative (fine sands, gentle slopes) beaches [3,4,6,7,9]. These studies showed that physical variables inevitably drive the composition of the macroinvertebrate community. Therefore, the number of macroinvertebrate species, their abundance and/or their biomass, generally increases towards the more benign morphodynamic beach type [4]. Beach type thus represents a dynamic spectrum of environmental gradients where both, local and regional factors constrain and modify macroinvertebrate distributions. However, the processes that explain macroinvertebrate species distribution patterns across different spatial scales using several morphodynamic conditions and beach characteristics have not been examined.

There are a number of environmental and biological variables that influence the distribution and abundance of beach macroinvertebrates, including sediment grain size, hydrodynamic regime, biotic interactions and food supply [6,7,10–12]. Some of these variables affect beach species patterns at a local scale (e.g. slope), other variables have an influence over a broader scale (e.g. food availability); while other variables have both local and large scale effects (e.g. sediment characteristics). Thus, species-environment relationships can change with the scale of observation, because at different spatial scales different processes may influence species distributions [13]. For example, a species may be locally limited to beaches with fine grain sizes, but at a larger geographical scale species can also be limited by other variables, such as food availability. Therefore, over a broad spatial scale, the same species may show preferences for beaches with coarser sediment types, but also beaches that are food-enriched. The analysis of scale dependent relationships can help to distinguish whether specific variables are direct drivers of species patterns or are mediated by other variables [13].

In this study, we examined the role of different environmental variables for describing the regional and the local scale distributions of representative macrobenthic species across 39 beaches along the North coast of Spain. This shoreline is characterised by extensive, intermediate morphodynamic beaches that differ in size, orientation and physical characteristics across more than 2000 km of coastline. This large continuum of beaches provides an opportunity to assess the drivers of species distributions at contrasting-scales, i.e., regional versus local responses of macroinvertebrates over a gradient of intermediate beaches. To identify which environmental variables explain the regional distribution of the macrobenthic species we used estimates of chlorophyll a, temperature and morphodynamic variables that broadly characterise a beach. We expect that at a regional scale, the macroinvertebrate abundance and richness would follow a longitudinal gradient related to primary productivity [14]. On the other hand, for each beach a specific abiotic variable might be the driver limiting or shaping a local species response. To test whether the responses of individual species to the same environmental variable differ at a local scale, we examined the fitted responses of species within a beach to local variables (e.g., slope, sediment).

Different techniques have been used for testing beach species environment relationships (e.g., [4,6,7,15,16]), but these have mainly focused on contrasting morphodynamic gradients at a single spatial scale. To examine the role of environmental variables for describing macroinvertebrate species distributions at different spatial scales we used gradient-boosted regression trees (BRTs). BRT stems from machine learning [17], and can automatically model complex functions and the interactions between variables without making assumptions about the shape of the fitted functions or the interactions between variables [18,19]. It thus provides a powerful tool for analysing complex ecological datasets, and examining the distribution of beach species over different spatial scales.

Methods

Study Site and Sampling Procedure

Thirty nine exposed sandy beaches were sampled from 1995 to 1999 along the North coast of Spain (Figure 1). Sampling was carried out during spring tides in September and October to reduce biotic and abiotic variability linked to seasonal cycle in exposed beaches [7,8,14]. This coast encompasses *ca.* 2000 km of shoreline between 42° 07′N, 1° 46′ W and 43° 22′ N, 8° 49′ W (data compiled from [8,14], and Rodil unpublished data). All beaches are mesotidal, with maximum tidal ranges close to 4 m. This coast is affected by upwelling events along the North margin of the Iberian Peninsula (from West to East) stimulating primary production during the spring and summer seasons [20], related to cold sea water temperature and high chlorophyll a concentrations (Figure 1).

At each beach, macroinfauna samples were collected at 10 equally spaced intervals (distance variable depending on the total width of each beach) from above the drift line to the low swash zone with one 0.05 m^2 core per interval to a depth of 15 cm (see [8,14] for further details). All macroinvertebrate samples were taken along three randomly chosen transects during low tide. A total of 1170 macrobenthos samples were collected in the study area. All the sediment samples were sieved through 1-mm mesh and the residue was preserved in 4% formalin. The individuals were later sorted from the sediments, identified and counted in the laboratory. We obtained abundance (individuals/m^2) and species richness (number of species) as the main beach macroinvertebrate community descriptors. Sediment samples for the determination of the granulometric measures and morphodynamic beach characteristics were collected in triplicate at the same sites as the macroinvertebrates (see supporting Table S1 for a detailed explanation of all the environmental variables). No specific permits were required for the described field studies.

The beach environment from the northern coast of Spain comprises a full set of environmental variables (see Table S1). Typical beach morphodynamic indices, such as Dean's parameter (Ω), Beach State Index (BSI) and Beach Index (BI) [3] were used to characterize the beach morphodynamic type (see Table S1 for an index description). These indices classified all the beaches as a series of intermediate tidally dominated environments (92% and 97% of the beaches according to Ω and BSI, respectively); considered the most varied and widespread type of beach worldwide [3] falling between the two main morphodynamic extremes (reflective and dissipative). The rating system proposed by [21] was used to estimate the wave exposure rate (ER, Table S1). Estimates of average sea surface temperature and chlorophyll a concentrations were obtained from Bio-Oracle (http://www.oracle.ugent.be/) [22]. Specifically, annual average chlorophyll a (chla) and sea surface temperature (sst), proxies of seasonality and temporal variation in nutrient supply, were derived from satellite images (Aqua-MODIS and SeaWiFS; http://oceancolor.gsfc.nasa.gov/) summarizing information across several years and were then used to generate three relevant metrics: annual maximum, minimum and mean (Figure 1). The estimates of chla and sst were extracted from the GIS layers to the sample sites in this study using the intersect points tool from Hawths Tools in ArcGIS 9.2. These two metrics showed a specific geographic gradient with decreasing temperature and high chla in waters from East to West (Figure 1).

All variables (Table S1) were checked for collinearity using a Pearson correlation. Where the correlation was greater than 0.7, the variable within the pair that had the best functional relevance (i.e., the most important morphodynamic influence according to the literature) of the variables was kept in the model.

Machine Learning Analyses

We used boosted regression trees analysis (BRT) to examine the association between total macroinvertebrate abundance and species richness, with respect to variables describing the beach environment across the northern Spanish coastline. BRT, a machine learning approach [17,23], combines the regression tree algorithm and the boosting algorithm to produce an ensemble of regression trees. The boosting algorithm improves standard regression tree modelling by adding a stochastic component to the model, which continuously emphasizes the poorly explained part of the data space [17,19]. BRT can be considered an advanced form of regression, which also uses a link function to examine a range of response types including binomial, Poisson and Gaussian [24]. Although the ecological uses of BRT have only been recently applied, it has shown to be a useful technique for dealing with complex biological data sets [25–28].

The predictive performance of the BRT models is optimized by means of the learning rate (lr) and tree complexity (tc). The lr is used to shrink the contribution of each tree as it is added to the model, and tc determines the number of nodes in a tree and should reflect the true interaction order on the response being

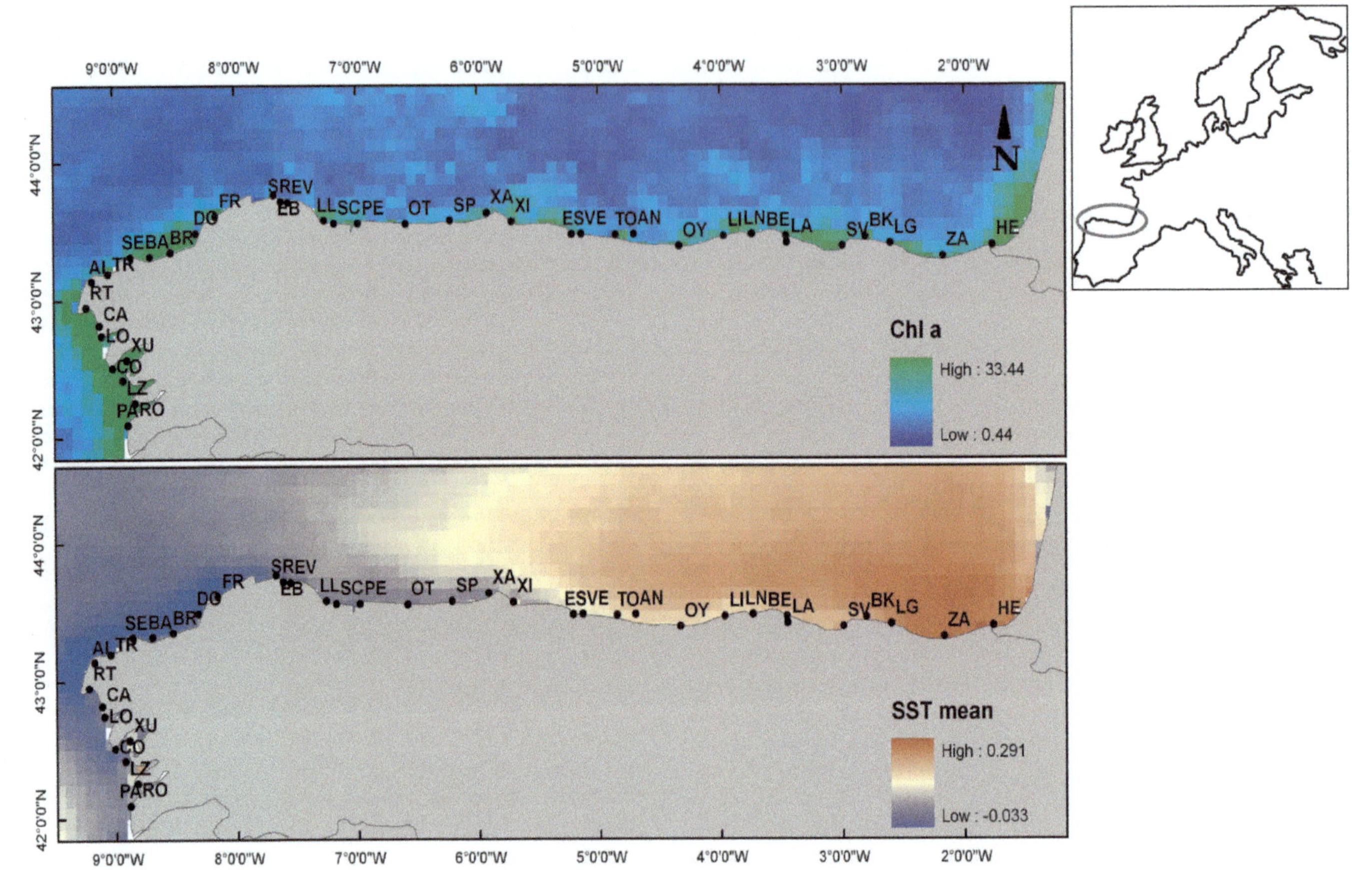

Figure 1. The Spanish northern shoreline, showing the location of the 39 beaches. (from East to West; HE: Hendaya; ZA: Zarautz; LG: Laga; BK: Bakio; SV: Salvaje; LA Laredo; BE: Berria; LN: Langre; LI: Liencres; OY: Oyambre; AN: Andrín; TO: Toranda; VE: Vega; ES: Espasa; XI: Xivares; XA: Xagó; SP: San Pedro; OT: Otur; PE: Peñarronda; SC: San Cosme; LL:Llás; EV: Viveiro; SR: San Román; EB: Bares; FR: Frouxeira; DO: Doniños; BR: Barrañán; BA: Baldaio; SE: Seiruga; TR: Traba; AL: Area Longa; RT: Rostro; CA: Carnota; LO: Louro; XU: Xuño; CO: Corrubedo; LZ: Lanzada; RO: Rodas; PA: Playa América). Estimates of mean sea surface temperature (sstmean) and chlorophyll a (chla) concentrations are shown.

modelled [23]. All BRT models had a tc of 3 and were optimised for their lr so that a minimum of 1000 trees was fitted for each model [19]. Model performance was assessed using ten-fold cross-validation (cv). This procedure compares fitted values from ten individual models, each derived from a random subset of the full data, against the portion of the data withheld from the model [29]. Interactions between predictors are modelled automatically with BRT because the structure of a tree means that the response to one predictor variable depends on values of predictors higher in the tree [27]. All BRT analyses were carried out in R (version 2.14.0) [30] using the 'gbm' library [31], supplemented with functions from [19].

Regional Scale Analysis

Across all the 39 beaches, variation in macroinvertebrate abundance and species richness, and the occurrence of specific species was examined with respect to six predictor variables: Dean morphodynamic index (Ω), exposure rate (ER), sediment mean grain size (MGS) and shear strength (SS), sea surface temperature (sst), and chlorophyll a (chla) (see Table S1). Morphodynamic indices, and also ER are traditionally used to make beach comparisons [3], rather than to find different results within a beach. As sediment characteristics (MGS and SS) can show an effect over different spatial scales, they were included in both, the local and regional BRT models.

First of all, we explored the variation in species richness and macroinvertebrate abundance with respect to the beach variables across the geographic gradient using a BRT model with a Poisson link function. We further examined the distributions of the most common macroinvertebrate species at beaches along the northern Spanish coastline using a BRT model with a Bernoulli link function to fit the presence/absence response with respect to the predictor variables. Model performance was examined with the area under the receiver-operating characteristic curve (AUC) [32]. The AUC provides a measure of the degree to which the fitted values discriminate the observed presences and absences, where a value of 1 indicates that the presences and absences were perfectly discriminated while a value of 0.5 indicates that a model cannot discriminate the presences better than random.

To assess the relevance of each variable to the BRT models, the variable contribution was used. This variable gives a measure of how often the predictor is selected, and the improvement to the model as the result of a variable being selected [27]. This measure assesses the number of times a variable is selected for splitting, weighted by the squared improvement to the model as a result of each split, and averaged over all trees [19]. The relative contribution of each variable to a model was assessed using functions provided by [19]. In the cases where the predictor variables played a small contribution (<3%) to explaining variation in the final BRT model, these variables were eliminated

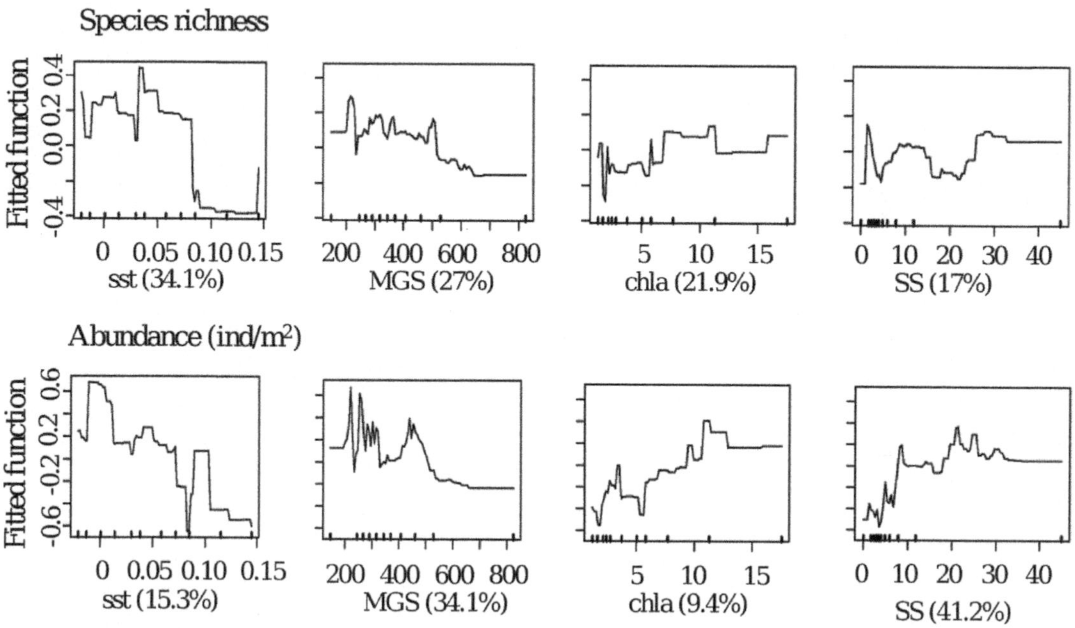

Figure 2. Variation in species richness and macroinvertebrate abundance (ind/m^2) predicted by a BRT model. The relative importance of the most influential variables is shown as %. Ticks across the bottom of each plot show the distribution of deciles for each predictor variable. A common scale is used on the vertical axis (centred to have zero mean over the data distribution). Mean grain size (MGS), shear strength (SS), chlorophyll a (chla) and mean sea surface temperature (sstmean).

from the final models. The probability of occurrence at a regional scale was examined using perspective plots ([19], gbm.perspec).

Local Scale Analysis

At a local scale, the species distributions were examined with respect to exposure time (ET), sediment water content, slope, sediment characteristics (MGS and SS), sediment temperature, and Beach (Table S1). Beach is added as a factor in the BRT models, to determine whether individual species distributions displayed different responses between the different beaches. Note that in this analysis we added the condition that recorded observations had to exceed an N = 5 before a species at a particular beach could be included in this analysis, thus not all 39 beaches in the regional analysis ended up in the species models for the local analysis. Less than N = 5 would mean that we don't have enough data to describe the distribution of a species with respect to an environmental gradient.

To identify whether individual species had different environmental associations among beaches, we examined the abundance of the most common species (in terms of abundance and occurrence; Table S2, supporting information) with respect to the environmental variables using a BRT model with a Poisson link function. This model represented the data better than a log-normal model with a Gaussian error distribution. The model performance was assessed using the cross-validated (cv) explained deviance, which provides a measure of the goodness-of-fit between the predicted and raw values, and the cross-validated correlation (cvCor), which provides a measure of correlation between the recorded observations and the model fitted values. The cvCor is calculated as a Pearson correlation coefficient and thus takes into account how far the prediction varies from the observed data [33].

To visualize the fitted functions from the local BRT model, partial dependence plots were used ([19], gbm.unipred). These functions show the effect of a variable on the response, while controlling for the average effect of all other variables in the model [19]. The partial dependence plots for the most important predictors of the main species were presented, and interactions were explained in the text.

Results

The environmental predictors from the 39 studied beaches displayed different physical and morphodynamic characteristics (Figure S1, supporting information) along the Spanish coastline (Figure 1), but there was no significant trend in harshness or a shift in the morphodynamic conditions across the geographic stretch. Sediment temperature was significantly related to beach location (Figure S1), with lower sediment temperatures at the most western beaches ($y = 25.2–0.02x$; $R^2 = 0.47$, $p<0.001$). There were strong correlations (>0.7) between some of the main environmental variables, such as MGS, with sediment water content, Dean and ER. In this case, only MGS was selected for the final BRT regional models (including MGS, SS, chla and sst). Sediment water content and ET were highly correlated with beach slope, thus ET was excluded from the final local model (including MGS, SS, slope and beach) (see Table S1 for code explanation).

A total of 71 species were collected through all the 39 beaches. Macroinvertebrate community was dominated by Crustacean species (41%), Isopoda and Amphipoda being the main components of this group (71%). Polychaetes and molluscs were also present but in lower percentages (31 and 3%, respectively). In the supralittoral zone, a variety of insects (13 species) were found above the drift line (18%). Five different species (7%) belonging to different groups (one nemertean, one oligochaeta and three fish species) were also found (Table S3, supporting information).

Regional Scale

Using 4 environmental variables to describe the beach conditions across the northern Spanish coastline, the BRT models described species richness (0.51 ± 0.02 cvCor) and macroinverte-

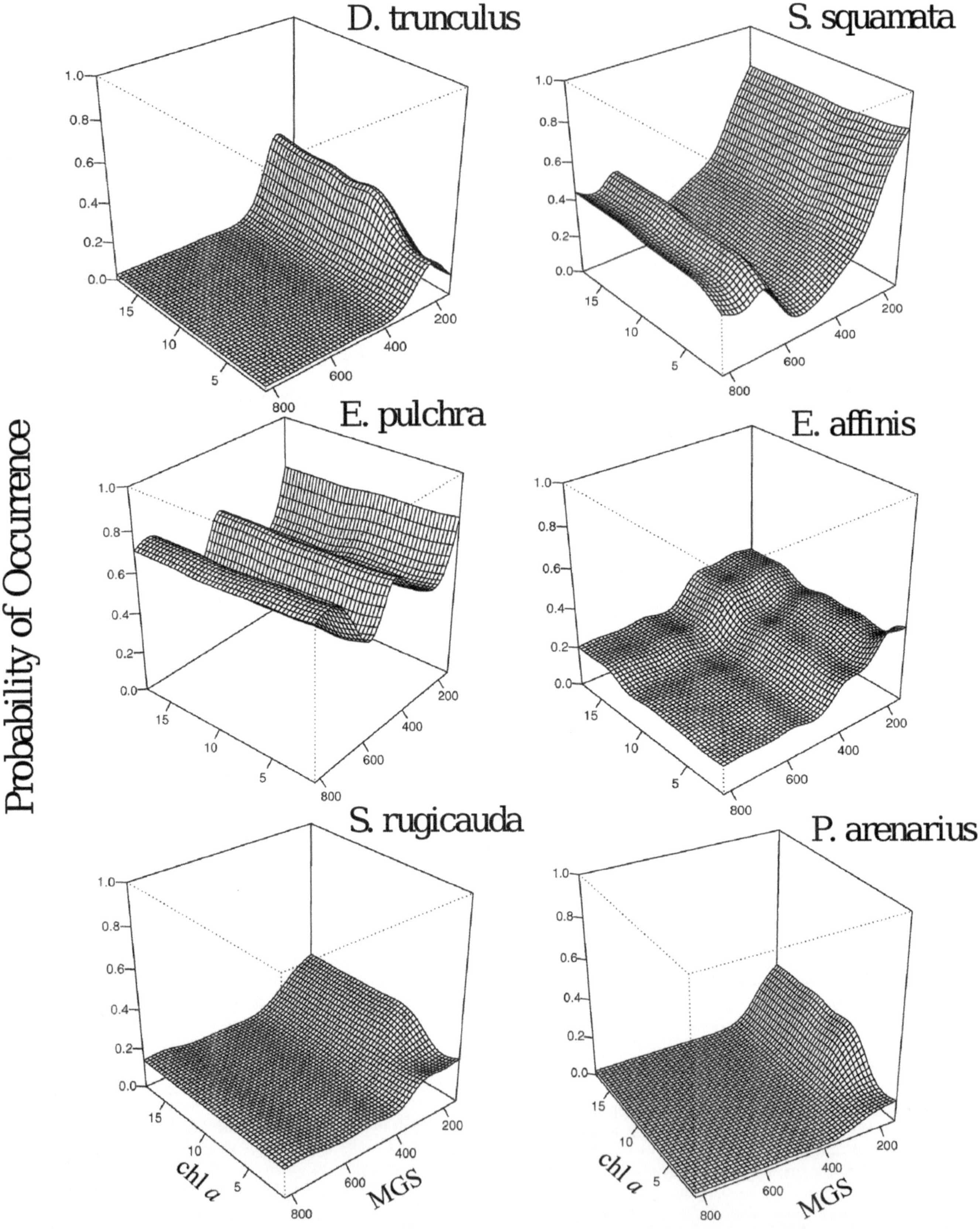

Figure 3. Three-dimensional partial dependence plots. Plots show the strongest pairwise interactions in the beach model for the main species (*Donax trunculus, Scolelepis squamata, Eurydice pulchra, Eurydice affinis, Sphaeroma rugicuda, Pontocrates arenarius*). Mean grain size (MGS), and chlorophyll a water concentration (chl*a*) as the main predictor variables.

Table 1. Predictive performance of the BRT model relating the beach species distributions to environment, averaged across the species (presence/absence) over the regional scale.

					Cross-validation*			
Main species	**pa**	**lr**	**n.trees**	**Null deviance**	**Unexplained deviance (se)**	**Explained deviance (se)**	**AUC (se)**	**Correlation (se)**
B. pelagica	78	0.001	1250	0.490	0.439 (0.01)	10.3 (0.03)	0.751 (0.03)	0.257 (0.04)
D. trunculus	88	0.001	1200	0.534	0.446 (0.03)	16.4 (0.05)	0.815 (0.03)	0.325 (0.06)
E. affinis	213	0.001	1200	0.949	0.877 (0.01)	7.6 (0.01)	0.709 (0.02)	0.276 (0.02)
E. pulchra	588	0.003	1000	1.386	1.356 (0.01)	2.2 (0.01)	0.6 (0.01)	0.176 (0.02)
P. arenarius	332	0.001	1050	1.193	1.146 (0.01)	4 (0.01)	0.636 (0.02)	0.22 (0.03)
S. squamata	368	0.001	1700	1.245	1.164 (0.01)	6.5 (0.01)	0.677 (0.01)	0.284 (0.02)
S. rugicauda	186	0.0003	2150	0.876	0.857 (0.01)	2.2 (0.01)	0.61 (0.02)	0.147 (0.02)
T. brito	49	0.000	1350	0.348	0.338 (0.01)	2.7 (0.02)	0.661 (0.04)	0.106 (0.03)
T. saltator	61	0.0003	1050	0.410	0.406 (0.01)	1 (0.01)	0.608 (0.04)	0.068 (0.03)

*Mean and se estimated within model building. Standard error (se).
Estimates of predictive performance were calculated using 10-fold cross validation and a tree complexity of 3. Learning rate (lr), the number of trees (n.trees), and the number of positive observations (pa) are shown. AUC: area under the receiver operating characteristic curve. The final environmental variables included in the models are shown in Table 2. Beach species included in the final model: *Bathyporeia pelagica, Donax trunculus, Eurydice affinis, Eurydic pulchra, Pontocrates arenarius, Scolelepis squamata, Sphaeroma rugicauda, Talorchestia brito* and *Talitrus saltator.*

brate abundance (0.48±0.07 cvCor) with good predictive performances (Figure 2). Furthermore, the models explained 77.1 and 66.6% of the cross validated deviance for species richness and abundance, respectively. Inspection of the contributions of the predictor variables indicated that sst (34.1%), MGS (27%), chla (21.9%) and SS (17%) were the most important variables for explaining species richness (Figure 2). Species richness tended to increase with increasing chla and to decrease with increasing MGS, SS and sst (Figure 2). The variables most important for macroinvertebrate abundance were sediment shear strength (41.2%) and MGS (34.1%). Abundance was also related to high chla (9.4%) and low sst (15.3%) (Figure 2).

The BRT regional model showed a good performance for all the species (cvAUC >0.6), and for *B. pelagica, D. trunculus* and *E. affinis* (cvAUC >0.7) the model showed a very good degree of discrimination among the predictor variables (Table 1). The distribution of the dominant species along this coastline was described mainly by sediment variables (MGS and SS) and by chla and sst (Table 2). MGS explained the greatest amount of variability in describing the probability of occurrence across the beaches for *B. pelagica* (60.4%), *S. rugicauda* (60.1%) and *D. trunculus* (53.5%) (Table 2). However, sediment predictor variables were associated with chla (Figure 3). Thus, although the probability of occurrence of the most common species was higher with lower MGS, the fitted probabilities of occurrence were always associated with high chla concentrations (Figure 3). *E. pulchra* showed a wide geographic distribution across all the beaches with different probabilities of occurrence no matter the MGS-chla association (Figure 3).

Table 2. The relative contribution (%) of the predictor variables to the BRT models over the regional scale.

Main species	**MGS**	**SS**	**sst**	**chla**
B. pelagica	60.4	17.7	11.2	10.7
D. trunculus	53.5	31.9	4.5	10.1
E. affinis	40.1	47.7	3.8	8.3
E. pulchra	50.6	35	9.7	4.7
P. arenarius	20.7	64	11.1	4.1
S. squamata	52	34.6	6.2	7.1
S. rugicauda	60.1	21.1	7	11.9
T. brito	17.5	60.5	12.8	9.2
T. saltator	39.4	40.7	12.2	7.6

The predictor variables, by relevance were mean grain size (MGS), sediment shear strength (SS), mean sea surface temperature (sst), and chlorophyll *a* (chla). The predictor variables with the highest contribution described the recorded observations of the dominant beach species (same as in Table 1). Beach species included in the final model: *Bathyporeia pelagica, Donax trunculus, Eurydice affinis, Eurydic pulchra, Pontocrates arenarius, Scolelepis squamata, Sphaeroma rugicauda, Talorchestia brito* and *Talitrus saltator.*

Local Scale

The abundance distribution of the most common species within a beach was explained with a high level of predictive performance (cvCor =0.3–0.76, Table 3), but some of the species had a relatively lower model performance (e.g. the model for *B. pelagica* explained only 20% of the predictive deviance, compared to 65% for *S. squamata*, or 64% for *E. affinis* and *E. pulchra*) (Table 3).

The predictor variables with the highest contribution to explaining the abundance distributions of individual species within the beaches are showed in Table 4. The sediment characteristics (MGS and SS) showed the highest overall contribution. The factor Beach was also an important variable for some species such as, *B. pelagica* (54.9%), *E. pulchra* (31.7%), *P. arenarius* (50.5%), *S. squamata* (42.6%), and *S. rugicauda* (34.4%). Other distributions of species, such as *T. brito* and *T. saltator* were explained by beach slope (63.2 and 97.2%, respectively) or MGS (*T. brito*, 31.1%) (Table 4).

The fitted responses of species to the predictor variables within a beach were generally repeatable between beaches (Figures 4, 5, 6). In some of the beaches not only was the shape of the predicted species response similar, but also the species abundance did not change very much between beaches (Figure 4, 5, 6). Thus, within the northern Spanish beaches, the bivalve *D. trunculus* was predicted to occur maximally in fine-grained sediments (MGS

Table 3. Comparison of the predictive performance of statistical models relating dominant species distributions to environment, averaged for the beach species in a local scale.

					Cross-validation*		
Main species	**pa**	**lr**	**n.trees**	**Null deviance**	**Unexplained deviance (se)**	**Explained deviance (se)**	**Correlation (se)**
B. pelagica	1170	0.0003	2200	3.90	3.1 (0.7)	0.2 (0.1)	0.331 (0.11)
D. trunculus	1170	0.001	3650	4.02	2.4 (0.3)	0.41 (0.09)	0.612 (0.06)
E. affinis	1170	0.005	2000	7.07	2.6 (0.2)	0.64 (0.03)	0.71 (0.08)
E. pulchra	1170	0.005	7300	18.52	6.7 (1.1)	0.64 (0.06)	0.764 (0.06)
P. arenarius	1170	0.005	2700	3.40	2.1 (0.15)	0.4 (0.04)	0.531 (0.04)
S. squamata	1170	0.005	5100	8.87	3.1 (0.4)	0.65 (0.05)	0.745 (0.04)
S. rugicauda	1170	0.005	1450	3.25	2.5 (0.5)	0.23 (0.1)	0.46 (0.06)
T. brito	1170	0.001	1900	5.72	3.6 (1.8)	0.4 (0.3)	0.614 (0.13)
T. saltator	1170	0.0001	1600	1.89	1.5 (0.3)	0.2 (0.17)	0.631 (0.06)

*Mean and SE estimated within model building. Values indicate the cross-validated unexplained and % explained deviances, and the cross-validated correlation (cvCor) and the standard errors (se).
Estimates of predictive performance were calculated using 10-fold cross validation and a tree complexity of 3. Learning rate (lr), the number of trees (n.trees), and the number of positive observations (pa) are shown. Environmental variables tested in the final model are showed in Table 4. Beach species included in the final model: *Bathyporeia pelagica, Donax trunculus, Eurydice affinis, Eurydic pulchra, Pontocrates arenarius, Scolelepis squamata, Sphaeroma rugicauda, Talorchestia brito* and *Talitrus saltator.*

<300 μm) and low sediment shear strength (SS <25 Kpa) (Figure 4). *S. squamata* showed a similar sediment trend (Figure 4), as this polychaete also revealed maximum peaks of predicted abundance in fine-grained sediments (MGS <300 μm) and low sediment shear strength (SS <10 Kpa). *D. trunculus* and *S. squamata*, showed an expected higher occurrence in the lower parts of the beach face slope (<0.2; below the drift line). However, these species showed differences in the abundances between some of the beaches, with the highest abundances in the most western beach locations (Figure 4).

Some species, such as *P. arenarius*, and *S. rugicauda* showed a preference for low SS values (<10 Kpa), and a widespread preference for a wide range of grain sizes (Figure 5). These two species were found maximally in low intertidal levels (Slope ≤0.2) and, although with a similar fitted response to any environmental gradient, they showed different abundances between some of the beaches. The highest abundances were generally found in the western beaches (Figure 5). *T. brito* showed the highest peaks of abundance in fine-grain sizes (200–400 μm) and very low compacting sediment forces (<4 Kpa) (Figure 5). The highest predicted sand hopper abundances were found at the mid-high intertidal levels (Slope >0.3; above the drift line), and the relative contribution of SS (5%) and beach (1%) were low (Table 4).

Table 4. The relative contribution (%) of the predictor variables to the BRT interaction models over the local scale.

Main species	Beach	MGS	Slope	SS
B. pelagica	54.9	38.3	2.7	4.1
D. trunculus	28.6	17.3	31.8	22.3
E. affinis	27.6	10.9	37.7	23.9
E. pulchra	31.7	20.2	13.8	34.3
P. arenarius	34.4	13.7	38	13.8
S. squamata	50.5	9	29.5	11
S. rugicauda	42.6	18.7	29.1	9.6
T. brito	0.7	31.1	63.2	5
T. saltator	0.2	1.3	97.2	1.3

The predictor variables were Beach, mean grain size (MGS), beach slope, and sediment shear strength (SS). The predictor variables with the highest contribution described best the recorded observations of the main beach species. Beach species included in the final model: *Bathyporeia pelagica, Donax trunculus, Eurydice affinis, Eurydic pulchra, Pontocrates arenarius, Scolelepis squamata, Sphaeroma rugicauda, Talorchestia brito* and *Talitrus saltator.*

Finally, the two sympatric cirolanid isopods *E. affinis* and *E.pulchra* showed repeatable response curves between beaches with respect to the main predictor variables (Figure 6); although they showed different abundances between some of the beaches. Both species were found more commonly in the lower parts of the intertidal (Slope <0.3), but also with some peaks of abundance in higher levels of the beach face slope (Slope >0.3, Figure 6). Interestingly, there was a contrasting response of the two species when the same variables were compared. *E. affinis* showed a specific association with low MGS (~300 μm) and SS (5–10 Kpa). By contrast, *E. pulchra* showed an association with a broad range of median grain sizes (100–500 um), different levels of sediment compaction (5–30) and different slope environments (Figure 6), charting saw-shaped relationships with the environmental variables.

Discussion

The distribution of the macroinvertebrate community is here explained with respect to 18 predictor variables across and within 39 beaches from the North coast of Spain. The current status and theories accept physical factors as the predominant predictor of community characteristics across beach types [3–9]. Gradually, however, a greater number of studies is showing the relevance of biotic interactions and food supply on exposed beaches (e.g., [11,12,14,34]).

Over a Large Scale: the Geographic Gradient

From a regional scale perspective, an ensemble of predictor variables including MGS, SS, and surrogates of food supply (chlorophyll a and sea surface temperature) showed the highest

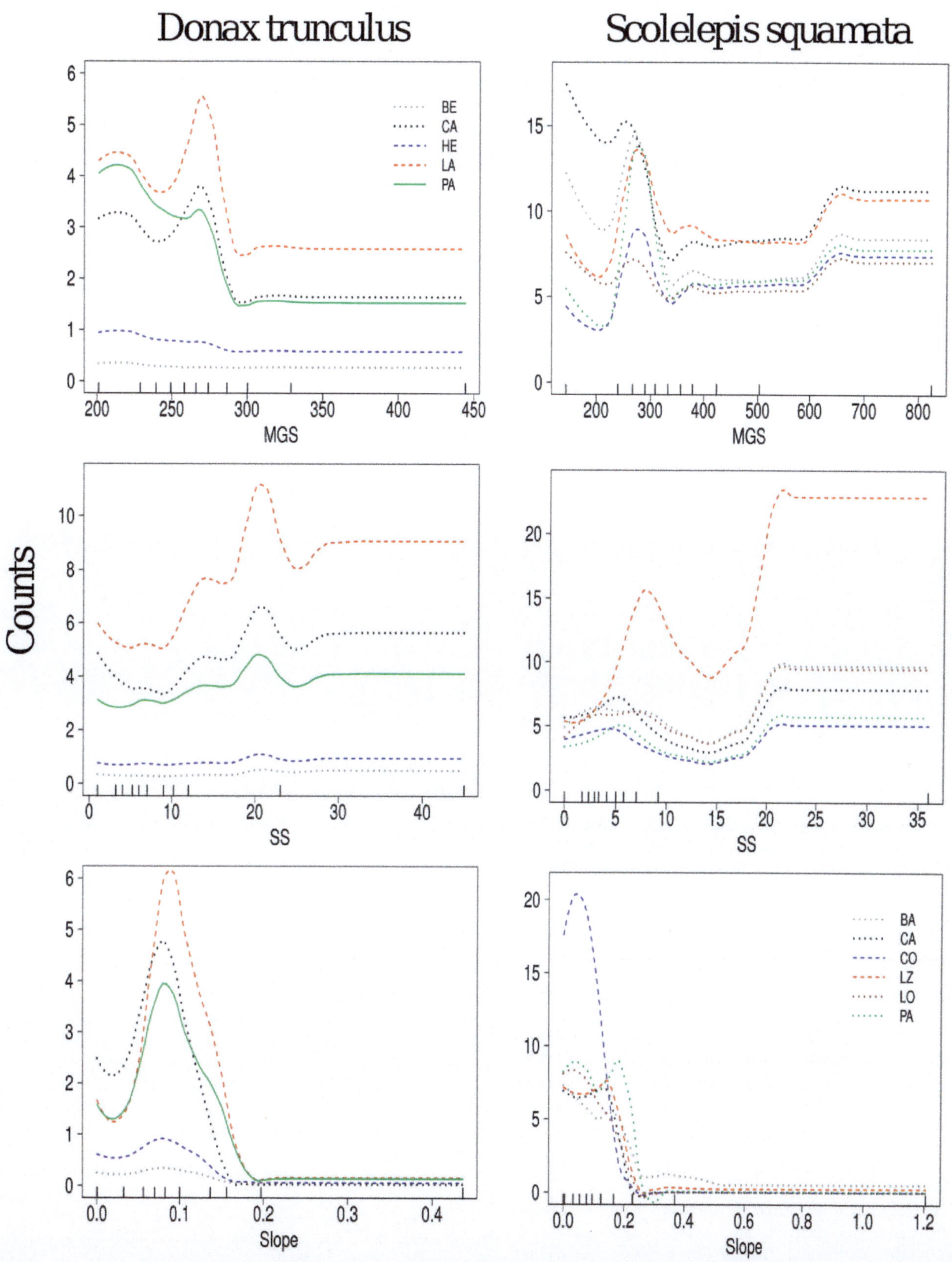

Figure 4. The abundance of *D. trunculus* and *S. squamata* predicted by the main beach variables. The species abundance (counts) was plotted over those beaches presenting a frequency >5 individuals. Ticks across the bottom of each plot show the distribution of deciles for each predictor variable. Main variables within a beach: mean grain size, shear sediment strength, and slope. Beach names as in Figure 1.

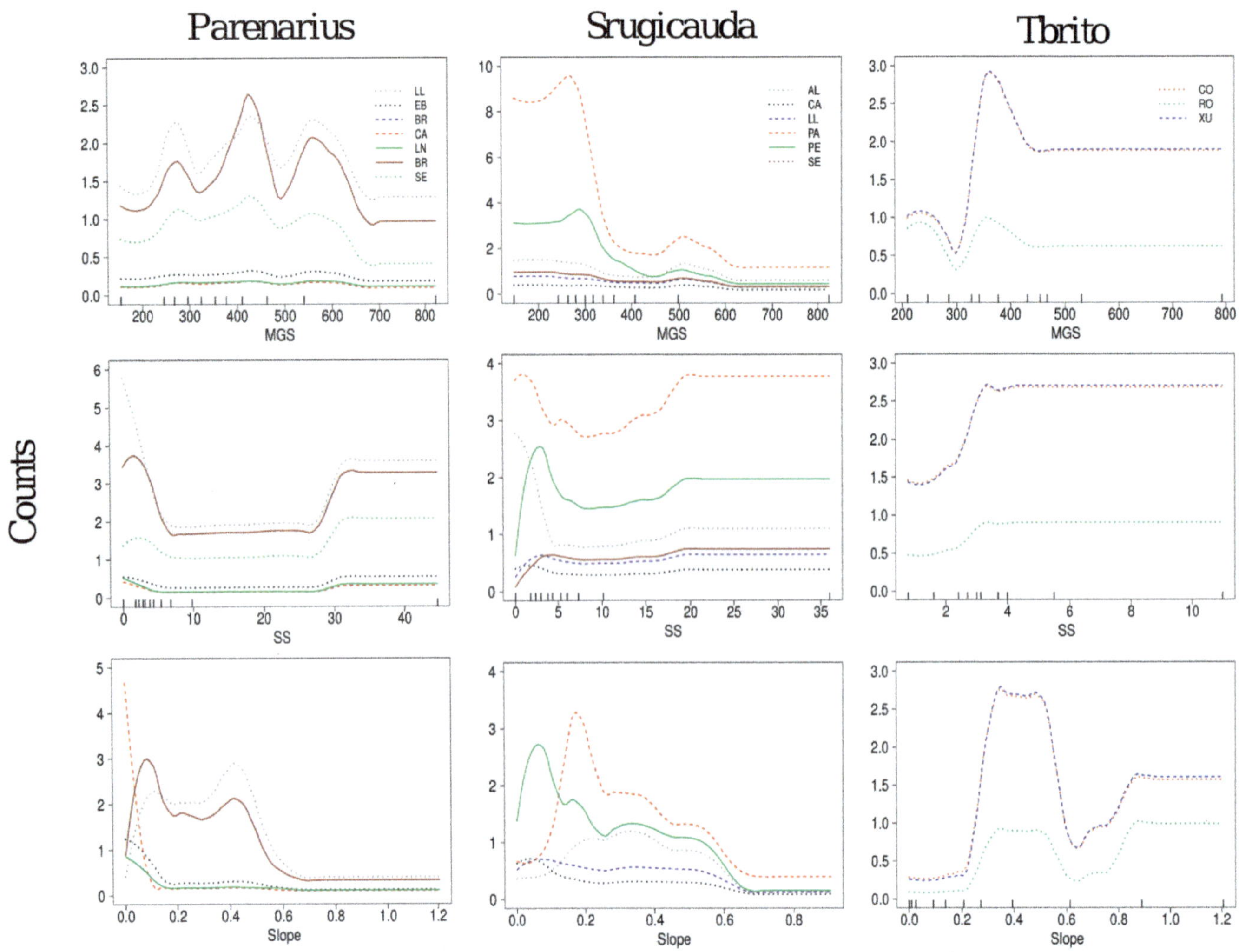

Figure 5. The abundance of *Pontocrates arenarius, Sphaeroma rugicauda* and *Talorchestia brito* predicted by the main variables. The species abundance (counts) was plotted over those beaches presenting a frequency >5 individuals. Predictor variables names as in Figure 4. Ticks across the bottom of each plot show the distribution of deciles for each predictor variable. Beach names as in Figure 1.

contribution to the models. Macroinvertebrate abundance and richness were higher where food supply was higher along the geographic gradient. Moreover, the spatial variation of the macroinvertebrate species reflects the extent to which other variables, apart from beach morphodynamics, shape the macroinvertebrate community from these locations. The contribution of the sediment characteristics to the dominant species from this region interacted with proxies of food availability, i.e., chla. This interaction was well described by the pairwise partial dependence plots, where the dominant species revealed a specific preference for finer sediments, but the highest probability of occurrence were always associated with chla, no matter the sediment traits. Thus, macroinvertebrate community was not subjected only to physical control, and those variables associated with the beach environment did not provide an exclusive description of macroinvertebrate distribution through the regional scale. This refers to a pattern in the NW coast of Spain that relates the low sea water temperature and high chla concentrations with periodic upwelling events, stimulating primary production and shaping a West to East gradient of food supply [14,20]. Surprisingly, although macroinvertebrate species generally occur in the highly productive areas, they did not show differences in their type of response to the environmental predictors through the whole geographic continuum.

In a Local Scale: from within a Beach

At a local scale, predictor variables such as sediment characteristics and beach face slope played the highest contribution to the models, performing almost an exclusive role explaining most of the macroinvertebrate community variation on the beaches. The response of the individual species to these predictor variables followed a repeatable pattern across a single predictor variable between beaches. Thus, the occurrence of truly marine species [3], such as *D. trunculus* or *S. squamata* increased in the lower parts of the beach (i.e., closer to the sea), and decreased with increasing slope (i.e., terrestrial conditions). On the other hand, a semi-terrestrial species, such as *T. brito* occurs mainly close to the higher parts of the beach, where normally finer sediments are also found. Sediment traits, such as MGS and SS described the observed beach occurrences, but the Beach itself played a main role explaining the species variation in the models. Since the responses of the species to the variables were similar, the relevance of beach as a predictor variable indicates differences in beach abundances.

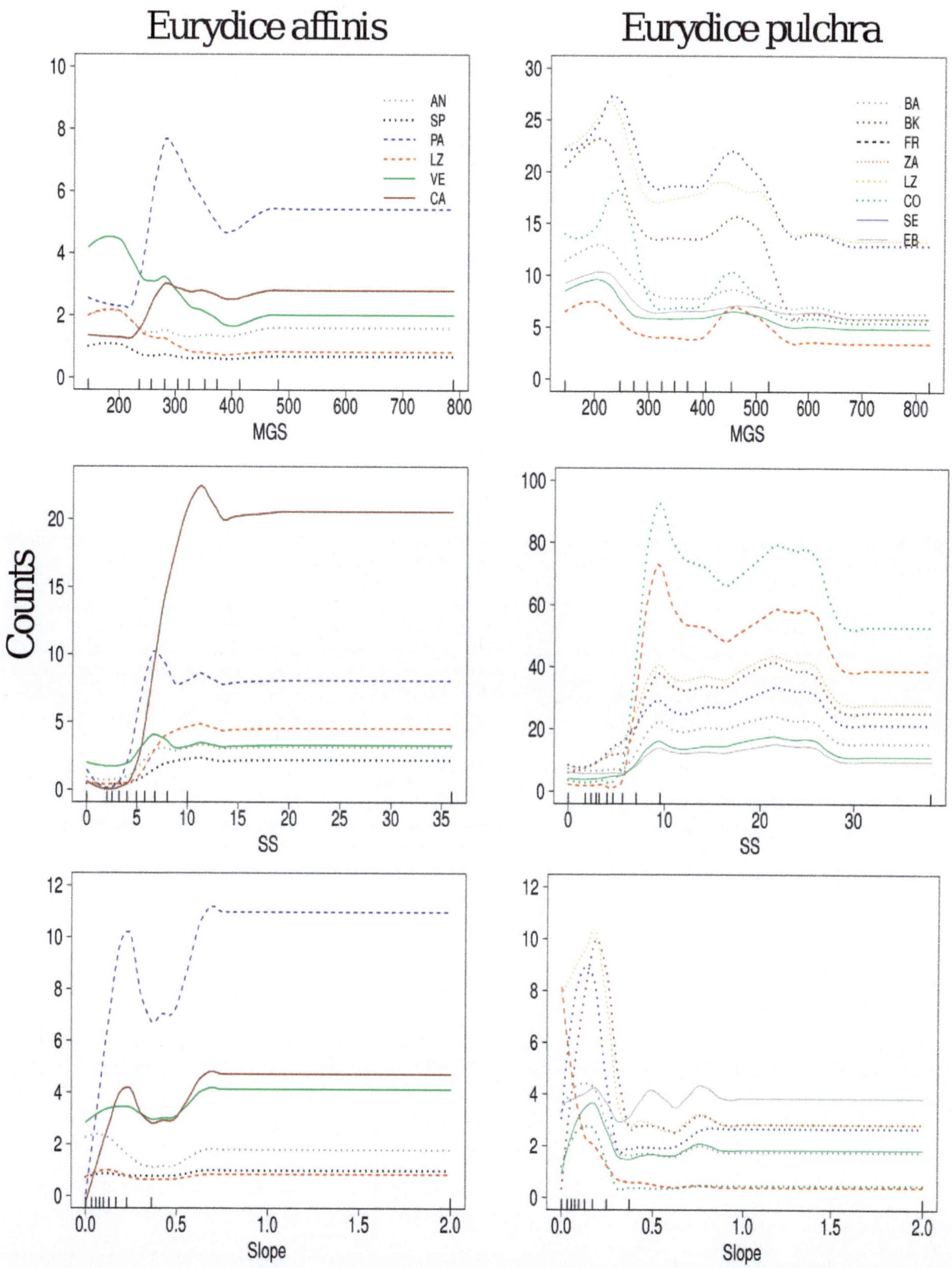

Figure 6. The abundance of *E. affinis* and *E. pulchra* predicted by the main beach variables. The species abundance (counts) was plotted over those beaches presenting a frequency >5 individuals. Predictor variables names as in Figure 4. Ticks across the bottom of each plot show the distribution of deciles for each predictor variable. Beach names as in Figure 1.

All these beaches are mesotidal, intermediate and show variable physical characteristics through the geographic continuum with no significant morphodynamic trends. The only difference among all the beaches was related to the increasing gradient of food supply, which can explain the differences in abundance among beaches.

A Contrasting Scale-approach: Large vs. Local

One of the most interesting results was that the fitted occurrence of the most common species across an environmental gradient was similar between beaches (e.g., decreases, increases or unimodal). In some of the cases, although the species response to the predictor variable was similar, the absolute occurrence observed changed between beaches. For instance, the bivalve *D. trunculus* and the polychaete *S. squamata* showed different abundances among some of the beaches. There are no cues in our data showing a shift in the environmental severity over the geographic stretch, but polychaete and mollusc species are truly marine species and are expected to be more sensitive to the harsh beach climate. Therefore, these species may display specific adaptations to the beach swash climates [3,4]. *D. trunculus* is a filter feeder depending on suspended matter and food particles in water. This suggests that, as scale moves through beaches at different longitudes, other factors become important overlaying any response to the main predictor variable. In the same way, *T. brito*, *P. arenarius* and *S. rugicauda* relationships with the variables shaped models with the same types of response curves, but with different abundances between some of the beaches. This might be due to complex interactions between life history, feeding, mobility, growth and reproduction, and other processes that would require detailed study to resolve. However, based on the history of upwelling events in this region we hypothesize that the food supply may affect the geographic abundance distribution of these species. This model type revealed that large-scale species-environment relationships arise from interactive processes between variables, rather than local responses to a specific abiotic variable being the dominant factor limiting distribution [13].

Exposed beaches in this coast are dominated by crustaceans performing tidal migrations throughout the entire beach looking for food, but also for a more stable habitat far away from the harsh wave climate [4,12,35,36]. In our locations, despite the high abundance of talitrid amphipods and tylid isopods, BRT models were not able to predict their occurrence related to any of the physical variables, except for *T. brito*. The relationships indicated a strong influence of a specific predictor variable on local processes for this sand hopper, i.e., similar form of response and occurrence across scales. These semi-terrestrial species, although highly mobile, spend their whole life cycle in the supratidal zone, burrowed under arid conditions and performing nocturnal feeding (on wrack and carrion) migrations [36,37]. They find a more stable environment in the supralittoral zone, which is independent of the dominant physical intertidal conditions [3,4]. Thus, neither the beach nor the sediment traits showed a high contribution for talitrid species models. In general terms, this may reflect some biological processes like predation or feeding competition without relation to the abiotic variables included in the models [38,39]. In fact, the availability of limiting resources in the beach ecosystem can potentially influence all these interactions, including cannibalism episodes reported for some talitrid species [40].

Some of the intraspecific differences found in this study might be the result of different densities and beach levels occupied, i.e., niche segregation, for some species, such as the conspicuous isopods *E. pulchra* and *E. affinis*. Both species display a tidal cycle of migration into the water column and dominated from mid tide level downwards [36], coping with the harsh beach conditions and remaining buried during daylight. *E. pulchra* was the most abundant species, inhabiting a broader intertidal spectrum and plotting a saw-shaped chart across the intertidal and over the geographic stretch. Therefore, this species is exposed to different environmental conditions where a flexible behaviour is mandatory to withstand short-term unpredictable changes in the swash zone. *E. affinis* had more specific trends and environmental preferences and its main occurrence was in mid-high levels of the beach, where a circadian rhythm related to tidal cycles is supposed to be an advantage for beach crustacean species in avoiding surface desiccation, predation or immersion [3,4]. These species were the most abundant in all the beaches studied, and the highest occurrences were once again related to the food-enriched western coast.

In some cases, the models derived from different beaches revealed similar species occurrences, indicating a strong influence of a specific variable on local processes. This means that relationships produced over any of those beaches may provide a gross qualitative indication of the abiotic effect [13]. Almost all the species were sometimes directly limited by a specific variable, and in these occasions data from any beach location should provide enough information to determine the specific species' general reaction to the variable measured. This situation can be extrapolated to other parts of the world, since the intermediate beach is the most common beach type [3]. The results obtained in this study revealed a mixed response to the predictor variables over the two contrasting scales, so important information can be gathered from examining differences between BRT models at different scales in order to build general beach species-environment relationships.

In summary, beach species dwell in a highly demanding ecosystem controlled by abiotic factors, but evidence of the importance of external food availability was shown. Our results also revealed that the local scale is a feasible way to construct general predictive species-environmental models, since relationships derived from different beaches predicted similar responses for certain macroinvertebrate species. However, additional information on aspects of specific species distribution and abundance can be derived by checking models at larger scales. Actual studies (e.g., [11,14,34]), including the present one, showed contrary evidence to the traditionally concept of exposed beaches as only physically stressed environments with negligible influence of the biotic compound. Finally, BRT added insight elucidating the importance of interactions between predictors for identifying the relative contributions of the predictors and species responses. This technique has proven useful for exploration, explanation and prediction offering both opportunities and challenges to sandy beach ecologists.

Supporting Information

Figure S1 Boxplots representing the main beach variables (mean ± standard deviation) from the northern Spanish coastline. Mean grain size (MGS), sediment water content and temperature (T), exposure to the air (ET), morphodynamic Dean index, shear strength (SS), species richness and exposure rate (ER). The dots represent outliers. See Figure 1 for beach names (n = 39) and Table S1 for further explanation on the variables.

Table S1 Environmental variables measured across the geographic beach continuum from the Spanish northern

shoreline. Description, range of values, units, and main references are included for better comprehension.

Table S2 Main species captured in the 39 exposed beaches from the North coast of Spain. (Bp: *Bathyporeia pelagica*, Dt: *Donax trunculus*, Ea: *Eurydice affinis*, Ep: *E. pulchra*, Gs: *Gastrosaccus sanctus*, Ha: *Haustorius arenarius*, Nc: *Nephtys cirrosa*, Ob: *Ophelia bicornis*, Pa: *Pontocrates arenarius*, Ss: *Scolelepis squamata*, Sr: *Sphaeroma rugicauda*; Ts: *Talitrus saltator*; Tb: *Talorchestia brito*, Te: *Tylos europaeus*).

Table S3 List of the beach species captured in the 39 exposed beaches from the North coast of Spain. This list shows the name and taxonomic composition of 59 truly sandy beach species. We also collected 13 species of uncertain origin, mainly insects: Aranei (1 species), Coleoptera (9), Diptera (2), and Hymenoptera (1).

Acknowledgments

Many thanks to all the benthos team from UVIGO for the hard work and enthusiasm during all these years. We thank A.M. Lohrer and three anonymous reviewers for helpful comments which improved the manuscript.

Author Contributions

Conceived and designed the experiments: IFR TJC ML. Performed the experiments: IFR ML. Analyzed the data: IFR TJC. Wrote the paper: IFR TJC.

References

1. Schlacher TA, Dugan J, Schoeman DS, Lastra M, Jones A, et al. (2007) Sandy beaches at the brink. Divers Distrib 13: 556–560.
2. Dugan JE, Defeo O, Jaramillo E, Jones AR, Lastra M, et al. (2010) Give beach ecosystems their day in the sun. Science 329 (5996): 1146.
3. McLachlan A, Brown AC (2006) The Ecology of Sandy Shores, 2nd ed. Burlington, MA, USA: Academic press 373 pp.
4. Defeo O, McLachlan A (2005) Patterns, processes and regulatory mechanisms in sandy beach macrofauna: a multi-scale analysis. Mar Ecol-Prog Ser 295: 1–20.
5. Jaramillo E, McLachlan A (1993) Community and population response of the macroinfauna to physical factors over a range of exposed sandy beaches in South-central Chile. Estuar Coast and Shelf S 37: 615–624.
6. McLachlan A, Jaramillo E, Donn TE, Wessels F (1993) Sandy beach macrofauna communities and their control by the physical environments: a geographical comparison. J Coastal Res 15: 27–38.
7. Jaramillo E, Contreras H, Duarte C, Quijón P (2001) Relationships between community structure of the intertidal macroinfauna and sandy beach characteristics along the Chilean coast. Mar Ecol 22(4): 323–342.
8. Rodil IF, Lastra M (2004) Environmental factors affecting benthic macrofauna along a gradient of intermediate sandy beaches in northern Spain. Estuar Coast and Shelf S 61: 37–44.
9. Celentano E, Defeo O (2006) Habitat harshness and morphodynamics: life history traits of the mole crab *Emerita brasiliensis* in Uruguayan sandy beaches. Mar Biol 149: 1453–1461.
10. Defeo O, Brazeiro A, de Alava A, Riestra G (1997) Is sandy beach macrofauna only physically controlled?. Role of substrate and competition in isopods. Estuar Coast and Shelf S 45: 453–462.
11. Dugan JE, Jaramillo E, Hubbard DM, Contreras H, Duarte C (2004) Competitive interactions in macroinfaunal animals of exposed sandy beaches. Oecologia 139: 630–640.
12. Lastra M, Page HM, Dugan JE, Hubbard DM, Rodil IF (2008) Processing of allochthonous macrophyte subsidies by sandy beach consumers: estimates of feeding rates and impacts on food resources. Mar Biol 154: 163–174.
13. Thrush SF, Hewitt JE, Herman PMJ, Ysebaert T (2005) Multi-scale analysis of species-environment relationships. Mar Ecol-Prog Ser 302: 13–26.
14. Lastra M, De la Huz R, Sánchez-Mata AG, Rodil IF, Aerts K, et al. (2006) Ecology of exposed sandy beaches in northern Spain: environmental factors controlling macrofauna communities. J Sea Res 55: 128–140.
15. Celentano E, Gutiérrez NL, Defeo O (2010) Effects of morphodynamic and estuarine gradients on the demography and distribution of a sandy beach mole crab: implications for source-sink habitat dynamics. Mar Ecol Prog Ser 398: 193–205.
16. Lercari D, Bergamino L, Defeo O (2010) Trophic models in sandy beaches with contrasting morphodynamics: comparing ecosystem structure and biomass flow. Ecol Model 23: 2751–2759.
17. Friedman JH, Hastie T, Tibshirani T (2000) Additive logistic regression: a statistical view of boosting. Ann Stat 28: 337–407.
18. De'ath G (2007) Boosted trees for ecological modelling and prediction. Ecology 88 (1): 243–251.
19. Elith J, Leathwick JR, Hastie T (2008) A working guide to boosted regression trees. J Anim Ecol 77: 802–813.
20. Alvarez-Salgado XA, Gago J, Miguez BM, Pérez FF (2001) Net ecosystem production of dissolved organic carbon in a coastal upwelling system: the Ria de Vigo, Iberian margin of the North Atlantic. Limnol Oceanogr 46: 135–147.
21. McLachlan A (1980) The definition of sandy beaches in relation to exposure: simple rating system. S Afr J Sci 76: 137–138.
22. Tyberghein L, Verbruggen H, Pauly K, Troupin C, Mineur F, et al. (2012) Bio-ORACLE: a global environmental dataset for marine species distribution modelling. Global Ecol Biogeogr 21(2): 272–281.
23. Friedman JH (2001) Greedy function approximation: a gradient boosting machine. Ann Stat 29: 1189–1232.
24. Hastie T, Tibshirani R, Friedman JH (2001) The Elements of Statistical Learning: Data Missing, Inference, and Prediction. Springer-Verlag, New York, 763 pp.
25. Leathwick JR, Elith J, Chadderton WL, Rowe D, Hastie T (2008) Dispersal, disturbance and the contrasting biogeographies of New Zealand's diadromous and non-diadromous fish species. J Biogeogr 35: 1481–1497.
26. Compton TJ, Leathwick JR, Inglis GJ (2010) Thermogeography predict the potential global range of the invasive European green crab (*Carcinus maneas*). Divers Distrib 16: 243–255.
27. Buston PM, Elith J (2011) Determinists of reproductive success in dominant pairs of clownfish: a boosted regression tree analysis. J Anim Ecol 80: 528–538.
28. Rijkenberg MJA, Langlois RJ, Mills MM, Patey MD, Hill PG, et al. (2011) Environmental forcing of nitrogen fixation in the Eastern Tropical and Subtropical North Atlantic Ocean. PLoS ONE 6(12): e28989.
29. Elith J, Graham CH, Anderson RP, Dudik M, Ferrier SG, et al. (2006) Novel methods improve prediction of species' distributions from occurrence data. Ecography 29: 129–151.
30. R Development Core Team (2011) R: A Language and Environment for Statistical Computing. R Foundation for Statistical Computing, Vienna, Austria. Available at: http://www.R-project.org.
31. Ridgeway G (2009) Generalized boosted regression models. Documentation on the R package, 1.5 'gbm'. Available at: http://www.R-Project.org.
32. Hanley JA, McNeil BJ (1982) The meaning and use of the area under a receiver operating characteristic (ROC) curve. Radiology 143: 29–36.
33. Parviainen M, Luoto M, Heikkinen RK (2009) The role of local and landscape level measures of greenness in modelling boreal plant species richness. Ecol model 220: 2690–2701.
34. Cisneros KO, Smit AJ, Laudien J, Schoeman DS (2011) Complex, dynamic combination of physical, chemical and nutritional variables controls spatio-temporal variation of sandy beach community structure. PLoS ONE 6: e23724.
35. Dugan JE, Hubbard DM, McCrary MD, Pierson MO (2003) The response of macrofauna communities and shorebirds to macrophyte wrack subsidies on exposed sandy beaches of southern California. Estuar Coast and Shelf S 58S: 133–148.
36. Rodil IF, Lastra M, Sánchez-Mata AG (2006) Community structure and intertidal zonation of the macroinfauna in intermediate sandy beaches in temperate latitudes: North coast of Spain. Estuar Coast and Shelf 67: 267–279.
37. Lastra M, Schlacher TA, Olabarria C (2010) Niche segregation in sandy beach animals: an analysis with surface-active peracarid crustaceans on the Atlantic coast of Spain. Mar Biol 157: 613–625.
38. Orr M, Zimmer R, Jelinski DE, Mews M (2005) Wrack deposition on different beach types: spatial and temporal variation in the pattern of subsidy. Ecology 86: 1496–1507.
39. Olabarria C, Incera M, Garrido J, Rodil IF, Rossi F (2009) Intraspecific diet shift in *Talitrus saltator* inhabiting exposed sandy beaches. Estuar Coast and Shelf S 84: 282–288.
40. Duarte C, Jaramillo E, Contreras H, Acuña K (2010) Cannibalism and food availability in the talitrid amphipod *Orchestoidea tuberculata*. J Sea Res 64: 417–421.

17

Ecology of Caribbean Sponges: Are Top-Down or Bottom-Up Processes More Important?

Michael P. Lesser[1]*, Marc Slattery[2]

1 Department of Molecular, Cellular and Biomedical Sciences, University of New Hampshire, Durham, New Hampshire, United States of America, 2 Department of Pharmacognosy, University of Mississippi, University, Mississippi, United States of America

Abstract

Benthic-pelagic coupling and the role of bottom-up versus top-down processes are recognized as having a major impact on the structure of marine communities. While the roles of bottom-up processes are better appreciated they are still viewed as principally affecting the outcome of top-down processes. Sponges on coral reefs are important members of the benthic community and provide a critically important functional linkage between water-column productivity and the benthos. As active suspension feeders sponges utilize the abundant autotrophic and heterotrophic picoplankton in the water column. As a result sponges across the Caribbean basin exhibit a consistent and significant pattern of greater biomass, tube extension rate, and species numbers with increasing depth. Likewise, the abundance of their food supply also increases along a depth gradient. Using experimental manipulations it has recently been reported that predation is the primary determinant of sponge community structure. Here we provide data showing that the size and growth of the sponge *Callyspongia vaginalis* are significantly affected by food availability. Sponges increased in size and tube extension rate with increasing depth down to 46 m, while simultaneously exposed to the full range of potential spongivores at all depths. Additionally, we point out important flaws in the experimental design used to demonstrate the role of predation and suggest that a resolution of this important question will require well-controlled, multi-factorial experiments to examine the independent and interactive effects of predation and food abundance on the ecology of sponges.

Editor: Roberto Pronzato, University of Genova, Italy, Italy

Funding: This project was funded by grants from the Caribbean Coral Reef Ecosystems program at the Smithsonian Institute, NOAAs National Institute of Undersea Science and Technology and the National Science Foundation. The funders had no role in the design of study, data collection and analysis, decision to publish, or preparation of the manuscript. The views expressed herein are those of the authors and do not necessarily reflect the views of the Smithsonian Institute, National Oceanic and Atmospheric Administration, their sub agencies or the National Science Foundation.

Competing Interests: The authors have declared that no competing interests exist.

* E-mail: mpl@unh.edu

Introduction

Sponges are ecologically and functionally important members of the benthic community on coral reefs [1–6]. In the Caribbean over 80 species have been recorded on reefs in the Florida Keys [7] and close to 300 species have been identified on Bahamian reefs [8]. In addition to their efficient filtering of seawater during feeding [9–11], sponges provide essential reef ecosystem functions such as providing habitat for numerous reef species including fish, brittle stars, and shrimp [12,13]. It is also increasingly recognized that sponges are crucial members of benthic food webs because of their ability to couple water column productivity with the secondary productivity of benthic communities [11,14–16].

Many mobile predators consume sponges (e.g., sea stars, fish, sea turtles), and spongivores can influence competitive interactions between sponges and corals [17]. Sponges have been proposed to be generally well defended against predation by both physical (i.e., spicules) and chemical means [18–21]; but see Thoms and Schupp [22] regarding the experimental difficulties of demonstrating chemical defenses in sponges. While it has been shown that cryptic or mangrove sponge species may be limited to specific refugia by spongivores [18,23–25], sponges throughout the Caribbean show a repeatable pattern of increasing biomass and diversity with depth to 150 m [7,8,26–31].

There is continuing interest in the relative importance of bottom-up versus top-down processes in structuring marine communities generally [32,34], and coral reef sponge communities in particular [6,24]. Many studies examining the bottom-up, top-down dichotomy have been conducted on suspension feeding invertebrates in rocky intertidal and subtidal habitats [35]. These studies have generally supported the importance of bottom-up and top-down processes, and the interaction of the two, on the structure of these communities [33,34]. For sponges on coral reefs, factors that change with depth, such as macro-scale flow velocities, solar irradiance, food supply and water temperature, have all been shown to influence sponge biology and ecology [1,2,6,7,11,36–38]. Recently, a paper by Pawlik et al. [39] reported on the growth rates of several species of sponge and concluded that rather than bottom-up processes (i.e., food supply) being a primary determinant of sponge growth and abundance, predation, mediated by the presence or absence of chemical defenses, is solely responsible for controlling the ecology of sponge communities. Here we present a natural field experiment *sensu* Diamond [40] demonstrating the importance of food supply in the ecology of *C. vaginalis*, one of the species used in Pawlik et al. [39], and compare this study with studies by Lesser [11], Trussell et al. [15] and Pawlik et al. [39] on *C. vaginalis*. In addition, we discuss in detail the experimental design flaws of Pawlik et al. [39] that make it difficult to come to the conclusions that predation alone structures sponge communities. We do so without any *a priori* dismissal of the potential role that predation, competition or food supply has alone, or together, on the ecology of this species, and on sponges generally.

Materials and Methods

Ethics Statement

No endangered or protected species were involved in this field study. No vertebrates were collected in this study; surveys were through visual census only. No "taking" of any flora or fauna was conducted except for samples of seawater. This study was conducted under a research license from Belize Fisheries, which was obtained by the Smithsonian Institution's Caribbean Coral Reef Ecosystems Program.

Study Site

Callyspongia vaginalis Lamarck is a common tubular morphology sponge exhibiting a wide bathymetric and geographic distribution in the Caribbean. Sponge size frequency distributions, picoplankton availability, feeding, tube extension rates and assessments of the total number of sponge species and individual sponge species numbers were conducted at Carrie Bow Cay, Belize (16° 48.16′ N, 88° 04.94′ W) from 3 to 46 m between 2000 and 2001.

Abundance, Size Frequency and Tube Extension Rates

Transects at 3, 10, 17, 23, 30 and 46 m were conducted by placing a 30 m transect tape parallel to the reef at each depth. Along each transect ten 1.0 m^2 quadrats were randomly placed and all sponges within the quadrat were enumerated and identified to species where possible. The mean ± SE number of sponges, and number of sponge species, per m^2 were calculated. Tube length was measured to the nearest 0.1 cm on sponges (N = 10) at each depth, and tubes (N = 3) on individual sponges (N = 10) were tagged 1.0 cm below the lip of the osculum at each depth. Sponges were then left undisturbed for one year when all tagged sponges were re-measured, and tube extension rates (cm mo^{-1}) were calculated by subtracting 1 cm from the total change in tube length. In contrast to Pawlik et al. [39], these sponges were not caged so this growth data actually represents the most ecologically relevant condition (i.e., net growth in the presence of both biotic and abiotic factors) although we cannot directly separate the effects of food supply from predation.

Food Availability and Feeding

Sponge pumping was assessed using sodium fluorescein dye. A small amount of dye (~1 ml) was injected into the exterior wall of a tube at the base of the sponge (N = 3) with a syringe, and the time it took for the dye to move over a known distance was measured and used to estimate pumping velocity, at a resolution of 0.5 cm s^{-1}. The diameter of each osculum was directly measured after dye ejection, and its area estimated using the formula for the surface area of a circle or ellipse. Volume flow rate was estimated by multiplying pumping rate by the cross-sectional area of the osculum. Volume flux was then computed by multiplying ejection speed by the cross-sectional area. This computation assumes plug flow rather than laminar flow coming out of the tube [15]. Careful examination of the dye fronts indicated that plug flow was the better approximation.

After pumping measurements were determined for the same sponges (N = 3) at each depth, ambient seawater was collected approximately 5 cm from the sponge. Matched samples of the excurrent flow were collected from approximately 3 cm inside the osculum of each sponge. Both ambient and excurrent samples were collected with 10 ml Vacutainer® syringes. All samples were fixed at a final concentration of 0.5% electron microscopy grade glutaraldehyde in filtered (0.2) seawater and initially frozen at 0°C for transportation on dry ice to the University of New Hampshire. Samples were then sent frozen to the Bigelow Laboratory for Ocean Sciences Flow Cytometry Facility where they were stored in liquid nitrogen until analysis as described previously [11,41]. Briefly, each sample was analyzed for cell abundances using a Becton Dickinson FACScan flow cytometer equipped with a 15 mW, 488 nm, air-cooled Argon ion laser. Simultaneous measurements of forward light scatter (FSC, relative size), 90 degree light scatter (SSC), chlorophyll fluorescence (>650 nm), and phycoerythrin fluorescence (560–590 nm) were made on all samples. Differentiation of cyanobacteria from prochlorophytes was based on the presence of phycoerythrin fluorescence. Cell abundance of the heterotrophic bacteria was determined by the use of PicoGreen (Molecular Probes), a dsDNA specific dye, which stains all bacteria (emission fluorescence 515–525 nm). Subtraction of the chl *a* containing picoplankton from total bacteria yielded the heterotrophic bacterial component of the community.

Food availability was separated into cyanobacteria, prochlorophytes, phytoplankton, heterotrophic bacteria, and then total cells. Filtration efficiency was calculated as 1 – (concentration of cells in the ex-current stream/ambient concentration of cells). Cells filtered per second were then computed by multiplying filtration efficiency (dimensionless) x volume flow rate (ml s^{-1}) x ambient concentration (cells ml^{-1}). All filtered cells were converted to carbon equivalents using the following conversions; heterotrophic bacteria: 20 fg C $cell^{-1}$ [42], *Prochlorococcus*: 53 fg C $cell^{-1}$ [43] and *Synechococcus*: 470 fg C $cell^{-1}$ [44]. The total carbon acquired $sponge^{-1}$ d^{-1} was then converted to energetic units (J s^{-1}) using a conversion factor of 1 mg C = 23.03 Joules, assuming an RQ = 1.0 which is appropriate for ammonotelic animals such as sponges [45].

Spongivores

The diversity and abundance of spongivorous fishes was assessed on replicate 2×10 m band transects at three depths (10, 30, and 46 m) on the fore-reef of Carrie Bow Cay (n = 5 transects per depth). Since spongivores tend to be home ranging [46], on subsequent dives we were able to collect dietary data from replicate individuals (n = 5 from each depth) during 5 min observation periods of each fish. Specifically, we recorded the percent of bites on sponges during the observation period, the total number of bites on sponges during the time period, the number of bites on the same individual, and the number of sponges that each fish species grazed. In addition, the 3-dimensional cover (= volume) and identity of sponges within randomly-assigned replicate (n = 5) 1 m^2 quadrats on each of the transects was assessed. From this data, we were able to determine the ranked abundance of the resident sponges relative to feeding preference (i.e., the ranked bites on specific species).

Statistical Analysis

Abundance, size frequency, tube extension rates, food availability and spongivore observations were analyzed using one-way ANOVAs at a significance level of 5%, using depth as a fixed factor. No unequal variances were detected using the F_{max} test, and when significant treatment effects were detected individual treatment differences were assessed using the Tukeys HSD multiple comparison test. Where appropriate, ratios and percentages were arcsin or log transformed for analysis and back transformed for presentation. Because the feeding rate of *Callyspongia vaginalis* scaled linearly with sponge size in a previous study [11] an ANCOVA was run on all feeding data with sponge size as a covariate.

Results

At Carrie Bow Cay sponge density increases significantly (ANOVA: $p<0.0001$) with increasing depth, as does the number of species (ANOVA: $p<0.0001$, Fig. 1). Sponge density and species number were significantly different between shallower (7.5–23 m) and deeper (30–46 m) depths (Tukeys HSD: $p<0.05$). Additionally, the density of *Callyspongia vaginalis* increased significantly with depth (ANOVA: $p<0.0001$, Fig. 1) with significant differences between the shallowest and deepest depths, and the middle depths exhibiting overlapping differences (Tukeys HSD $p<0.05$).

Sponge size, measured as tube length, showed a significant (ANOVA: $p<0.0001$) effect of depth for *Callyspongia vaginalis* with larger sponges found at deeper depths (Fig. 2 a). Multiple comparison testing showed that tube length at each depth was significantly different (Tukeys HSD: $p<0.05$) from each other (Fig. 2 a). Measurements of tube extension rates (cm mo^{-1}) showed a significant effect (ANOVA: $p<0.0001$) of depth for *Callyspongia vaginalis* (Fig. 2 b). Multiple comparison tests showed that sponges from the deeper depths (23–46 m) grew significantly faster (Tukeys HSD: $p<0.05$) than sponges from shallow depths, and sponges at 46 m grew faster than sponges at all other depths (Fig. 2 b).

Total food availability for *C. vaginalis* increased significantly (ANOVA: $p<0.031$) with depth (Fig. 3 a) with the highest concentration of picoplankton found at 46 m (Tukeys HSD: $P<0.05$). As observed in previous studies [11], the density of cyanobacteria within the picoplankton decreased while both the prochlorophytes and heterotrophic bacteria increased with depth (data not shown). The calculated percentage clearance for *C. vaginalis* was also significantly different with depth (ANOVA: $p<0.0013$), and ranged from 53% at 7.5 m to 84% at 46 m. The clearance rate at the shallowest depth (7.5 m) was significantly different (Tukeys HSD: $p<0.05$) from all other depths that were not significantly different from each other (data not shown). Unlike previous observations from the Florida keys where only two depths were analyzed [11] the ANCOVA analysis for the effects of depth, with tube length as a covariate, on feeding rates as well as carbon and energy acquisition was not significant (ANCOVA: $p<0.05$) and was not considered further. Combining the available food with percentage clearance and the volume flux calculations showed that *C. vaginalis* consumed significantly more food with increasing depth (ANOVA: $p<0.0001$), and acquired more carbon as particulate organic carbon (ANOVA: $p<0.0001$) and more energy (J d^{-1}; ANOVA: $p<0.0001$), based on an assumption of constant pumping rate and feeding (Fig. 3 b). In all cases multiple comparison tests showed that sponges at shallower depths had significantly lower rates of feeding, total carbon (Fig 3 c) and total energy acquisition (Fig. 3 d) when compared to deeper depths (Tukeys HSD: $P<0.05$).

There were no significant differences in spongivorous fish densities between the shallow reefs (10 and 30 m) and the deep reef (46 m), with the exception of spadefish that were not observed shallower than 39 m (Table 1; ANOVA: $p = 0.0326$). The percent of sponges in spongivorous fish diets was consistent across the depth gradient (ANOVA: $p = 0.0897$), although the number of sponge species in the diet of spongivores increased significantly

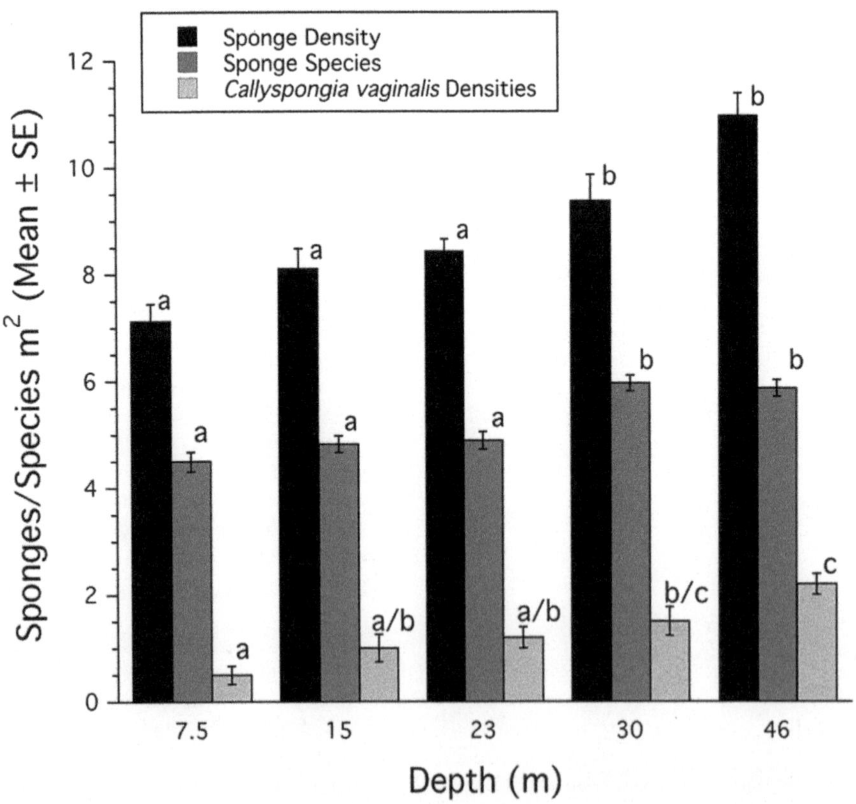

Figure 1. Sponge densities and diversity on Carrie Bow Cay, Belize (m^2, mean ± SE). Density of sponges, number of sponge species and density of *Callyspongia vaginalis* from different depths. Common superscripts indicate groups not significantly different from each other.

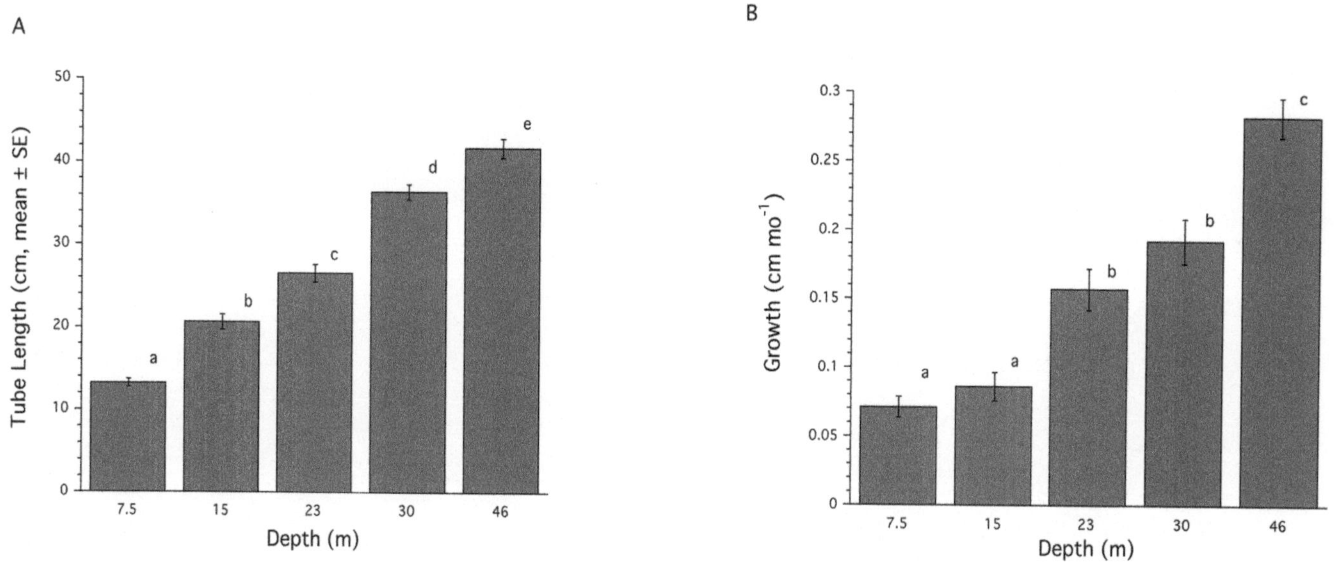

Figure 2. Tube length and growth for *Callyspongia vaginalis* on Carrie Bow Cay, Belize. Tube length (cm, mean ± SE) (A). Growth over a one-year period (cm, mean ± SE) (B). Common superscripts indicate groups not significantly different from each other.

with depth (ANOVA: $p<0.0001$). Finally, the number of bites per unit time was consistent across the depth gradient (ANOVA: $p=0.1121$), but spongivores on the deeper reef exhibited significantly more focal bites on individual sponges than did their conspecifics on the shallow reefs (ANOVA: $p<0.0001$). *Callyspongia vaginalis* was the 12th and 23rd most common sponge, by volume, on the shallow and deep reefs, respectively (data not shown). Additionally, *C. vaginalis* represented the 19th and 24th most common prey species on shallow and deep reefs respectively, based on focal bites.

Discussion

In tropical ecosystems neritic and pelagic productivity is dominated by picoplankton, including both photoautotrophic and heterotrophic bacteria [11,47,48]. The small size of these picoplankton (1.0 µm or less) makes them an ideal food resource for sponges that are active suspension feeders and consume a large fraction of these microbes [10,11,49]. Caribbean sponges such as *Callyspongia vaginalis*, *Agelas conifera*, and *Aplysina fistularis* have been shown to acquire their food heterotrophically [11,15], unlike their photoautotrophic counterparts typical of Pacific reefs such as the Great Barrier Reef [1,2].

Here we show that the bathymetric pattern of picoplankton distribution for cyanobacteria (*Synechococcus* sp.), prochlorophytes and heterotrophic bacteria (likely including various chemoautotrophs and Archaea) is similar to what has been previously reported for the waters adjacent to coral reefs across the Caribbean [11]. The carbon and energy acquired by sponges from these planktonic resources are substantial and result in a consistent and repeatable pattern of increasing sponge size and growth rates for both chemically defended and undefended species with increasing depth ([11], this study). The natural experiment described here used a chemically undefended sponge that is fully exposed to equivalent numbers of spongivores, specifically fishes [30,39,50], at depths up to 46 m, with no evidence for depth dependent differences in spongivory on this species. If anything, since spongivorous fishes are likely to follow their food supply [6] one would predict higher rates of sponge biomass removal by these predators with increasing depth that is not evident in this study or others [e.g., 11]. We did observe a greater dietary breadth at depth, and more focal bites on individual sponges, but this apparently had no effect on growth rates of sponges at depths greater than 30 m.

For *Callyspongia vaginalis* in particular, a well-controlled reciprocal transplant experiment was conducted on Conch Reef in the Florida Keys [15] the same reef where Pawlik et al. [39] conducted their studies. In that study *C. vaginalis* from the deep reef grew significantly (27%) more than conspecifics from the shallow habitat. To examine the effect of depth and the genetic or environmental basis of habitat-specific variation in sponge growth, Trussell et al. [15] conducted a reciprocal transplant experiment between deep (25 m) and shallow reefs (12 m). Sponges returned to their native site served as transplant controls (e.g., shallow —> shallow). A significant depth effect on sponge growth indicated that: 1) deep control sponges exhibited significantly higher growth rates than both shallow sponge controls and deep sponges transplanted to the shallow site, 2) there were no significant differences in growth between deep and shallow sponges maintained at the deep site, and 3) shallow sponges transplanted to the deep site showed only a trend of increased growth (0.36 ± 0.11 cm) over their controls (0.20 ± 0.11 cm). These data indicate that deep sponges generally grow faster than shallow sponges but are not able to maintain their high growth rates when transplanted to the shallow habitat. In addition, the large differences in growth observed between both control groups were not apparent between shallow sponges transplanted to the deep site and deep site controls. Hence, shallow sponges placed in the deep habitat and deep controls exhibited comparable growth rates. Moreover, without having tested directly the effects of predation, all of these results occurred while being exposed to the full suite of spongivores.

These results are also supported by simultaneous assessment of food availability, feeding and energetic budgets for sponges at each depth and the reciprocal transplants [15]. Deep reef habitats at Conch Reef are more conducive to higher growth rates than the shallow reef, which is consistent with the larger overall population size structure of sponge populations in the deep vs. shallow reefs on

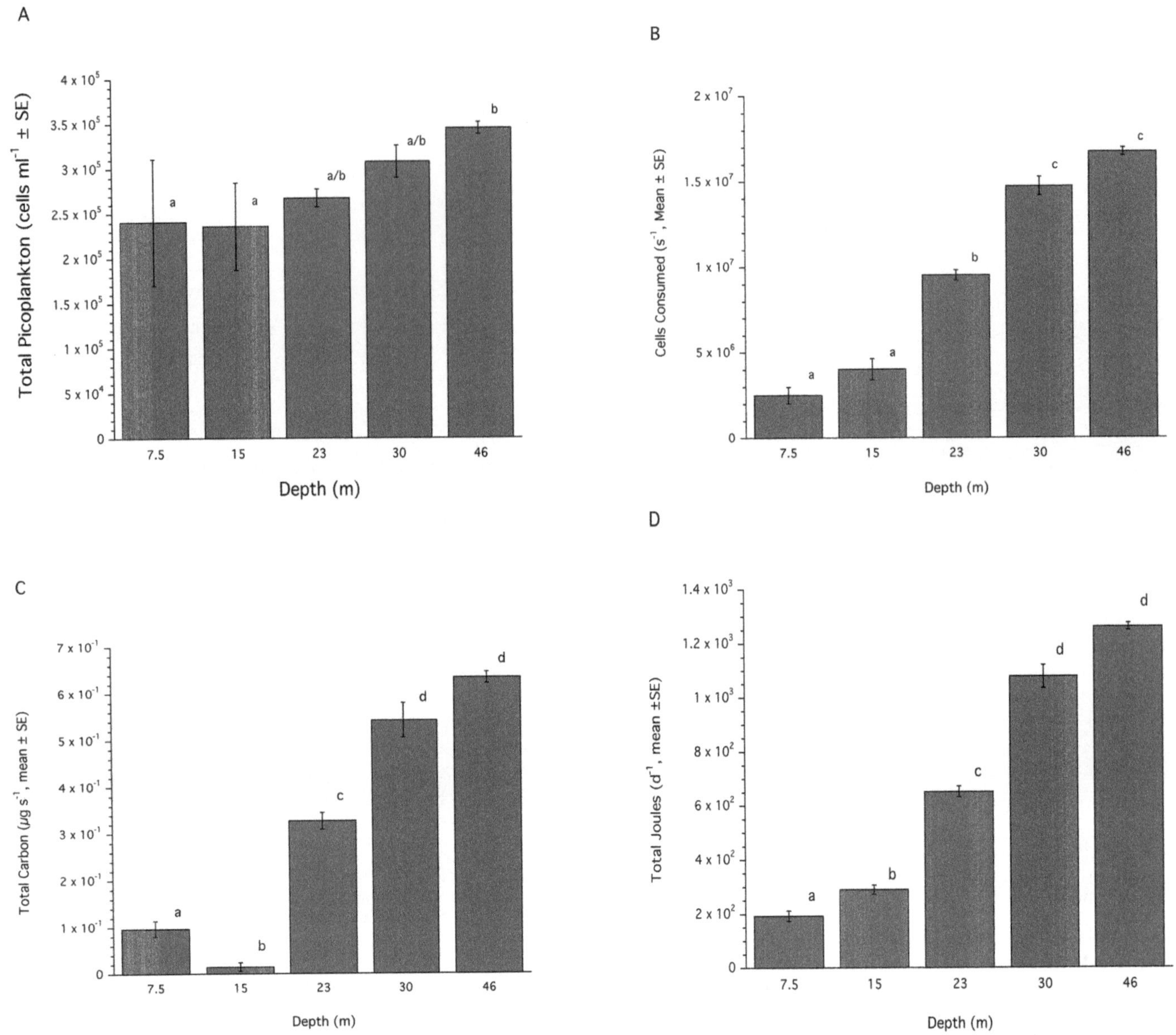

Figure 3. Picoplankton availability and feeding for *Callyspongia vaginalis* on Carrie Bow Cay, Belize. Depth-dependent concentrations of total picoplankton (A). Instantaneous total cells consumed per individual at each depth (B). Instantaneous intake of total particulate carbon per individual (C). Daily intake of total energy (J) per individual (D). Common superscripts indicate groups not significantly different from each other.

Conch Reef [11]. Moreover, the shift in the growth rates of deep reef sponges between the two habitats and the trend for shallow sponges suggests a phenotypically plastic response to different environments, rather than genetically different populations which has since been quantified for *C. vaginalis* [51]. The work on sponges at Conch Reef by Lesser [11] and Trussell et al. [15] is also consistent with transplant experiments done by Leichter et al. [52] that showed higher growth rates at 30 m versus 15 m for the coral *Madracis mirabilis* on Conch Reef. Additionally, long-term population studies on the giant barrel sponge *Xestospongia muta*, a sponge that exhibits highly variable chemical defenses [20], showed that sponge densities increased 2–5 times faster at 30 m compared to 10 m and 20 m, while sponge size frequency distributions shifted to smaller sponge sizes across all depths at Conch Reef [53]. McMurray et al. [53] attributed increases in sponge densities to episodic recruitment events and subsequent survival but did not explain the depth dependent density differences. If recruitment potential is the same over the depth range of 10–30 m, and the densities of spongivores are equivalent across depths at this location [39], then we would hypothesize recruitment survival could be explained by access to greater food resources at 30 m.

There are several flaws with the experimental design of Pawlik et al. [39] that reduce the impact of their conclusion regarding control by spongivorous fish for Caribbean sponge communities. First, in their *C. vaginalis* experiment Pawlik et al. [39] lost an entire treatment group which necessitated the use of multiple t-tests rather than the planned multifactorial ANOVA. However, the authors appear to use multiple t-tests without correcting the comparison-wise p value to control for the experiment-wise α of 0.05 (i.e., the 30 m uncaged treatment group was used in more than one comparison) and avoid making a Type I error [54]. In fact, it appears that correcting the p value for the number of

Table 1. Depth-specific spongivore densities and their diet on Carrie Bow Cay, Belize.

Spongivore	Density: 10–30 m $(2\times10\ m)^{-1}$ N = 10	Density: 46 m $(2\times10\ m)^{-1}$ N = 5	Dietary Breadth (% bites on sponges) [# sponge species eaten]	Bites $(15\ min)^{-1}$	Bites $(individual)^{-1}$
CHAETODONTIDAE					
Chaetodipterus jaber	0	2.6±1.6	na/31.4±4.5 [na/10.4±1.3]	0/10.2±3.1	0/4.2±0.9
MONOCANTHIDAE					
Cantherhines macrocerus	0.2±0.1	0.6±0.4	66.7±3.6/77.2±3.7 [7.9±0.5/10.2±1.3]	21.3±2.7/18.6±2.9	1.6±0.3/5.0±1.0
POMACANTHIDAE					
Holocanthus ciliarus	0.3±0.2	0.4±0.4	94.5±1.7/98.4±0.9 [24.7±1.0/35.2±2.7]	12.2±1.4/13.0±1.9	2.6±0.4/4.0±0.8
Holocanthus tricolor	0.6±0.3	0.4±0.2	92.8±1.5/97.0±1.5 [21.5±1.2/28.8±0.9]	17.8±1.2/19.0±2.1	1.6±0.2/10.2±1.5
Pomacanthus arcuatus	0.5±0.3	0.4±0.4	70.7±2.5/74.8±2.7 [21.2±1.3/28.8±0.9]	12.4±0.9/14.2±1.9	1.2±0.1/6.0±0.7
Pomacanthus paru	0.4±0.2	0.2±0.2	76.5±2.5/73.4±2.9 [19.7±0.7/30.4±1.4]	14.9±0.8/14.8±2.2	1.2±0.1/6.2±1.2
TETRODONTIDAE					
Canthigaster rostrata	4.8±1.5	4.4±1.8	20.3±2.4/45.4±2.1 [11.2±0.8/28.0±0.7]	47.9±12.1/48.8±9.1	1.9±0.3/12.8±2.9

Notes: Data presented are the mean number of fish ± SE per 2×10 m band transects (n = 5 replicates at each of three depths: 10 m, 30 m and 46 m). Percent of bites on sponges for each fish species are based on 15 min bite count observations of five fish on the shallow- and deep- reefs (na = not applicable: no fish observed at depth); the number of sponge species sampled during that period are included in parentheses. The average number of bites per 15 min observation period, and the average number of focal bites on individual sponges are recorded. In all cases where numbers are separated by a hash ("/"), the first number refers to the shallow depths and the second number refers to the deep depth.

multiple comparisons to an α of 0.025 shows that their results for caging effects ($p = 0.0486$) would not be significant. Also, there is no explanation for running a one-tailed *t*-test. Prior data in fact already shows greater growth at deeper depths [11] which *a priori* should have led the authors to conduct a two-tailed test to test for the possibility of greater growth in either shallow or deeper depths and the required doubling of those *p* values. Second, not having cage controls is a clear experimental design flaw, despite the report of no cage effects from Leong and Pawlik [55] where similar problems with *post hoc* multiple comparison analyses also occur; a study that itself did not report the cage control data. Cages will affect the hydrodynamics outside and inside the cage and therefore the flow speed and residency time of the water within the cage [56] with potential effects on food availability and sponge metabolism. Another underappreciated effect of caging would be a significant reduction in flow-induced feeding by sponges [57,58]. Furthermore, cages could also affect light levels within these structures (i.e., they will act as neutral density filters) that could negatively affect sponges in Pawlik et al. [39] that contain photoautotrophic symbionts (i.e., *Aplysina cauliformis*) [59,60]. There were, however, clearly no effects of cages or depth on the growth of *Amphimedon compressa and Aplysina cauliformis* and, not withstanding the lack of cage controls, this suggests that food supply, predation, and in the case of *A. cauliformis* light, did not significantly affect the growth of these sponges in these experiments so broad conclusions about the role of predation appear unsupported in this case. But controlling for cage effects in each experiment, even if the results agree with *a priori* anticipated outcomes, is required and when conducted appropriately can be very informative regarding the role of fish predation [61,62]. Third, sponges were collected from "10–30 m" and then divided into depth treatment groups (15 m and 30 m) with no regard to their prior history at the depths of collection. The appropriate design is to conduct reciprocal transplants, with disturbance controls [15,63], that match experimental depths with the sponge collection depths [15,64] such that any genetic, or prior environmental history differences or disturbance effects can be detected and accounted for. Finally, the assessment of sponge growth by Pawlik et al. [39] is by buoyant weight but the results are described on a percentage basis that has the potential to mask what are actually very small increases in biomass. While we appreciate the importance of volumetric parameters (i.e., wall-thickening) as a component of growth (e.g., [5]), in the specific case of *C. vaginalis* previous work has shown that both volume and tube length, in absolute units, increases with depth in *C. vaginalis* [11,15].

We also disagree with Pawlik et al. [39] that Conch Reef is typical of the coral reefs around the Caribbean basin. Pawlik et al. [39] note that the unique oceanography at Conch Reef (i.e., internal waves that drive a "plankton pump" [52]), should provide the strongest signal in the Caribbean basin for bottom-up processes, but the feeding and physiological experiments by Lesser [11] and Tussell et al. [15] were actually conducted during times when internal waves were absent. We do, however, appreciate how internal waves integrated annually would increase productivity and food availability to deeper coral reef communities *sensu* Leichter et al. [52,65,66] and effect long-term growth rates. It is therefore important to see similar results for *C. vaginalis* reported here from Carrie Bow Cay, Belize; the site is a Caribbean coral reef system that is not significantly affected, in terms of frequency or amplitude [67], by an internal wave driven plankton pump but there is a significant increase in picoplankton abundance with increasing depth, and the sponges on this reef are exposed to abundances of spongivores that are similar to what is found on Conch Reef ([6], Table 1).

Pawlik et al. [39] go on to argue that Conch Reef and other reefs throughout the Caribbean basin have a similar diversity of sponges and spongivores. In fact, sponge diversity on Conch Reef may be the lowest in the region (e.g., [7] vs. [8]), and even if diversity is similar function can be quite different [68]. Also, while spongivore diversity is much more similar across the region at depths up to 30 m, except for parrotfishes which arguably are not important spongivores [6,50,69], differences in spongivore abundance and diversity can have significant impacts on community structure, and likely function but mostly for reefs >50 m in depth ([30], Slattery and Lesser unpublished data).

When exposed to spongivorous fishes, *Callyspongia vaginalis*, a widely distributed and chemically undefended sponge ([20], Slattery unpublished data), maintains greater biomass and grows faster in proportion to the abundance of picoplankton ([11,15], this study). This abundant sponge is also of poor nutritional quality [70], and sponge nutritional quality has been shown to be more important than chemical defenses in determining whether fish consume different species of sponges [71]. Despite claims that differential predation, mediated by chemical defenses, is the major process controlling the ecology of sponges [19,39], this view is not universally supported [6,50] and the data on food availability and growth, differences in sponge nutritional quality, competition and disease all support roles for other processes in the ecology of sponges [5,72]. In fact, our data (Table 1) indicates that spongivory is consistent across depths from 10 to 46 m at Carrie Bow Cay, but the greater sponge biodiversity and biomass on deeper reefs has resulted in a more opportunistic feeding strategy that includes more sponge species as food and potentially less time searching for prey (i.e., more focal bites per individual). It is also possible that the latter behavior implies that these sponges are less defended than their shallow reef counterparts (e.g., Slattery et al. unpublished data), although this reinforces our point that predation has limited control on sponge growth rates on deep reefs since any tissue loss is compensated by enhanced food supply for either defended or undefended sponges [11]. It is interesting to note that based on their rank availability (12th) and their ranking as a prey species (19th) shallow reef *C. vaginalis* may have some feeding deterrent characteristics (i.e., chemical defenses and/or lowered nutritional content). In contrast, deep reef *C. vaginalis* are consumed at a level consistent with their availability in the population.

Understanding the ecology of sponges is more important than ever as there is increasing awareness that some tropical reefs could undergo a phase transition to sponge dominated reefs [73]. There is some evidence from long-term data that this is already happening [74], even in the presence of spongivorous fishes [75]. But evidence of sponge declines, putatively from disease outbreaks, also exist [5,76], and a recent cold-water event in the Florida keys resulted in significant decreases in both coral and sponge cover [77]. Importantly, Caribbean wide declines in coral reef fish have also been reported with significant declines in "invertivores" that include well-known spongivorous species [78]. Paddock et al. [78] conducted a meta-analysis of data from 1955–2007 and found that the most significant decline in fish populations occurred during 1996–2007. This begs the following question; if you believe, or can show, that spongivory by fishes is the primary process controlling sponge populations, is there a "tipping point" for sponge populations where predation pressure becomes so low that abiotic factors (e.g., food supply) become more important over time [79]? Additionally, the Caribbean wide invasion by lionfish has further reduced fish populations since 2005, including spongivores, on both shallow [80] and mesophotic [30] reefs.

Pawlik et al. [39], despite claims to the contrary, did not directly test the role of bottom-up factors, or food supply and feeding, on the ecology of their target sponges. Similarly, Lesser [11], Trussell et al. [15] and the data presented here did not directly test the role of predation on sponges. This is reminiscent of early discussions on the role of nutrients versus herbivory on the phase transition of reefs from coral- to algal-dominated reefs [75,81-84]. No one had yet carried out the multi-factor, multi-level experiments required to tease apart the respective roles of nutrients versus herbivory, and their interaction, in these phase transitions (*sensu* [85]). When that was finally accomplished it was found that herbivory controls algal production, but that nutrients can play an important role under certain circumstances [86]. In a similar manner, no one has yet conducted carefully controlled experiments that examine the independent and interactive effects of predation and food supply on the ecology of sponges. To do this requires a detailed and controlled experimental approach carried out on sponges under varying levels of food supply and predation pressure. We predict that, similar to algal phase shifts on coral reefs [75,84,87], the controlling factors will be much more complex than either predation or food levels in isolation.

Acknowledgments

We thank Deborah Gochfeld for constructive comments on the manuscript.

Author Contributions

Conceived and designed the experiments: MPL MS. Performed the experiments: MPL MS. Analyzed the data: MPL MS. Contributed reagents/materials/analysis tools: MPL MS. Wrote the paper: MPL MS.

References

1. Wilkinson CR (1983) Net primary productivity in coral reef sponges. Science 219: 410–412.
2. Wilkinson CR (1987) Interocean differences in size and nutrition of coral reef sponge populations. Science 236: 1654–1657.
3. Wilkinson CR, Cheshire AC (1990) Comparisons of sponge populations across the Barrier Reefs of Australia and Belize: evidence for higher productivity in the Caribbean. Mar Ecol Prog Ser 67: 285–294.
4. Bell JJ (2008) The functional biology of sponges. Est Coast Shelf Sci 79: 341–353.
5. Wulff JL (2006) Rapid diversity and abundance decline in a Caribbean coral reef sponge community. Biol Cons 127: 167–176.
6. Wulff JL (2012) Ecological interactions and the distribution, abundance, and diversity of sponges. Adv Mar Biol 61: 273–344.
7. Schmahl GP (1990) Community structure and ecology of sponges associated with four Southern Florida coral reefs. In: New Perspectives in Sponge Biology. Rützler K, editor. Smithsonian Institution Press, Washington, D.C, pp. 384–390.
8. Reed JK, Pomponi SA (1997) Biodiversity and distribution of deep and shallow water sponges in the Bahamas. Proc of the 8th Int Coral reef Symp 2: 1387–1392.
9. Reiswig HM (1974) Water transport, respiration and energetics of three tropical marine sponges. J Exp Mar Biol Ecol 14: 231–249.
10. Pile AJ (1997) Finding Reiswig's missing carbon: quantification of sponge feeding using dual-beam flow cytometry. Proc 8th Int Coral Reef Symp 2: 1403–1410.
11. Lesser MP (2006) Benthic-pelagic coupling on coral reefs: feeding and growth of Caribbean sponges. J Exp Mar Biol Ecol 328: 277–288.
12. Diaz MC, Rützler K (2001) Sponges: An essential component of Caribbean coral reefs. Bull Mar Sci 69: 535–546.
13. Fiore CL, Jutte PC (2010) Characterization of macrofaunal assemblages associated with sponges and tunicates collected off of the southeastern United States. Invert Biol 129: 105–120.
14. Gili JF, Coma R (1998) Benthic suspension feeders: their paramount role in littoral marine food webs. Trends Ecol Evol 13: 316–321.
15. Trussell GC, Lesser MP, Patterson MR, Genovese SJ (2006) Depth-specific differences in growth of the reef sponge *Callyspongi vaginalis*: role of bottom-up effects. Mar Ecol Prog Ser 323: 149–158.

16. Perea-Blázquez A, Davy SK, Bell JJ (2012) Estimates of particulate organic carbon flowing from the pelagic environment to the benthos through sponge assemblages. PLoS ONE 7: e29569.
17. Hill MS (1998) Spongivory on Caribbean reefs releases corals from competition with sponges. Oecologia 117: 143–150.
18. Pawlik JR (1997) Fish predation on Caribbean reef sponges: an emerging perspective of chemical defenses. Proc 8th Int Coral reef Symp 2: 1255–1258.
19. Pawlik JR (2011) The chemical ecology of sponges on Caribbean reefs: natural products shape natural systems. Bioscience 61: 888–898.
20. Pawlik JR, Chanas B, Toonen RJ, Fenical W (1995) Defenses of Caribbean sponges against predatory reef fish. I. chemical deterrency. Mar Ecol Prog Ser 127: 183–194.
21. Becerro MA, Thacker RW, Turon X, Uriz M, Paul VJ (2003) Biogeography of sponge chemical ecology: comparisons of tropical and temperate defenses. Oecologia 135: 91–101.
22. Thoms C, Schupp PJ (2007) Chemical defense strategies in sponges: a review. In: Porifera Research: Biodiversity, Innovation and Sustainability. Custódio MR, Lôbo-Hadju G, Hajdu Muricy G, editors. Série Livros. Museu Nacional, Rio de Janeiro, pp. 627–637.
23. Wulff JL (1997) Parrotfish predation on cryptic sponges of Caribbean coral reefs. Mar Biol 129: 41–52.
24. Wulff JL (2005) Ecological interactions of marine sponges. Can J Zool 84: 146–166.
25. Pawlik JR (1998) Coral reef sponges: Do predatory fishes affect their distribution? Limnol Oceanogr 43: 1396–1399.
26. Rützler K, Macintyre IG (1982) The habitat distribution and community structure of the barrier reef complex at Carrie Bow Cay, Belize. In: The Atlantic Barrier Reef Ecosystem at Carrie Bow Cay, Belize. 1 Structure and Communities. Rützler K, Macintyre IG, editors. Smithsonian Institution Press, Washington, pp. 9–45.
27. Suchanek TH, Carpenter RC, Witman JD, Harvell CD (1984) Sponges as important space competitors in deep Caribbean coral reef communities. In: The ecology of deep and shallow coral reefs. Reaka ML, editor. Symposia Series for Undersea Research, NOAA's Undersea Research Program. U.S. Department of Commerce, pp 55–60.
28. Liddell WD, Avery WE, Ohlhorst SL (1997) Patterns of benthic community structure, 10-250 m, the Bahamas. Proc 8th Int Coral Reef Symp 1: 437–442.
29. Lehnert H, Fischer H (1999) Distribution patterns of sponges and corals down to 107 m off north Jamaica. Mem Queensland Mus 44: 307–316.
30. Lesser MP, Slattery M (2011) Phase Shift to Algal Dominated Communities at Mesophotic Depths Associated With Lionfish (*Pterois* volitans) Invasion on a Bahamian Coral Reef. Biol Invasions 13: 1855–1868.
31. Slattery M, Lesser MP (2012) Mesophotic Coral Reefs: A Global Model of Community Structure and Function. Ecology of Mesophotic Coral reefs, Proc 12th Int Coral Reef Symp, Cairns, Australia, 9–13 July 2012, 5 pp.
32. Power ME (1992). Top-down and bottom-up forces in food webs: do plants have primacy. Ecology 73: 733–746.
33. Menge BA (1992) Community regulation: under what conditions are bottom-up factors important on rocky shores. Ecol Monogr 73: 755–765.
34. Menge BA (2000) Top-down and bottom-up community regulation in marine rocky intertidal habitats. J Exp Mar Biol Ecol 250: 257–289.
35. Coma R, Ribes M, Gili J-M, Zabala M (2000) Seasonality in coastal benthic ecosystems. Trends Ecol Evol 15: 448–453.
36. Wilkinson CR, Vacelet J (1979) Transplantation of marine sponges to different conditions of light and current. J Exp Mar Biol Ecol 37: 91–104.
37. Alcolado PM (1979) Ecological structure of the sponge fauna in a reef profile of Cuba. In: Biologie des Spongiaires. Levi C, Boury-Esnault N, editors. Colloques Internationaux du C.N.R.S., Paris, France, pp. 297–302.
38. Palumbi SR (1986) How body plan limits acclimation: Responses of a demosponge to wave force. Ecology 67: 208–214.
39. Pawlik JR, Loh T-L, McMurray SE, Finelli CM (2013) Sponge communities on coral reefs are controlled by factors that are top-down, not bottom-up. PLoS ONE 8: e62573.
40. Diamond J (1986) Overview: laboratory experiments, field experiments, and natural experiments. In: Community Ecology. Diamond J, Case TJ, editors. New York: Harper and Row, pp 3 – 22.
41. Lesser MP, Shumway SE, Cucci T, Smith J (1992) Impact of fouling organisms on mussel rope culture: Interspecific competition for food among suspension-feeding invertebrates. J Exp Mar Biol Ecol 165: 91–102.
42. Ducklow HW, Kirchman DL, Quinby HL, Carlson CA, Dam HG (1993) Stocks and dynamics of bacterioplankton carbon during the spring bloom in the eastern North Atlantic Ocean. Deep Sea Res 40: 245–263.
43. Morel A, Ahn YH, Partensky F, Vaulot D, Claustre H (1993) *Prochlorococcus* and *Synechococcus*: a comparative study of their optical properties in relation to their size and pigmentation. J Mar Res 51: 617–649.
44. Campbell L, Nolla HA, Vaulot D (1994) The importance of *Prochlorococcus* to community structure in the central North Pacific Ocean. Limnol Oceanogr 39: 954–960.
45. Parsons TR, Takahashi M, Hargrave B (1984) Biological Oceanographic Processes. 3rd Ed. Pergamon.
46. Hourigan TF, Stanton FG, Motta PJ, Kelley CD, Carlson B (1989) The feeding ecology of three species of Caribbean angelfishes (family Pomacanthidae). Environ Biol Fishes 24:105–116.
47. Partensky F, Hess WR, Vaulot D (1999) *Prochlorococcus*, a marine photosynthetic prokaryote of global significance. Microbiol Mol Biol Rev 63: 106–127.
48. Ting CS, Rocap G, King J, Chisolm S (2002) Cyanobacterial photosynthesis in the oceans: the origins and significance of divergent light harvesting strategies. Trends Microbiol 10: 134–142.
49. Ribes M, Coma R, Atkinson MJ, Kinzie III RA (2003) Particle removal by coral reef communities: picoplankton is a major source of nitrogen. Mar Ecol Prog Ser 257: 13–23.
50. Randall JE, Hartman WD (1968) Sponge-feeding fishes of the West Indies. Mar Biol 1: 216–225.
51. López-Legentil S, Erwin PM, Henkel TP, Loh T-L, Pawlik JR (2010) Phenotypic plasticity in the Caribbean sponge *Callyspongia vaginalis* (Porifera: Haplosclerida). Scientifica Marina 74: 445–453.
52. Leichter JJ, Shellenbarger G, Genovese SJ, Wing SR (1998) Breaking internal waves on a Florida (USA) coral reef: a plankton pump at work? Mar Ecol Prog Ser 166: 83–97.
53. McMurray SE, Henkel TP, Pawlik JR (2010) Demographics of increasing populations of the giant barrel sponge *Xestospongia muta* in the Florida Keys. Ecology 91: 560–570.
54. Sokal RR, Rohlf FJ (2012) Biometry: the principles and practice of statistics in biological research. 4th ed., W.H. Freeman and Co, New York.
55. Leong W, Pawlik JR (2010) Evidence of resource trade-off between growth and chemical defenses among Caribbean coral reef sponges. Mar Ecol Prog Ser 406: 71–78.
56. Miller LP, Gaylord B (2007) Barriers to flow: the effects of experimental cage structures on water velocities in high-energy subtidal and intertidal environments. J Exp Mar Biol Ecol 344: 215–228.
57. Vogel S (1977) Current-induced flow through living sponges in nature. Proc Natl Acad Sci 74: 2069–2071.
58. Leys SP, Yahel G, Reidenbach MA, Tunnicliffe V, Shavit U, et al. (2011) The sponge pump: the role of current induced flow in the design of the sponge body plan. PLoS ONE 6: e27787.
59. Erwin PM, Thacker RW (2007) Incidence and identity of photosynthetic symbionts in Caribbean coral reef sponge assemblages. J Mar Biol Ass UK 87: 1683–1692.
60. Maldonando M, Ribes M, van Duyl FC (2012) Nutrient fluxes through sponges: biology, budgets, and ecological implications. Adv Mar Biol 62: 113–182.
61. Steele MA (1996) Effects of predators on reef fishes: separating cage artifacts from effects of predation. J Exp Mar Biol Ecol 198: 249–267.
62. Connell SD (1997) Exclusion of predatory fish on a coral reef: the anticipation, pre-emption and evaluation of some caging artifacts. J Exp Mar Biol Ecol 213: 181–198.
63. Slattery M, Starmer J, Paul V (2001) Temporal and spatial variation in defensive metabolites of the tropical Pacific soft corals *Sinularia maxima* and *S. polydactyla*. Mar Biol 138: 1183–1193.
64. Wulff JL (2005) Trade-offs in resistance to competitors and predators, and their effects on the diversity of tropical marine sponges. J Animal Ecol 74: 313–321.
65. Leichter JJ, Wing SR, Miller SL, Denny MW (1996) Pulsed delivery of subthermocline water to Conch Reef (Florida Keys) by internal tidal bores. Limnol Oceanogr 41: 1490–1501.
66. Leichter JJ, Stewart HL, Miller SJ (2003) Episodic nutrient transport to Florida coral reefs. Limnol Oceanogr 48: 1394–1407.
67. Leichter JJ, Helmuth B, Fischer AM (2006) Variation beneath the surface: quantifying complex thermal environments on coral reefs in the Caribbean, Bahamas and Florida. J Mar Sci 64: 563–588.
68. Mouillot D, Graham NAJ, Villéger S, Mason NWH, Bellwood DR (2013) A functional approach reveals community responses to disturbances. Trends Ecol Evol. 28: 167–168.
69. Cardosa SC, Soares MC, Oxenford HA, Côte IM (2009) Interspecific differences in foraging behavior and functional role of Caribbean parrotfish. Mar Biod Rec 2: e148.
70. Chanas B, Pawlik JR (1995) Defenses of Caribbean sponges against predatory reef fish. II. Spicules, tissue toughness, and nutritional quality. Mar Ecol Prog Ser 127: 195–211.
71. Duffy JE, Paul VJ (1992) Prey nutritional quality and the effectiveness of chemical defenses against tropical reef fishes. Oecologia 90: 333–339.
72. Gochfeld DJ, Easson CG, Freeman CJ, Thacker RW, Olson JB (2012) Disease and nutrient enrichment as potential stressors on the Caribbean sponge Aplysina cauliformis and its bacterial symbionts. Mar Ecol Prog Ser 456: 101–111.
73. Bell JJ, Davy SK, Jones T, Taylor MW, Webster NS (2013) Could some coral reefs become sponge reefs as our climate changes. Global Change Biol doi: 10.1111/gcb.12212.
74. Colvard NB, Edmunds PJ (2011) Decadal-scale changes in abundance of non-scleractinian invertebrates on a Caribbean coral reef. J Exp Mar Biol Ecol 397: 153–160.
75. Norström AV, Nyström M, Lokrantz J, Folke C (2009) Alternative states on coral reefs: beyond coral-macroalgal phase shifts. Mar Ecol Prog Ser 376: 295–306.
76. Webster NS (2007) Sponge disease: a global threat? Environ Microbiol 9: 1363–1375.
77. Colella MA, Ruzicka RR, Kidney JA, Morrison JM, Brinkhuis VB (2012) Cold-water event of January 2010 results in catastrophic benthic mortality on patch reefs in the Florida Keys. Coral Reefs 31: 621–632.

78. Paddack MJ, Reynolds JD, Aguilar C, Appledoorn RS, Beets J, et al. (2009) Recent region-wide declines in Caribbean reef fish abundance. Current Biol 1: 590–595.
79. Hughes TP, Linares C, Dakos V, van de Leemput IA, van Nes EH (2013) Living dangerously on borrowed time during slow, unrecognized regime shifts. Trends Ecol Evol 28: 149–155.
80. Green SJ, Akins JL, Maljkovic'A, Côte IM (2012) Invasive lionfish drive Atlantic coral reef fish declines. PLoS ONE 7: e32596.
81. LaPointe BE (1997) Nutrient thresholds for bottom-up control of macroalgal blooms on coral reefs in Jamaica and Southeast Florida. Limnol Oceanogr 42: 1119–1131.
82. Hughes T, Szmant AM, Steneck R, Carpenter R, Miller S (1999) Algal blooms on coral reefs: what are the causes? Limnol Oceanogr 44: 1583–1586.
83. Smith JE, Hunter CL, Smith CM (2010) The effects of top-down versus bottom-up control on benthic coral reef community structure. Oecologia 163: 497–507.
84. Dudgeon SR, Aronson RB, Bruno JF, Precht WF (2010) Phase shifts and stable states on coral reefs. Mar Ecol Prog Ser 413: 201-216.
85. Worm B, Lotze HK, Hillebrand H, Sommer U (2002) Consumer versus resource control of species diversity and ecosystem functioning. Nature 417: 848–851.
86. Burkepile DE, Hay ME (2006) Herbivore vs. nutrient control of marine primary producers: context-dependent effects. Ecology 87: 3128–3139.
87. Bruno JF Sweatman H, Precht WF, Selig ER, Schutte VGW (2009) Assessing evidence of phase shifts from coral to macroalgal dominance on coral reefs. Ecology 90: 1478–1484.

18

Physiological Condition of Juvenile Wading Birds in Relation to Multiple Landscape Stressors in the Florida Everglades: Effects of Hydrology, Prey Availability, and Mercury Bioaccumulation

Garth Herring[1]*, Collin A. Eagles-Smith[1], Dale E. Gawlik[2], James M. Beerens[2], Joshua T. Ackerman[3]

1 United States Geological Survey, Forest and Rangeland Ecosystem Science Center, Corvallis, Oregon, United States of America, **2** Department of Biological Sciences, Florida Atlantic University, Boca Raton, Florida, United States of America, **3** United States Geological Survey, Western Ecological Research Center, Dixon Field Station, Dixon, California, United States of America

Abstract

The physiological condition of juvenile birds can be influenced by multiple ecological stressors, and few studies have concurrently considered the effects of environmental contaminants in combination with ecological attributes that can influence foraging conditions and prey availability. Using three temporally distinct indices of physiological condition, we compared the physiological response of nestling great egrets (*Ardea alba*) and white ibises (*Eudocimus albus*) to changing prey availability, hydrology (water depth, recession rate), and mercury exposure in the Florida Everglades. We found that the physiological response of chicks varied between species and among environmental variables. Chick body condition (short-term index) and fecal corticosterone levels (medium-term) were influenced by wetland water depth, prey availability, region, and age, but not by mercury contamination. However, mercury exposure did influence heat shock protein 70 (HSP70) in egret chicks, indicating a longer-term physiological response to contamination. Our results indicate that the physiological condition of egret and ibis chicks were influenced by several environmental stressors, and the time frame of the effect may depend on the specialized foraging behavior of the adults provisioning the chicks.

Editor: Chang-Qing Gao, Central South University, China

Funding: Financial or inkind support was provided by the U.S. This work received support from the Fish and Wildlife Service (grant/contract number: 401815G028) and U.S. Geological Survey. The funders had no role in study design, data collection and analysis, decision to publish, or preparation of the manuscript.

Competing Interests: The authors have declared that no competing interests exist.

* Email: gherring@usgs.gov

Introduction

Environmental stressors, such as contaminants, prey availability, and adverse weather, can elicit strong physiological responses from wild birds in order to help them overcome short-term challenges. These responses can be particularly harmful during the early stages of life if the response exceeds physiological norms [1–3]. Physiological [4] and behavioral [5] responses can manifest throughout the annual cycle by many avian species in order to cope with seasonal environmental stochasticity. Understanding avian responses to multiple environmental stressors, when one of those stressors is an environmental contaminant, is poorly understood, yet this situation is prevalent in many anthropogenically-impacted environments [6,7].

Mercury (Hg) contamination and subsequent bioaccumulation in waterbirds is problematic in wetlands throughout the world. Inorganic Hg is converted into the toxic and bioaccumulative form methylmercury (MeHg) under biogeochemical conditions that are common in wetland ecosystems [6,7]. Once methylated, MeHg biomagnifies as it is transferred through the food chain [8–10]. Environmentally relevant Hg concentrations in birds have been associated with a suite of impaired physiological and reproductive responses, including altered reproductive hormone levels [11,12], compromised hypothalamic-pituitary-adrenal axis [13–15], and reduced reproductive performance [16].

The decline of several species of Everglades' wading bird between the 1930s and 2001 has been related to changes in prey availability and alterations to the ecosystem's hydrology [17–19] and possibly due to Hg exposure [20,21]. Great egrets (*Ardea alba*; hereafter egret) and white ibis (*Eudocimus albus*; hereafter ibis) are two of the most common breeding waterbirds in the Everglades, and they exhibit different foraging strategies that influence their exposure to ecological stressors. Ibis are more limited in their use of habitats and available prey than egrets [22]. Specifically, Ibis select high-quality foraging patches that they then abandon relatively quickly once prey availability drops, whereas egrets tend to remain at foraging sites even when prey densities are low [22]. Although both species select different foraging habitats [23] nest survival for both species is similarly influenced by hydrological conditions and prey availability [24]. Furthermore, Hg concentrations differ between egrets and ibises in the Florida Everglades [10,25–28], which is likely a function of their foraging ecology and prey selection [28].

Herein, we assessed the physiological response of juvenile egrets and ibis to changing prey availability, hydrology (water depth, recession rate), and mercury exposure. We measured physiological biomarkers and body condition for egret and ibis nestlings in two consecutive years that differed greatly in hydrologic conditions, prey availability, and Hg exposure. Additionally, we measured physiological biomarkers of environmental stressors that manifest across a range of different time frames. In particular, we used measures of physiological condition that we expected would represent a temporal continuum of responses relative to the age of wading bird chicks in our study, including a body-condition index (short-time frame; 1–2 days; [29]), fecal corticosterone metabolites [FCORT] (medium-time frame; 2–7 days; [30,31]), and heat shock proteins 60 and 70 [HSP60, HSP70] (long-time frame: weeks - >1 month; [28,32–34]). Corticosterone (CORT) is a hormone that serves as a signal to modify both behavior and metabolism during a period of acute stress. Corticosterone is released by the hypothalamo-pituitary-adrenal axis into the blood stream when birds experience acute stress, facilitating a rapid response to overcome the stressor [2,35]. Variation in CORT levels in waterbirds have been correlated with a variety of factors including prey availability [31,36], Hg exposure [12,15], and changing hydrological conditions [37]. In contrast, heat shock proteins are highly conserved molecular chaperones that function to maintain optimal cell function and homeostasis [38,39] by being amplified through up-regulation during periods of stress to minimize cell protein damage [40,41]. Elevated heat shock proteins are commonly associated with responses to damage associated with heavy metals including Hg [42,43], decreased food availability [31], and rapid changes in hydrological conditions at foraging sites for birds [37]. In order to better understand the potential conservation implications of exposure to multiple environmental stressors we simultaneously tested those effects on these physiological biomarkers and chick body condition.

Methods

Ethics Statement

We made every attempt to reduce disturbance, stress, and other impacts to target and non-target species during the course of this research. Florida Atlantic University Institutional Animal Care and Use Committee (Protocol A0534) approved this study, and we conducted the research under U.S. Geological Survey banding permit 23354 and Florida Fish and Wildlife Conservation Commission Scientific Research Permit WX04487.

Study Area

In 2006 and 2007, we monitored egret and ibis nestlings in Florida Everglades wading bird colonies at the Arthur R. Marshall Loxahatchee National Wildlife Refuge (Lox; $n = 8$ colonies) in Palm Beach County, Water Conservation Area (WCA) 2A ($n = 2$ colonies) in Broward County, and WCA 3A ($n = 5$ colonies) in Broward and Miami-Dade Counties, Florida, USA (Figure S1). Colonies were located by using a combination of ground and aerial surveys [24], by tracking radio-tagged adults [23], and by visiting known colony sites (e.g., South Florida Wading Bird Reports, South Florida Water Management District). Although we attempted to sample colonies in which both species nested, only one species nested in five of the twelve colonies.

Chick sampling

We surveyed nests within egret and ibis colonies that were associated with our radio-tagged adults ([23,24], Figure S1). We revisited nests and resampled chicks (n = 61 egret and 65 ibis nests) approximately every 7 days, and determined the hatching order and post-hatch age of chicks.

We used FCORT as an index of stress because this approach minimizes handling-induced bias that results during bird sampling. We collected fecal material as described in [37], and fecal samples were stored in labeled cryovial tubes and placed on ice. We recorded mass for both species to the nearest 5 g using a spring scale and recorded all other morphometric measurements to the nearest 1 mm. We sampled all nestlings during the morning to minimize heat stress and time of day effects on physiological metrics.

We sampled blood from chicks to measure both HSPs and Hg concentrations. From each chick, we collected up to 2 ml of blood from the brachial vein using a 27-gauge needle. Blood samples were stored on ice in heparinized cryovials until transport to the laboratory. Following The Ornithological Council guidelines [44], we limited the total blood volume of blood collection to <1% of bird mass and thus had limited blood volume available for determining physiological parameters and Hg concentrations. Therefore, we separated plasma and RBC fractions in the laboratory via centrifugation (15 min, 5,000 rpm) and determined HSPs and THg concentrations in RBCs. Each fraction was stored frozen at −20°C until analyses could be determined.

Laboratory Methods

Fecal samples were homogenized in the laboratory, divided into 2 equal 1-ml wet portions, and dried using a Labconco CentriVap Concentro (Labconco, Kansas City, MO). Dried samples (~0.25 g) were then mixed with 5 ml of 95% ethanol and vortexed for 30 min. After centrifugation (15 min, 2500 × g) the supernatant was transferred to a new vial, and evaporated under a stream of nitrogen gas. Corticosterone metabolites were then resuspended in diluted extraction buffer and measured using the Correlate-EIA Corticosterone Enzyme Immunoassay Kit (EIA [45];) following the manufacturer's instructions (Assay Design, Inc., Ann Arbor, MI). Inter- and intra-assay coefficients of variation for FCORT internal standards were 7% and 11%, and 8% and 9% respectively for egrets and ibises. We validated that FCORT levels did not change after freezing as in the case of mammals [46] by freezing and measuring FCORT levels monthly for 6 months [45].

A subset of RBCs was washed three times using phosphate buffered saline, centrifuged and the pellet was removed after the final wash. The RBC pellet was then mixed with 1× extraction reagent and a protease inhibitor cocktail (Sigma), vortexed for 5 min and then sonicated for 1 min. Samples were again centrifuged (15 min, 2500 g) and the supernatant removed. We measured HSP60 (HSPD1) and HSP70 (HSP72/HSPA1A) in the supernatant using EIA kits specific to just those stress proteins and not all other HSP60 and HSP70 family members EIA kits (Assay Designs, Inc., Ann Arbor, MI). Inter- and intra-assay coefficients of variation for HSP60 and HSP70 internal standards were 5% and 7% and 6% and 7% respectively for egrets and ibis. All samples were run in duplicate, and means of duplicates were used in all analyses. All EIA kits were validated using serial dilutions and spike tests to determine percent recovery [45].

Mercury Determination

We analyzed the remaining portion of RBCs for total mercury (THg) concentrations because the majority of Hg in blood is in the

RBC fraction [47,48] and RBC Hg concentrations are correlated with whole blood Hg concentrations [49,50]. Further, almost all (>95%) of the mercury in avian blood is in the methylmercury form and THg concentrations are highly correlated with methylmercury levels [51,52]. To minimize any potential variability in THg concentrations associated with processing and differential moisture loss during long-term storage of samples, RBC fractions were dried (48 hours at 50°C) to a constant weight before THg analysis. We measured up to 0.05 g of dried blood into nickel sample vessels (weighed to the nearest 0.00001 g; Mettler Toledo model XS105, Mettler Toledo, Columbus, Ohio, USA). Following U.S. Environmental Protection Agency method 7473 [53], we determined THg concentrations in each RBC sample on a Milestone DMA-80 direct mercury analyzer (Milestone, Monroe, Connecticut, USA). Analytical equipment was calibrated using certified standard solutions prior to analysis, and accuracy and precision were evaluated within each analytical batch through the inclusion of certified reference materials (either dogfish muscle tissue [DORM-3] or dogfish liver [DOLT-4] by the National Research Council of Canada, Ottawa, Canada), calibration verifications (liquid standards), duplicates, and blanks. Recoveries averaged 102.4±0.06% ($n = 17$) and 97.7±0.04% ($n = 33$) for certified reference materials and calibration checks, respectively. Absolute relative percent difference for all duplicates averaged 1.7%±0.08% ($n = 16$).

Hydrological Conditions

We used the Everglades Depth Estimation Network (EDEN [54];) to estimate water depths and water level recession rates within foraging ranges of adults associated with breeding colonies where chicks were sampled. The EDEN uses a system of water level gauges to produce a water surface model coupled with a ground elevation model to estimate water depth for the entire freshwater portion of the Greater Everglades [55]. The EDEN calculates water level depths in 400 m×400 m grid cells at daily time steps accounting for evapotranspiration, rainfall, and sheet flow. The estimated water depths have been found to be accurate to within 5 cm [55].

To estimate recession rates and water depths, we used protocols developed during previous studies [23,24,28]. Briefly we used fixed radius buffers (hereafter foraging range) associated with breeding colonies based on mean distances flown plus one standard error by radio tagged adult egrets and ibises in a concurrent study. We defined foraging range as the average distance that radio tagged adults flew from breeding colonies to a foraging site throughout the breeding season. We extracted the water depth from each EDEN grid cell within the foraging ranges for each specific colony, year, and species combination. For each day that we sampled chicks, we calculated water recession rate and water depth as the mean value for the preceding 2-week period [24,28]. Positive recession rates indicate decreasing water levels. Surface water could still be present when estimated cell depth <0 cm; however, the 5-cm error in EDEN depths suggests that a majority of the ground surface would be exposed in a given 400 m×400 m grid cell.

Statistical procedures

We used linear mixed-effects models in JMP using maximum likelihood estimation (JMP 2001 [56];) to test the effects of environmental stressors (water depth, recession rate, prey availability, and Hg) on the physiological condition (chick body condition index, FCORT, HSP70, and HSP60) of egret and ibis chicks. We included age of chicks in models to account for variation in prey selection behavior as chicks aged [57–59]. The hatch order of chicks within the nest was included to account for competition for food within nests [24,28]. Lastly, we included region because our previous research found differences in adult physiological condition among Everglades' regions [37]. We did not include a year effect because it was confounded with large differences in prey biomass observed between years. For egret chick models, we statistically nested the chick identification code within the nest identification code and added it as a random effect to the model to account for the non-independence of sampling chicks multiple times and different numbers of chicks within the same nest. For ibis chick models, we included the nest identification code as a random effect to account for the non-independence of sampling multiple chicks within a nest. We used the maximum likelihood estimation method in our linear models. Because we intentionally studied two species with contrasting foraging strategies, we ran separate models for each species. Natural log transformations where used to improve normality of residuals for prey biomass, THg, FCORT, HSP60, and HSP70 data. We assessed the level of multicollinearity using the variance inflation factor (VIF; 60); across all models the VIFs ranged from 1.04–2.41 suggesting no significant multicollinearity. We calculated a marginal and conditional coefficient of determination (R^2) for each model to examine goodness-of-fit in R [GLMM R^2; 61]. Briefly, for mixed effects models, the marginal R^2 ($R^2_{GLMM(m)}$) reports the goodness-of-fit for just the fixed effects, while the conditional R^2 ($R^2_{GLMM(c)}$) incorporates both the random and fixed effects as a measure of explained variation [61].

In the case of chick mass, we did not have repeated measurements of all chicks across time and subsequently could not calculate growth rates. Instead, we calculated a chick body condition index that allowed us to examine the potential effects of multiple landscape stressors on a relative measure of the condition of the chick that was independent of the mass of a chick [29,62]. To do so, we first we calculated a structural body size index using principal components analysis on the morphometric measurements (e.g. tarsus, bill length, wing length) for each species separately. We then regressed chick mass on tarsus length (the variable our principal components analysis identified as the best predictor of structural size for both egrets and ibis), including the first principal component as a covariate. We then used the residuals expressed as a percentage of the predicted mass, to obtain our chick body condition index [29,62]. Benson *et al.* [29] found that measurements taken during a single visit to a colony were useful to detect variation in chick body condition across time.

Results

During the 2006 and 2007 nesting seasons we collected blood samples from 254 chicks (169 egret; 2006 = 27, 2007 = 142, and 85 ibis; 2006 = 30, 2007 = 55). The geometric mean across all dates, regions, and hatch order was 6.99±0.63 (SE) in egrets and 4.57±0.83 in ibises for FCORT concentration (μg/g), 13.73±0.96 in egrets and 14.73±2.06 in ibis for HSP60 concentration (ng/ml), 4.44±0.44 in egrets and 2.53±0.35 in ibis for HSP70 concentration (ng/ml), and 4.13±0.16 in egrets and 0.58±0.07 in ibis for THg concentration (μg/g dw).

Chick body condition index

Egret chick body condition was influenced by water depth ($F_{1,\ 128.8} = 6.60$, $P = 0.01$), but not region ($F_{2,\ 114.7} = 2.30$, $P = 0.10$), water recession rate ($F_{1,\ 119.2} = 2.28$, $P = 0.14$), prey biomass ($F_{1,\ 142.7} = 0.38$, $P = 0.53$), age ($F_{1,\ 52.4} = 0.43$, $P = 0.51$), hatch order ($F_{1,\ 79.9} = 1.95$, $P = 0.17$), nor Hg concentration ($F_{1,\ 150.8} = 0.02$, $P = 0.88$). The low marginal R^2 value of our egret chick body condition model ($R^2_{GLMM(m)} = 0.10$, $R^2_{GLMM(c)} = 0.21$) suggests that the model was a relatively poor predictor of body condition. However, egret chick body condition increased with increasing water depth ($\beta = 0.20 \pm 0.13$–95% confidence interval) on average by 23% across the range of water depths observed (range = −46 to 17 cm; Figure 1A).

Conversely, ibis chick body condition was influenced by both region ($F_{1,\ 37.6} = 9.09$, $P = 0.004$) and age ($F_{1,\ 75} = 5.26$, $P = 0.02$), but not by water depth ($F_{1,\ 68.8} = 0.0001$, $P = 0.99$), water recession rate ($F_{1,\ 72.5} = 0.005$, $P = 0.98$), prey biomass ($F_{1,\ 55.6} = 0.06$, $P = 0.80$), hatch order ($F_{1,\ 38.6} = 33$, $P = 0.56$), nor Hg concentration ($F_{1,\ 75} = 0.10$, $P = 0.75$). The ibis chick body condition model marginal R^2 value ($R^2_{GLMM(m)} = 0.23$), suggested that the model was a modest predictor of ibis chick body condition. Similarity between the marginal R^2 value and the conditional R^2 ($R^2_{GLMM(c)} = 0.23$) suggested that little additional variance was explained by the random effect associated with sampling multiple chicks within nests. Ibis chick body condition was 1.9 times higher in Lox (mean = 5.36±2.68 SE) than WCA3A Lox (−10.02±4.31 SE; $\beta = 7.69 \pm 4.99$) and decreased by 26% across the range of ages ($\beta = -0.64 \pm 0.54$, range = 3–30 days; Figure 1B).

Fecal corticosterone metabolites

Egret FCORT metabolite concentrations were influenced by age ($F_{1,\ 13.5} = 13.50$, $P < 0.001$), water depth ($F_{1,\ 7.5} = 7.50$, $P = 0.007$), and region ($F_{1,\ 108.6} = 3.06$, $P = 0.007$), but not by water recession rate ($F_{1,\ 106.8} = 0.01$, $P = 0.97$), prey biomass ($F_{1,\ 124.9} = 2.82$, $P = 0.09$), hatch order ($F_{1,\ 80.5} = 0.78$, $P = 0.38$), nor Hg concentration ($F_{1,\ 133.6} = 1.65$, $P = 0.20$). The similarity between the marginal and conditional R^2 values ($R^2_{GLMM(m)} = 0.34$, $R^2_{GLMM(c)} = 0.37$) suggests that the fixed effects alone were considerably more influential than the random effects associated with sampling multiple chicks within the same nest on more than one occasion. Egret FCORT metabolite concentrations decreased with increasing age ($\beta = -0.04 \pm 0.02$) and water depth ($\beta = -0.02 \pm 0.01$) on average by 79% across the range of ages (range = 3–42; Figure 2A) and 72% across the range of water depths (range = −46 to 17 cm; Figure 2B). Egret FCORT metabolite concentrations were higher in WCA3A (11.74±2.45 SE) than either WCA2A ($t_{118.3} = 2.09$, $P = 0.04$; mean = 4.71 ±1.60 SE) or Lox ($t_{102.3} = 2.25$, $P = 0.03$; mean = 6.34±0.76 SE) respectively, but similar between WCA2A and Lox ($t_{114.5} = -0.83$, $P = 0.40$).

Similar to egrets, region influenced ibis FCORT metabolite concentrations ($F_{1,\ 54.8} = 4.40$, $P = 0.04$), however unlike egrets, prey biomass also influenced ibis FCORT metabolites ($F_{1,\ 58.4} = 19.41$, $P < 0.001$). Ibis FCORT metabolites were not influenced by water depth ($F_{1,\ 59.6} = 0.65$, $P = 0.42$), water recession rate ($F_{1,\ 60.4} = 0.36$, $P = 0.55$), age ($F_{1,\ 60.8} = 0.25$, $P = 0.62$), hatch order ($F_{1,\ 46.1} = 1.43$, $P = 0.23$), nor Hg concentration ($F_{1,\ 59.1} = 0.001$, $P = 0.96$). The ibis FCORT model had a moderate marginal R^2 value ($R^2_{GLMM(m)} = 0.36$), but the conditional R^2 value ($R^2_{GLMM(c)} = 0.71$) suggested that a similar amount of the variability in ibis FCORT concentrations were associated with just the random effect of sampling multiple chicks within the same nest. FCORT metabolite concentrations were 3.3 times higher in Lox (mean = 5.26±1.68 SE) than WCA3A (1.60±0.70 SE; B = 0.59±0.54), and decreased by 98% (B = −1.40±0.62), across the range of prey biomass (range = 1.8–26.7 g/m^2; Figure 2C).

Heat shock protein 70

Egret HSP70 concentrations were influenced by Hg concentration ($F_{1,\ 109.9} = 4.09$, $P = 0.04$), but not region ($F_{2,\ 100.3} = 1.56$, $P = 0.21$), water depth (F_1, $101.2 = 0.16$, $P = 0.68$), water recession rate ($F_{1,\ 91.0} = 0.37$, $P = 0.54$), prey biomass ($F_{1,\ 99.36} = 0.23$, $P = 0.62$), age ($F_{1,\ 110.8} = 0.02$, $P = 0.87$), nor hatch order ($F_{1,\ 76.6} = 1.63$, $P = 0.20$). The overall amount of variation in egret HSP70 explained by the fixed effects was relatively low ($R^2_{GLMM(m)} = 0.13$), however very little of that variation was explained by our random effects ($R^2_{GLMM(c)} = 0.16$) associated sampling multiple chicks from the same nest or sampling those chicks on several occasions. The beta coefficient estimate for Hg ($\beta = -0.48 \pm 0.22$) indicated a decrease in HSP70 concentrations of 83% across the range Hg values (range = 0.26–11.01 μg/g dw; Figure 3).

In the case of ibis, neither water depth ($F_{1,\ 30.9} = 0.12$, $P = 0.72$), region ($F_{1,\ 28.6} = 0.11$, $P = 0.73$), water recession rate ($F_{1,\ 32.9} = 0.0009$, $P = 0.97$), prey biomass ($F_{1,\ 32.2} = 0.12$, $P = 0.72$), age ($F_{1,\ 31.4} = 0.49$, $P = 0.48$), hatch order ($F_{1,\ 29.5} = 0.02$, $P = 0.87$), nor Hg concentration ($F_{1,\ 13.9} = 0.0001$, $P = 0.99$) influenced HSP70 concentrations. Correspondingly, the fixed effects ($R^2_{GLMM(m)} = 0.06$) explained very little of the variation in HSP70, however the conditional effects ($R^2_{GLMM(c)} = 0.74$) associated with sampling chicks from the same nest played a considerably larger role in ibis HSP70 concentrations.

Heat shock protein 60

Egret chick HSP60 concentrations were not influenced by water depth ($F_{1,\ 118.9} = 0.0002$, $P = 0.98$), region ($F_{2,\ 111.4} = 0.01$, $P = 0.98$), water recession rate ($F_{1,\ 119.7} = 0.27$, $P = 0.60$), prey biomass ($F_{1,\ 86.6} = 0.91$, $P = 0.34$), age ($F_{1,\ 134.5} = 1.81$, $P = 0.18$), hatch order ($F_{1,\ 81.7} = 0.05$, $P = 0.81$), nor Hg concentration ($F_{1,\ 120.2} = 0.86$, $P = 0.34$). The low R^2 values associated with both the fixed and random effects suggested that variation in egret HSP60 results from factors other than those measured in this study ($R^2_{GLMM(m)} = 0.04$, $R^2_{GLMM(c)} = 0.04$).

In the case of Ibis HSP60 concentrations were influenced by age ($F_{1,\ 43.4} = 4.32$, $P = 0.04$), but not prey biomass ($F_{1,\ 45.5} = 0.89$, $P = 0.35$), water depth ($F_{1,\ 42.0} = 0.11$, $P = 0.74$), water recession rate ($F_{1,\ 46.0} = 0.04$, $P = 0.83$), hatch order ($F_{1,\ 31.6} = 0.02$, $P = 0.87$), nor Hg concentration ($F_{1,\ 46.6} = 1.61$, $P = 0.21$). The overall amount of variation in ibis HSP60 explained by the fixed effects was relatively low ($R^2_{GLMM(m)} = 0.15$), but the random effect ($R^2_{GLMM(c)} = 0.89$) associated with sampling multiple chicks from the same nest explained considerably more of the variation in ibis HSP60 concentrations. The beta coefficient estimates for age ($\beta = -0.05 \pm 0.04$) indicated HSP concentrations declined by 74% across the range of ages (range = 3–30 days; Figure 4).

Discussion

Different environmental stressors were associated with different time frames and biomarkers of physiological condition in egret and ibis chicks. The short and moderate time frame markers, chick body condition and FCORT, respectively, were correlated with landscape variables such as prey biomass and water depth, whereas there was no measurable effect of Hg. In contrast, the long-term physiological markers were not influenced by any of the measured landscape stressors, but one (HSP70 in egrets) was influenced by Hg concentration. One potential confounding aspect in our study is the fact that the same hydrology variables used to define the landscape stressors that we tested to determine their influence on physiological condition are also know to influence concentrations of Hg in egret and ibis adults and chicks [28]. Previously, Herring *et al.* [28] found that water depths and recession rates, while not the primary variables that influenced Hg concentrations, did play a secondary role in influencing Hg concentrations in egrets and ibis. Disentangling those effects is

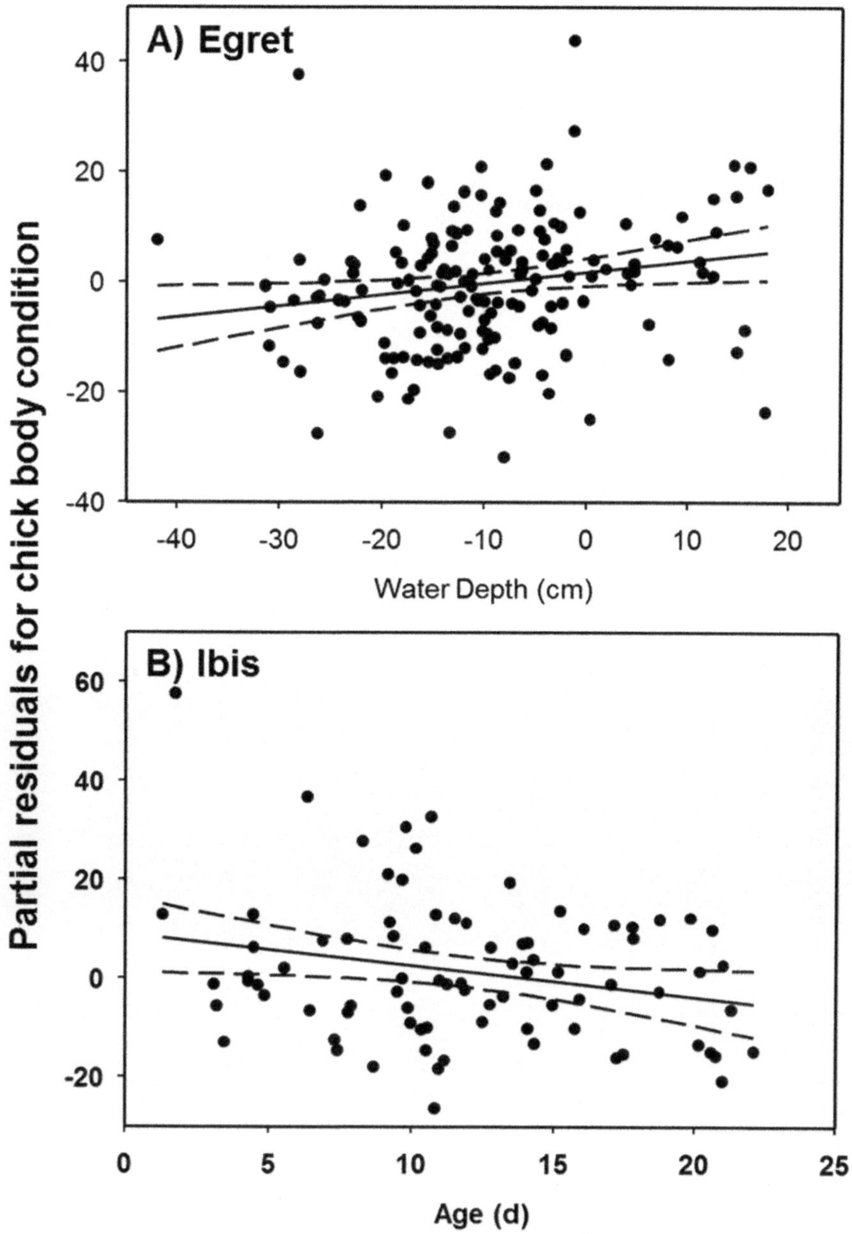

Figure 1. Relationship between partial residuals of chick body condition index and water depth (A; cm/^{2}) for great egret chicks and age (B; days) for white ibis chicks in the Florida Everglades, after accounting for other model variables. Dashed lines indicated 95% confidence.

beyond the scope of this study, but illustrates the complexity of examining different environmental stressors on the physiological condition of birds.

Egret HSP70 decreased by 83% across the range (low-high) of Hg concentrations, suggesting that Hg may have down-regulated these heat shock proteins. Previous research on heat shock protein regulation in relation to both inorganic Hg and MeHg have found that heat shock protein regulation can be either inhibited (down-regulated) or induced (up-regulated) depending on the chemical form of Hg [63]. Up-regulation of heat shock proteins are associated with protective stress responses to minimize cell protein damage [40,41,64]. Methylmercury exposure in birds is associated with oxidative stress in liver, kidney, and brain tissue [65], and up-regulation of heat shock proteins has recently been identified as a potential biological defense mechanism against MeHg toxicity [66,67]. However, it is unclear what the consequences are for down-regulation of heat shock proteins as we observed with egrets. Presumably, a reduction of heat shock proteins could result in decreased cell function and homeostatic imbalance, given that this is the primary role heat shock proteins when an individual is not in a stressful situation [38,39]. Opposite response patterns of other stress biomarkers in relation to Hg exposure have been documented in other birds (e.g., [15]). Herring *et al.* [15] found that Hg exposure in Forster's tern (*Sterna forsteri*) chicks depressed FCORT concentrations by up to 81% via the downregulation of the HPA axis. These results suggest that the physiological stress response systems of birds are dynamic and that Hg exposure can negatively influence critical physiological processes.

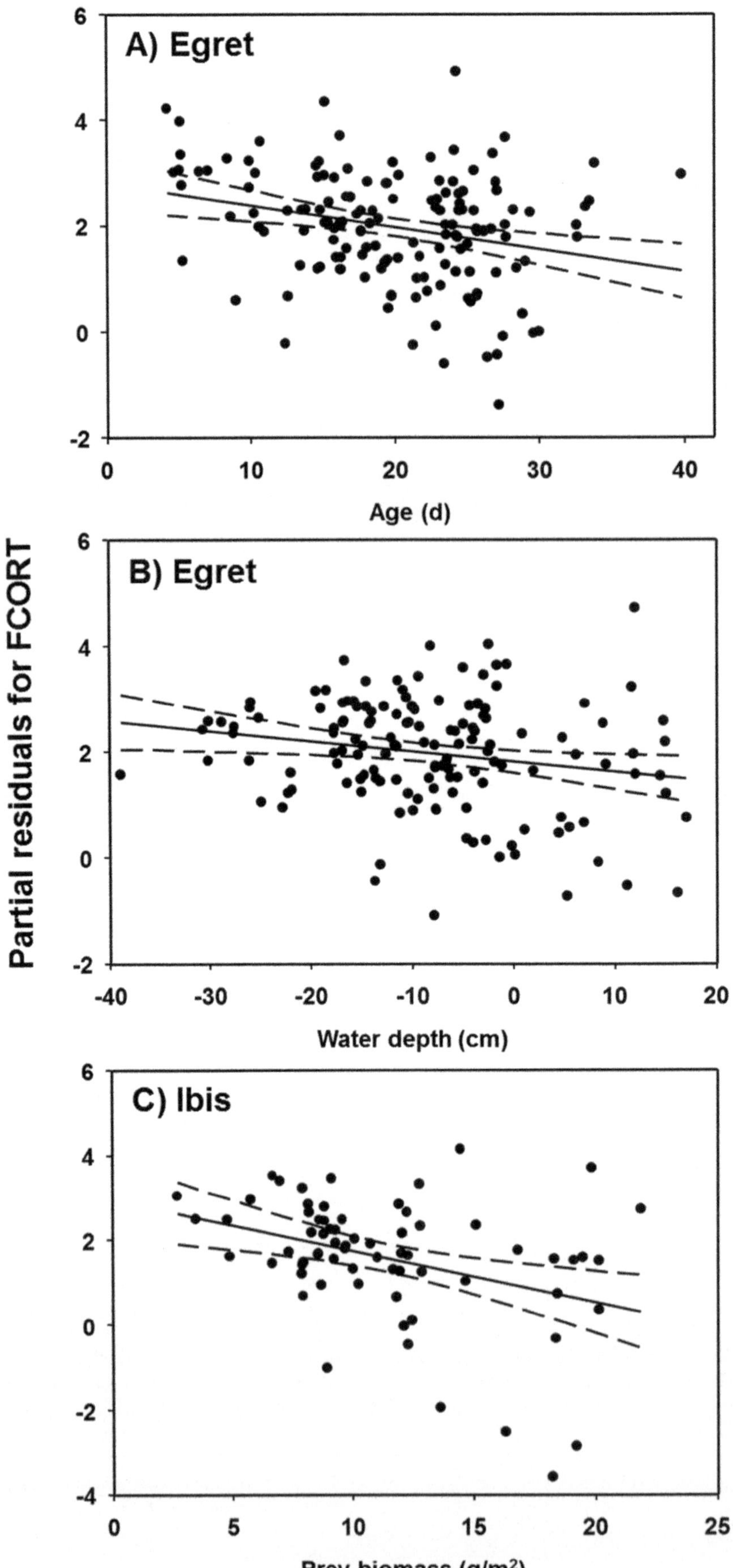
A) Egret
Age (d)
B) Egret
Water depth (cm)
C) Ibis
Prey biomass (g/m2)
Partial residuals for FCORT

Figure 2. Relationship between partial residuals of fecal corticosterone metabolite (μg/g dw) and chick age (A; days) and water depth (B; cm) for great egret chicks and prey biomass (C; g/m^2) for white ibis in the Florida Everglades, after accounting for other model variables. Dashed lines indicated 95% confidence interval.

Current Hg exposure levels of egrets and ibises throughout much of the Florida Everglades are low relative to historical reports [28,68]. However, the fact that for egrets, one of their HSPs was influenced by Hg, demonstrates that the current Hg levels could still influence the physiology of this species. Moreover, Hg hotspots still exist within the Florida Everglades (e.g., interior portions of Everglades National Park not sampled in this study) where Hg levels in fish far exceed levels measured anywhere else within the ecosystem, and still pose a risk to wading birds [10,21]. Ibis heat shock proteins were not influenced by Hg exposure in this study, which was likely a function of Hg exposure levels being approximately 7 times lower than those of egrets.

We found that ibis but not egret FCORT levels were influenced by prey biomass levels, and that they declined by 98% across low-to-high range of prey biomass at foraging sites. This is consistent with responses in pre-breeding adult birds [37], suggesting that this relationship may be consistent across life stages. Similarly, ibis nest success was also more sensitive to changes in prey biomass than that of egrets [24], demonstrating the importance of prey biomass to ibis. Egret FCORT metabolite concentrations also declined by 79% across the age range (young to old) of chicks. Other studies of CORT stress levels in chicks have found that concentrations often increase as chicks get older, in preparation for fledging [3,69,70]. However, there is some evidence that elevated CORT levels can result in decreased protein deposition in growing feathers [71], which may lead to poorer feather quality [72]. Increased baseline CORT levels can also impair growth rates and immune function [73]. Alternatively egret CORT levels may be higher in younger chicks to facilitate improved begging to adults [74,75] and promote growth at a critical stage in the chicks' development since CORT allows the organism to mobilize needed energy reserves [76].

We found support for the influence of hydrological variables on egret physiology, in particular egret chick body condition increased with increasing water depth. Critical to understanding these responses is the fact that the range of water depths associated with our study did not exceed the observed foraging range and optimal depths for egrets [23] and generally would be considered to be shallow for egret foraging [22]. As such, the increase in chick body condition is likely driven by the fact that egrets prefer larger prey items [57] found in deeper depths relative to those very shallow sites where large prey die rapidly. We expect that body condition would have dropped had depths increased beyond the normal foraging range of egrets.

Neither water depths nor recession rates influenced ibis chick physiological condition in the current study. Although recession rates were found to influence the pre-breeding physiological condition of ibis previously [37], recession rates during the nestling

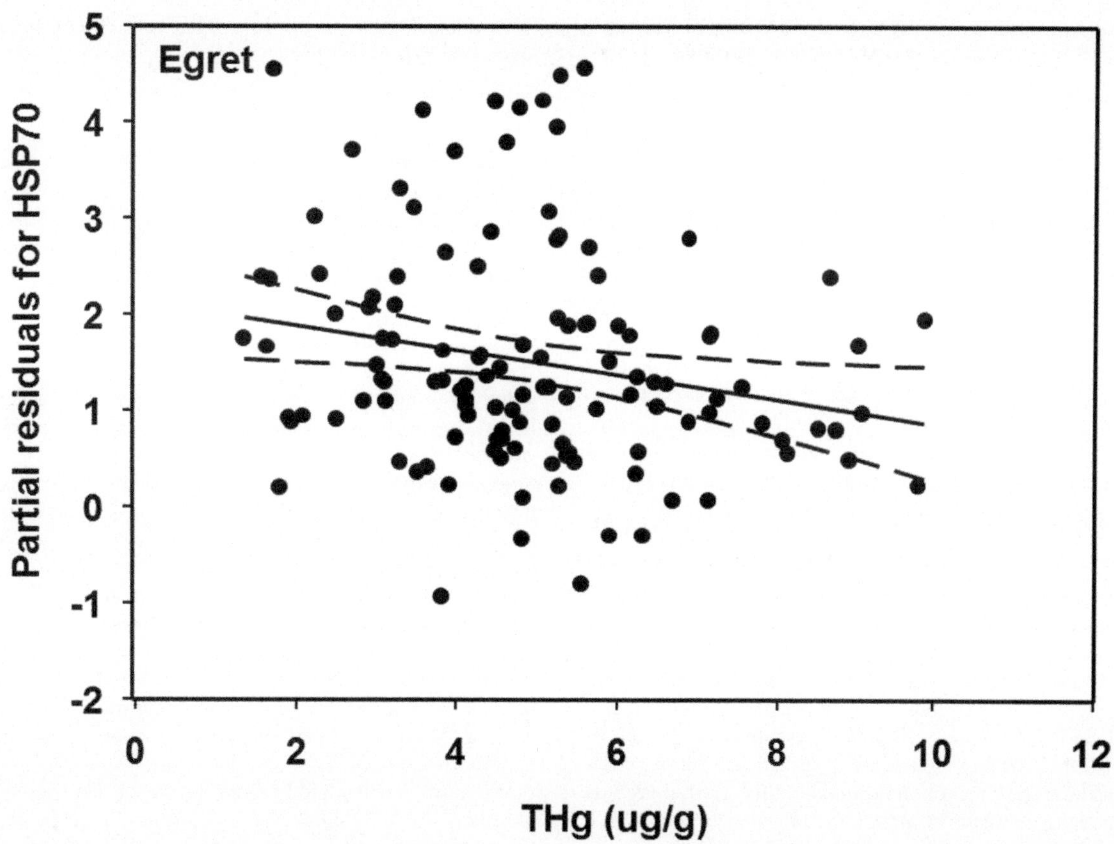

Figure 3. Partial residuals of HSP70 concentrations (ng/ml) and red blood cell THg (μg/g dw) for great egret chicks in the Florida Everglades, after accounting for other model variables. Dashed lines indicated 95% confidence interval.

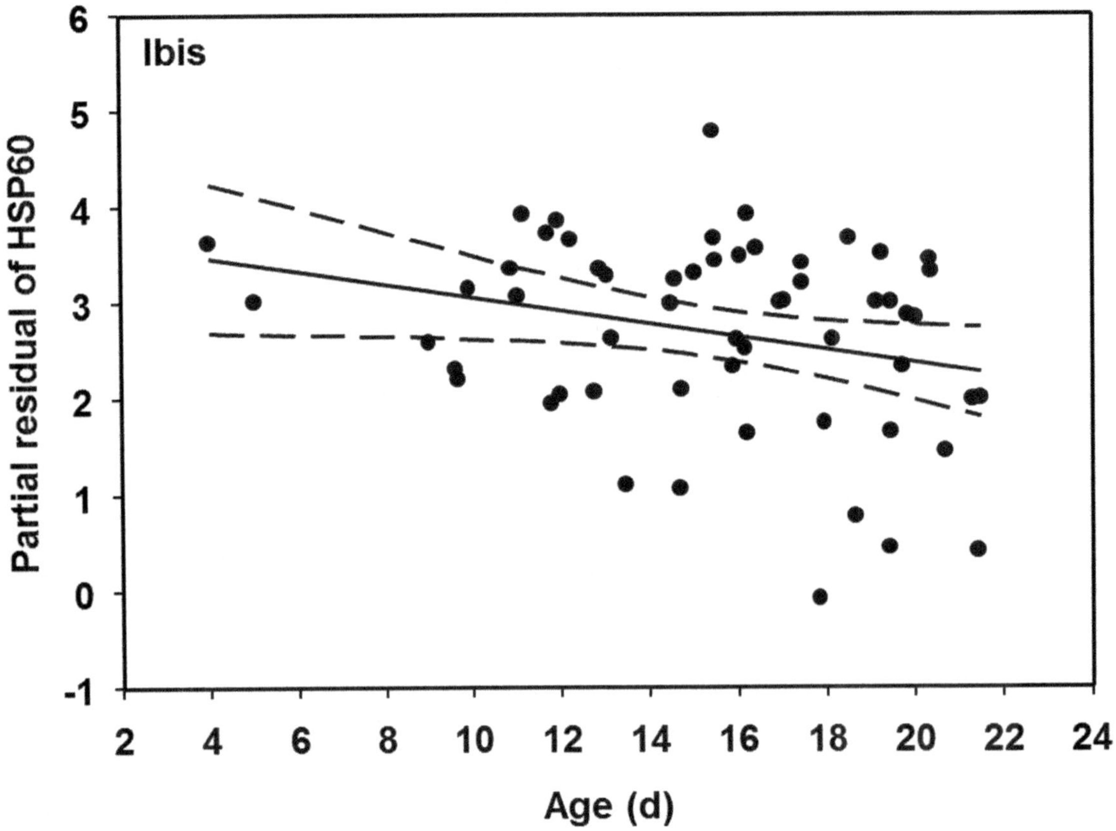

Figure 4. Relationship between partial residuals of HSP60 (ng/ml) and chick age (days) for white ibis chicks in the Florida Everglades, after accounting for other model variables. Dashed lines indicated 95% confidence interval.

stage may play a smaller role in chick physiological condition because water depths have largely reached optimal foraging depths for adults around breeding colonies [22,23]. Thus, small variation in water depth and recession rate likely have little influence on accessibility of foraging sites or the availability of prey for ibis. However, if rainfall events occur, rapid negative recession rates that result in increased water depths may result in the loss of foraging site access and reduced prey availability. Although these conditions did not occur in the present study, they can have a substantial effect on nestlings [77] and would be expected to decrease chick physiological condition. Similarly, water depths were largely within the foraging range and were near optimal within the proximity of breeding colonies during this study. In fact, water depths associated with ibises were less variable than those observed in egrets (coefficient of variation = 20% vs 56% respectively). Under conditions with deeper or more variable water depths, we expect that ibises may have an increased physiological response due to their dependence on high quality (shallow water with high prey densities) foraging sites.

We also found that chick age was an influential variable in explaining physiological responses, including ibis body condition and HSP60, and egret FCORT. The body condition of ibis chicks declined 26% from the youngest to the oldest chicks. Although it is unclear why chick body condition declined as they aged, there was not a concomitant increase in stress biomarkers such as FCORT or HSPs. In fact, HSP60 levels actually declined by 74% across the range of ibis chick ages. One previous study found that chickens (*Gallus gallus domesticus*) that were experimentally fed lower energy diets, resulting in lower body masses, also had lower levels of HSPs relative to those with higher energy diets and higher masses [78]. Ibis body condition may have declined as the chicks prepared to fledge, a period where many bird species reduce their body mass/condition [79,80]. It is believed that this reduction in overall mass allows birds to change from plump immobile chicks to slender and proficient flying juveniles [80–82].

Conservation and Management Implications

There were two primary findings of this study that have the potential to benefit conservation of wading bird species. First, we found support for chick physiology to be influenced by hydrology variables such as water depths. Water managers have the potential to regulate the hydrology throughout much of the Everglades through a system of canals, levees, and pumps, theoretically allowing them to adjust water depths for wading birds. This hydrology variable has also been previously linked to pre-breeding adult physiological condition [37], nesting success [24], and foraging site selection of egrets and ibises [23]. Optimal water depths for foraging wading birds differ by species [22,23], foraging strategies (searchers and exploiters), and years depending on prey biomass conditions; however maintaining water depths between −22 and 21 cm has been suggested to provide suitable foraging conditions for both species and foraging groups [23].

We also found that current levels of Hg exposure in the Everglades may influence the expression of heat shock proteins in egrets. Understanding what the consequences are for these changes in heat shock protein concentrations may provide insight as to how Hg may have impacted Everglades' wading bird populations in the past. This understanding may be relevant to future patterns of wading bird nesting because there are portions of the Everglades that still contain high levels of Hg in prey species of wading birds [10].

Supporting Information

Figure S1 Location of study area indicating wading bird nesting colonies sampled during 2006 and 2007. The primary Water Management Areas (WCAs), Arthur R. Marshall Loxahatchee National Wildlife Refuge (Lox), and Everglades National Park (ENP) are indicated.

Table S1 Data used in analyses of the Physiological condition of juvenile wading birds in relation to multiple landscape stressors in the Florida Everglades: effects of hydrology, prey availability, and mercury bioaccumulation.

Acknowledgments

Disclaimer: Any use of trade, product, or firm names is for descriptive purposes only and does not imply endorsement by the U.S. Government. We appreciate the support and cooperation of the Arthur R. Marshall Loxahatchee National Wildlife Refuge staff. We thank our field research crews and fellow researchers that assisted in collection of field data and laboratory analyses: T. Anderson, N. Baker, T. Beck, B. Botson, K. Donner, H. Herring, N. Hill, A. Horton, B. Imdieke, B. Johnson, S. Lantz, and J. Pierce.

Author Contributions

Conceived and designed the experiments: GH CES DEG JTA. Performed the experiments: GH JEB. Analyzed the data: GH. Contributed reagents/materials/analysis tools: GH DEG CES. Contributed to the writing of the manuscript: GH CES DEG JEB JTA.

References

1. Wingfield JC (1988) Changes in reproductive function of free-living birds in direct response to environmental perturbations. In: Stetson, J. H. (ed) Processing of environmental information in vertebrates. Berlin; Springer-Verlag: 121–148.
2. Astheimer LB, Buttemer WA, Wingfield JC (1992) Interactions of corticosterone with feeding, activity and metabolism in passerine birds. Ornis Scandinavia 23: 355–365.
3. Kitaysky AS, Kitaiskaia EV, Piatt JF, Wingfield JC (2003) Benefits and costs of increased levels of corticosterone in seabird chicks. Hormone Behav 43: 140–149.
4. Hahn TP (1998) Reproductive seasonality in an opportunistic breeder, the Red Crossbill, *Loxia curvirostra*. Ecology 79: 2365–2375.
5. Kacelnik A (1984) Central place foraging in starlings (*Sturnus vulgaris*) I: Patch residence time. J Anim Ecol 53: 283–299.
6. Marvin-DiPasquale MC, Agee JL, Bouse RM, Jaffe BE (2003) Microbial cycling of mercury in contaminated pelagic and wetland sediments of San Pablo Bay, California. Environ Geol 43: 260–267.
7. Hall BD, Aiken GR, Krabbenhoft D, Marvin-DiPasquale M, Swarzenski CM (2004) Wetlands as principal zones of methylmercury production in southern Louisiana and the Gulf of Mexico region. Environ Pollut 154: 124–134.
8. Sundlof SF, Spalding MG, Wentworth JD, Steible CK (1994) Mercury in livers of wading birds (Ciconiiformes) in southern Florida. Arch Environ Contam Toxicol 27: 299–305.
9. Frederick PC, Spalding MG, Dusek K (2002) Wading birds as bioindicators of mercury contamination in Florida: annual and geographic variation. Environ Toxicol Chem 21: 163–167.
10. Rumbold DG, Lange TR, Axelrad DM, Atkeson TD (2008) Ecological risk of methylmercury in Everglades National Park, Florida, USA. Ecotoxicology 17: 632–641.
11. Adams EM, Frederick PC, Larkin ILV, Guillette LJ (2009) The effects of chronic methylmercury exposure upon fecal testostosterone, estradiol, and corticosterone levels in juvenile White Ibises (*Eudocimus albus*). Environ Toxicol Chem 28: 982–989.
12. Jayasena N, Frederick PC, Larkin ILV (2011) Endocrine disruption in white ibises (*Eudocimus albus*) caused by exposure to environmentally relevant levels of methylmercury. Aquat Toxicol 105: 321–327.
13. Franceschini MD, Lane OP, Evers DC, Reed JM, Hoskins B, et al. (2009) The corticosterone stress response and mercury contamination in free-living tree swallows, *Tachycineta bicolor*. Ecotoxicology 18: 514–521.
14. Wada HD, Cristol A, McNabb FMA, Hopkins WA (2009) Suppressed adenocortical responses and thyroid hormone levels in birds near a mercury-contaminated river. Environ Sci Technol 43: 6031–6038.
15. Herring G, Ackerman JT, Herzog M (2012) Mercury exposure suppresses baseline stress levels in juvenile birds. Environ Sci Technol 46: 6339–6346.
16. Jackson AK, Evers DC, Etterson MA, Condon AM, Folsom SB, et al. (2011) Mercury exposure affects the reproductive success of a free-living terrestrial songbird, the Carolina wren (*ThryothorusLudovicianus*). Auk 128: 759–769.
17. Kushlan JA (1986) Response of wading birds to seasonally fluctuating water levels: strategies and their limits. Waterbirds 9: 155–162.
18. Frederick PC, Spalding MG (1994) Factors affecting reproductive success of wading birds (Ciconiiformes) in the Everglades ecosystem. Pages 659–692 *in* Everglades: The Ecosystem and Its Restoration (Davis SM, Ogden JC, Eds.). St. Lucie Press, Delray Beach, FL.
19. Ogden JC (1994) A comparison of wading bird nesting colony dynamics (1931–1946 and 1974–1989) as an indication of ecosystem conditions in the southern Everglades. Pages 533–570 *in* Everglades: The Ecosystem and Its Restoration (Davis SM, Ogden JC, Eds.). St. Lucie Press, Delray Beach, FL.
20. Heath JA, Frederick PC (2005) Relationships among mercury concentrations, hormones, and nesting effort of White Ibises in the Florida Everglades. Auk 122: 255–267.
21. Frederick P, Jayasena N (2010) Altered pairing behavior and reproductive success in white ibises exposed to environmentally relevant concentrations of methylmercury. Proc Royal Soc Lond B 278: 1851–1857.
22. Gawlik DE (2002) The effects of prey availability on the numerical response of wading birds. Ecol Monog 72: 329–346.
23. Beerens JM, Gawlik DE, Herring G, Cook MI (2011) Dynamic habitat selection by two wading bird species with divergent foraging strategies in a seasonally fluctuating wetland. Auk 128: 651–661.
24. Herring G, Gawlik DE, Cook MI, Beerens JM (2010) Sensitivity of Great Egret and White Ibis nesting to reduced prey density. Auk 127: 660–670.
25. Frederick PC, Spalding MG, Sepulveda MS, Williams Jr GE, Nico L, et al. (1998) Exposure of great egret nestlings to mercury through diet in the Everglades of Florida. Environ Toxicol Chem 18: 1940–7.
26. Rumbold DG, Niemczyk SL, Flink LE, Chandrasekhar T, Harkanson B, et al. (2001) Mercury in eggs and feathers of great egrets (*Ardea alba*) from the Florida Everglades. Arch Environ Contam Toxicol 41: 501–507.
27. Herring G, Gawlik DE, Rumbold DA (2009) Feather mercury concentrations and physiological condition of Great Egret and White Ibis nestlings in the Florida Everglades. Science of the Total Environment 407: 2641–2649
28. Herring G, Eagles-Smith C, Ackerman J, Gawlik D, Beerens J (2013) Landscape factors and hydrology influence mercury concentrations in wading birds breeding the Florida Everglades, USA. Sci Total Environ 460: 637–646.
29. Benson J, Suryan RM, Piatt JF (2003) Assessing chick growth from a single colony visit to a seabird colony. Marine Ornithol 31: 181–184.
30. Wasser SK, Hunt KE, Brown JL, Cooper K, Crockett CM (2000) A generalized fecal glucocorticoid assay for use in a diverse array of nondomestic mammalian and avian species. Gen Comp Endocrinol 120: 260–275.
31. Herring G, Cook MI, Gawlik DE, Call EM (2011) Food availability is expressed through physiological stress indicators in nestling white ibis: a food supplementation experiment. Func Ecol 25: 682–690.
32. Sørensen JG, Kristensen TN, Loeschke V (2003) The evolutionary and ecological role of heat shock proteins. Ecol Letters 6: 1025–1037.
33. Tomás G, Martínez J, Merino S (2004) Collection and analysis of blood samples to detect stress proteins in wild birds. J Field Ornith 75: 281–278.
34. Herring G, Gawlik DE (2007) The role of stress proteins in the study of allostatic overload in birds, use and applicability to current studies in avian ecology. Sci World J 7: 1596–1602.
35. Wingfield JC (1994) Modulation of the adrenocortical response to stress in birds. *In* Davey KG, Peter RE, Tobe SS (eds). Perspectives in Comparative Endocrinology National Research Council of Canada, Ottawa, Ontario, Canada.
36. Kitaysky AS, Piatt JF, Wingfield JC, Romano M (1999) The adrenocortical stress response of black-legged kittiwake chicks in relation to dietary restrictions. J Comp Phys B 69: 303–310.
37. Herring G, Gawlik DE (2012) Differential physiological response to landscape prey availability by the great egret and white ibis. J Wild Manage 77: 58–67.
38. Linquist S (1986) The heat shock response. Ann Rev Biochem 55: 1151–1191.
39. Bierkens JGEA (2000) Applications and pitfalls of stress-proteins in biomonitoring. Toxicology 153: 61–72
40. Locke M, Noble EG (1995) Stress proteins: the exercise response. Can J Appl Phys 20: 155–167.
41. Råberg L, Grahn M, Hasselquist D, Svensson E (1998) On the adaptive significance of stress-induced immunosupression. Proc Royal Soc Lon B 265: 1637–1641.

42. Papaconstantinou AD, Brown KM, Noren BT, McAlister T, Fisher BR, et al. (2003) Mercury, cadmium, and arsenite enhance heat shock protein synthesis in chicken embryos prior to embryotoxicity. Birth Defects Res B 68: 456–464.
43. Liu YR, Zheng YM, Zhang LM, Luan YX, He JZ (2010) Effects of mercury on reproduction, avoidance, and heat shock protein gene expression of the soil springtail *Folsomia candida*. Environ Toxicol Chem 29: 654–659.
44. Gaunt AS, Oring LW (1997) Guidelines for the Use of Wild Birds in Research. Washington: Ornithological Council Special Publication.
45. Herring G, Gawlik DE (2009) Stability of avian fecal corticosterone metabolite levels in frozen avian feces. J Wild Manage 73: 1010–1013.
46. Khan MZ, Altmann I, Isani SS, Yu J (2002) A matter of time: evaluating the storage of fecal samples for steroid analysis. Gen Comp Endocrinol 128: 57–64.
47. Myers GJ, Davidson PW (1998) Prenatal methylmercury exposure and children: neurological development, and behavioral research. Environ Health Perspect 106: 841–847.
48. Day RD, Segars AL, Arendt MD, Lee AM, Peden-Adams MM (2007) Relationship of blood mercury levels to health parameters in the loggerhead sea turtle (*Caretta caretta*). Environ Health Perspect 115: 1421–1428.
49. Chen JW, Chang CC, Wang SL (2011) Mercury concentration of whole blood and red blood cell in Taiwan residents. Epidemiology 22: S289.
50. Lieske CL, Moses SK, Castellini JM, Kleka J, Hueffer K, et al. (2011) Toxicokinetics of mercury in blood compartments and hair of fish-fed sled dogs. Acta Vet Scanda 2011: 53: 66.
51. Thompson DR, Furness RW (1989) Comparison of the levels of total and organic mercury in seabird feathers. Mar Pollut Bull 20: 577–579.
52. Fournier F, Karasov WH, Kenow KP, Meyer MW, Hines RK (2002) The oral bioavailability and toxicokinetics of methylmercury in common loons (*Gavia immer*) chicks. Comp Biochem Physiol A 133: 703–714.
53. U.S. Environmental Protection Agency (2000) Method 7473, Mercury in solids and solutions by thermal decomposition, amalgamation, and atomic absorption spectrophotometry, test methods for evaluating solid waste, physical/chemical methods SW 846, Update IVA. U.S. Government Printing Office, Washington, DC.
54. U. S. Geological Survey (2006) The Everglades Depth Estimation Network (EDEN) for Support of Ecological and Biological Assessments. U. S. Department of the Interior. Fact Sheet 2006–3087.
55. Liu Z, Volin JC, Owen VD, Pearlstine LG, Allen JR (2009) Validation and ecosystem applications of the EDEN water-surface model for the Florida Everglades. Ecohydrology 2: 182–194.
56. JMP (2001) A guide to statistics and data analysis using JMP and JMP IN software. Second Edition. SAS Institute, Cary, North Carolina, USA.
57. Smith JP (1997) Nesting season food habits of 4 species of herons and egrets at Lake Okeechobee, Florida. Waterbirds 20: 198–220.
58. Loftus WF (2000) Accumulation and fate of mercury in an Everglade aquatic food web; Doctoral Thesis. Florida International University: Miami, FL.
59. Dorn NJ, Cook MI, Herring G, Boyle R, Nelson J, et al. (2011) Diet variation among White Ibis (*Eudocimus albus*) chicks: prey composition depends on recent hydrologic conditions. Ibis 153: 323–335.
60. Neter J, Kutner MH, Nachtsheim CJ, Wasserman W (1996) Applied linear regression models. Third Edition. Times Mirror Higher Education Group, Chicago, Illinois, USA.
61. Nakagawa S, Schielzeth H (2013) A general and simple method for obtaining R^2 from generalized linear mixed-effects models. Method Ecol Evol 4: 133–142.
62. Beauliea M, Thierry AM, Raclot T, Maho YL, Robert-Coudert Y, et al. (2009) Sex-specific paraetal startegies according to the sex of offspring in the Adélie penguin. Behav Ecol 20: 878–883
63. Franzellitti S, Fabbri E (2006) Cytoprotective responses in the Mediterranean mussel exposed to Hg^{2+} and $CH3Hg^{+}$. Biochem Biophys Res Comm 351: 719–725.
64. Sørensen JG (2010) Application of heat shock protein expression for detecting natural adaptation and exposure to stress in natural populations. Curr Zool 56: 703–713.
65. Hoffman DJ, Eagles-Smith CA, Ackerman JT, Adelsbach TL, Stebbins KR (2011) Oxidative stress response of Forster's terns *(Sterna forsteri)* and Caspian terns *(Hydroprogne caspia)* to mercury and selenium bioaccumulation in liver, kidney, and brain. Environ Toxicol Chem 30: 920–929.
66. Brkljačić J, Perišić T, Dundjerski J, Matić G (2007) Interaction of rat renal glucorticoid receptor with Hsp90 and Hsp70 upon stress provoked by mercury. J Appl Toxicol 27: 43–50.
67. Hwang GW (2012) Role of intracellular defense factors against methylmercury toxicity. Biol Pharm Bull 35: 1881–1884.
68. Sepulveda MS, Frederick PC, Spalding MG, Williams GE (1998) Mercury contamination in free-ranging great egret nestlings (*Ardea albus*) from southern Florida. Environ Toxicol Chem 18: 985–992.
69. Schwabl H (1999) Developmental changes and among-sibling variation of corticosterone levels in an altricial avian species. Gen Comp Endocrinol 116: 403–408.
70. Kern M, Bacon W, Long D, Cowie RJ (2001) Possible roles for corticosterone and critical size in the fledging of nestling pied flycatchers. Physiol Biochem Zool 74: 651.
71. Romero LM, Strochlic D, Wingfield JC (2005) Corticosterone inhibits feather growth: potential mechanism explaining seasonal down regulation of corticosterone during molt. Comp Biochem Phys A 142: 65–73.
72. Desrochers DW, Reed JM, Awerman J, Kluge JA, Wilkison J (2009) Exogenous and endogenous corticosterone alter feather quality. Comp Biochem Phys A 152: 46–52.
73. Butler MW, Leppert LL, Duffy AL (2010) Effects of small increases in corticosterone levels on morphology, immune function, and feather development. Phys Biogeo Zool 83: 78–86.
74. Kitaysky AS, Wingfield JC, Piatt JF (1999) Corticosterone facilitates begging and affects resource allocation in the Black-legged Kittiwake. Behav Ecol 12: 619–625.
75. Loiseau C, Sorci G, Dano S, Chastel O (2008) Effects of experimental increase of corticosterone levels on begging behavior, immunity and parental provisioning rate in house sparrows. Gen Comp Endocrinol 155: 101–108.
76. Wingfield JC (2005) Modulation of the adrenocortical response to acute stress in breeding birds. In: Dawson A, Sharp PJ (eds.), Functional Avian Endocrinology. Narosa Publishing House, New Delhi, India.
77. Frederick PC, Collopy MW (1989) Nesting success of Ciconiiform species in relation to water conditions in the Florida Everglades. Auk 106: 625–634.
78. Gabriel JE, Ferro MIT, Ferro JA, Silva MM, Givisiez PEN, et al. (2000) Influence of dietary energy level on hepatic 70-kDa heat shock protein expression in broiler chicks submitted to acute heat stress. Braz J Poult Scie 2: 259–266.
79. Gray CM, Hamer KC (2001) Prefledging mass recession in Manz shearwaters: parental desertion or nestling anorexia? Anim Behav 62: 705–709.
80. Mauck RA, Ricklefs RE (2005) Control of fledging age in Leach's storm petrel *Oceanodroma leucorhoa*: chick development and pre-fledging mass loss. Func Ecol 19: 73–80.
81. Martins TLF (1997) Fledging in common swift, *Apus apus*; weight watching with a difference. Anim Behav 54: 99–108.
82. Wright J, Markham S, Denney SM (2014) Facultative adjustment of pre-fledging mass loss by nestling swifts preparing for flight. Proc Royal Soc Lon B 273: 1895–1900.

19

Disproportional Representation of Primates in the Ecological Literature

Eckhard W. Heymann[1]*, Dietmar Zinner[2], Jörg U. Ganzhorn[3]

1 Abt. Verhaltensökologie & Soziobiologie, Deutsches Primatenzentrum, Göttingen, Germany, 2 Abt. Kognitive Ethologie, Deutsches Primatenzentrum, Göttingen, Germany, 3 Abt. Tierökologie, Zoologisches Institut, Universität Hamburg, Hamburg, Germany

Abstract

We address the question why papers dealing with the ecology of primates are so sparsely represented in the general ecological literature. A literature analyses based on entries in Web of Science and PrimateLit reveals that despite a large number of papers published on primates in general and on the ecology of primates, only a very small fraction of these papers is published in high-ranking international ecological journals. We discuss a number of potential reasons for the disproportion and highlight the problems associated with experimental research on wild primates and constraints on sample size as major issues.

Editor: Joanna E. Lambert, The University of Texas at San Antonio, United States of America

Funding: The authors have no support or funding to report.

Competing Interests: The authors have declared that no competing interests exist.

* E-mail: eheyman@gwdg.de

Introduction

Primatological studies have played a major role in the development of conceptual frameworks in evolutionary ecology. The comparative approach of J. H. Crook on the social organization of birds and primates was instrumental to the development of the fields of "socio-biology" and "behavioral ecology" in general [1,2]. His work stimulated a number of primatological papers that contributed to the development of general ecological questions (e.g., [3,4]). Since then, the comparative approach seems to have shifted away from primate ecology, and primate ecology seems to have lost importance for the academic community at large, while analyses of social systems still remain prominent (e.g., [5]). Many ecological studies in primatology are designed or considered "case studies" rather than studies of general interest that contribute to the understanding of ecological and evolutionary questions in general. This perception or reality has important consequences as, first, academic funding agencies are reluctant to fund case studies, and second, the scientific community and primatologists themselves are unaware of the wealth of information for many distinct species (probably unmatched by any other taxon) and the potential to contribute to the understanding of general evolutionary questions in ecology. With the present analysis we wanted to test whether this perception on the lack of importance of primatological studies in the ecological literature is justified.

This perceived lack of importance is even more intriguing when considering that the number of publications on primate ecology is very high when compared to other taxa with a much larger number of species. A search in Web of Knowledge^{SM} v.5.9 (search date: 18 June 2013) with terms "primate* and ecology", "mammal* and ecology", "bird* and ecology" and "insect* and ecology" yielded 1646, 5947, 8979 and 9749 hits, respectively. Comparing these figures with the number of species (primates: 479, mammals: ~6000, birds: ~10.000, insects: >1.000.00) reveals an obvious disproportion: in relation to the number of species, there are many more papers on primates than on the other taxa.

Therefore, in this paper we examine the following questions:

(1) Which fraction of papers published on primates (independent of the type of publication, i.e. journal articles, book chapters etc.) deals with aspects of ecology?

(2) Which fraction of papers on primate ecology is published in high-ranking ecological journals?

(3) Which fraction of papers published in high-ranking ecological journals deals with primates and how does this compare with other taxa?

We then discuss potential reasons for the underrepresentation of primate papers in ecological journals.

Methods

To examine the significance of ecological research questions in primatology, we searched in PrimateLit (http://primatelit.library.wisc.edu) with "keyword = ecology and feral". The term "feral" was used to exclude papers dealing with "lab ecology", a keyword used in PrimateLit for papers dealing with housing of and environmental enrichment for captive primates. We restricted this search to the period 1980–2009 for two reasons: (a) some of the currently high-ranking ecological journals only started in the 80 s or 90 s; (b) PrimateLit ceased to be updated after November 2010. We plotted 5-year running means of the number of hits per year against the time in years.

To examine the representation of papers dealing with primate ecology in ecological journals, we performed two different analyses. First, we searched in PrimateLit, with "keyword = ecology and feral" and "publication = <journal name>" and restricted this search to the period 1980–2009. We used the following journals on the basis of their scope (which should make a journal potentially feasible for submitting primate ecology papers) and their impact factors: Ecology, Ecology Letters, Functional Ecology, Journal of Animal Ecology, Journal of Ecology, Oecologia, Oikos. We then determined the total number of articles dealing with any taxon or topic in the journals mentioned above by searching the Web of Knowledge[SM] [v.5.9] with the journal name in the field "publication name" (search date: 15 April 2013). On the basis of these search results, we calculated the percentage of papers from the search in PrimateLit that were published in the above mentioned journals. We also examined, on the basis of values summed up over 5-year periods, whether there is a temporal trend in the representation of primate ecology papers in ecological journals.

Second, we compared the primate data with other taxa, namely insects, all classes of vertebrates, and major orders of mammals. We searched the Web of Knowledge Knowledge[SM] v.5.9 (search date: 18 June 2013) first with the search term "<taxon> and ecology" (e.g. "bird* and ecology") in the field "topic" and then with this term plus the journal name in the field "publication name". The search period comprised 1945–2013. We then calculated, for each taxon and over the entire search period, the percentage of papers in the above mentioned ecological journals. To estimate how primate papers perform in comparison to other taxa, we calculated the mean percentage and SD over all taxa except primates and compared this mean to the value for primates with a single sample t-test in Statistica 10.0 [6].

Results

Significance of ecological research questions in primatology

Over the period 1980–2009, 7.6% of papers listed in PrimateLit dealt with aspects of primate ecology. The percentage almost doubled between 1980 and 2009, with a steep increase starting around 1990 (Fig. 1).

Representation of primate papers in ecological journals

Papers dealing with aspects of primate ecology are only marginally represented in high-ranking ecological journals, accounting on average for only 0.3% of all papers in these journals (Fig. 2). For the period 2010–2013, this remains unchanged (0.3% of a total of 4980 papers).

The representation of papers dealing with primate ecology in ecological journals dropped drastically from >3% in 1980–84 to the currently low value with some fluctuation over time, despite the strong increase in the total number of papers dealing with primate ecology (Fig. 3). The absolute number of primate papers in ecological journals varied relatively little between 1990 and 2004 (10–20 papers/5 year period), and increased to 33 in the period 2005–2009. The total number of primate papers with "keyword = ecology and feral" increased from 644 (1990–94) to 3027 (2005–09). Over the entire 30-year period, the percentage of primate ecology papers published in ecological journals is below 1%.

Comparison with other taxa

Compared to other taxa, primates are poorly represented in ecological journals. The proportion of papers that matched the search term "primate* and ecology" (1.5%) was lower than for most other taxa except for cetaceans that had an even lower value (1.4; Fig. 4). The mean percentage ± SD over all taxa is 5.1±2.8, and only primates and cetaceans lie outside one SD from the mean. The percentage for primates (1.5) is significantly smaller than the mean percentage over all other taxa (5.4±2.7; $t = 4.878$, d.f. $= 11$, $p = 0.000487$).

Discussion

Our analyses revealed that despite a strong increase in the number of papers dealing with primate ecology over the last decades, there is a disproportionately low representation of such papers in general ecological journals. The general increase in primatological papers was not matched by a similar increase in the number of primate papers published by ecological journals. Their percentage actually dropped (Fig. 3). Furthermore, the representation is similar to the representation of cetaceans and considerably lower than for most other taxa. There are several potential explanations for this which we discuss below:

(a) "Bad ecologist" hypothesis

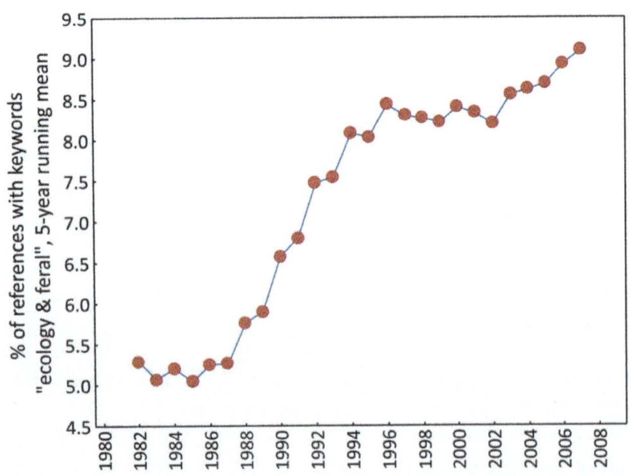

Figure 1. Temporal trend in the percentage of primatological publications dealing with aspects of primate ecology over the period 1980–2009. Dots represent 5-year running means; therefore, 1980, 1981, 2008 and 2009 are not represented by a dot.

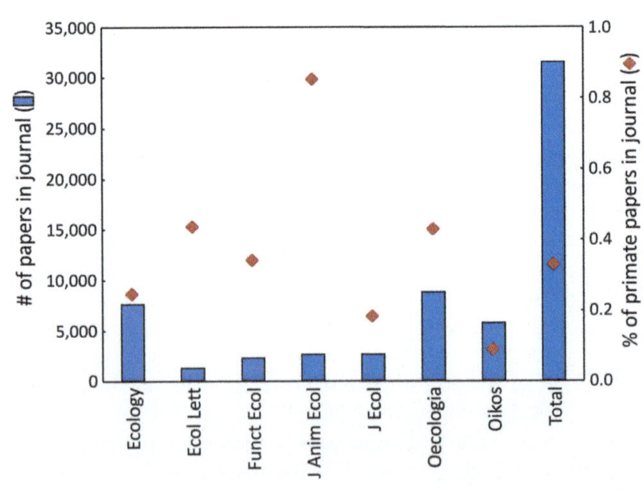

Figure 2. Total number of papers published in different ecological journals and percentage of these papers that relate to primates. Ecol Lett = Ecology Letters, Funct Ecol = Functional Ecology, J Anim Ecol = Journal of Animal Ecology, J Ecol = Journal of Ecology.

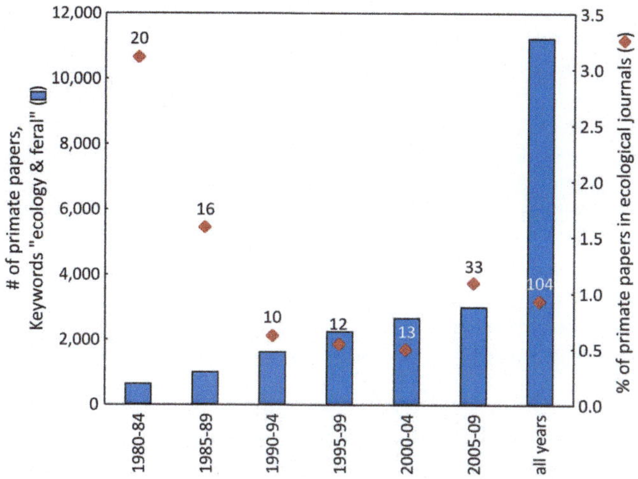

Figure 3. Number of primate papers over 5-year periods and percentage of primate papers in ecological journals. Numbers above diamonds indicate the total number of primate papers in ecological journals in the respective period.

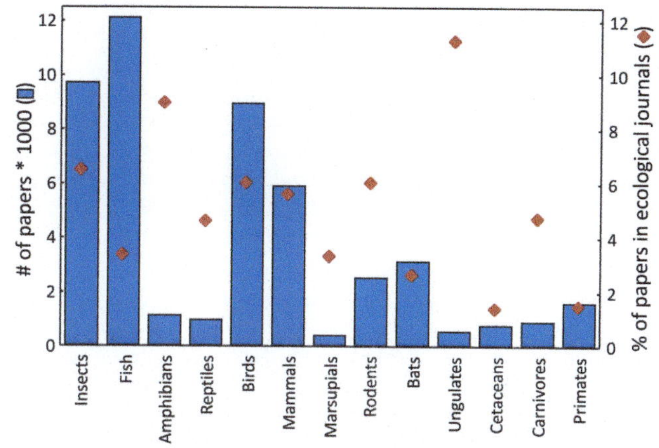

Figure 4. Number of papers on the ecology of insects, different classes of vertebrates and major orders of mammals, and the percentage of these papers that are published in high-ranking ecological journals.

Many primatologists may be interested in primates per se rather than in ecological problems and lack a sound theoretical background in general ecology. Thus, they may either not attempt publication outside the primatological circle, or they may not ask the "right" questions, or they cannot put their findings into a perspective that makes the results of their research interesting for a broader ecological readership. This is perhaps related to the history of primate ecological research. As Clutton-Brock [7, p. vii] stated in the preface to one of the first books dedicated to primate ecology, "interest in primate ecology has been stimulated by investigation of the adaptive significance of social systems". Furthermore, as Janson [8, p. 521] pointed out, "field primatology emerged largely from a descriptive anthropological tradition" which led to primate field studies focusing on a more descriptive primate ecology ("The diet of the ...", "Habitat use in a population of ..." etc.). This is not necessarily "bad ecology", but it clearly makes findings less appealing to a wider, ecologically interested audience. The relatively minor fluctuation in the total number of primate papers in ecological journals over the years, despite an almost 5-fold increase in the number of primate papers with keyword "ecology" and "feral" from 1990 to 2009, probably reflects this rather descriptive approach in primatology.

Carmel et al. [9] recently analysed trends in ecological research by examining the topics and approaches of papers published in all ecological journals and in "core ecological journals" (partially overlapping with our journal list) between 1980 and 2010. They found that a majority of studies deals with single species (71% and 66% of papers in all journals and core journals, respectively) and use observational approaches (59% and 45%). Therefore, the principal setup of most primate studies – observational on single species – should not preclude publication in ecological journals (although the percentage of observational studies in core ecological journals is considerably lower than in all ecological journals).

(b) Lack of experimental confirmation

Many research questions in ecology require experimental work to confirm observational findings and to test hypotheses. Lack of experimental confirmation of findings may result in lower acceptance of primate ecological research to ecologists who advocate more rigid experimental approaches. However, many experiments are impossible with primates on ethical and/or on practical grounds.

Ethical limitations

Primatologists have used playbacks of conspecific or allospecific vocalizations [10–12], platforms with food and/or sensory stimuli [13,14], cognitive test apparatuses [15,16], and provisioning with tools [17,18] to manipulate the behaviour and study the cognitive and sensory capabilities of primates. These kinds of experiments raise few ethical concerns, if any. However, many ecological experiments may require much stronger manipulation. E.g., to test whether a species exerts a competitive effect on another species or whether the lack of pollinators or seed dispersers affects plant populations requires the removal of complete groups or populations. For primate ecology studies, this would be absolutely unacceptable for ethical reasons.

As a proxy, pseudo-experimental situations where primate species have been extinguished, occur in reduced population densities due to hunting, or are absent for other reasons can provide relevant information. A comparison of hunted and non-hunted sites in Amazonia revealed competitive effects of large-sized on medium-sized primate species [19]. Similarly, a comparison of sites with low and high hunting pressure demonstrated that the lack of primate seed dispersal affects patterns of plant recruitment [20]. The absence of spider monkeys resulted in higher fruit set in a population of *Symphonia globulifera* compared to a population where these primates heavily preyed upon flowers of this plant [21].

Given that the majority of primates are group-living [5] experiments that require the temporal removal or isolation of individuals from a group are also ethically problematic (apart from practical problems). Experiments like those of Kawai [22] where entire groups were captured and released at a different place would no longer receive approval. Consequently, ecological experiments that require temporal removal of animals from their natural setting have been performed with solitary foragers like mouse lemurs, *Microcebus* [23,24], but not with group-living primates.

Practical limitations

(i) Sample size limitations

The majority of primates live in groups that range together and show a high degree of co-ordination and synchronization of activities [25,26]. Individuals within such groups thus cannot be considered as statistically independent samples. Observing a sufficiently large number of different groups to obtain high statistical power is in most cases practically impossible (due to limitations of funding, time constraints, limited size of study areas etc.).

Sample size can also be limited due to the high behavioural and ecological flexibility and high intra-specific and intra-population variability of most primates (see (ii)).

(ii) Flexibility and variability

Most primates are behaviourally and ecologically highly flexible [27–29]. This requires either many replica or the control of a large number of potentially or actually confounding variables; something that is rarely done in primate ecology [8]. Most primates also have slow life histories and long generation times [30] which can introduce an additional major source of variation [31]. Together with the issue of sample size limitations discussed above, this results in low statistical power.

(iii) Primate intelligence and cognition

Primates – by virtue of their higher cognitive capabilities compared to most other animals – can rapidly habituate to experimental stimuli and settings. They may even manipulate or destroy the experimental setup (see [32] for an example from captive primates). Even if planned very well and based on previous knowledge and experience, experiments may fail, simply because primates respond in an unanticipated manner. Janson [8] provides first-hand examples on failures of ecological and behavioural field experiments with primates. E.g., a setup that had worked very well with some groups, did not work with another group, or experiments turned out not "to fit within the lifestyle" [8, p. 531] of the animals. Experimental feeding platforms may not be attractive enough if alternative food resources become available, a factor that may be uncontrollable in the often large home ranges and complex tropical habitats of primates.

From all other taxa, practical problems, particularly for experimentation, and ethical concerns are probably most similar in cetaceans. The low representation of this taxon in high-ranking ecological journals supports our hypotheses outlined above.

There are a few other conceivable hypotheses that may apply.

(a) "Boring primates" hypothesis: Most primates live in tropical forest and occupy only a few, relatively similar niches. Therefore, ecological questions are similar for different primate species.

(b) "Geographic origin of researcher" hypothesis: Many researcher study ecology of species living in their home countries. Since still most researchers come from northern industrialized countries, species living in these regions could be overrepresented in the ecological literature.

(c) "Inbreeding" hypothesis: Primatologists may tend to publish their results – even if they could be of more general interest – in taxon-specific journals. Conversely, editors of general-interest ecology journals may be more willing to decide that 'there are other more appropriate journals for this manuscript' simply because in fact there are many primate journals relative to any other comparable order of animals (e.g. there is no 'Journal of Woodpeckerology'). However, the latter seems to be unlikely. This can be illustrated by the case of the representation of papers on birds and rodents. Both taxa have a similar representation in ecological journals (Fig. 4), but while there are 130 or more ornithological journals [33], we are aware of only one journal dedicated to rodents.

Conclusions

We have identified some potential causes for the low representation of primate papers in the general ecological literature, but do not pretend that this is an exhaustive listing. Conceptual problems are probably the easiest to overcome by modifying the curricula in primatology classes. Most practical problems could be overcome in an ideal world with less restricted funding and less time pressure for publications. However, in the real world, these problems constrain ecological research on primates and consequently may lead to results of primate ecological research being less attractive for a general ecological audience. Nevertheless, we hope that this paper stimulates discussions, both within the primatological community and between primate ecologists and general ecologists. Primate ecology has the potential to make important contributions to general ecology, particularly with regard the long-term ecological variation [34] and consequences of personality for ecological interactions and processes [35].

Acknowledgments

We thank Dr. Charlie Janson and two anonymous reviewers for commenting and very helpful suggestions on an earlier version of this manuscript.

Author Contributions

Analyzed the data: EWH. Wrote the paper: EWH DZ JUG.

References

1. Crook JH (1965) The adaptive significance of avian social organization. Symposia of the Zoological Society of London 14: 181–218.
2. Crook JH, Gartlan JS (1966) Evolution of primate societies. Nature 210: 1200–1203.
3. Clutton-Brock TH (1974) Primate social organisation and ecology. Nature 250: 539–542.
4. Milton K, May ML (1976) Body weight, diet, and home range area in primates. Nature 259: 459–462.
5. Mitani JC, Call J, Kappeler PM, Palombit RA, Silk JB (2012) The evolution of primate societies. Chicago: University of Chicago Press. 730 p.
6. StatSoft, Inc. (2011). STATISTICA für Windows (Software-System für Datenanalyse), Version 10.0. Available: www.statsoft.com.
7. Clutton-Brock TH (1977) Primate ecology. Studies on feeding and ranging behaviour in lemurs, monkeys and apes. London: Academic Press.
8. Janson CH (2012) Reconciling rigor and range: observations, experiments, and quasi-experiments in field primatology. International Journal of Primatology 33: 520–541.
9. Carmel Y, Kent R, Bar-Massada A, Blank L, Liberzon J, et al. (2013) Trends in Ecological Research during the Last Three Decades – a systematic review. PLoS ONE 8(4): e59813.
10. Robinson JG (1979) Vocal regulation of use of space by groups of titi monkeys *Callicebus moloch*. Behavioral Ecology and Sociobiology 5: 1–15.
11. Raemaekers JJ, Raemaekers P (1985) Field playback of loud calls to gibbons (*Hylobates lar*): territorial, sex-specific and species-specific response. Animal Behaviour 33: 481–493.
12. Zuberbühler K (2000) Interspecies semantic communication in two forest primates. Proceedings of the Royal Society of London B 267: 713–718.
13. Janson CH (1998) Experimental evidence for spatial memory in foraging wild capuchin monkeys, *Cebus apella*. Animal Behaviour 55: 1229–1243.
14. Bicca-Marques JC, Garber PA (2004) Use of spatial, visual and olfactory information during foraging in wild nocturnal and diurnal anthropoids: a field experiment comparing *Aotus*, *Callicebus*, and *Saguinus*. American Journal of Primatology 62: 171–182.

15. Pesendorfer MB, Gunhold T, Schiel N, Souto A, Huber L, et al. (2009) The maintenance of traditions in marmosets: individual habit, not social conformity? A field experiment. PLoS ONE 4: e4472.
16. Schnoell A, Fichtel C (2012) Wild redfronted lemurs (*Eulemur rufifrons*) use social information to learn new foraging techniques. Animal Cognition 15: 505–516.
17. Matsuzawa T (1994) Field experiments on use of stone tools by chimpanzees in the wild. In: Wrangham RW, McGrew WC, De Waal FBM, Heltne PG, editors. Chimpanzee cultures. Cambridge: Harvard University Press. pp. 351–370.
18. Fragaszy D, Pickering T, Liu Q, Izar P, Ottoni E, et al. (2010) Bearded capuchin monkeys' and a human's efficiency at cracking palm nuts with stone tools: field experiments. Animal Behaviour 79: 321–332.
19. Peres CA, Dolman PM (2000) Density compensation in neotropical primate communities: evidence from 56 hunted and nonhunted Amazonian forests of varying productivity. Oecologia 122: 175–189.
20. Effiom EO, Nuñez-Iturri G, Smith HG, Ottosson U, Olsson O (2013) Bushmeat hunting changes regeneration of African rainforests. Proceedings of the Royal Society B: Biological Sciences 280: 20130246.
21. Riba-Hernández P, Stoner KE (2005) Massive destruction of *Symphonia globulifera* (Clusiaceae) flowers by Central American spider monkeys (*Ateles geoffroyi*). Biotropica 37: 274–278.
22. Kawai M (1960) A field experiment on the process of group formation in the Japanese monkey (Macaca fuscata), and the releasing of the group at Ôhirayama. Primates 2: 181–255.
23. Goerlitz HR, Siemers BM (2007) Sensory ecology of prey rustling sounds: acoustical features and their classification by wild grey mouse lemurs. Functional Ecology 21: 143–153.
24. Thorén S, Linnenbrink M, Radespiel U (2011) Different competitive potential in two coexisting mouse lemur species in northwestern Madagascar. American Journal of Physical Anthropology 145: 156–162.
25. King AJ, Cowlishaw G (2009) All together now: behavioural synchrony in baboons. Animal Behaviour 78: 1381–1387.
26. Boos M, Kolbe M, Kappeler PM, Ellwart T (2011) Coordination in human and primate groups. Heidelberg: Springer. 288 p.
27. Chapman CA, Fedigan LM (1990) Dietary differences between neighboring *Cebus capucinus* groups: local traditions, food availability or responses to food profitability? Folia Primatologica 54: 177–186.
28. Jones CB (2005) Behavioral flexibility in primates: causes and consequences. New York: Springer.
29. Codron D, Lee-Thorp JA, Sponheimer M, de Ruiter D, Codron J (2006) Inter- and intrahabitat dietary variability of chacma baboons (Papio ursinus) in South African savannas based on fecal δ13C, δ15N, and %N. American Journal of Physical Anthropology 129: 204–214.
30. Martin RD (1990) Primate origins and evolution. London: Chapman and Hall. 804 p.
31. Nowell AA, Fletcher AW (2008) The development of feeding behaviour in wild western lowland gorillas (*Gorilla gorilla gorilla*). Behaviour 145: 171–193.
32. Vitale AF, Visalberghi E, De Lillo C (1991) Responses to a snake model in captive crab-eating macaques (*Macaca fascicularis*) and captive tufted capuchins (*Cebus apella*). International Journal of Primatology 12: 277–286.
33. Wikipedia website. Available: http://en.wikipedia.org/wiki/List_of_ornithology_journals. Accessed: 25 Sep 2013.
34. Kappeler PM, Watts DP (2012) Long-term field studies of primates. Heidelberg: Springer. 495 p.
35. Wolf M, Weissing FJ (2012) Animal personalities: consequences for ecology and evolution. Trends in Ecology & Evolution 27: 452–461.

20

Ecological Conditions Favoring Budding in Colonial Organisms under Environmental Disturbance

Mayuko Nakamaru[1]*, Takenori Takada[2], Akiko Ohtsuki[1,3], Sayaki U. Suzuki[1,3¤], Kanan Miura[1], Kazuki Tsuji[4]

1 Department of Value and Decision Science, Tokyo Institute of Technology, Tokyo, Japan, **2** Graduate School of Environmental Earth Science, Hokkaido University, Sapporo, Japan, **3** The Graduate University for Advanced Studies, Kanagawa, Japan, **4** Faculty of Agriculture, University of the Ryukyus, Okinawa, Japan

Abstract

Dispersal is a topic of great interest in ecology. Many organisms adopt one of two distinct dispersal tactics at reproduction: the production of small offspring that can disperse over long distances (such as seeds and spawned eggs), or budding. The latter is observed in some colonial organisms, such as clonal plants, corals and ants, in which (super)organisms split their body into components of relatively large size that disperse to a short distance. Contrary to the common dispersal viewpoint, short-dispersal colonial organisms often flourish even in environments with frequent disturbances. In this paper, we investigate the conditions that favor budding over long-distance dispersal of small offspring, focusing on the life history of the colony growth and the colony division ratio. These conditions are the relatively high mortality of very small colonies, logistic growth, the ability of dispersers to peacefully seek and settle unoccupied spaces, and small spatial scale of environmental disturbance. If these conditions hold, budding is advantageous even when environmental disturbance is frequent. These results suggest that the demography or life history of the colony underlies the behaviors of the colonial organisms.

Editor: Ahmed Moustafa, American University in Cairo, Egypt

Funding: This work is supported by JSPS KAKENHI Grant Number 21770076 (M.N.), 23370006 (T.T.), 25340115 (T.T.), 20033015 (K.T.), 21247006 (K.T.), and Inamori Foundation (M.N.). The funders had no role in study design, data collection and analysis, decision to publish, or preparation of the manuscript.

Competing Interests: The authors have declared that no competing interests exist.

* E-mail: nakamaru@valdes.titech.ac.jp

¤ Current address: National Agricultural Research Center, Ibaraki, Japan

Introduction

Dispersal distance varies widely among organisms. Dispersal determines the genetic structure of the meta-population and influences the evolution of individual characteristics such as mortality, reproductive efforts, sex allocation and altruism [1–3]. Since dispersal is a costly activity, why organism disperse has become an important issue in evolutionary ecology [4,5]. In the pioneering meta-population model of Hamilton and May (1977) it was shown that dispersal can evolve even when the dispersal cost is so high that nearly all dispersers are destined to die. Subsequent studies identified specific factors that favor dispersal over non-dispersal, such as the avoidance of competition among relatives and inbreeding [1,2]. Disturbances such as predation, drought and flooding are regarded as important general triggers of dispersal, encouraging organisms to hedge the risks [6].

However, many organisms do not disperse long distances. For example, some organisms often adopt an "budding" strategy, where a colony splits into two or more parts which subsequently move away from each other. The budding by clones, buds, or propagules has been studied theoretically and experimentally from the viewpoint of the evolution of altruism and sex ratio [7–9]. However, those studies did not focus on the difference in dispersal distance, while in many real organisms the budding strategy is typically characterized by shorter-range dispersal in comparison to non-budding dispersal. Contrary to the conventional view, organisms reproducing by budding often survive in environments undergoing frequent disturbance. For example, many invasive clonal plants that are specialists of disturbed habitats predominantly use vegetative reproduction, despite retaining the seed-producing capability that would enhance their spread [10,11]. New colonies of branching gorgonian coral (*Plexaurella* sp.) exposed to waves are predominantly founded by fragments of broken branches, rather than by inseminated gametes that can migrate long distances [12]. In some invasive ants, such as *Linepithema humile* (the Argentine ant) and *Wasmannia auropunctata* (the tiny fire ant), winged queens do not engage in nuptial (dispersal) flights. Instead they mate within or near natal nests, and new colonies are founded by colony-splitting (fission or budding), in which queens search for a new nest site on foot, accompanied by some workers [13,14]. These invaders are recognized as specialists of disturbed habitats [13,15,16]. However, the ecological conditions favoring budding remain obscure.

All of the above examples (clonal plants, corals and ants) are colonial or super organisms that must maintain a spatially fixed life following initial settlement. Assuming such colonial life, Nakamaru et al. (2007) pioneered the investigation of ecological conditions favoring budding under disturbance in a spatially explicit computational model (lattice model) [17]. They identified an important trade-off between dispersal distance and offspring size (table 1); small reproductive units, such as plant seeds, spawned coral eggs and winged ant queens, travel long-distance, whereas large units, such as the propagules of clonal plants, broken branches of corals and dependently-founded ant colonies remain within their local environment. This difference in dispersal affects the survival and growth of offspring; small offspring suffer low survival and reach maturity comparatively slowly, while large offspring (buds) benefit from high survival and rapid growth to maturity. Nakamaru et al. (2007) concluded that the short distance dispersal strategy (budding) is advantageous over long distance dispersal (non-budding) under conditions of (i) very high mortality of a small-sized colony, or (ii) relatively frequent environmental disturbance over a small spatial scale.

However, Nakamaru et al. did not clarify the effect of spatial structure and the life history on the above-mentioned advantage. For each colony, they specifically assumed that growth is logistic and that death rate is a negative exponential function of colony size. However, other survivorship patterns and other growth functions are possible in nature. The authors failed in their quest for a mathematical model that could accommodate relatively complicated assumptions and therefore could not obtain analytical results. These drawbacks have precluded a precise understanding of the adaptive significance of budding. To more generally understand the ecological conditions that induce budding dispersal, here we investigate the characteristics of an organism's life history functions, such as survivorship and colony growth, that favor budding dispersal under disturbance.

We adopt the general view that life history parameters (survival, growth and reproduction functions) directly determine the evolved dispersal strategy [18,19]. Environmental conditions, such as spatial structure and disturbance, influence the above parameters in a complicated manner through environment vs. phenotype interactions. In this paper we make a simple generalized assumption that is applicable to colonial organisms adopting similar resource allocation strategies. We first focus on colonial life history using a model without spatial structure, called the baseline model. This model assumes discrete colony size, assigning different death and growth probabilities to each colony size. Under this assumption, we can construct discrete equations or a matrix model in the baseline model that permit exact solutions. We clarify the effects of life histories or demographics (such as survivorship and growth rate) on the competition between the two strategies, i.e. budding vs. non-budding. We also examine the conditions under which budding is advantageous over non-budding when realistic and possible assumptions are added to the baseline model.

Methods and Results

Model 1: The baseline model in the completely mixed population

It is common that an individual is assumed to be a basic unit in the agent-based simulation model. In contrast, here the basic unit in our agent-based simulation model is the clonal (super) organism or the colony of individuals, and the model is called the colony-based model. Gardner and Grafen (2009) proved that there is no mathematical difference between an individual organism and a clonal group if social groups have no intracolonial conflicts [20].

We consider only non-spatial structures or completely mixed populations, ignoring the effect of dispersal distance. Instead, we aim to clarify the effect of colonial life history patterns on the population dynamics. The colony size-structured matrix model assumes four colony sizes, designated size 1, size 2, size 3 and size 4. The first three denote immature colonies, while size 4 represents the mature colony. Life history pattern is described by two parameters; the baseline death probability and the growth probability of the colony (Fig. 1 A). The baseline death probability of a colony of size i is defined as d_i ($0 \leq d_i \leq 1$, $i = 1, 2, 3$ and 4.). An additional death probability, independent of the baseline death, is imposed by disturbance. The survival probability of a colony of size i is $1-d_i$, defined as p_i ($0 \leq p_i \leq 1$, $i = 1, 2, 3$ and 4). Let g_i be the probability that a colony of size i expands to size i+1 ($0 \leq g_i \leq 1$, $i = 1, 2, 3$). When a colony of size i survives with probability p_i, it expands to size i+1 with probability g_i ($i = 1, 2, 3$); otherwise it remains at its present size with probability $1-g_i$. When the colony has reached size 4, it can divide into two colonies, one of which will colonize an empty site. Therefore the chance of division is proportional to the density of empty sites in the population multiplied by the division rate, h, where h controls the division speed. A size 4 colony deprived of the opportunity to divide will survive with probability p_4.

The colony divides by one of two strategies: designated 2:2 division and 1:3 division (Fig.1 B and C). A size 4 colony adopting the 2:2 division strategy divides into two colonies of size 2, one of which moves to an empty site. Alternatively, a size 4 colony adopting the 1:3 division strategy divides into two unequally sized colonies (of size 1 and size 3), the smaller of which moves to an empty site. This situation mimics nature; if a biological colony splits into two equal parts, one of the daughter colonies moves to a new site close to the natal colony; if the colony is unequally split, the smaller colony disperses far from the natal colony. The 2:2 and 1:3 division strategies correspond to budding and seed-like dispersal, respectively. To purely clarify the effect of division ratio on the population dynamics, we ignore spatial structure and distance.

The size-structured matrix model for colonies adopting the 2:2 division strategy in the population is constructed as

$$x(t) = S(t)x(t), \tag{1}$$

Table 1. Life history characteristics of long and short dispersal units.

Dispersal units	Seeds, spawned eggs, and independently founding queens	vegetative propagule, broken coral braches, or dependently-founded ant colonies
Dispersal distance	long	short
Size	small	large
Time to reproduction	long	short
Number	many	few
Survival rate	low	high

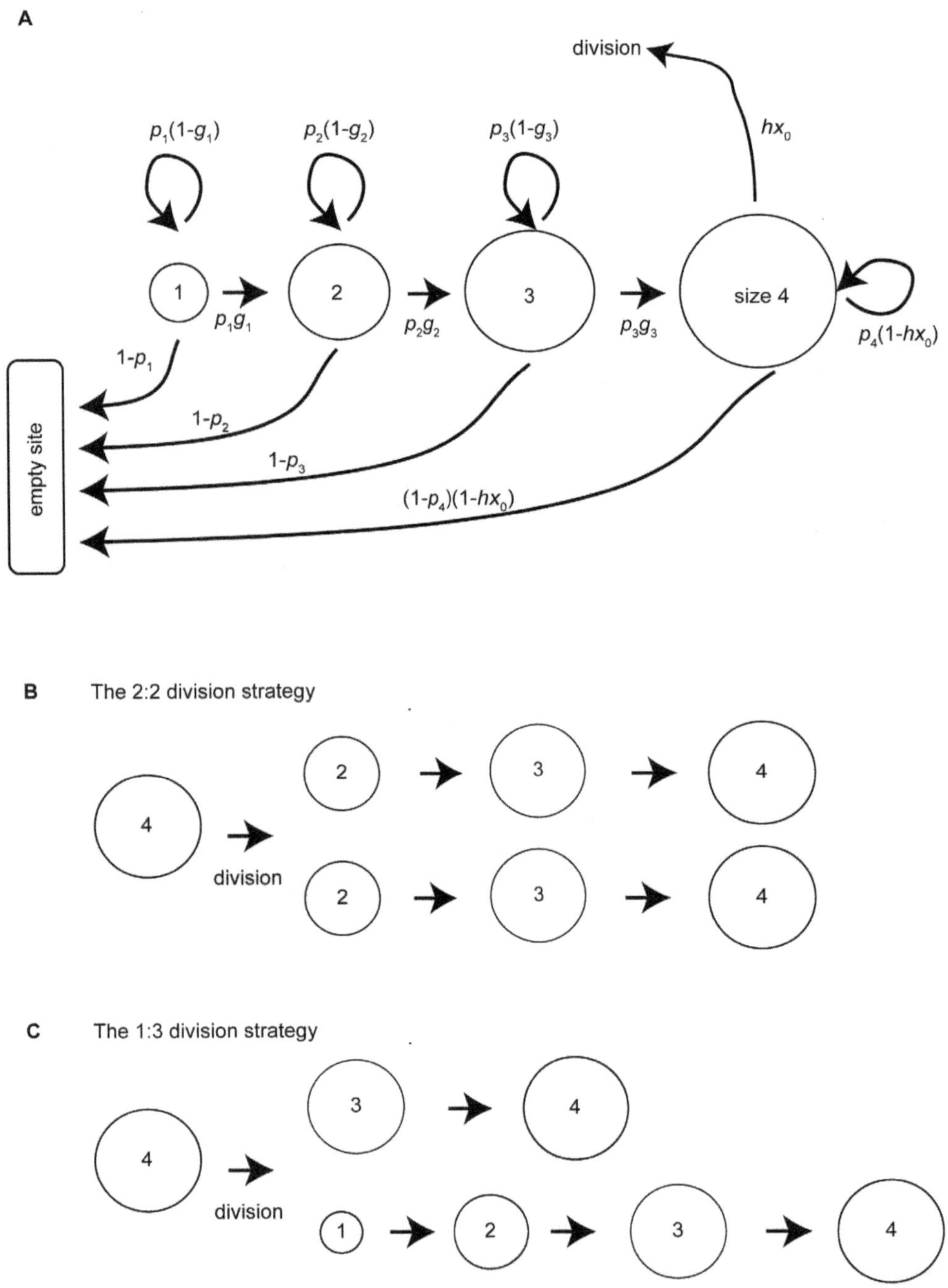

Figure 1. Life history of a colony and its division strategies. (A) Life history of a colony. (B) The 2:2 division strategy, in which a mature (size 4) always divides into two size 2 daughter colonies. The new colonies grow to maturity. (C) The 1:3 division strategy, in which a mature colony always divides into a small (size 1) and a large (size 3) daughter colony. The size 1 matures slowly via size 2 and size 3, while the size 3 colony matures rapidly.

where $x(t) = (x_0(t), x_1(t), x_2(t), x_3(t), x_4(t))^t$ and

$$S(t)=\begin{bmatrix} 1-hx_4(t) & 1-p_1 & 1-p_2 & 1-p_3 & (1-p_4)(1-hx_0(t)) \\ 0 & p_1(1-g_1) & 0 & 0 & 0 \\ hx_4(t) & p_1g_1 & p_2(1-g_2) & 0 & hx_0(t) \\ 0 & 0 & p_2g_2 & p_3(1-g_3) & 0 \\ 0 & 0 & 0 & p_3g_3 & p_4(1-hx_0(t)) \end{bmatrix}.$$

Let $x_0(t)$ be the density of empty sites at time t and $x_i(t)$ ($i = 1, 2, 3$ and 4) be the density of colonies of size i adopting the 2:2 division strategy at time t, subject to $\Sigma_i\ x_i(t) = 1$. The values of $hx_4(t)$ in element (3, 1) and $hx_0(t)$ in element (3, 5) show that, when a size 4 colony divides, one of the two resulting size 2 colonies disperses to an empty site while the other remains in its natal habitat.

Similarly, the size-structured matrix model for colonies adopting the 1:3 division strategy in the population is constructed as

$$y(t) = L(t)y(t), \tag{2}$$

where $y(t) = (y_0(t), y_1(t), y_2(t), y_3(t), y_4(t))^t$ and

$$L(t)=\begin{bmatrix} 1-hy_4(t) & 1-p_1 & 1-p_2 & 1-p_3 & (1-p_4)(1-hy_0(t)) \\ hy_4(t) & p_1(1-g_1) & 0 & 0 & 0 \\ 0 & p_1g_1 & p_2(1-g_2) & 0 & 0 \\ 0 & 0 & p_2g_2 & p_3(1-g_3) & hy_0(t) \\ 0 & 0 & 0 & p_3g_3 & p_4(1-hy_0(t)) \end{bmatrix}.$$

Let $y_0(t)$ be the density of empty sites at time t and $y_i(t)$ ($i = 1, 2, 3$ and 4) be the density of colonies of size i adopting the 1:3 division strategy at time t, subject to $\Sigma_i\, y_i(t) = 1$. When a size 4 colony divides, the probability of the resulting size 1 colony successfully migrating to an empty site is determined by $hy_4(t)$ in element (2, 1), while the probability that the size 3 colony remains in its natal habitat is $hy_0(t)$ in element (4, 5).

The dynamics of competition between colonies adopting 1:3 and 2:2 division strategies can be described using the competition matrix, $F(w(t))$:

$$w(t{+}1) = F(w(t))w(t), \tag{3}$$

in which the vector $w(t)$ consists of $x_i(t)$ and $y_i(t)$ ($i = 1, 2, 3, 4$) and $z_0(t)$ which is the density of empty sites, and the matrix F represents the colony size transition. Detailed information about Eq. (3) is given in Appendix S4.

Results when all growth probabilities are one

Mathematical results of the colony size-structured matrix model. Here we consider the simplest case in which all growth probabilities are one ($g_i = 1$). By local stability analysis of Eq. (1), the equilibrium point where colonies adopting the 2:2 division strategy become extinct ($x_0^*(t) = 1.0$) is locally unstable when

$$2p_2p_3 > p_4 + (1-p_4)/h, \tag{4}$$

otherwise the equilibrium point $x_0^*(t) = 1.0$ is locally stable (see appendix S1 and S2-(i)). This result implies that colonies adopting the 2:2 division strategy could stably survive in the local environment under conditions satisfying the inequality (4). Inequality (4) is called the viability condition of the 2:2 division strategy. When $p_1 \leq p_2 \leq p_3 \leq p_4$, inequality (4) is locally stable at the equilibrium $x_1^*(t)+x_2^*(t)+x_3^*(t)+x_4^*(t)>0$ and $x_0^*(t)<1.0$ (see appendix S1 and S2-(ii)).

Similarly, by local stability analysis of Eq. (2), the equilibrium point where colonies adopting the 1:3 division strategy become extinct ($y_0^*(t) = 1.0$) is locally unstable when

$$(1+p_1p_2)p_3 > p_4 + (1-p_4)/h. \tag{5}$$

Under this condition, colonies adopting the 1:3 division strategy can stably survive; hence, inequality (5) is called the viability condition of the 1:3 division strategy (appendix S1 and S3-(i)). When $p_1 \leq p_2 \leq p_3 \leq p_4$, inequality (5) defines the locally stable condition at equilibrium ($y_1^*(t) + y_2^*(t) + y_3^*(t) + y_4^*(t) > 0$ and $y_0^*(t) < 1.0$) (see appendix S1 and S3-(ii)).

We also analyze the matrix model in which two strategies compete for empty sites (see appendix S4). The notation of parameters and variables is unaltered, apart from the density of empty sites at time t, denoted $z_0(t)$. As shown in appendix S4-(ii), the equilibrium point of the pure 2:2 division strategy, ($x_1^*(t) + x_2^*(t) + x_3^*(t) + x_4^*(t) > 0$ and $y_i^*(t) = 0$ ($i = 1$–4)), is locally stable when inequalities (4) and (6) are satisfied, where (6) is given by

$$2p_2 > 1 + p_1p_2. \tag{6}$$

Appendix S4-(iii) also shows that the equilibrium point of the pure 1:3 division strategy, ($y_1^*(t) + y_2^*(t) + y_3^*(t) + y_4^*(t) > 0$ and $x_i^*(t) = 0$ ($i = 1$–4)), is locally stable when inequalities (5) and (7) are satisfied, where (7) is given by

$$2p_2 < 1 + p_1p_2. \tag{7}$$

Inequalities (6) and (7) are called the competition conditions.

These analyses indicates that the 2:2 division strategy wins against the 1:3 division strategy when inequalities (4) and (6) are satisfied, while upholding inequalities (5) and (7) favors the 1:3 division strategy. Inequalities (4)–(7) indicate that the division rate h does not affect the competition condition, but does affect the viability conditions. Under the viability conditions, small h restrains colony survivorship. The competition conditions are influenced by the survival probabilities of colonies of size 1 and 2 (see inequalities (6) and (7)), but not by those of older colonies.

Equation (6) can be rewritten as $p_2 > 1/(2-p_1) \geq 1/2$, indicating that colonies adopting the 2:2 division strategy can overcome their 1:3 counterparts when p_2 exceeds 0.5. Inequality (7) can be rewritten as $p_2 < 1/(2-p_1) \leq 1$. Equations (6) and (7) also indicate that relatively high p_2 can favor colonies adopting the 2:2 division strategy.

When $d_1 = d_2 = d_3 = d_4$, i.e. the dynamics are size-independent, inequality (7) always holds. This result implies that, in size-independent dynamics, the 1:3 division strategy is always advantageous over the 2:2 division strategy.

To investigate whether inequalities (4) – (7) hold for any values of p_1, p_2, p_3 and p_4, we conduct the numerical simulations of Eq. (D1) presented in appendix S4, and also run the colony-based simulations.

Simulation results of competition for empty sites between two reproductive strategies (2:2 division and the 1:3 division). We conducted colony-based simulations based on the assumptions in the previous section. We assume 2,500 sites, either unoccupied or occupied by colonies adopting one of two reproductive strategies. Each parameter set was run 100 times.

Three forms of the baseline death probability function are considered in the simulations. Death type (i) assumes that $d_2 = d_3 = d_4$. If d_1 is higher than the other rates, this function is similar to the exponential decreasing function adopted in Nakamaru et al. [17]. In death type (ii), the death probability function is a linear function of size, and defining d_1 and d_4 automatically determines d_2 and d_3. In death type (iii), we assume that $d_1 = d_2 = d_3$. If $d_1 = d_4$, the four death probabilities are identical. In this scenario, we can investigate the size-independent ecological dynamics and compare them with the size-dependent dynamics.

To understand the effect of colony division and death probability on the ecological dynamics described by Eqs. (1), (2) and (D1), the theoretical results (computed from inequalities (4) – (7)) and the simulation results are plotted in Fig. 2. We now compare the results of inequalities (4) – (7) with those of the colony-based simulations and the numerical calculations of eq. (3), and show that these approaches are consistent. Figure 2 A shows how the outcomes are affected by two death probabilities (d_1 and d_4) in death type (i). Each parameter set was tested in 100

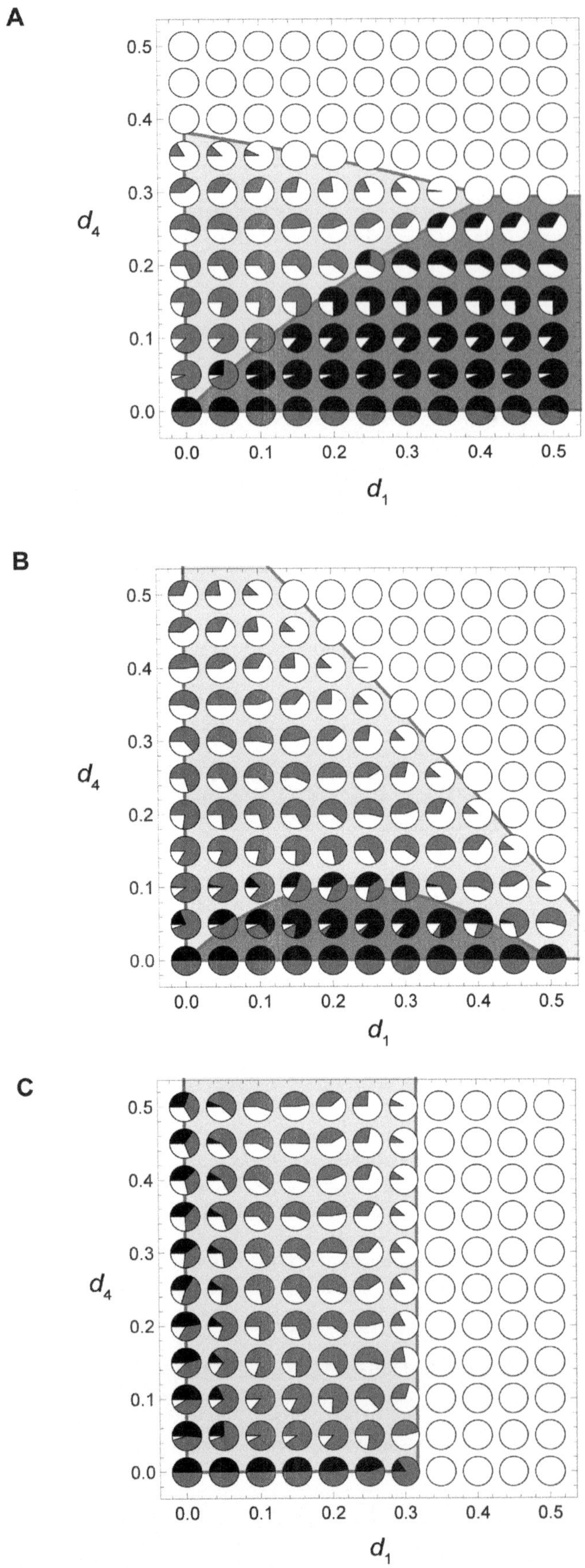

Figure 2. The d_1-d_4 graphs of the theoretical and simulated solutions in a completely mixed population. The horizontal and vertical axes represent d_1 and d_4, respectively The lines in the graphs delineate the viability and competition conditions in the completely mixed population. Theoretical analysis predicts that the 2:2 division strategy wins in the dark gray area, while the 1:3 division strategy wins in the light gray area. Each pie chart shows the average densities of sites occupied by either of the two strategies, as well as the empty sites (open circles) after 10,000 iterations in 100 simulation runs. The black and gray portions of the filled circles indicate the average density of sites occupied by the 2:2 and 1:3 division strategies, respectively. In $d_4=0.0$, matured colonies adopting both strategies ultimately survive and neutrally coexist. Initial densities are $z_0=0.2$ and $x_1=x_2=x_3=x_4=y_1=y_2=y_3=y_4=0.1$. Other parameters are: 2,500 sites in the population, $u=0$ and $h=1$. (A) Death probability function assumes that $d_2=d_3=d_4$. (B) Death probability function is linear. (C) Death probability function assumes that $d_1=d_2=d_3$.

simulation runs. Since the simulated and theoretical outcomes are consistent, we conclude that inequalities (4) – (7) can predict the simulated ecological dynamics even though $p_1 \leq p_2 \leq p_3 \leq p_4$ is violated (see appendix S2). When inequality (7) is satisfied (roughly, $d_1 \leq d_4$), colonies adopting the 1:3 division strategy dominate the population. Conversely, if inequality (6) is upheld (roughly, $d_1 > d_4$), the population is dominated by colonies undergoing 2:2 division. When d_4 is high or neither viability condition is satisfied, size 4 colonies die out before dispersing their daughter colonies and all colonies become extinct.

Figure 2 B shows the simulation outcomes in the d_1- d_4 graph in death type (ii). In parameter sets satisfying $d_1 > d_4$ and small d_4, colonies adopting the 2:2 division strategy dominate the population in some cases. When d_1 or d_4 is high, all colonies become extinct. Otherwise, the population is dominated by colonies adopting the 1:3 division strategy. As also shown in Fig. 2 B, inequalities (4) – (7) can predict the simulated ecological dynamics even when $p_1 \leq p_2 \leq p_3 \leq p_4$ is violated (See appendix S2). In death type (iii), colonies adopting the 2:2 division strategy are always overwhelmed by those undertaking the 1:3 division strategy (figure 2 C).

Figure 2 displays the outcomes of special cases of the death functions. One fixed ($d_4=0.2$) and three variable (d_1, d_2 and d_3) death probabilities were assumed in the additional simulations (figures are not shown). The outcome of this scenario also support inequalities (4) – (7), in which the competition between the two strategies depends on both d_1 and d_2, and colony survivorship is affected by all death probabilities. The high d_1 ensures that colonies adopting the 2:2 division strategy ultimately win, while low d_2 favors the 2:2 division strategy.

We conclude that the 2:2 division strategy is advantageous over the 1:3 strategy when the death probability of size 1 colonies is relatively higher than that of size 2 colonies (see Eq. (6) and Fig. 2). Otherwise, the 1:3 strategy is favorable. The reasons for these findings are now discussed: Fig. 1 C shows that the 1:3 strategy divides a mature colony into two daughter colonies of unequal size (1 and 3). The size 1 colony matures much more slowly than the size 3 colony, which rapidly (by comparison) reaches size 4 and divides. By contrast, the 2:2 strategy divides a mature colony into two half-sized colonies (figure 1 B). A size 2 colony matures more rapidly than a size 1 colony but less rapidly than one of size 3. Therefore, if all colonies share the same death probability, the 1:3 division strategy is advantageous over the 2:2 strategy because the probability of reproduction is higher for size 3 colonies than for size 2 colonies (see Fig. 1 B and C).

Now consider the scenario in which the death probability of a size 1 colony is prohibitively high. Once colonies adopting the 1:3

division strategy have divided, the size 1 daughter colonies die out, while the size 3 colonies survive and mature. Colonies adopting the 2:2 division strategy are not affected by the small survival chance of size 1 colonies, and both daughter colonies will likely survive.

Next, we assign a high death probability to a size 2 colony, and observe the effects on the ecological dynamics. Figure 1 B indicates that colonies adopting the 2:2 division strategy can scarcely increase their numbers because most of the daughter colonies dies out before reaching maturity. Meanwhile, although new size 1 colonies also die out, the size 3 colonies can survive to maturity. Therefore, imposing a high death probability on size 2 colonies favors colonies adopting the 1:3 division strategy. This discussion is supported by the competition conditions (Eqs. (6) and (7)).

The results when all growth probabilities are less than one

The following analysis concerns the effect of growth probability on the ecological dynamics. Applying the methods of appendices S1–S3 to Eqs. (1) and (2), the viability condition of the 2:2 and 1:3 division strategies are respectively obtained as

$$2p_2p_3g_2g_3>(p_4+(1-p_4)/h)(1-p_2(1-g_2))(1-p_3(1-g_3)), \quad (8)$$

and
$$(p_1p_2g_1g_2+(1-p_1(1-g_1))(1-p_2(1-g_2)))\,p_3g_3 >(p_4+(1-p_4)/h)\,(1-p_1(1-g_1))\,(1-p_2(1-g_2))(1-p_3(1-g_3)). \quad (9)$$

Applying the methods of appendix S4 to Eq. (3) and to the general equation (D1), we can calculate the conditions under which the 2:2 division strategy wins against the 1:3 strategy when inequalities (8) and (10) are satisfied, where (10) is given by

$$2p_2g_2(1-p_1(1-g_1)) > (1-p_1(1-g_1))(1-p_2(1-g_2)) + p_1p_2g_1g_2. \quad (10)$$

On the other hand, if inequalities (9) and (11) hold, the 1:3 division strategy can defeat the 2:2 strategy, where (11) is given by

$$2p_2g_2(1-p_1(1-g_1)) < (1-p_1(1-g_1))(1-p_2(1-g_2)) + p_1p_2g_1g_2. \quad (11)$$

Inequalities (10) and (11) are called the competition conditions. The competition conditions are independent of the division rate h, but h does affect the viability conditions. In essence, small h restrains colony survival. The competition conditions are influenced by the survival and growth probabilities of size 1 and 2 colonies (see also inequalities (6) and (7)), but not by those of older colonies.

When $d_1=d_2=d_3=d_4$ and $g_1=g_2=g_3$, interpreted as no colony dynamics, inequality (11) is always upheld. This result implies that, in static colonies, the 1:3 division strategy always confers survival at the expense of the 2:2 division strategy.

We now compare the results of inequalities (8)-(11) with those of the colony-based simulations and the numerical calculations of eq. (3), and show that these approaches are consistent. First, we examine the effect of growth probabilities on the system using the outcomes derived from inequalities (8) – (11). Figure 3 shows the effect of g_2 and g_1 ($=g_3$) on the competition and viability conditions defined in inequalities (8) – (11). If $g_2>g_1=g_3$, the fastest growing colonies are those increasing from size 2 to 3. This scenario mimics logistic growth. When $d_1=0.35>d_2=d_3=d_4=0.15$, and also $g_1=g_2=g_3=1$ (see Figures 2 A and 3 A), the population is dominated by colonies adopting the 2:2 division strategy. When

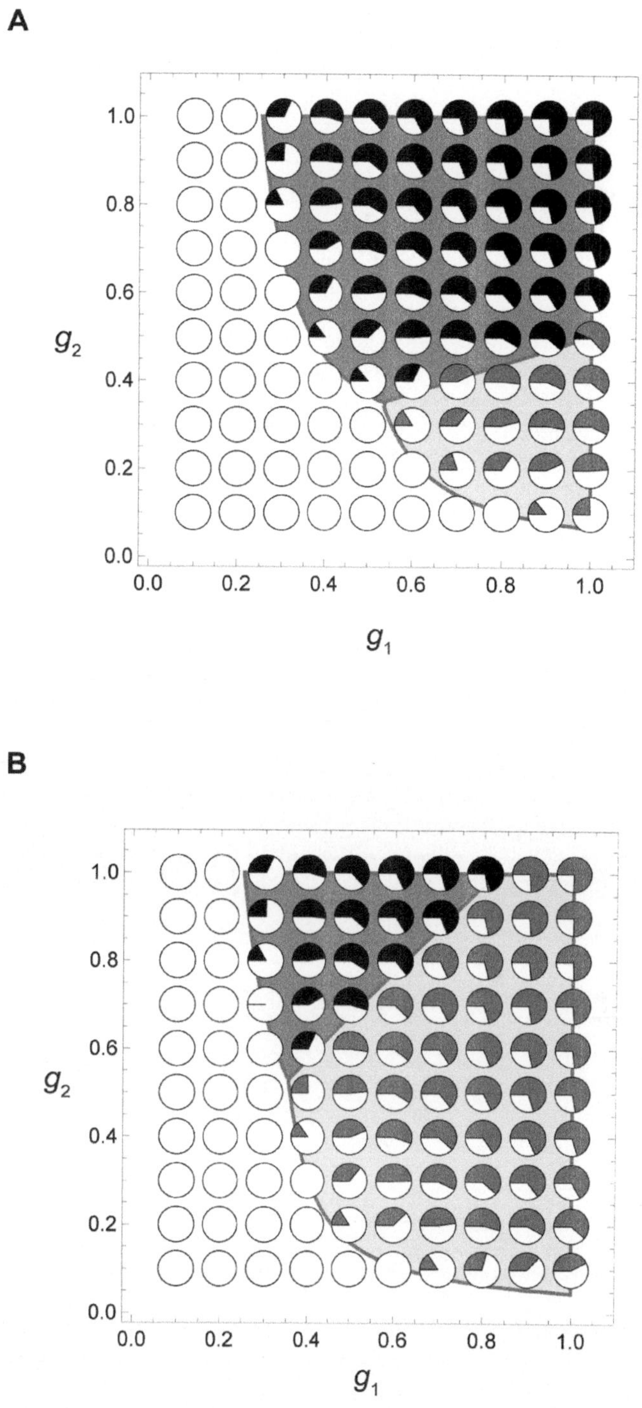

Figure 3. The g_1-g_2 plot of simulated and theoretical results in a completely mixed population. Simulations are iterated 6,250 times. Parameters are $d_2=d_3=d_4=0.15$, $h=1$, $u=0$. The horizontal and vertical axes represent g_1 ($=g_3$) and g_2, respectively. Initial densities are $z_0=0.6$ and $x_1=x_2=x_3=x_4=y_1=y_2=y_3=y_4=0.05$. More information is provided in the caption to Fig. 2. (A) $d_1=0.35$. (B) $d_1=0.15$.

$g_2>g_1=g_3$, colonies adopting the 2:2 division strategy can always defeat their unequally dividing counterparts. Conversely, lower g_2 favors colonies adopting the 1:3 division strategy. When $d_1=d_2=d_3=d_4=0.15$, and also $g_1=g_2=g_3=1$, the 1:3 division strategy becomes favorable (see figure 3 B). When $g_2<g_1=g_3$, colonies adopting the 1:3 division strategy outcompete those

adopting the 2:2 division strategy. The reverse occurs when $g_2 > g_1 = g_3$, especially when g_2 is high. According to Figure 3, low and high g_2 promote the 1:3 division strategy and the 2:2 division strategy, respectively. These results are supported by the simulation outcomes.

Model 2 and results: Direct competition between a dispersing colony and settled colonies

Thus far, the model has assumed no direct competition between colonies; dispersing colonies avoid sites occupied by existing colonies, and seek unoccupied sites in which to settle. However, direct competition is the norm in nature. In this section, we examine the effect of direct competition imposed by dispersing colonies settling at new sites.

Competition is introduced to the baseline model with $g_i = 1$. Once a colony has divided into two, one of the daughter colonies migrates from the natal colony site to a new site chosen randomly from the site population. If the site is empty, the dispersing colony settles there. If the site is occupied by another colony, the dispersing colony competes with the settled colony for the space. Let w_{ij} be the probability that a dispersing colony of size i invades and annihilates a settled colony of size j. We investigate three types of direct competition: (i) $w_{ij} = 0.0$, (ii) $w_{ij} = 1.0$, (iii) $w_{ij} = 1$ if $i > j$, $w_{ij} = 0.5$ if $i = j$, and $w_{ij} = 0.0$ if $i < j$. In situation (i), the settled colony can overcome the dispersing colony regardless of its size, and the dispersing colony dies out. In (ii), the dispersing colony always combats the settled colony, the settled colony dies out, and the dispersing colony occupies the site. Situation (iii) implies that a larger colony always overtakes a smaller one.

If all colonies in the population adopt the 2:2 division strategy, the matrix construct is

$$x(t) = S_1(t)x(t), \tag{12}$$

where $x(t) = (x_0(t), x_1(t), x_2(t), x_3(t), x_4(t))^t$ and

$$S_1(t) = \begin{bmatrix} 1-hx_4(t) & 1-p_1 & 1-p_2 & 1-p_3 & (1-p_4)(1-h) \\ 0 & -hx_4(t)w_{21} & 0 & 0 & 0 \\ hx_4(t) & p_1+hx_4(t)w_{21} & 0 & hx_4(t)w_{23} & h+hx_4(t)w_{24} \\ 0 & 0 & p_2 & -hx_4(t)w_{23} & 0 \\ 0 & 0 & 0 & p_3 & p_4(1-h)-hx_4(t)w_{24} \end{bmatrix}.$$

At equilibrium, $x^* = (1, 0, 0, 0, 0)^t$ is locally unstable when $2p_2p_3 > p_4 + (1-p_4)/h$ regardless of w_{ij}, which corresponds to inequality (4).

Similarly, if all colonies in the population adopt the 1:3 division strategy, the matrix construct is

$$y(t) = L_1(t)y(t), \tag{13}$$

where $y(t) = (y_0(t), y_1(t), y_2(t), y_3(t), y_4(t))^t$ and

$$L_1(t) = \begin{bmatrix} 1-hy_4(t) & 1-p_1 & 1-p_2 & 1-p_3 & (1-p_4)(1-h) \\ hy_4(t) & 0 & hy_4(t)w_{12} & hy_4(t)w_{13} & hy_4(t)w_{14} \\ 0 & p_1 & -hy_4(t)w_{12} & 0 & 0 \\ 0 & 0 & p_2 & -hy_4(t)w_{13} & h \\ 0 & 0 & 0 & p_3 & p_4(1-h)-hy_4(t)w_{14} \end{bmatrix}.$$

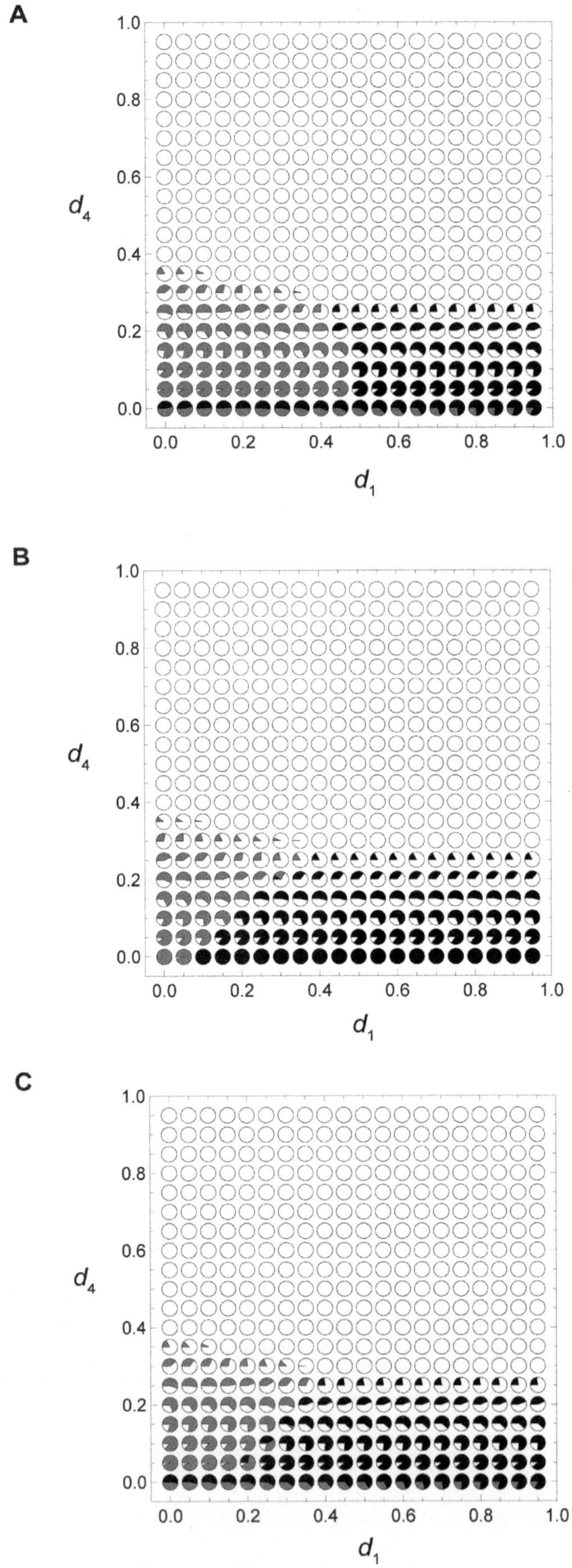

Figure 4. The d_1-d_4 graph of the colony-based simulations in which a dispersing colony competes with settled colonies. The horizontal and vertical axes represent d_1 and d_4, respectively. Death type (i), in which $d_2 = d_3 = d_4$, is assumed. Each pie chart shows the average densities of sites occupied by either of the two strategies, as well as the empty sites (open circles) after 5,000 iterations of (A) and (B) and after 20,000 iterations of (C) in 100 simulations. The black and gray portions of each circle indicate the average densities of sites occupied by colonies undertaking the 2:2 division strategy and the 1:3 division strategy, respectively. The open portion represents the average density of empty sites. Initial densities are $z_0 = 0.6$ and $x_1 = x_2 = x_3 = x_4 = y_1 = y_2 = y_3 = y_4 = 0.05$. The other parameters are: 2,500 sites, $u = 0$ and $h = 1$. (A) Direct competition (i), in which $w_{ij} = 0.0$. (B) Direct competition (ii), in which $w_{ij} = 1.0$. (C) Direct competition (iii), in which a bigger colony always combats a smaller one.

At equilibrium, $y^* = (1, 0, 0, 0, 0)^t$ is locally unstable when $(1+p_1p_2)p_3 > p_4 + (1-p_4)/h$ regardless of w_{ij}, which corresponds to inequality (5).

Interestingly, the baseline model with $g_i = 1$ and the direct competition model share the same local stability condition; an equilibrium in which all colonies become extinct. This arises because a newly-divided dispersing colony encountering many empty sites at equilibrium will more likely settle at a site not already occupied. In this situation, direct competition exerts little influence on the population dynamics.

To investigate the effect of direct competition among colonies adopting either reproductive strategy, we run the colony-based simulations under death type (i), which most strongly favors the 2:2 division strategy among the three death categories. As shown in Figure 4, among the three categories of direct competition, the 2:2 division strategy is most favored in category (ii).

Next, we discuss why direct competition category (ii), rather than category (i), confers advantage to the 2:2 division strategy, focusing on the disadvantage of the 1:3 division strategy in this scenario. The 1:3 division strategy disperses a size 1 colony, while the larger colony remains at the site. Recall that, in this strategy, a size 3 colony matures more rapidly than a colony of size 1. In direct competition category (ii), a size 3 colony dies out, whereas a size 1 colony disperses and settles at a new site. Since a size 1 colony matures more slowly than a size 2 colony, the 2:2 division strategy is advantageous over the 1:3 division strategy. Conversely, in direct competition category (i), a size 3 colony can survive but a size 1 colony cannot colonize a new site. In this scenario, the 1:3 division strategy is preferable.

The assumptions of the baseline model and direct competition category (i) are similar in that dispersing colonies can settle only at empty sites. However the two scenarios generate completely different outcomes (see Figures 2 A and 4 A). In Figure 2 A, the 2:2 division strategy is more advantageous than in Figure 4 A, because dispersing colonies always die out under direct competition (i), unless they find an unoccupied site to colonize. Conversely, in the baseline model, colonies do not disperse unless empty sites are available. Although the dispersing colonies die out, the 1:3 division strategy confers less damage than the 2:2 division strategy because the size 3 colonies remaining at the natal site after division compensate for the loss of the smallest colonies in direct competition category (i). Populations of colonies adopting the 2:2 division strategy lose their dispersing size 2 members, which is more damaging to the population dynamics than the loss of size 1 colonies. Therefore the 2:2 division strategy cannot outcompete the 1:3 division strategy in this scenario.

Although direct competition (ii) most favors the 2:2 division strategy among the three direct competitions, it disadvantages the 2:2 division strategy relative to the baseline model with $g_i = 1$ (See Figures 2 A and 4 B). Therefore we conclude that the 2:2 division strategy is advantageous if newly-divided colonies disperse solely to empty sites and never directly compete with colonies to claim an occupied site. In the following section, we investigate the effect of spatial structure and environmental disturbance on the population dynamics by imposing spatial structure on the baseline model.

Model 3 and results: Spatial structure and environmental disturbance

In a previous section, we found that imposing death type (i) and the logistic growth of colonies favor the 2:2 division strategy over the 1:3 strategy when dispersing colonies seek and settle at unoccupied spaces in a completely mixed population. In this section, we examine the environmental factors that favor the 2:2 division strategy over the 1:3 division strategy. To this end, we use the lattice-structured population as a spatial structure, assuming death type (i) and that a dispersing colony seeks and settles at an unoccupied site. Although the 2:2 division strategy is advantageous when $g_2 > g_1$ and g_3, here we set $g_1 = g_2 = g_3 = 1$ for simplicity. We then run simulations to investigate how the population dynamics are affected by spatial structure and environmental disturbance.

We first present the assumptions of the spatial structured model without environmental disturbance. We use a lattice model as a spatial structure model. Each colony distributed in each lattice site, possesses two traits; colony division and dispersal distance. Each colony adopts either the 2:2 or the 1:3 division strategy (Figure 1), and disperses either short- or long-distance.

This setup yields four possible division/dispersal strategies: 1:3 long, 1:3 short, 2:2 long and 2:2 short. The 1:3 long strategy disperses the size 1 colony to a random site in the lattice-structured population. The 1:3 short strategy displaces the size 1 colony to an empty site among the nearest neighbors. The 2:2 long disperses one of the size 2 colonies to a random site in the lattice-structured population. The 2:2 short strategy and deposits one of the divided size 2 colonies in a nearest-neighbor empty site. Among these strategies, the 1:3 long and 2:2 short are most commonly found in nature. Thus, we focus on the competition between the 1:3 long and 2:2 short strategies in the lattice-structured population.

Simulated results in the absence of environmental disturbance. First, to clarify the effect of the lattice-structured population on the competition between the two strategies, we assume no environmental disturbance. Computer simulations were conducted using death type (i) and $g_1 = g_2 = g_3 = 1$. The outcomes of the simulations are presented in Figure 5. When the colony is static ($d_1 = d_2 = d_3 = d_4$), and the death probabilities relatively low, the 1:3 long overwhelms 2:2 short. Colonies adopting the 2:2 short strategy can outcompete their 1:3 long counterparts if the death probability of size 1 colonies is sufficiently high. The 1:3 long strategy is favorable over the 2:2 short because the latter strategy places a daughter colony only at a nearest neighbor site. Thus, the 2:2 short strategy cannot search for unoccupied sites when the local colony density is high, and colonies adopting this strategy are prevented from increasing their number. On the other hand, the 1:3 long strategy disperses a new colony to a random site, releasing it from local density effects. Provided that empty sites exist in the population, colonies adopting the 1:3 long strategy can sustain population growth. We also conducted simulations in which death probabilities d_1, d_2 and d_3 were variable while d_4 was fixed at 0.2. The outcomes of these simulations are consistent with other results (see Figure S1).

When all of g_1, g_2, and g_3 are less than one, the lattice-structured population displays logistic growth, which favors the 2:2 short strategy over the 1:3 long (see Figure S2).

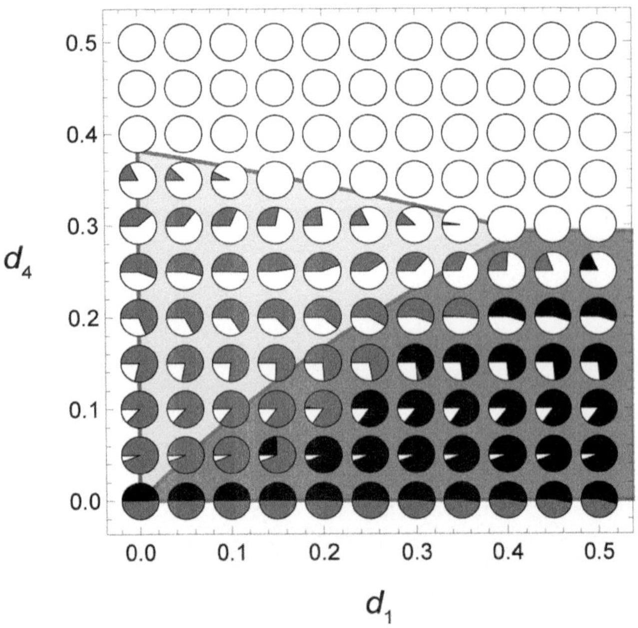

Figure 5. The d_1-d_4 graph of the two-dimensional lattice-structured population. The horizontal and vertical axes represent d_1 and d_4, respectively. Each circle shows the average densities of sites occupied by either of the two strategies, as well as the empty sites (open circles) after 10,000 iterations in 100 simulations on the two-dimensional lattice. The black and gray portions of each circle indicate the average density of the sites occupied by colonies undertaking the 2:2 division and 1:3 division strategies, respectively. The open portion represents the average density of the empty sites. The lines in the graphs delineate the theoretical solutions in a completely mixed population. In the dark gray area, the 2:2 division strategy theoretically wins over the 1:3 strategy. In the light gray area, the predicted winning strategy is 1:3. In $d_4 = 0.0$, mature colonies always survive and a neutral relationship develops between the strategies, which then coexist. Initial densities are $z_0 = 0.2$ and $x_1 = x_2 = x_3 = x_4 = y_1 = y_2 = y_3 = y_4 = 0.1$. The other parameters are: 2,500 lattice sites, $u = 0$ and $h = 1$. The death probability assumes death type (i).

Furthermore, if competition is introduced among all four division/dispersal strategies, colonies adopting either short-distance strategy rapidly become extinct because competition strongly favors long dispersal distance. After just a few iterations, the population comprises colonies adopting the 1:3 long strategy and the 2:2 long strategy. Since the 1:3 and 2:2 long-dispersing colonies in the lattice-structured population behave identically to their counterparts adopting the 1:3 and the 2:2 division strategies in the completely mixed population, the final competition outcomes among all four division/dispersal strategies are those of the 1:3 and the 2:2 division strategies in the completely mixed population (see Figure S3).

Simulated results in the presence of environmental disturbance. We now introduce two parameters of environmental disturbance, u and q, defined as the frequency of disturbance and the disturbance scale, respectively [17]. These parameters are independent of the baseline natural death. q can be interpreted as the degree of spatial autocorrelation. The probability that a colony of size i is removed from a site by environmental disturbance (leaving the previously occupied site empty) is defined as u_i $(0.0 \leq u_i \leq 1.0)$. Disturbance is assumed independent of colony size and density $(u_1 = u_2 = u_3 = u_4 = u)$. A disturbance kills a colony occupying a lattice site with probability u. The disturbance then spreads to a nearest neighbor site (selected randomly among eight neighbors) with probability q and kills the residential colony (if present). If the site is unoccupied, it is merely disturbed. The disturbance is repeated with probability q until halted with probability $1 - q$. For example, if u is low and q is high, the lattice contains a few large or clustered disturbed areas. If u is high and q is low, many small areas are disturbed in the lattice.

Figure 6 shows the effect of the disturbance parameters u and q on the competition between 1:3 long and 2:2 short when $g_1 = g_2 = g_3 = 1$. This simulation assumes death type (i). As demonstrated in Figure 6, low q and relatively high u confer an advantage to the 2:2 strategy when d_1 is higher and the difference between d_1 and d_2 $(= d_3 = d_4)$ is larger (see figure 6 C, D and E). Increasing u and q favors the 1:3 long strategy at certain death probabilities favoring the 2:2 short strategy in the absence of disturbance. These outcomes are supported by standard dispersal theory. The simulation outcomes are influenced by disturbance when the growth probabilities are less than 1 (see figure S4). According to this figure, the 2:2 short strategy is advantageous over 1:3 long when size 2 colonies grow rapidly (high g_2) relative to g_1 (which is fixed; see figure S4 A–C and D–F). Figure S4 A and D, together with Figure S4 B and E, indicate that the 2:2 short strategy is advantageous when g_1 is low relative to g_2.

Discussion and Conclusions

We examined the ecological conditions under which budding can be advantageous over migration in colonial organisms. The main conclusions are summarized as follows: (1) when the life history parameters are independent of colony size, budding colonies are overwhelmed by dispersive colonies. (2) When the death probability of the smallest colonies is relatively higher than that of more mature colonies, survival favors the budding strategy. The death function can be modeled as a negative exponential function of the continuous lifespan. (3) Budding wins against dispersal when the growth rates of size 1 and size 3 colonies are lower than that of size 2 colonies. In this scenario, colonies undergo logistic growth, which ensures their survival if they adopt the budding strategy. (4) When a newly-divided dispersing colony does not compete with a settled colony for space, but instead colonizes open spaces, the budding strategy is advantageous over dispersal. This corresponds to the situation in which founding colonies leave their natal site, and seek and settle at a nearby open space.

We then examine the effects of spatial structure and environmental disturbance on the population dynamics. Relative to dispersal, the budding strategy proved less favorable in spatial structured populations than in completely mixed populations. This occurs because long distance dispersal prevents local overcrowding. In the event of environmental disturbance, long distance dispersal confers a survival advantage over budding species because migrating species can escape the risk. However, when the scale of environmental disturbance is small, budding can retain an advantage even under relatively high frequency of environmental disturbance.

Our current paper roughly supports the main results presented in Nakamaru *et al.* (2007). Among the various growth and death functions, logistic colony growth (corresponding to g_1, $g_3 < g_2$) and an exponentially decreasing death function (corresponding to $d_1 >> d_2$, d_3, d_4) favor budding over dispersal. However, the assumptions of our current model differ from those in Nakamaru et al (2007). For example, the previous study assumed continuous colony size, and that each colony obeys logistic growth and an exponentially decreasing death function. Here, to more easily investigate the effect of different growth and death functions on the

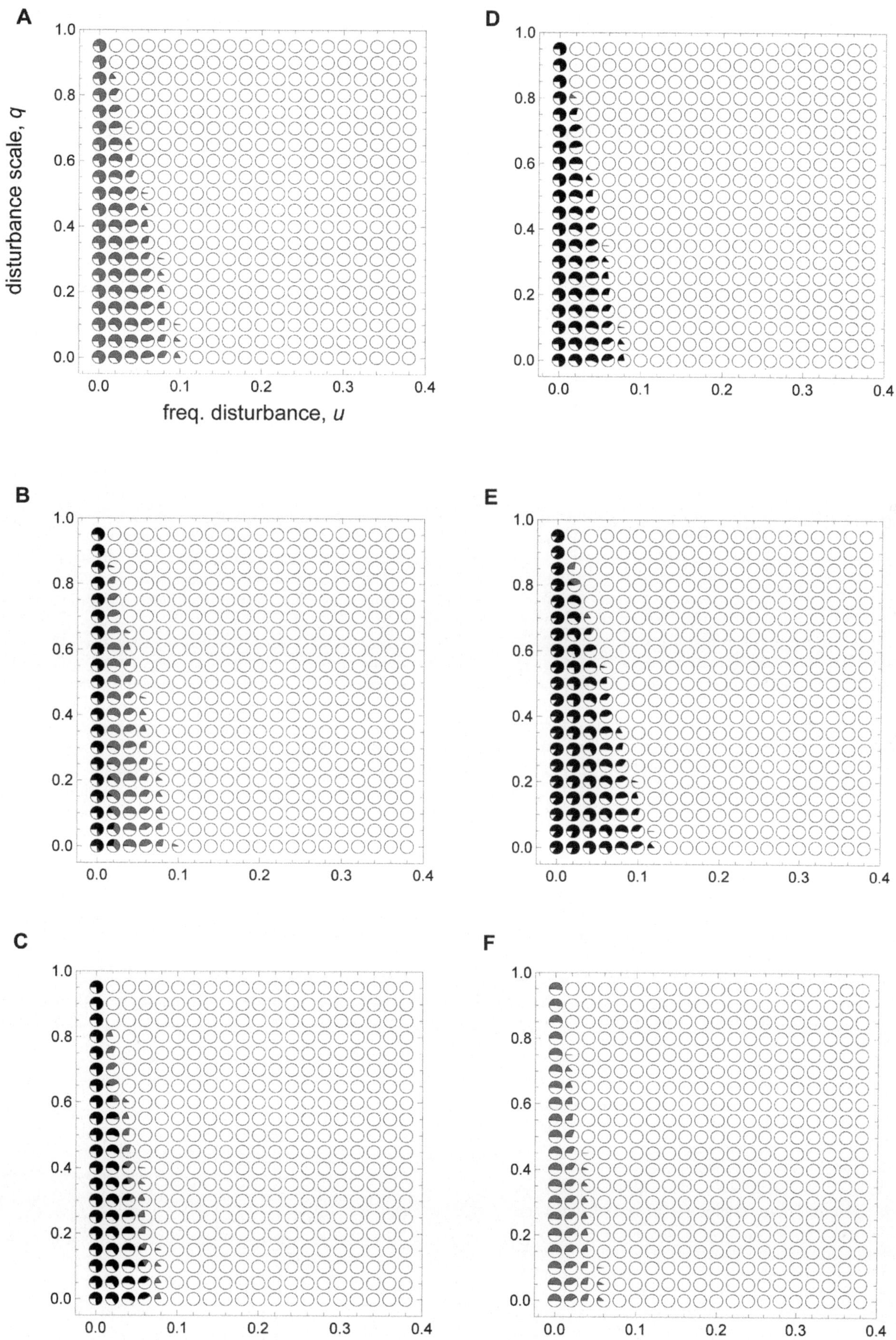
A
B
C
D
E
F
1.0
0.8
0.6
0.4
0.2
0.0
0.0
0.1
0.2
0.3
0.4
disturbance scale, q
freq. disturbance, u

Figure 6. The *u*-*q* graph of colony-based simulations in the two-dimensional lattice-structured population after 5,000 iterations. Details are provided in the caption to Figure 5. The death probability assumes death type (i). (A) $d_1 = 0.25$ and $d_2 = 0.15$, (B) $d_1 = 0.3$ and $d_2 = 0.15$, (C) $d_1 = 0.35$ and $d_2 = 0.15$, (D) $d_1 = 0.4$ and $d_2 = 0.15$, (E) $d_1 = 0.35$ and $d_2 = 0.1$ and (F) $d_1 = 0.35$ and $d_2 = 0.2$. Other parameters are: $g_1 = g_2 = g_3 = 1$ and $h = 1$.

evolutionary population dynamics, we discretized the colony size. Despite these differences in model assumptions, the results are consistent between the two studies, which strengthens our conclusions.

Budding likely evolves when very small individuals (or colonies) undergo logistic growth and are subject to relatively high mortality. Furthermore, our models suggest that budding is favored when dispersing offspring (colonies) can seek unoccupied spaces in which to settle without inciting conflict. Ant or coral colonies (or colonies of ramets in plants) usually exhibit type III survivorship curves (negative exponential mortality functions) [21–23] and logistic growth [21,24–26]. Furthermore, ants are known to seek suitable empty sites when their colony disperses [27,28]. Thus, these organisms are naturally predisposed to budding. Although our results also suggest that environmental disturbance promotes long-dispersing by its risk hedging effect and discourages budding, the benefits of budding may outweigh the disadvantage of short-distance dispersal in some of the above mentioned organisms (some corals, plants and ants) that reproduce by budding in frequently disturbed environments.

Rigorous empirical testing of our hypothesis is limited by the difficulty of distinguishing the baseline death and growth rates from the changes in those parameters induced by disturbance. However, we consider that our model predictions could be tested on realized average death and growth rates that include both baseline and additional (or reduced) rates caused by disturbance. As already mentioned, life history parameters such as survival and growth functions appear to directly determine which strategy should evolve. If the average death and growth rates of fission-dividing colonial organisms fulfill the predicted conditions favoring the 2:2 division strategy described in Figure 2 and Eqs. (4) and (6), our models would be supported. The life history of species or closely related species adopting both budding and non-budding tactics must be investigated in empirical studies.

Previous studies on the evolution of body-condition dependency in dispersal have not directly addressed the budding phenomenon [29–31]. Instead, these models assume semelparity or an annual lifecycle. Budding is possible only when a parent (or part of the parent's body) survives after reproduction, coexists with the offspring (perennial) and re-reproduces (iteroparity). Therefore, our results are not directly comparable with the results of these studies.

We now discuss the factors that should be incorporated into our current model to render it more realistic. We assumed that a colony reproduces by dividing in two. We also assumed that (not always, however) large offspring disperse over a smaller distance (size-dispersal distance trade off). However, another reproductive constraint has been reported in the literature; the trade-off between size and number of offspring (e.g. [32]). We consider that the size-number trade-off is a sensible proposition when the parental effort or total parental resource that can be allocated to reproduction is constant. For generality in our current study, the total reproductive effort was assumed labile (for instance, the dispersive strategy (3:1 strategy) invests more resource in survival than budding, given that the dispersed portion is the offspring). We now introduce the 2:1:1: division strategy, in which a size 4 colony divides into one size 2 colony and two size 1 colonies that then disperse, and the 1:1:1:1 division strategy, in which a size 4 colony divides into four size 1 colonies. We constrained the reproduction effort to be constant (mimicking a size–number trade off) in the competition between 2:1:1 vs. 2:2 strategies. However, biological strategies might be more precisely described by 2:1:1 and 1:1:1:1 strategies; for example; plants may release many seeds and ant colonies disperse many winged queens simultaneously. We have partially tackled this situation. In well-mixed populations, the local stable conditions of the 1:1:1:1 and the 2:1:1 division strategies are $4p_1p_2p_3 > p_4 + (1-p_4)/h$ and $p_2p_3(1+2p_1) > p_4 + (1-p_4)/h$, respectively (method presented in the Appendices). The left hand side of both inequalities specifies the probability that dividing colonies mature to size 4. This result implies that (i) budding wins against the other three strategies when the survival probability of size 1 colonies (p_1) is very small, and (ii) if the converse is true (p_1 is high), the 1:1:1:1 division strategy may be favored. The latter condition might simulate the evolution of semelparity. However, the strongest strategy when members of the population adopt all four strategies has yet to be elucidated.Furthermore, one must consider relaxing the assumption on the timing of reproduction. Size 4 colonies only reproduce in the current model. If we assume that size 2 and 3 colonies can reproduce, we can analyze the conditions under which annual or perennial life cycles evolve (regarding the reproductive strategy of size 2 colonies as annual semelparity).

The current model does not include colony size effects in disturbance. For instance, if smaller colonies are more vulnerable to environmental disturbance than larger ones, the budding might be more advantageous over dispersal than is predicted in this paper. Finally, our study considers neither explicit competition among individuals within each colony nor direct interference induced by competition between neighboring colonies. In future work, we will incorporate these factors into a more realistic model.

Budding has been considered to play a key role in evolutionary ecology. Theoretically, budding increases the genetic variance among groups and/or the genetic relatedness within groups, thus promoting the evolution of group-advantageous traits such as altruism and female-biased sex allocation [33–35]. On the other hand, restricted migration (which frequently accompanies fission) intensifies the local competition among related individuals and hinders the evolution of altruism, despite the apparent enhancement of genetic relatedness of interactants [36–38]. However, recent theoretical and empirical studies suggest that budding can solve this "viscous population dilemma" and enable the evolution of altruism, because budding increases the number of more related individuals in the background local population [7,8,39–45]. All of the above-mentioned theoretical studies focus on the effect of budding on evolved social traits. Thus, budding is assumed as the given condition, without considering the environments in which budding evolves. The latter problem is the focus of our current paper. On the other hand, the ecological life history models introduced here neglect issues such as relatedness and intra-colonial conflicts. Therefore, our current study is complementary to existing studies. In future studies, we hope to incorporate relatedness and intra-colonial conflicts into the model scheme, to better understand the ecological drivers of coevolving life history and social traits.

Supporting Information

Figure S1 Effect of life history parameters on competition between the two strategies in the two-dimensional lattice. Black spheres indicate that the 2:2 short strategy overcomes the 1:3 long strategy, while gray spheres indicate the opposite. Where no spheres are presented, colonies have become extinct. The volume of each sphere denotes the average density of the strategy after 1,000 iterations. This graph shows the effect of imposing the death probabilities d_1, d_2, d_3. The parameters are $d_4 = 0.2$, $g_1 = g_2 = g_3 = 1$, $h = 1$. Initial densities are $z_0 = 0.6$ and $x_1 = x_2 = x_3 = x_4 = y_1 = y_2 = y_3 = y_4 = 0.05$.

Figure S2 Effect of life history parameters on competition between the two strategies in the two-dimensional lattice. Black spheres indicate that the 2:2 short strategy overcomes the 1:3 long strategy, while gray spheres indicate the opposite. Where no spheres are presented, the colonies have become extinct. The volume of each sphere denotes the average density of the strategy after 6,250 iterations. This graph illustrates the effect of growth probabilities g_1, g_2, and g_3. Other parameters are: $d_1 = 0.35$, $d_2 = d_3 = d_4 = 0.15$. Initial densities are $z_0 = 0.6$ and $x_1 = x_2 = x_3 = x_4 = y_1 = y_2 = y_3 = y_4 = 0.05$.

Figure S3 Competition among four possible strategies in a structured population residing on a two dimensional lattice, after 10,000 iterations. The parameters are $g_2 = g_3$, $d_1 = 0.35$, $d_2 = d_3 = d_4 = 0.15$ and $h = 1$. The initial population density of each strategy is the same and the initial density of vacant sites is 0.6.

Figure S4 The *u*-*q* graph of colony-based simulations in the two-dimensional lattice structured population after 6,250 iterations. Details are provided in the caption to Figure 5. (A) $g_1 = 0.8$ and $g_2 = 0.8$, (B) $g_1 = 0.8$ and $g_2 = 0.6$, (C) $g_1 = 0.8$ and $g_2 = 0.4$, (D) $g_1 = 0.6$ and $g_2 = 0.8$, (E) $g_1 = 0.6$ and $g_2 = 0.6$ and (F) $g_1 = 0.6$ and $g_2 = 0.4$. Other parameters are: $d_1 = 0.35$, $d_2 = d_3 = d_4 = 0.15$, $h = 1$ and $g_3 = 0.9$. Initial density of each size of each strategy is identical, and the initial density of vacant sites is 0.6.

Appendix S1 Corollary of Jury's criterion.

Appendix S2 Local stability analysis of positive equilibrium on the 2:2 division strategy.

Appendix S3 Local stability analysis of positive equilibrium on the 1:3 division strategy.

Appendix S4 Local stability analysis in competitive system between two strategies.

Acknowledgments

We thank Evan Economo for his helpful comments.

Author Contributions

Conceived and designed the experiments: MN. Performed the experiments: AO SUS KM. Analyzed the data: AO SUS KM MN. Wrote the paper: MN TT KT. Made the mathematical model: MN. Proved the mathematical model: TT. Analyzed the mathematical model: TT MN.

References

1. Clobert J, Danchin E, Dhonde AA, Nichols JD (2001) Dispersal. Oxford: Oxford University Press.
2. Levin SA, Muller-Landau HC, Nathan R, Chave J (2003) The ecology and evolution of seed dispersal: A theoretical perspective. Annu Rev Ecol Syst 34: 575–604.
3. Clobert J, Baguette M, Benton TG, Bullock JM (2012) Dispersal ecology and evolution. Oxford: Oxford University Press.
4. Wynee-Edwards V (1962) Animal dispersion in relation to social behaviour. New York: Hafaer.
5. Williams GC (1966) Adaptation and natural selection. Princeton: Princeton University Press.
6. Ronce O, Olivieri I, Clobert J, Danchin E (2001) Perspectives on the study of dispersal evolution. In: Clobert J, Danchin E, Dhonde AA, Nichols JD, editors. Dispersal. Oxford: Oxford University Press. pp. 341–357.
7. Gardner A, West SA (2006) Demography, altruism, and the benefits of budding. J Evol Biol 19: 1707–1716.
8. Kummerli R, Gardner A, West SA, Griffin AS (2009) Limited dispersal, budding dispersal, and cooperation: an experimental study. Evolution 63: 939–949.
9. Gardner A, Arce A, Alpedrinha J (2009) Budding dispersal and the sex ratio. Journal of Evolutionary Biology 22: 1036–1045.
10. Liu J, Dong M, Miao SL, Li ZY, Song MH, et al. (2006) Invasive alien plants in China: role of clonality and geographical origin. Biol Invasions 8: 1461–1470.
11. Honnay O, Jacquemyn H (2010) Clonal plants: beyond the patterns—ecological and evolutionary dynamics of asexual reproduction. Evol Ecol 24: 1393–1397.
12. Coffroth MA, Lasker HR (1998) Population structure of a clonal gorgonian coral: The interplay between clonal reproduction and disturbance. Evolution 52: 379–393.
13. Passera L (1994) Characteristics of tramp species. In: William DF, editor. Exotic ants: biology, impact, and control of introduced species. Colorado: Westview Press. pp. 23–43.
14. Holway DA, Lach L, Suarez AV, Tsutsui ND, Case TJ (2002) The causes and consequences of ant invasions. Annu Rev Ecol Syst 33: 181–233.
15. Carpintero S, Reyes-Lopez J, de Reyna LA (2003) Impact of human dwelling on the distribution of the exotic Argentine ant: a case study in the Doñana National Park. Biol Conserv 115: 279–289.
16. Tsuji K (2010) Unicolonial ants: loss of colony identity. In: Breed M, Moore J, editors. Encyclopedia of Animal Behavior. Oxford: Academic Press. pp. 469–473.
17. Nakamaru M, Beppu Y, Tsuji K (2007) Does disturbance favor dispersal? An analysis of ant migration using the colony-based lattice model. J Theor Biol 248: 288–300.
18. Caswell H (2000) Matrix population models 2nd edition. Sunderland: Sinauer Associates, Inc., Publishers.
19. Stearns SC (1992) The Evolution of Life Histories. Oxford: Oxford University Press.
20. Gardner A, Grafen A (2009) Capturing the superorganism: a formal theory of group adaptation. J Evol Biol 22: 659–671.
21. Hölldobler B, Wilson EO (1990) The ants. Cambridge: Belknap Press of Harvard University Press.
22. Babcock RC (1991) Comparative demography of three species of scleractinian corals using age- and size-dependent classifications. Ecol Monogr 61: 225–244.
23. Harper JL (1977) Population Biology of Plants. Waltham: Academic Press.
24. Yamaguchi M (1983) Growth data analysis in the reef-building coral {Pocillopora} {damicornis} (Linnaeus). Galaxea 2: 21–27
25. van Woesik R, Sakai K, Ganase A, Loya Y (2011) Revisiting the winners and the losers a decade after coral bleaching. Mar Ecol Prog Ser 434: 67–76.
26. Suzuki JI, Hucthings MJ (1997) Interactions between shoots in clonal plants and the effects of stored resources on the structure of shoot population. In: de Kroon H, van Groenendael J, editors. The Ecology and Evolution of Clonal Plants. Leiden: Backhuys Publishers.
27. Stroeymeyt N, Giurfa M, Franks NR (2010) Improving decision speed, accuracy and group cohesion through early information gathering in house-hunting ants. PLoS One 5: e13059.
28. Cronin AL (2012) Consensus decision making in the ant Myrmecina nipponica: house-hunters combine pheromone trails with quorum responses. Anim Behav 84: 1243–1251.
29. Gyllenberg M, Kisdi E, Utz M (2008) Evolution of condition-dependent dispersal under kin competition. J Math Biol 57: 285–307.
30. Gyllenberg M, Kisdi E, Utz M (2011) Body condition dependent dispersal in a heterogeneous environment. Theor Popul Biol 79: 139–154.
31. Gyllenberg M, Kisdi E, Utz M (2011) Variability within families and the evolution of body-condition-dependent dispersal. J Biol Dyn 5: 191–211.
32. Smith CC, Fretwell SD (1974) The optimal balance between size and number of offspring. Am Nat 108: 499–506.
33. Slatkin M, Wade MJ (1978) Group selection on a quantitative character. Proc Natl Acad Sci U S A 75: 3531–3534.

34. Goodnight CJ, Stevens L (1997) Experimental studies of group selection: what do they tell us about group selection in nature? Am Nat 150: S59–S79.
35. Ono S, Misawa K, Tsuji K (2003) Effect of Group Selection on the Evolution of Altruistic Behavior. J Theor Biol 220: 55–66.
36. Wilson DS, Pollock GB, Dugatkin LA (1992) Can sltruism evolve in purely viscous populations? Evol Ecol 6: 331–341.
37. Taylor PD (1992) Altruism in viscous populations - an inclusive fitness model. Evol Ecol 6: 352–356.
38. Taylor PD (1992) Inclusive fitness in a homogeneous environment. Proc Biol Sci 249: 299–302.
39. Sharp SP, Simeoni M, Hatchwell BJ (2008) Dispersal of sibling coalitions promotes helping among immigrants in a cooperatively breeding bird. Proc Biol Sci 275: 2125–2130.
40. Slatkin M (1977) Gene flow and genetic drift in a species subject to frequent local extinctions. Theor Popul Biol 12: 253–262.
41. Goodnight KT (1992) The effect of stochastic variation on kin selection in a budding-viscous population. Am Nat 140: 1028–1040.
42. Lehmann L, Perrin N, Rpusset F (2006) Population demography and the evolution of helping behaviors. Evolution 60: 1137–1151.
43. Bradley BJ, Doran-Sheehy DM, Vigilant L (2007) Potential for female kin associations in wild western gorillas despite female dispersal. Proc Biol Sci 274: 2179–2185.
44. Heinsohn R, Dunn P, Leggs S, Double M (2000) Coalitions of relatives and reproductive skew in cooperatively breeding white-winged choughs. Proc Biol Sci 267: 243–249.
45. Williams DA, Rabenold KN (2004) Male-biased dispersal, female philopatry, and routes to fitness in a social corvid. J Anim Ecol 74: 150–159.

Large Spatial Scale of the Phenotype-Environment Color Matching in Two Cryptic Species of African Desert Jerboas (Dipodidae: *Jaculus*)

Zbyszek Boratyński[1,2]*, José Carlos Brito[2,3], João Carlos Campos[2,3], Maija Karala[1], Tapio Mappes[1]

1 Division of Ecology and Evolutionary Biology, Department of Biological and Environmental Science, University of Jyväskylä, Jyväskylä, Finland, **2** CIBIO/InBIO, Centro de Investigação em Biodiversidade e Recursos Genéticos, Universidade do Porto, Vairão, Portugal, **3** Departamento de Biologia, Faculdade de Ciências da Universidade do Porto, Porto, Portugal

Abstract

We tested the camouflage hypothesis, or the linkage between animal (Saharan rodent) and habitat coloration, on the largest geographical scale yet conducted. We aimed to determine whether phenotypic variation is explained by micro-habitat variation and/or genetic polymorphism to determine 1) the strength of linkage between fur color and local substrate color, and 2) the divergence in fur coloration between two genetic clades, representing cryptic species, throughout the complete range of the African desert jerboas (*Jaculus jaculus*). We used a combination of museum and field-collected specimens, remote sensing tools, satellite and digital photography and molecular genetic and phylogenetic methods to investigate the above hypotheses. Along with showing that the two divergent genetic clades of jerboas occur sympatrically throughout their African distribution, we showed significant covariation between dorsal fur coloration of the animals and the color of their habitat. We also described significant phenotypic divergence in fur color, consistent with genetic divergence between the sympatric clades. The linkage between environment and phenotype supports the idea that the selection promoting cryptic coloration is persistent in contemporary populations of jerboas, however the phenotypic divergence indicates that it has different strengths (or optima) in the two clades. The mosaic distribution of micro-habitats occupied by geographically sympatric clades suggests that it may influence both ecological and evolutionary dynamics between these two cryptic species.

Editor: Axel Janke, BiK-F Biodiversity and Climate Research Center, Germany

Funding: Fieldwork was supported by National Geographic Society (7629-04 and 8412-08) and Fundação para a Ciência e Tecnologia (PTDC/BIA-BEC/099934/2008), through the EU program COMPETE. Laboratory work was supported by the Academy of Finland (132190 to TM), the University of Jyväskylä (mobility grant to ZB) and the Centre of Excellence in Evolutionary Research. Logistic support for fieldwork was given by Pedro Lopes Lda. (Trimble GPS), Off Road Power Shop and A. Araújo (P.N. Banc d'Arguin), and S.M.O. Lehlou (Ministry of Environment, Mauritania). Museum samples were kindly provided by the Smithsonian Institution, National Museum of Natural History, Division of Mammals, Washington D.C., USA. ZB and JCC are financed by Fundação para a Ciência e Tecnologia (SFRH/BPD/84822/2012 and SFRH/BD/87885/2012, respectively), and JCB by project "Biodiversity, Ecology and Global Change" cofinanced by North Portugal Regional Operational Programme 2007/2013 (ON.2 – O Novo Norte), under the National Strategic Reference Framework (NSRF), through the European Regional Development Fund (ERDF). The funders had no role in study design, data collection and analysis, decision to publish, or preparation of the manuscript.

* E-mail: boratyns@jyu.fi

Introduction

Spatial distribution and covariation between individual and environmental variability is a key issue in modern evolutionary ecology [1] [2]. Broad geographic distribution is expected to generate genetic patterns, if isolation by distance or geographic barriers is strong enough. With such isolation, differences can evolve between populations by drift alone, and can be mitigated by migration, gene flow and subsequent genome homogenization. However, in situations where local habitats form mosaics, geographic barriers are usually absent and the distance between populations is negligible. In these cases, divergence between populations evolves in response to local habitat variability, and diversification can be driven by ecological rather than spatial processes, even if the gene flow is present [3–5].

Coloration has been an important model for ecological and evolutionary studies since the early days of evolutionary theory [6] [7]. Darwin provided several examples of camouflage coloration, as evidence for adaptation, such as the color difference between black and red grouses [8]. Coloration quickly became a model system in evolutionary research [9–14] with the evolution of camouflage, or the covariation between the coloration of animals and their habitats, being one of the most striking examples of adaptation [15–20]. Theoretical studies of the evolution of camouflage coloration suggested that in areas of variably colored habitats, prey either reach a compromise between the two or more different micro-habitats or specialize in one of them, optimizing survival [10]. Compromise results in coloration that is intermediate between the micro-habitats, as in the desert lizard, *Sceloporus magister* [21] and a number of moth species [16], and may be the best strategy in some circumstances [22]. On the other hand, specialization results in color polymorphism, as in the mouse, *Peromyscus polionotus*, which consists of a dark-coated inland form and light-colored coastal populations [12] [23]. The trade-off between these two evolutionary pathways is shaped by resource availability and predation risk, features that often vary spatially across local micro-habitats and broad geographic areas.

The presence of coat color polymorphism in the model species studied here, the lesser Egyptian jerboa, *Jaculus jaculus*, has led to the description of a number of subspecies and even the proposition of separate species (i.e., *J. deserti*) [24]. It has been suggested that in *J. jaculus* dorsal fur color matches the background substrate and that animals living in gravel rock deserts are darker than populations inhabiting sandy desert areas and dunes [24]. This implies that camouflage adaptation to avoid predation has led to speciation, although substrate types and micro-habitats are assorted in the Sahara [25]. Since obvious topographic barriers that might promote population isolation and geographical genetic structure are missing within the African range of this species [26], migration is thought to homogenize adaptive responses.

The main aim of this study was to rigorously test on a large geographical scale if phenotypic variation, namely the polymorphism in dorsal fur color, can be predicted by variation in the color of the local micro-habitat, and if this phenotypic variation is linked to the geographic patterns of genetic polymorphism in lesser African jerboas [26]. We predicted that if natural selection has indeed favored camouflage in this species, we should detect (1) a significant correlation between dorsal fur and local substrate colors. Furthermore, if genetic clades (i.e., cryptic species) specialized to divergent micro-habitats and are still under selection promoting camouflage, then (2) a significant difference in fur coloration between them will be observed [27]. Whether these clades represent separate species or not, the ability to explore different, although assorted, micro-habitats would allow them to occur in sympatry over a large geographic scale. These different clades might also reflect variation in the level of the specialization to the micro-habitats, as indicated by strength of the correlation. To test the above predictions in a comprehensive way for these clades over such a large geographical scale, we developed a study based on satellite photography and remote sensing methods.

Materials and Methods

Specimens

The lesser Egyptian jerboa (*Jaculus jaculus*) is mainly a nocturnal animal hunted by predators (birds of prey, owls) that use also visual clues for prey detection [15] [28] [29]. Samples of jerboas (85 specimens) were collected at the Smithsonian Natural History Museum, Washington, D.C., USA (69) and during field expeditions to Mauritania (16 road-killed individuals; see Appendix S1). Museum samples were originally collected as part of the African Mammal Project (1961–1972) and were preserved and stored without any chemical treatment as dried skins and skulls [30]. These samples represent the complete African distribution of the species (Fig. 1). Additional specimens (46) were included in phylogenetic analyzes (GenBank, *J. jaculus*: JN214504-JN214546, NC_005314, *J. orientalis*: JN652663, *Dipus sagitta*: AM407909 and *Allactaga elater*: AJ389534).

Ethics statement

Ethical approval was not required for this study as all sampled animals originated from a museum collection or road-killed individuals. Museum sampling was conducted at the Smithsonian Institution, National Museum of Natural History, Division of Mammals, Washington, D.C., USA. Fieldwork was carried out with permission from the Ministére Délégué auprès du Premier Ministre Chargé de l'Environnement, Nouakchott, Mauritania (Permit: 460/MDE/PNBA). This permit was valid for all of Mauritania and no specific permissions were required for any of the sampled localities. Analyzes were done at the CITES registered laboratory: 13PT0065/S. IACUC approval was not requested as no animal was sacrificed and there were no animal husbandry, experimentation or care/welfare concerns.

Molecular laboratory procedures

DNA from museum (tissues from inside skins and skulls) samples was extracted and purified using QIAamp (Qiagen) protocol. Polymorphic fragment of the cytochrome b gene, marker commonly used in phylogenetic and phylogeographic studies in mammals, was amplified and sequenced using primers designed for *Jaculus*: Jac1R (5′-TGCTGGTTTACAAGACCA-3′) and Jac4F (5′-CAAACCCACTTAATACGC-3′). The laboratory procedures were conducted in four stages with variable methods to account for the reduced quality of DNA in the museum tissues. DNA was amplified using the DreamTag kit. The 19 µl mixes contained 2 µl of DNA, 0.1 µl of DreamTag polymerase, 1.2 µl of each primer (1R and 4F), 0.4 µl of $MgCl_2$ (3 mM), 2 µl of dNTP, 2 µl of reaction buffer and 10.1 µl of water. A PCR program with denaturation at 95°C, annealing (Ta) at 53°C and elongation at 72°C was maintained for 40 cycles. The results were confirmed with agarose gel electrophoresis. The PCR products were purified using the Exo-Sap procedure. Sequencing mixes contained 2 µl of purified PCR product, 1 µl of 3.2 µM primer, 0.5 µl of 5× BigDye sequencing buffer and 3.75 µl of 25× Ready Reaction Premix in 18 µl volume. Sequencing was conducted in Applied BioSystems 3130xl Genetic Analyzer. Unsuccessful samples were re-amplified with FailSafe. The 20 µl mixes contained 2 µl of DNA, 10 µl of FailSafe Premix F, 0.2 µl of FailSafe Entzyme Mix, 1.2 µl of 1R and 4F primers, and 5.4 µl of water. For amplification, a Touchdown PCR was used. Ta gradually lowered from 56 to 53°C during the first six PCRs to allow selective amplification. The following 31 cycles were kept at constant Ta. To ensure a sufficient concentration of DNA this step was repeated twice, with the end product of the first PCR used as the template for next one. The bands from agarose gels were cut, purified and sequenced with the Applied BioSystems analyzer as described above. The last 8 unsuccessful samples were amplified in 10 µl PCR mix (1 µ of DNA, 5 µl FailSafe Premix, 1 µl FailSafe Enzyme Mix, 0.6 µl of each primer and 2.7 µl of water) and a Touchdown PCR (56–52 °C) during the first six cycles with a total of 41 cycles. The PCR reaction was repeated using the PCR product as template. Sixteen field-collected samples were analyzed following published protocols [31].

Phylogenetic analyzes

The sequences were aligned and analyzed using SeqScape (Applied BioSystems) and ClustalX (www.clustal.org). Phylogeny was conducted with the Maximum Likelihood method (ML in PhyML 3.00 [32]), with the K80 model and the proportion of invariable sites (0.644) was selected with jModelTest 2 [33], and with the Bayesian method (BI) using GTR model with prior uniform distributions for the proportion of invariable sites (0.00, 1.00) and gamma shape parameter (0.00, 200.00; MrBayes 3.1.2; Fig. 2) [34]. Four Markov Chain Monte Carlo simulations were run for 10 million generations, starting from a random tree. The trees were sampled every 100th generation with the first 30% of the trees discarded as burn-in. The remaining trees were used to construct a 50% majority consensus tree. The robustness of tree nodes was assessed by 1000 bootstrap replicates (ML) and posterior probabilities (BI). Descriptive statistics, tests of selective neutrality and mismatch haplotype distributions were calculated with Arlequin 3.11 (http://cmpg.unibe.ch/software/arlequin3).

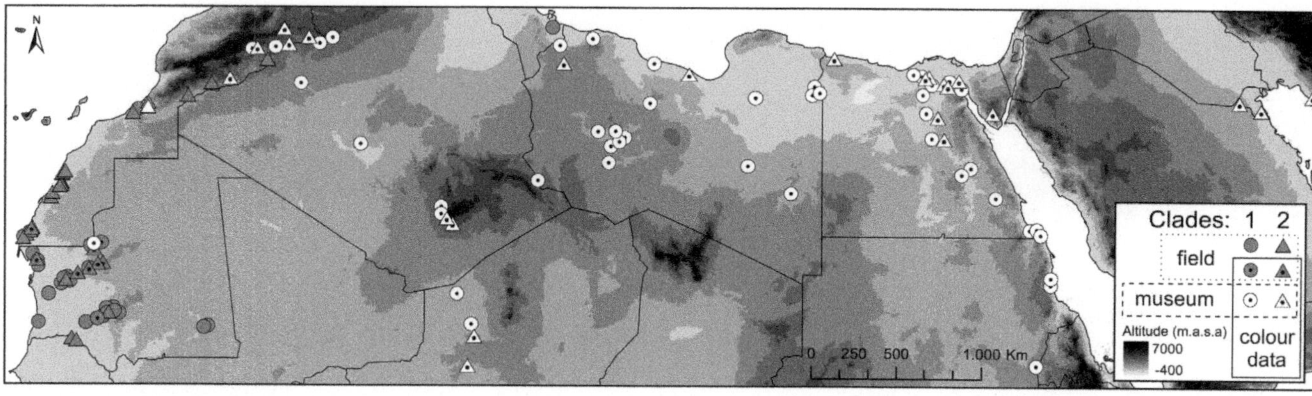

Figure 1. Sample locations of *Jaculus* specimens included in the study. Circles and triangles refer to clade 1 and 2, respectively. Grey and white figures refer to field and museum collected samples, respectively. Figures with dots refer to genetic samples accompanied by phenotypic data.

Animal coloration

Digital photography analysis was used to estimate animal coloration [35] only on specimens in suitable conditions (i.e., clean, without excessive dust). Museum (68) and field-collected (9) specimens were photographed alongside a ColorChecker (X-Rite, Michigan, USA) with a Canon EOS 350 digital camera. The high-resolution photographs (3888×2592 pixels) were saved in TIFF format and white balance was corrected using GIMP 2.0 (www.gimp.org) with the white reference of the ColorChecker. Fujifilm Hyper-Utility 2 (www.fujifilm.com) was employed to quantify coloration from a square-shaped area on each animal's uniformly colored back twice, independently. The squares, selected independently, and randomly, for two measurement trials, were as large as possible (2 cm^2) while excluding the darker fur near the tail and the lighter fur occasionally showing from the sides or undersides of the animals. Luminosity and reflectance of red, blue and green were measured as a Red-Green-Blue (RGB, 8-bit) standard values (from 0 to 255). The museum archives provided geographic coordinates or descriptions of the collection location of the museum samples [30]. Field samples were georeferenced using the global positioning system (GPS).

Environmental coloration

Remote sensing (RS) techniques were applied to estimate environmental coloration of sample locations from available NASA Landsat satellite images (landsat.gsfc.nasa.gov). A total of 90 satellite images from the Landsat 5 Thematic Mapper (TM) and Landsat 7 Enhanced Thematic Mapper (ETM$^+$) were obtained through the Global Visualization Viewer (GLOVIS; glovis.usgs.gov; see Appendix S1). The time period of the 90 images (with spatial resolution of 30 m) corresponded to the dry seasons of 2003. To quantify repeatability we also used 90 images from 1999–2002 (see Appendix S1). Only images with cloud cover lower than 10% were analyzed. To eliminate atmospheric effects and obtain the real surface reflectance, an absolute atmospheric correction was performed through two steps: 1) conversion of 8-bit satellite-quantized calibrated digital numbers to at-satellite radiance; and 2) conversion of at-satellite radiance to atmospherically-corrected surface reflectance [36]. This atmospheric correction accounts for variations in Earth-Sun distance, sun angle, light diffuse irradiance, atmospheric transmittances and biases associated to the measurements of each satellite sensor (see detailed information in Supporting Information). Luminosity was calculated as a RGB weighted average using ITU-R BT.601 standard coefficients (International Telecommunications Union 2011: www.itu.int/rec/R-REC-BT.601-7-201103-I/en). The landscape reflectance was measured as average pixel value from a 10 kilometer radius around the sample origin. This accounted for species movement abilities (around 10 km per night) and for possible error emerging from inaccurate coordinates of pre-GPS samples. The radii of samples collected prior to local intensive infrastructure development were moved slightly to prevent man-made, irrigated areas or open waters from being included. All RS analyzes were developed in ArcGIS 9.3 [37]. The most intensive predation on jerboas is expected during evenings/mornings. Satellite images were taken in daylight and animal skins were photographed in full light conditions. Full light photography thus gives a relative, rather than direct, measurement of animal-habitat match [15] [19] [20], even if much efforet has been put to normalize variations in color attributes between satellite images and direct digital images of the animals.

Statistical analyzes

Repeatability of fur and habitat coloration for the two independent measurements was estimated with the intraclass correlation coefficient (ICC) based on between and within group's mean squares from ANOVA [38]. Correlations between fur and environmental colorations were tested with partial Pearson correlation that included the origin of samples (either museum or field-collected). Variables potentially affecting animal coloration, including coloration of habitat, were tested with the mixed model procedure. The model importance and statistical significance of the effects were estimated with mixed model and Akaike information criterion (AIC) with *nlme* and *MuMIn* packages in R 3.0.1 (R Development Core Team, 2011). Fur luminosity and reflectance of red, blue and green were tested in independent analyzes as response variables, although colors have, with some exceptions [39], less biological meaning for nocturnal predators than overall luminosity. Individual ID was included in models as a random predictor, clade affiliation (with two levels of two genetic clades) as a fixed predictor, environmental luminosity and environmental reflectance of red, blue and green as continuous predictors, and factorial interaction between fixed effects and continuous predictors were included to construct the most complicated starting models. Fixed effects of continental affiliation (Africa versus Middle East), environment modification (unmodified versus urban, agricultural or near water, as detected from satellite images), sampling type (museum versus field) and their interactions with continuous predictors were also tested.

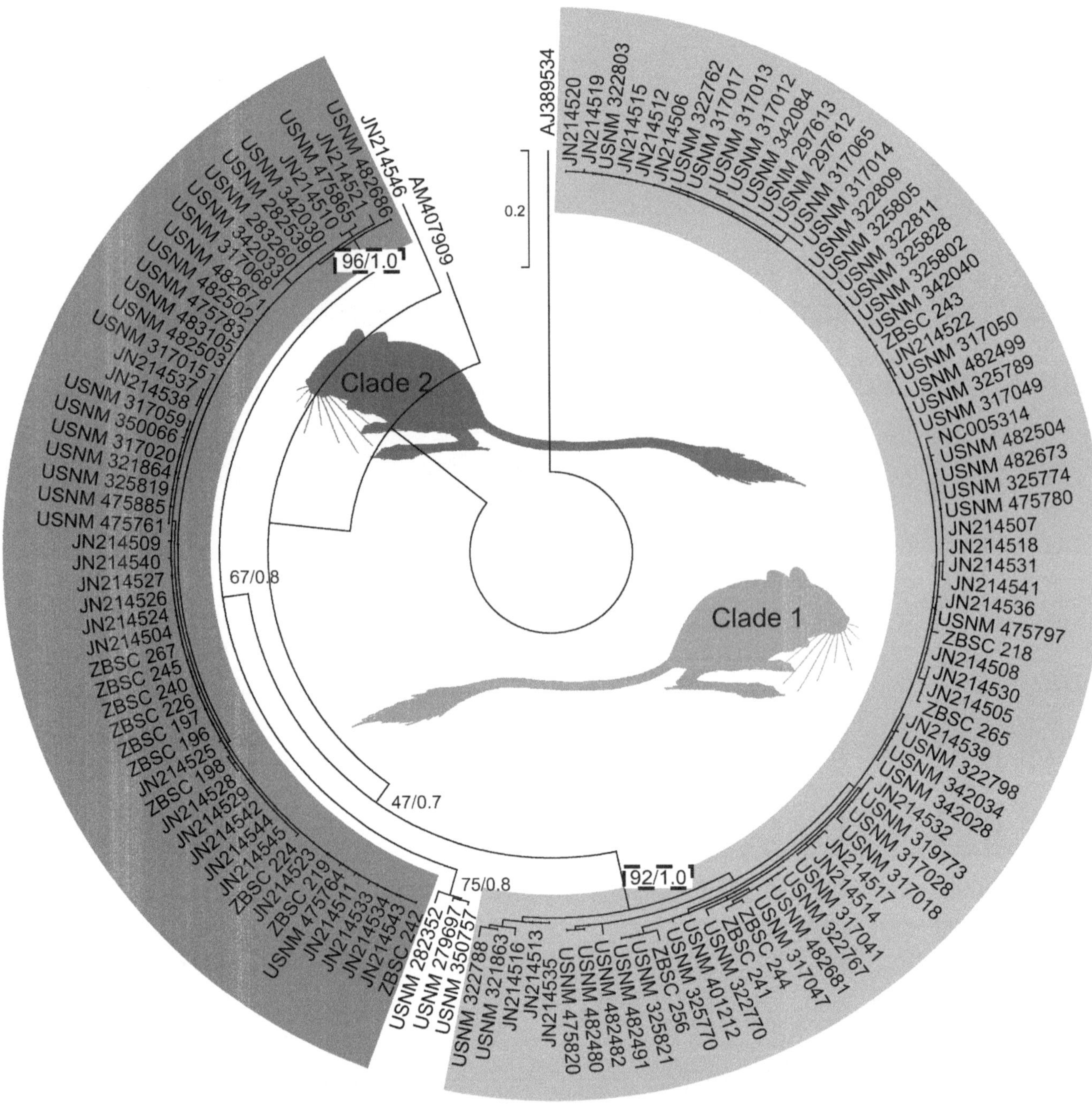

Figure 2. Maximum likelihood (ML) tree of a portion of the cytochrome b gene. Numbers in the external nodes refer to codes of the samples (see Appendix S1). Numbers in the internal nodes represent the percentage of bootstrap support for ML and posterior probabilities for Bayesian methods. Framed values represent statistics for the two main clades. The intensity of grey color of shadings and drafted animals is based on the overall dorsal fur luminosity of individuals pictured on the appended photograph (see Appendix S2).

Results

Genetic variation

A total of 131 sequences (at least 352 base pairs long) covering the entire species range in North Africa were analyzed. Barcodding ML and BI methods were consistent in showing two well-supported monophyletic clades (Fig. 2) with the broad distribution overlap (Fig. 1). The mismatch haplotype distributions for the African clades were closer to unimodality (Appendix S1), consistent with the Sudden Expansion Model ($p > 0.4$). Tajima's D and Fu's F neutrality tests were negative, with the exception of Tajima's D for the combined data set (Table 1). Estimated diversity indices (Table 1) and demographic parameters were slightly higher for African clade 1 [τ(CI) = 3.4(0.7–12.5); Θ_0(CI) = 1.7(0–8.1); Θ_1(CI) = 24.7(6.9–∞)] than for clade 2 [τ (CI) = 1.4(0–5.9); Θ_0(CI) = 1.0(0–5.9); Θ_1(CI) = ∞(3.7–∞); based on 1000 replicates] but their confidence intervals overlapped. The clades differed by 16 fixed mutations, with an average number of nucleotide substitutions per site of 0.10.

Table 1. Sequence diversity and neutrality tests for cytochrome b sequences of lesser Egyptian jerboas. n_i = number of samples, n_h = number of haplotypes, n_p = number of polymorphic sites, H = haplotype diversity, π% = nucleotide diversity.

	n_i	n_h	n_p	H	π (%)	Tajima's D	p	Fu's F_s	p
all	127	64	78	0.97	6.09(2.99)	1.15	0.91	−13.95	0.015
Clade 1	72	40	42	0.96	1.31(0.72)	−1.55	0.04	−25.70	<.0001
Clade 2	52	21	22	0.90	0.70(0.41)	−1.67	0.02	−14.27	<.0001
Clade $2_{\text{including Middle East}}$	55	24	38	0.91	1.26(0.70)	−1.62	0.03	−10.36	0.002

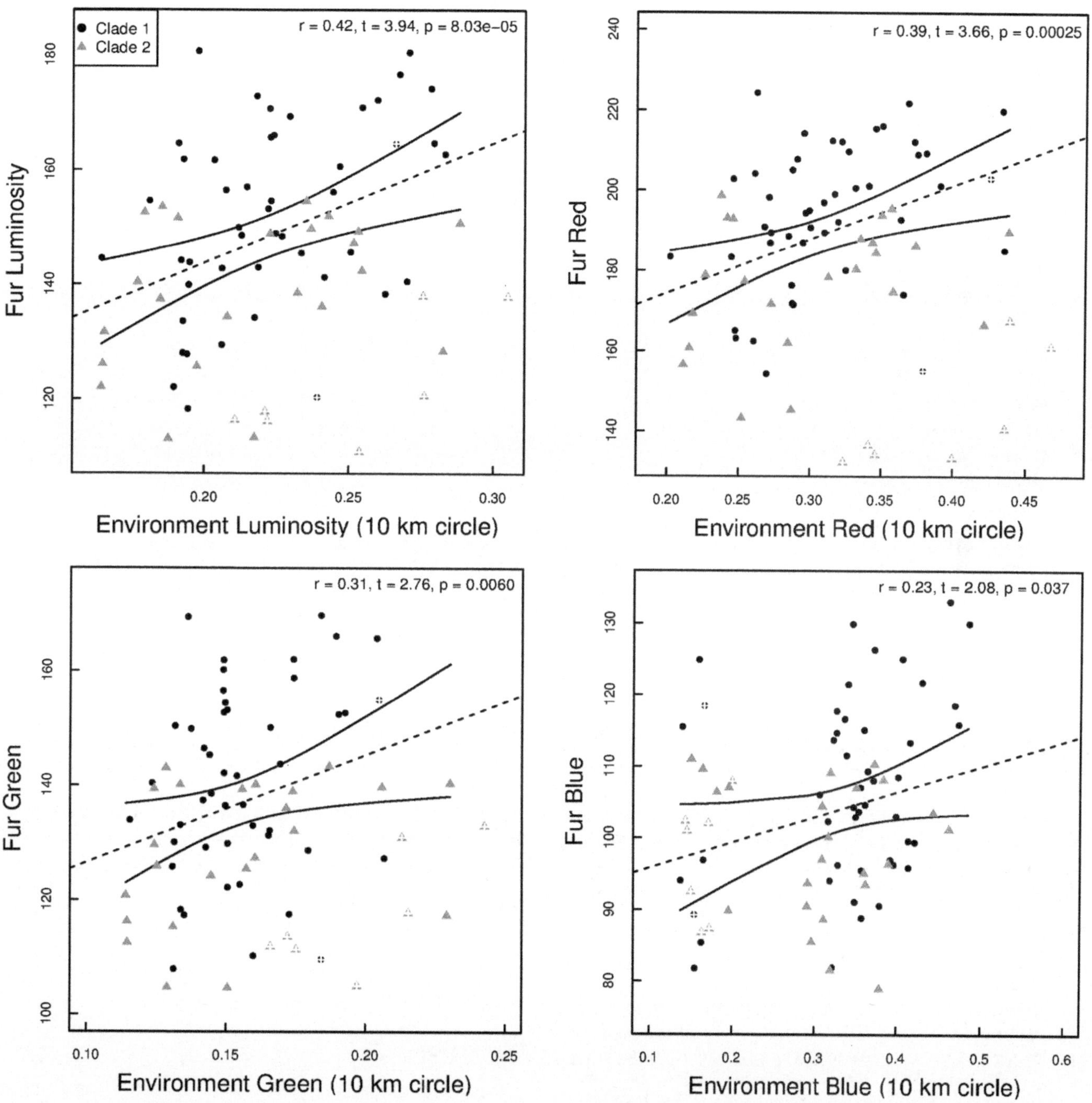

Figure 3. The association between individuals' dorsal fur and environmental luminosity. Circles (black) and triangles (grey) refer to clade 1 and 2, respectively. Figures with crosses refer to field collected samples. Statistics are based on partial Pearson correlations accounting for sample type, i.e., museum or field specimens.

Table 2. Model components and parameters from the AIC model selection procedure.

	Models parameters				
Dependent and independent variables	**df**	**logLik**	**AIC**	**Delta**	**Weight**
Fur Luminosity					
Clade+M/F+Red+Green	7	−290.5	628.8	0.00	0.29
Clade+M/F+Red+Blue+Luminosity	8	−285.3	629.8	1.00	0.18
Clade+M/F+Red+Blue+Green	8	−285.9	629.8	1.00	0.18
Clade+M/F+Red+Green+Luminosity	8	−283.7	629.8	1.00	0.18
Clade+M/F+Blue+Green+Luminosity	8	−284.7	629.8	1.00	0.18
Fur Red					
Clade+M/F+Red+Green	7	−299.4	647.8	0.00	0.27
Clade+M/F+Red+Green+M/F*R.	8	−292.7	648.0	0.23	0.24
Clade+M/F+Red+Green+M/F*G.	8	−292.2	648.4	0.64	0.2
Clade+M/F+Red+Green+C.*R.+C.*G.+M/F*R.	10	−278.9	649.1	1.28	0.14
Clade+M/F+Red+Green+C.*R.+M/F*R.	9	−287.0	649.2	1.37	0.14
Fur Blue					
Clade+Green+Luminosity	6	−277.9	595.2	0.00	0.29
Clade+Green+Luminosity+C.*G.+C.*L.	8	−264.2	595.8	0.55	0.22
Clade+Green+Luminosity+C.*L.	7	−271.9	596.1	0.94	0.18
Clade+Red+Blue	6	−281.1	596.2	0.94	0.18
Clade+Red+Blue+Luminosity	7	−273.9	596.8	1.53	0.13
Fur Green					
Clade+M/F+Red+Green	7	−290.2	628.3	0.00	0.23
Clade+M/F+Red+Blue	7	−292.9	628.5	0.19	0.21
Clade+M/F+Green+Luminosity	7	−289.9	628.5	0.24	0.21
Clade+M/F+Red+Blue+Luminosity	8	−284.9	628.9	0.58	0.17
Clade+M/F+Red+Blue+Green	8	−285.4	628.9	0.58	0.17

Fur luminosity and RGB colors, the dependent variables, were analyzed in separate mixed models that included clade (two clades) and museum/field (M/F) affiliations as fixed factors, luminosity (Luminosity) and RGB environmental coloration as continuous predictors with factorial interactions between them, and a random factor of individual code.

Repeatability and correlations

The estimated luminosity and RGB colors of dorsal fur were highly and significantly repeatable, with an ICC of 0.98 ($p < 0.001$; separate RGB colors: 0.99, 0.98, 0.97; $p < 0.001$). ICC for luminosity of the environment was 0.75 ($p < 0.001$; separate RGB colors: 0.65, 0.68, 0.82; $p < 0.001$). Dorsal fur luminosity was significantly correlated with environmental luminosity, both for the complete dataset (Fig. 3) and for datasets including only animals belonging to clade 1 (partial correlations: $r = 0.45$, $t = 3.36$, $df = 47$, $p = 0.0008$) or clade 2 ($r = 0.40$, $t = 2.24$, $df = 30$, $p = 0.025$). Dorsal fur red, green and blue color bands were also significantly correlated ($r \geq 0.23$, $p \leq 0.037$; except for the blue band within clade 2 ($p > 0.4$) with environmental red, green and blue colors bands (see Appendix S1).

Phenotypic and environmental variation and covariation

The mixed models showed consistent results among analyzes and highlighted the importance of clade affiliation and environmental coloration (RGB colors and luminosity) in predicting variability in the animals' dorsal fur luminosity (Table 2; see also: see Tables S3–S5 in Appendix S1 for consistent results of reduced datasets). The models explaining variability in dorsal fur coloration were consistent and always included the clade affiliation of the specimen and environmental coloration. The estimated relative importance of traits showed that clade affiliation and red and green bands of environmental coloration, as well as the type of the sample (either field collected or museum), were the traits strongest related to luminosity of dorsal fur (Table S5 in Appendix S1). As the differences between models were not strong ($DELTA < 2$; likely caused by correlations between color predictors) here we focused on the simplest, best models to satisfy parsimony. Analyzes with these mixed models showed red coloration of the environment and animals clade affiliation as the most consistent and significant predictors among different variables and analyzes (Table 3; Table S6 in Appendix S1). When the best models were analyzed within the clades separately, the significant effects of environmental coloration and the effect of sample origin (field versus museum) were maintained (for clade 1: red: $t = 3.37$, $p = 0.0016$, origin: $t = -2.09$, $p = 0.042$; for clade 2: red: $t = 2.30$, $p = 0.0297$, origin: $t = -4.02$, $p = 0.0004$).

Discussion

It has been shown in previous studies on rodents that adaptive dorsal fur coloration can evolve rapidly [40] in response to selection pressures caused by predation [19] and that the molecular and structural mechanisms for coloration seem to be relatively labile [20] [41] [42] [43]. Though there are studies on

Table 3. Results for best models, for all significant effects, of mixed model analyzes for variation in dorsal fur luminosity and RGB colors.

Dependent	and independent var.	Value	SE	df	t	p
Fur Luminosity						
	Intercept	131.0	10.27	73	12.8	<.0001
	Clade	−13.1	3.35	73	−3.91	0.0002
	Field/Museum	−25.4	5.72	73	−4.44	<.0001
	Red	113.0	28.6	73	3.95	0.0002
Fur Red						
	Intercept	181.5	11.8	72	15.4	<.0001
	Clade	−16.8	3.8	72	−4.45	<.0001
	Field/Museum	−39.5	6.35	72	−6.21	<.0001
	Green	−396.1	184.4	72	−2.15	0.035
	Red	296.4	86.8	72	3.42	0.0010
Fur Blue						
	Intercept	80.0	9.6	73	8.29	<.0001
	Clade	−6.5	2.9	73	−2.44	0.017
	Green	−266.1	113.4	73	−2.35	0.022
	Luminosity	334.5	97.4	73	3.44	0.0010
Fur Green						
	Intercept	117.4	10.2	73	11.5	<.0001
	Clade	−11.6	3.3	73	−3.50	0.0008
	Field/Museum	−20.9	5.7	73	−3.68	0.0004
	Red	113.9	28.4	73	4.01	0.0001

Best models were depicted from the AIC model selection procedure.

geography of fur color polymorphisms, they have focused mainly on few local populations. Here, we developed a novel remote sensing methodology (based on publicly available resources and therefore easily applicable in other systems) to study camouflage over a large geographic scale (roughly the size of the Australian continent) that covered the complete African/Saharan range of our model species, the African desert jerboa (Figs 1 and 2) [26]. Previous research based on both mitochondrial and nuclear sequences showed no signs of introgression implying reproductive isolation between the two sympatric jerboa mitochondrial clades [26] [31]. These previous studies hypothesized, on one hand, that the clades coexist in geographical sympatry across the entire Sahara, but also that they segregate to distinct micro-habitats [24]. The implied adaptive mechanism behind the origin of these putative species is the evolution of camouflage, as a response to avoid avian predators [28] [29]. Both the differences in fur luminosity between the clades and the animal-habitat color match detected here support this mechanism (Tables 2 and 3). Alternatively color polymorphism between clades could be related to thermoregulation, as the differences of mean annual temperature between the two clades were documented [31], although it seems to be less likely as jerboas regulate temperature internally [44].

The detected linkage between consistent (repeatable) environmental and phenotypic variability suggests that natural selection may be promoting cryptic coloration of dorsal fur and that this mechanism may have generated phenotypic divergence between the clades (Fig. 3). The observed correlation ($r = 0.42$) implies that relatively strong selection pressures are shaping this phenotypic polymorphism (Tables 2 and 3), a result consistent with previous experimental studies on other rodents [15] [19]. Although the statistical interactions between clade and environmental effects were not significant (Table 2), the correlation between phenotypic and environmental luminosity in clade 1 ($r = 0.45$) was stronger compared to clade 2 (0.40). Also, the mismatch distribution between individual and environmental coloration was more skewed within clade 1 compared to clade 2 (Fig. 4). Both results suggest tighter micro-habitat preferences within clade 1. It is also possible that clade 2 is competitively excluded by the more dominant clade 1 from the optimum micro-habitat, which may be responsible for the less skewed distribution for clade 2 (Fig. 4). Nonetheless, within both clades habitat-phenotype covariation was persistent. Thus, our results confirmed that in the admixture of the rocky (darker) and sandy (lighter) desert micro-habitats over North Africa, these sympatric clades may persist in ecological separation, with clade 1 being associated with brighter, sandy, areas and clade 2 with darker, rockier locations. Indeed, the phenotypic divergence strongly suggests that, to some extent, barriers to gene flow between the two putative species may have evolved, perhaps even as a result of ecologically-based differential selection. The role of geographic isolation in shaping this pattern is not obvious and cannot be completely excluded, but the persistence of both clades in sympatry, and the persistence of phenotype-environment association within both clades, suggests that geographic isolation might not be crucial for their coexistence [45]. Yet the distribution of the differences between mitochondrial haplotypes (Figure S1 in Appendix S1) and the observed overall genetic polymorphism (Table 1) suggest a slightly more recent demographic history of clade 2 versus clade 1, as already suggested in previous works [26] [31]. This might imply significant differences in the history of

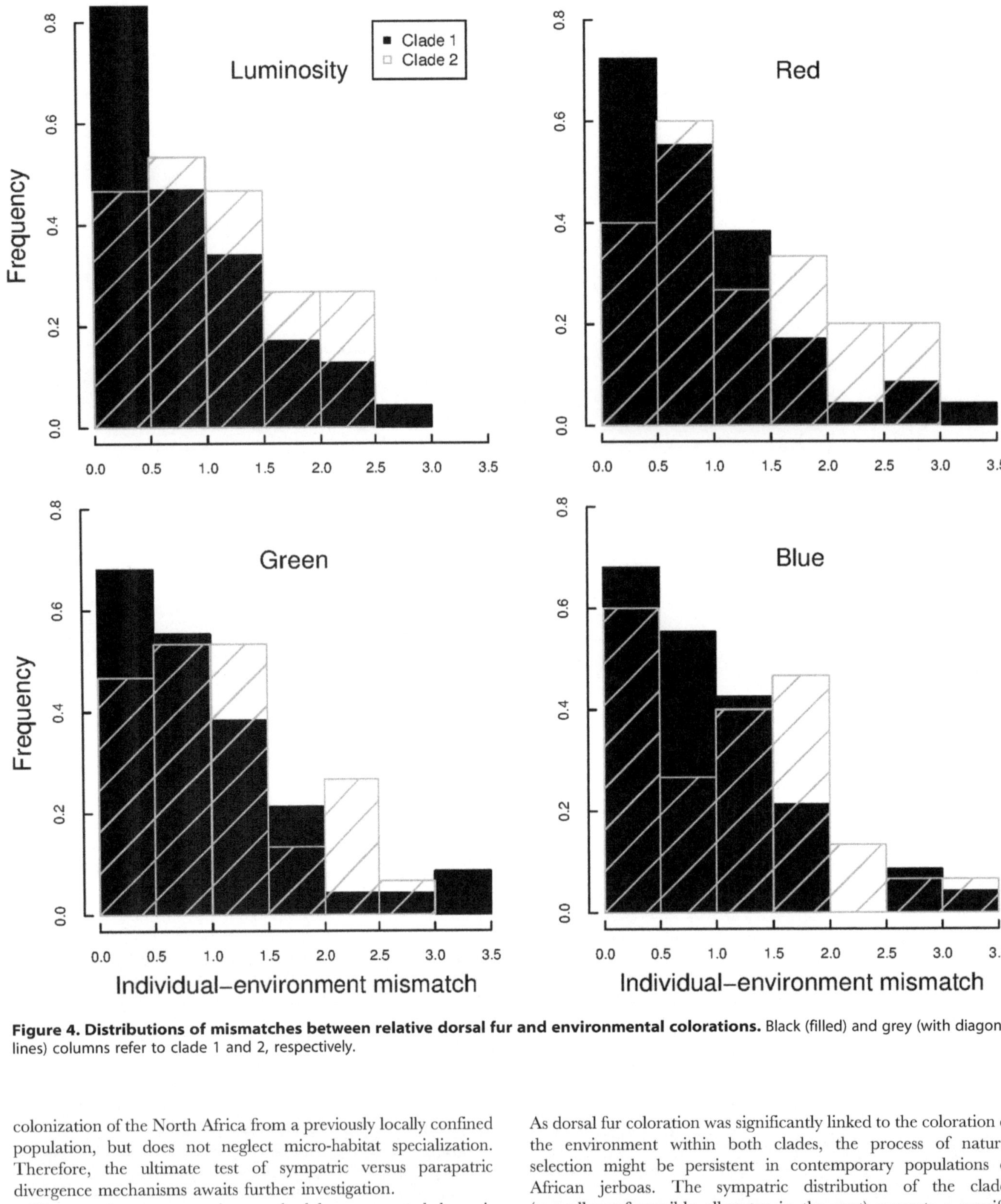

Figure 4. Distributions of mismatches between relative dorsal fur and environmental colorations. Black (filled) and grey (with diagonal lines) columns refer to clade 1 and 2, respectively.

colonization of the North Africa from a previously locally confined population, but does not neglect micro-habitat specialization. Therefore, the ultimate test of sympatric versus parapatric divergence mechanisms awaits further investigation.

The novel remote sensing methodology presented here is applicable to other systems intended to investigate animal-habitat color covariation over large geographical scales. Using this technique, we have shown here the environmental basis of animal-habitat color matching as well as consistent phenotypic differences between sympatric genetic clades of African desert jerboas over a large geographical scale (the whole of North Africa). As dorsal fur coloration was significantly linked to the coloration of the environment within both clades, the process of natural selection might be persistent in contemporary populations of African jerboas. The sympatric distribution of the clades (regardless of possible allopatry in the past) suggests a specific micro-habitat preference, which likely has affected reproductive isolation, allowing the persistence of separate demes. This highlights the importance of an ecological mechanism, the evolution of camouflage, as a process affecting the evolution of distinct African desert jerboas. Tests on the extent of the reproductive isolation between these putative species will help to validate a speciation hypothesis. Further analyzes will also evaluate whether predation pressures are indeed differential between, and disruptive within, the clades, and whether or not it is based on ground color variation, which could cause an increase in the dorsal fur color polymorphism observed here.

Supporting Information

Appendix S1 Detailed information about data (samples GenBank affiliations, satellite images), remote sensing method and additional (partial Pearson correlations, mixed models, haplotypes mismatches) results included in the study.

Appendix S2 Representatives of genetic clades, upper individual belongs to clade 1 and lower to clade 2.

Acknowledgments

We acknowledge Sami Kyröläinen for molecular laboratory efforts, Jackson H. Jennings for language correction and Johanna Mappes, Nina Seren, Anne Roulin and anonymous reviewer for helpful comments on the manuscript.

Author Contributions

Conceived and designed the experiments: ZB JC MK JCB. Performed the experiments: JC MK ZB. Analyzed the data: ZB JC MK TM JCB. Contributed reagents/materials/analysis tools: TM. Wrote the paper: ZB.

References

1. Novembre J, Di Rienzo A (2009) Spatial patterns of variation due to natural selection in humans. Nature Rev Genet 10: 745–55.
2. Ng J, Landeen EL, Logsdon RM, Glor RE (2013) Correlation between *Anolis* lizard dewlap phenotype and environmental variation indicates adaptive divergence of a signal important to sexual selection and species recognition. Evolution 67: 573–82.
3. Rice WR, Salt GW (1990) The evolution of reproductive isolation as a correlated character under sympatric conditions: experimental evidence. Evolution 44: 1140–1152.
4. Coyne JA, Orr HA (2004) Speciation. SunderlandSinauer.
5. Schluter D (2009) Evidence for ecological speciation and its alternative. Science 323: 737–741.
6. Blair WF (1947) Variation in shade of pelage of local populations of the cactus-mouse (*Peromyscus eremicus*) in the Tularosa Basin and adjacent areas of southern New Mexico. Laboratory Vertebrate Biology, University of MichiganContribu-Contribution37: 1–7.
7. Dice L (1947) Effectiveness of selection by owls of deer mice (*Peromyscus maniculatus*) which contrast with their background. ContributionLaboratory Vertebrate Biology, University of Michigan50115
8. Darwin CR (1859) On the origin of species. LondonJohn Murray
9. TurnerJRG (1967) Some early works on heliconiine butterflies and their biology (Lepidoptera, Nymphalidae). J Linn Soc London Zool 46: 255–266.
10. Merilaita S, Tuomi J, Jormalainen V (1999) Optimization of cryptic coloration in heterogeneous habitats. Biol J Linn Soc 67: 151–161.
11. Caro TM (2005) The adaptive significance of coloration in mammals. BioScience 55: 125–136.
12. Hoekstra HE, Hirschmann RJ, Bundey RA, Insel PA, Crossland JP (2006) A single amino acid mutation contributes to adaptive color pattern in beach mice. Science 313: 101–104.
13. Mullen LM, Vignieri SN, Gore JA, Hoekstra HE (2009) Adaptive basis of geographical variation: genetic, phenotypic and environmental differences among beach mouse populations. Proc R Soc London Ser B 276: 3809–3818.
14. Stevens M, Merilaita S (2009) Animal camouflage: current issues and new perspectives. Proc R Soc London Ser B 364: 423–427.
15. Kaufman DW (1974) Adaptive coloration in *Peromyscus polionotus*: experimental selection by owls. J Mammal 55: 271–283.
16. Endler JA (1988) Frequency-dependent predation, crypsis and aposematic coloration. Philos Trans R Soc London Ser B 319: 505–523.
17. Heth G, Beiles A, Nevo E (1988) Adaptive variation of pelage color within and between species of the subterranean mole rat (*Spalax ehrenbergi*) in Israel. Oecologia 74: 617–622.
18. Bond AB, Kamil AC (1998) Apostatic selection by blue jays produces balanced polymorphism in virtual prey. Nature 395: 594–596.
19. Vignieri SN, Larson JG, Hoekstra HE (2010) The selective advantage of crypsis in mice. Evolution 64: 2153–2158.
20. Linnen CR, Poh YP, Peterson BK, Barrett RD, Larson JG, et al (2013) Adaptive evolution of multiple traits through multiple mutations at a single gene. Science 339: 1312–1316.
21. Norris KS, Lowe CH (1964) An analyzis of background color-matching in amphibians and reptiles. Ecology 45: 565–580.
22. Merilaita S, Lyytinen A, Mappes J (2001) Selection for cryptic coloration in a visually heterogeneous habitat. Proc R Soc London Ser B 268: 1925–1929.
23. Haldane JBS (1948) The theory of a cline. J Genet 48: 277–284.
24. Gharaibeh BM (1997) Systematics, distribution, and zoogeography of mammals of Tunisia. Ph.D. thesisTexas Tech University
25. Brito JC, Godinho R, Martínez-Freiría F, Pleguezuelos JM, Rebelo H, et al. (2013) Unravelling biodiversity, evolution and threats to conservation in the Sahara-Sahel. Biol Rev Early Online DOI: 10.1111/brv.12049.
26. Ben Faleh AR, Granjon L, Tatard C, Duplantier JM, Dobigny G, et al. (2012) Phylogeography of two cryptic species of African desert jerboas (Dipodidae: *Jaculus*). Biol J Linn Soc 107: 27–38.
27. Hoekstra HE (2006) Genetics, development and evolution of adaptive pigmentation in vertebrates. Heredity 97: 222–34.
28. Hendrie CA, Weiss SM, Eilam D (1998) Behavioural response of wild rodents to the calls of an owl: a comparative study. J Zool 245: 439–446.
29. Leonardi G, Dell'Arte GL (2006) Food habits of the Barn Owl (*Tyto alba*) in a steppe area of Tunisia. J Arid Environ 65: 677–681.
30. Schmidt DF, Ludwig CA, Carleton MD (2008) The Smithsonian Institution African Mammal Project (1961–1972). An annotated gazetteer of collecting localities and summary of its taxonomic and geographic scope. Washington DCSmithsonian Institution Scholarly Press.
31. Boratyński Z, Brito JC, Mappes T (2012) The origin of two cryptic species of African desert jerboas (Dipodidae: *Jaculus*). Biol J Linn Soc 105: 435–445.
32. Guindon S, Dufayard JF, Lefort V, Anisimova M, Hordijk W, et al. (2010) New algorithms and methods to estimate maximum-likelihood phylogenies: assessing the performance of PhyML 3.0. Syst Biol 59: 307–321.
33. Darriba D, Taboada GL, Doallo R, Posada D (2012) jModelTest 2: more models, new heuristics and parallel computing. Nature Methods, 9: 772.
34. Ronquist F, Teslenko M, van der Mark P, Ayres D, Darling A, et al. (2012). MrBayes 3.2: Efficient Bayesian phylogenetic inference and model choice across a large model space. Syst Biol 61: 539–542.
35. Stevens M, Párraga C, Schmidt A, Cuthill IC, Partridge JC, et al. (2007) Using digital photography to study animal coloration. Biol J Linn Soc 90: 211–237.
36. Chander G, Markham BL, Helder DL (2009) Summary of current radiometric calibration coefficients for Landsat MSS, TM, ETM+, and EO-1 ALI sensors. Remote Sens Environ 113: 893–903.
37. ESRI (2008) Arcmap 9.3. Redlands, CaliforniaEnvironmental Systems Research Institute, Inc.
38. Boratyński Z, Alves PC, Berto S, Koskela E, Mappes T, et al. (2011) Introgression of mitochondrial DNA among *Myodes* voles: consequences for energetics? BMC Evol Biol 11: 355.
39. Avilés JM, Parejo D (2013) Color also matters for nocturnal birds: owlet bill coloration advertises quality and influences parental feeding behaviour in little owls. Oecologia173399408.
40. Hoekstra HE, Krenz JG, Nachman MW (2005) Local adaptation in the rock pocket mouse (*Chaetodipus intermedius*): natural selection and phylogenetic history of populations. Heredity 94: 217–28.
41. Nachman MW, Hoekstra HE, D'Agostino SL (2003) The genetic basis of adaptive melanism in pocket mice. Proc Natl Acad Sci USA 100: 5268–73.
42. Nachman MW (2005) The genetic basis of adaptation: lessons from concealing coloration in pocket mice. Genetica 123: 125–36.
43. Singaravelan N, Pavlicek T, Beharav A, Wakamatsu K, Ito S, et al. (2010) Spiny mice modulate eumelanin to pheomelanin ratio to achieve cryptic coloration in "evolution canyon," Israel. PloS ONE 5: e8708.
44. Hooper ET, El Hilali M (1972) Temperature regulation and habits in two species of Jerboa, genus *Jaculus*. J Mammal 53: 574–593.
45. Rundle HD, Nosil P (2005) Ecological speciation. Ecol Lett 8: 336–352.

22

Production and Robustness of a Cacao Agroecosystem: Effects of Two Contrasting Types of Management Strategies

Rodolphe Sabatier*¤, Kerstin Wiegand, Katrin Meyer

Department of Ecosystem Modelling, Büsgen-Institute, Georg-August-University of Göttingen, Göttingen, Germany

Abstract

Ecological intensification, i.e. relying on ecological processes to replace chemical inputs, is often presented as the ideal alternative to conventional farming based on an intensive use of chemicals. It is said to both maintain high yield and provide more robustness to the agroecosystem. However few studies compared the two types of management with respect to their consequences for production and robustness toward perturbation. In this study our aim is to assess productive performance and robustness toward diverse perturbations of a Cacao agroecosystem managed with two contrasting groups of strategies: one group of strategies relying on a high level of pesticides and a second relying on low levels of pesticides. We conducted this study using a dynamical model of a Cacao agroecosystem that includes Cacao production dynamics, and dynamics of three insects: a pest (the Cacao Pod Borer, *Conopomorpha cramerella*) and two characteristic but unspecified beneficial insects (a pollinator of Cacao and a parasitoid of the Cacao Pod Borer). Our results showed two opposite behaviors of the Cacao agroecosystem depending on its management, i.e. an agroecosystem relying on a high input of pesticides and showing low ecosystem functioning and an agroecosystem with low inputs, relying on a high functioning of the ecosystem. From the production point of view, no type of management clearly outclassed the other and their ranking depended on the type of pesticide used. From the robustness point of view, the two types of managements performed differently when subjected to different types of perturbations. Ecologically intensive systems were more robust to pest outbreaks and perturbations related to pesticide characteristics while chemically intensive systems were more robust to Cacao production and management-related perturbation.

Editor: Nicholas J. Mills, University of California, Berkeley, United States of America

Funding: This study was part of the DFG-funded research project "Environmental and land-use change in Sulawesi, Indonesia: Socioeconomic and ecological perspectives" (PAK 569, WI1816/12). KMM was partly funded by the State of Lower Saxony (Ministry of Science and Culture; Cluster of Excellence "Functional Biodiversity Research"). The funders had no role in study design, data collection and analysis, decision to publish, or preparation of the manuscript.

Competing Interests: The authors have declared that no competing interests exist.

* E-mail: rodolphe.sabatier@agroparistech.fr

¤ Current address: INRA, UMR 1048 SADAPT, Paris, France

Introduction

New paradigms in agriculture based on ecological intensification such as natural farming [1], agroecology [2,3], the evergreen revolution [4], or the doubly green revolution [5] are presented as challenging alternatives to more conventional farming relying on a high level of chemical input. They are presented as being more respectful of the environment while ensuring a high level of production. A central idea behind these paradigms is that associated biodiversity is strongly impacted by the use of chemicals in conventional farming [6,7] while the associated biodiversity could provide a large range of ecosystem services that often have the same effect as the chemical used (e.g. pest regulation, [8]). According to these paradigms, reducing the amount of chemicals used would maintain biodiversity and ecosystem services at a high level and ensure high yields with lower economic costs. Moreover, based on the ecological concept of stability of ecosystems [9], agroecosystems with high levels of biodiversity are considered more robust, stable and resilient toward perturbations. In other words, such systems have the advantages of a high productivity due to the maintenance of high levels of ecosystem services [10] and of a strong capacity to resist to perturbations [11].

The three concepts of resilience, stability and robustness, although related, slightly differ: resilience is "the persistence of relationships within a system and is a measure of the ability of these systems to absorb changes of state variables, driving variables, and parameters, and still persist" [12], stability is "the ability of a system to return to an equilibrium state after a temporary disturbance" [12] and robustness is: "the ability to maintain performances in the face of perturbation and uncertainty" [13]. In this study we consider the issue of robustness that addresses the performance of the system when the perturbation occurs.

The overall production of an agroecosystem is quite easy to quantify but its behavior in face of a perturbation is much more difficult to assess as it strongly depends on the perturbation considered [14].

In this study our aim is to assess productive performance and robustness toward diverse perturbations of an agroecosystem managed with two contrasting strategies. More precisely, we address the following two questions:

1: What is the shape of the relationship between robustness to perturbations and yield of an agroecosystem under a broad range of different management schemes?

2: Where are the management strategies based on ecological processes (hereafter called ecological strategies, ES) and the ones based on pesticide use (hereafter called chemical-based strategies, CBS) located within the range of outcomes found in 1) and which of these two types of strategies performs better with respect to yield and robustness?

To answer these questions, we developed a model based on a case study of the Cacao agroecosystem in Central Sulawesi (Indonesia). The aim of this model was to capture some general patterns of the interactions between an agroecosystem, its environment and its management by the farmer. Hence, the product of this study is not intended for direct application in the field, but rather for informing management decisions on a general basis. To calibrate and parameterize this model, the Cacao case study was chosen as it captures the main above-ground ecosystem services and disservices and therefore represents a wide variety of agroecosystems well. Cacao crops were introduced in South East Asia 200 years ago. First records of Cacao production in Indonesia date back to 1848 (15) but production remained low (<5000 $t.year^{-1}$ until the late 1970's; [15,16]). Production strongly increased in the last decades to reach more than 800 000 $t.year^{-1}$ in 2010 [16]. In Indonesia, Cacao production is continuous but yield is not constant through the year and shows a main peak in January and in some cases a minor peak six months later. Two main pests impact Cacao production: *Helopeltis theobromae* and the Cacao Pod Borer *Conopomorpha cramerella* [17]. Cacao pollination is done by midges (Ceratopogonidae) and is a limiting factor for production [18].

After briefly describing the model that we used, we determine the relationship between ecosystem functioning and the number of spraying events and define two groups of strategies, ES and CBS. Then we determine the relationship between yield and robustness for the set of all possible management strategies and six types of perturbations. We then look more precisely at the relative positions of the two subsets of management strategies ES and CBS on the production-robustness relationship. Our results show that ecosystem functioning is strongly negatively correlated to the number of spraying events and show how both production and robustness of the two extreme types of strategies depend on both the type of pesticide used and the type of perturbation.

Materials and Methods

Model overview

Robustness of agroecosystems is difficult to address in the field due to the high complexity of agroecosystems and their low reproducibility for experimental purposes. In this context, modeling approaches are powerful tools. *In silico* experiments can cope with the complexity and their high level of reproducibility makes them useful frameworks to represent complex agroecosystems and to study their robustness to perturbations.

The study that we present here is based on a model developed and presented in [19]. This model gives an agroecological representation of a Cacao plantation (Figure 1). It is a discrete time model with a time step of one month and a time horizon of 20 years (T = 240 months). It links the Cacao pod dynamics to the population dynamics of a pest species (the Cacao Pod Borer, *Conopomorpha cramerella*) and two characteristic but unspecified beneficial insect populations (a pollinator of Cacao and a parasitoid of the Cacao Pod Borer). The Cacao Pod Borer and parasitoid parts of the model were inspired by both the Cacao model of [20] and the more general Nicholson and Bailey host-parasitoid model [21].

The insects impact the Cacao yield in several ways. The pollinators N_{Pol} positively affect the number of pods of age 0 $Pods_0$ by pollinating the obligately outcrossing Cacao plants. The Cacao Pod Borer population reduces the amount of Cacao beans eventually harvested. The Cacao Pod Borer population N_{CPB} is regulated by a parasitoid N_{Par}. All three insect populations are affected by the use of pesticides by the farmer. We distinguish two effects of spraying: the efficiency η (the effect on the Cacao Pod Borer) and its selectivity θ (the ratio of effects on beneficial and on pest populations). Efficiency and selectivity range from 0 to 1, $\eta = 1$ means that 100% of the Cacao Pod Borer are killed by the pesticides, $\theta = 1$ means that the effect of pesticide application on beneficial insects is as strong as its effect on the Cacao Pod Borer. Since the aim of pesticide application in the field is to control the Cacao Pod Borer, we limited the study to pesticides having a stronger effect on the target species (Cacao Pod Borer) than on the other species (beneficial insects). We therefore implicitly assume that farmers would not use pesticides that have a net negative impact. In this sense, we avoided the trivial situation where pesticides should be banned due to net negative effects. The model computes the Cacao yield dynamics Y through time as well as the dynamics of the three insect populations for different types of pesticides (characterized by their selectivity and efficiency), and for different timings of spraying.

The model can be described by the following system:

$$\begin{cases} Pods_0(t+1) = f_1(t, N_{Pol}(t)) \\ Y(t) = f_2(Pods_0(t-5), N_{CPB}(t-2)) \\ N_{CPB}(t+1) = (1 - \mathbb{1}_{\{Spray\}}(t)\eta) f_3(Pods_0(t-3), N_{Par}(t), N_{CPB}(t)) \\ N_{Par}(t+1) = (1 - \mathbb{1}_{\{Spray\}}(t)\eta(1-\theta)) f_4(N_{Par}(t), N_{CPB}(t)) \\ N_{Pol}(t+1) = (1 - \mathbb{1}_{\{Spray\}}(t)\eta(1-\theta)) f_5(N_{Pol}(t)) \end{cases} \quad (1)$$

Cacao Pod Borers preferentially attack pods of age 3, pods are harvested at age 5. $\mathbb{1}_{\{Spray\}}$, the characteristic function related to

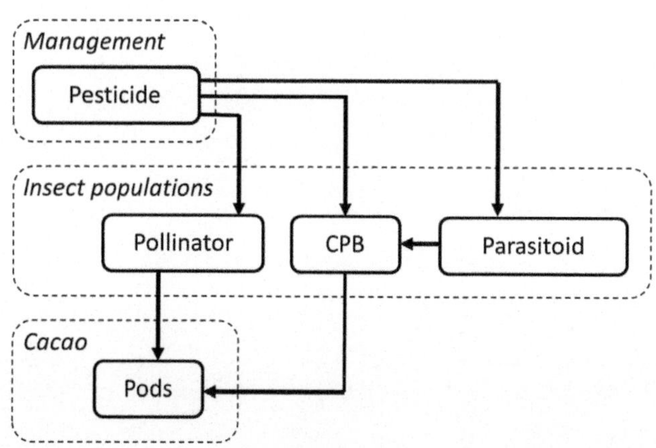

Figure 1. Overview of the model structure. CPB stands for Cacao Pod Borer. Bold black arrows stand for the processes considered in the ecosystem functioning index (EFI; eqn 4).

the spraying event is defined as follows:

$$\begin{cases} 1_{\{Spray\}}(t)=1 \text{ if spraying is true at time } t \\ 1_{\{Spray\}}(t)=0 \text{ if spraying is false at time } t \end{cases} \quad (2)$$

f_1–f_5 are the functions related to the different natural dynamics (see Appendix S1, eqn S.1.1, S.1.11, S.1.8, S.1.9 and S.1.6)

Model calibration and sensitivity analysis

Most parameter values were taken from the literature, when possible from studies conducted in central Sulawesi (Table 1); [19].

Data on pollinators of the Cacao (ceratopogonid midges) are very scarce and we could not calibrate this dynamics on data from any of the pollinating species. Therefore we relied on available data from another tropical species of the same family: *Leptoconops albiventris*. Although it concerned a different species than the pollinators of our system, we considered these data suitable enough regarding the ambitions of the model. These data [22] provide a 6-months survey of a population of midges subjected to regular pesticide applications and were used to calibrate the equation corresponding to pollinator dynamics. Pesticide applications are followed by an instantaneous drop of the midge population followed by a fast recovery of the population. Given

Table 1. List of parameters.

Name	Description	Value	Unit	Reference
Intra-annual dynamic of Cacao pod production (eqn S.1.2)				
α_1	*Mean sinusoid 1*	7.176	pods.ha^{-1}	[23]*,[24]*
β_1	*Amplitude sinusoid 1*	1.425	pods.ha^{-1}	[23]*,[24]*
Ω	*Time period sinusoids*	12	months	[23]*,[24]*
Inter-annual trend of Cacao pod production (eqn S.1.3)				
α_A	*Age dependence parameter 1*	3.82	-	[42]
β_A	*Age dependence parameter 2*	0.086	-	[42]
γ_A	*Age dependence parameter 3*	1.33	-	[42]
A_0	*Initial age*	120	months	[42]
φ_A	*Standardization coefficient*	2.5 10^{-3}	-	[42]*
Pollination effect (eqn S.1.4, eqn S.1.5)				
α_P	*Pollination factor 1*	−0.83	kg. ha^{-1}.	[18]
β_P	*Pollination factor 2*	0.34	kg. ha.$^{-1}$ pollinated flower^{-1}	[18]
γ_P	*Standardization coefficient*	2.36	ha.kg^{-1}	[18]*
α_ψ	*Pollinated flowers per pollinator unit*	40	pollinated flower.pollinator^{-1}	[18]
Effects of management (eqn S.1.6)				
η	*Spraying effect on pests (efficiency)*	$\in[0,1]$	-	-
θ	*Spraying effect on beneficial insects (selectivity)*	$\in[0,1]$	-	-
Pollinator dynamics (eqn S.1.6)				
λ_P	*Growth rate*	100	-	[22]*
c_P	*Competition rate*	99	pollinator^{-1}	[22]*
Cacao Pod Borer (CPB) dynamics (eqn S.1.8)				
α_{CPB}	*Number of eggs per female*	130	egg.female CPB^{-1}	[20]
β_{CPB}	*Egg predation rate*	0.15	-	[20]
S_0	*Larvae survival*	0.1	CPB.egg^{-1}	[20]*
σ_{CPB}	*Adult sex-ratio*	0.5	female CPB. CPB^{-1}	Wielgoss & Clough (unpub)
δ_{CPB}	*Adult predation rate*	0.41	-	[20]
μ	*Density dependence coefficient 1*	0.69	pod.CPB−^{1}	[20]
ρ	*Density dependence coefficient 2*	1.92	-	[20]
Parasitoid dynamics (eqn S.1.9, eqn S.1.10)				
a_{Par}	*Parasitism probability*	0.003	-	[20]*
Yield function (eqn S.1.12)				
ω	*Weight of dry beans per pod*	0.03	kg.pod^{-1}	Clough (unpub)*
α_Y	*Yield loss parameter 1*	0.01	CPB.pod^{-1}	[20]
β_Y	*Yield loss parameter 2*	6.33	-	[20]

*refers to parameters calibrated from the cited reference.

that we do not aim at quantitative prediction but at qualitative understanding, we only modeled relative abundances and standardized the midge population in the absence of spraying to 1000 at equilibrium.

A sensitivity analysis was then conducted to test the effect of a variation of +/− 10% of each parameter of the model on the average Cacao yield and on the average population sizes of each of the three insects (for details, see [19]). The sensitivity analysis showed a high sensitivity of the model to the two parameters of the intra-annual dynamics of Cacao pod production (parameters α_I, and β_I in appendix S1). Calibration of these key parameters was then adjusted using data from a survey of the Cacao yield of two plots in central Sulawesi [23,24]. Initially conducted to compare rainfall treatments, this database provided us with six control subplots that we used for calibration. The survey was conducted from January to December 2007 with a two-week time step. Calibration was obtained by minimizing the Root Mean Square Error (RMSE) of the full model compared to the Cacao yield data. To ensure a periodic pattern of the dynamics, calibration was made on two successive years. This sensitivity analysis made it possible to isolate the most sensitive parameters. Once the sensitivity to input data had been tested, all simulations were run in a deterministic manner, to keep the number of simulation within feasible limits.

Simulations

Indices. The model that we used made it possible to compute the agroecosystem dynamics for the whole set of possible management strategies. For each simulated spraying strategy, we recorded the average yearly yield and the number of spraying events as well as two specific indices to record information on robustness and ecological functioning of the system. The Robustness index records the average deviation (in absolute values) of the productive output for a given set of perturbations. It reads as follows:

$$Robustness = \frac{Y_M - M(|Y - y_i|)}{Y_M} \quad (3)$$

with Y_M the maximum reachable yield (750 kg.ha^{-1}), Y the yield without perturbation and y_i the yield with perturbation i (i belongs to the set of studied perturbations). Function M stands for the mean and | | for the absolute value.

The Ecosystem Functioning Index (EFI) relates to the ecological functioning of the agroecosystem. It synthesizes the different ecological processes at stake and reads as follows:

$$EFI = P_F + P_{CPB} + P_{Pods} \quad (4)$$

with P_F the pollination rate, P_{CPB} rate of parasitism of the Cacao Pod Borers and P_{Pods} the infestation rate of the Cacao pods. This index synthesizes all ecological functions at stake in our modeled agroecosystem and does not only focus on ecosystem services. We decided to consider all functions regardless of their effects on production to reflect information available in real systems. Indeed, in real systems it is difficult to put a number on services specifically, especially when the distinction between services and disservices is not clear [25] or when the services are not known well.

Typology of the management strategies. To identify spraying strategies that combine extreme intensities of management and ecosystem functioning we built two contrasting groups of strategies, depending on their position in the ecosystem functioning (*EFI*) - number of spraying events (N_S) - plane. The set of Ecological Strategies (ES) encompass all strategies that are within both the 10% *EFI* upper quantile and the 10% N_S lower quantile. The set of Chemical Based Strategies, i.e. the CBS-management strategies encompass all strategies that are within both the 10% *EFI* lower quantile and the 10% N_S upper quantile.

Production and robustness of the management strategies. We analyzed the relationship between yield and robustness of the system under six types of perturbations related to management or environmental conditions. More precisely, we subjected the system to the following perturbations.

Perturbations related to modifications of the environmental conditions.

- Pest outbreak: For each management strategy, we simulate two alternative perturbations via a sudden increase in the Cacao Pod Borer population by 500 and 1000 individuals.ha^{-1} at month 3 (time of the year where the number of pods sensitive to this pest is the highest).
- Variation in Cacao production: For each management strategy, we simulate four alternative situations with each of the two parameters of the Cacao pod dynamics (mean and amplitude of the pod production sinusoid) increased or decreased by 10%. This perturbation could reflect diverse environmental variations, including climatic ones.

Perturbations related to modifications of the management strategies.

- Event shift: For each management strategy, we simulate 12 alternative, modified strategies involving the shift of one of the 12 spraying/non-spraying events (i.e. 1 spraying event was replaced by a non-spraying event and *vice versa*).
- Temporal shift: For each management strategy, we simulate two alternative strategies by shifting the entire spraying sequence by one month (either shifted one month earlier or one month later).

Perturbations related to the pesticide characteristics.

- Pesticide efficiency: This scenario corresponds to a modification of the pesticide efficiency. For each management strategy, we simulate two alternative situations with a pesticide whose efficiency is either increased or decreased by 20%.
- Pesticide selectivity: This scenario corresponds to a modification of the pesticide selectivity. For each management strategy, we simulate two alternative situations with a pesticide whose selectivity is either increased or decreased by 20%.

These six perturbations were tested on the 2^{12} possible management strategies. Then the relative performances of the ES and CBS defined were assessed. To compare the performances of ES and CBS, we compared all possible pairs of strategies (always one from the ES set and one from the CBS set) and computed the proportion of ES that performed better than CBS (hereafter called Comparison Index, *CInd*).

We first give a detailed analysis of the response of the Cacao agroecosystems to these perturbations for a single type of pesticide (Efficiency = 0.5, Selectivity = 0.5). Then, we conduct the same analysis for 81 different pesticide types (Efficiency and Selectivity ranging from 0.1 to 0.9 with a step of 0.1) so as to test the range of validity of these first findings.

Numeric computations and statistical analyses were performed with Python 2.7.2 (http://www.python.org/).

Results

Calibration

After calibration, our model gave a reasonable visual fit (Figure 2). Comparison of the model outputs with the Cacao yield data showed that we managed to capture the general behavior of the system.

Typology of the management strategies

We logically observed a strong relationship between the number of spraying events N_S and the functioning of the agroecosystem *EFI* (Figure 3; $EFI = 1.63 - 0.05\ N_S$, $p < 10^{-3}$, $R^2 = 0.68$). The management strategies built a continuum. At the two extremes of this continuum, we distinguished two subsets of trajectories: the Chemical Based Strategies (CBS) that corresponded to a high number of spraying events and a low functioning of the ecosystem and the Ecological Strategies (ES) that corresponded to a low number of spraying events and a high ecosystem functioning (Figure 3).

Production and robustness of the management strategies

The relationship between yield and robustness differed between the different types of perturbations (Table 2). Negative relationships were found for environment- and management-related perturbations while a positive relationship was found for pesticide-related perturbations.

On average, the yield obtained with ES was higher than the yield obtained with CBS (Figure 4, *CInd* = 0.71). The robustness of CBS was lower than the one of the ES under perturbations due to a pest outbreak (Figure 4.a.; *CInd* = 1.00) as well as under pesticide-related perturbations (Figure 4e, *CInd* = 0.99 and Figure 4f, *CInd* = 0.80). However, robustness of CBS was higher than the robustness of ES under perturbations related to management (Figure 4c, *CInd* = 0.12 and Figure 4d, *CInd* = 0.32) or to production (Figure 4b, *CInd* = 0.02).

These results can be qualitatively explained in the following way. Systems managed with ES have a high level of ecosystem functioning. This gives a higher capacity of self-regulation to the system, which explains its high robustness to pest outbreaks and to the pesticide characteristics. However ES are very specific strategies aimed at "driving" the system instead of "controlling" it. This explains their low robustness to both production- and management-related perturbations.

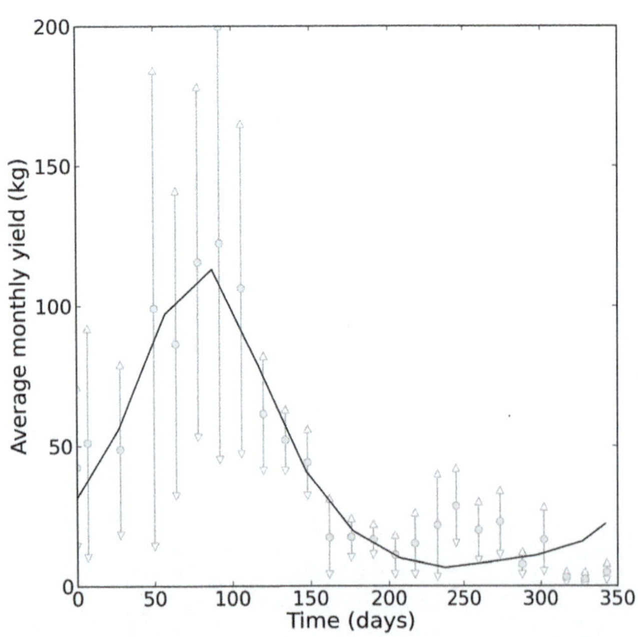

Figure 2. Model output and data from the study area (zoom on year 10). The solid black curve corresponds to the model output, the gray dots to the average monthly yields observed in the six plots of the study area. The grey bars around these dots correspond to the average values +/− standard deviation.

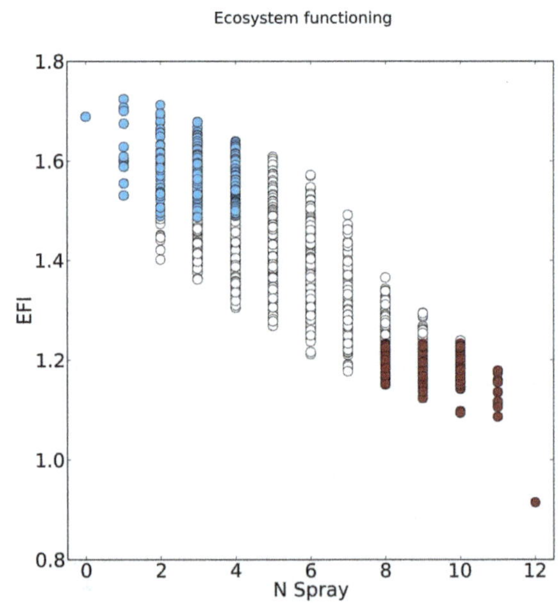

Figure 3. Intensity of spraying and ecosystem functioning of the different spraying strategies. The x-axis is the Ecosystem Functioning Index (EFI) and the y-axis is the number of spraying events per year. Each symbol represents one of the possible spraying strategies. Black symbols stand for the subset of Chemical Based Strategies (CBS; high spraying intensity and a low ecosystem functioning), white symbols stand for the subset of Ecological Strategies (ES; low spraying intensity and a high ecosystem functioning). Gray symbols stand for all other strategies.

Effect of the type of pesticide

Comparison of the robustness of CBS and ES showed similar results for the different types of pesticides as for the first pesticide detailed in the former section (Figure 5). Due to the non-linearity of model dynamics, the patterns observed are not necessarily

Table 2. Relationship between yield and robustness for four types of perturbation (linear models).

Perturbation	Trend	Intercept	Slope	R^2	P-value
Environment (pest outbreak)	Negative	600	−160	0.03	<10-3
Environment (production)	Negative	1158	−1018	0.65	<10-3
Management (event shift)	Negative	2478	−2053	0.14	<10-3
Management(temporal shift)	Negative	859	−413	0.03	<10-3
Pesticide (selectivity)	Positive	−270	747	0.37	<10-3
Pesticide (efficiency)	Positive	−478	944	0.19	<10-3

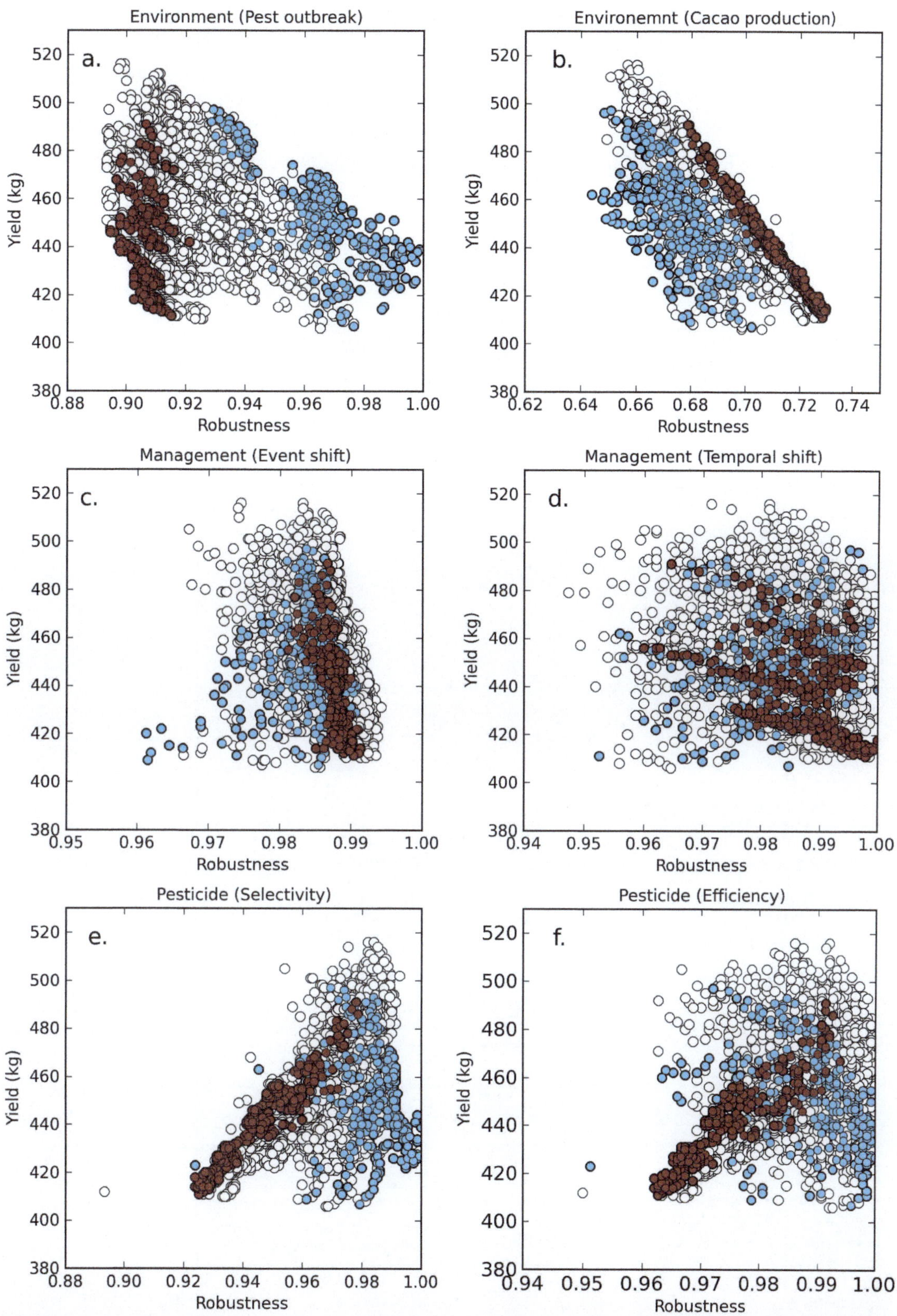

Figure 4. Relationship between yield and robustness under different perturbations. Under environmental perturbation (a: pest outbreak; b: production variation), under a modification of the management strategy (c: shift in one of the spraying/non spraying events; d: temporal shift) or under a perturbation linked to the type of pesticide (e: selectivity; f: efficiency). Each white symbol represents one of the 4096 possible spraying strategies. Dark red symbols stand for the subset of Chemical Based Strategies (CBS, high spraying intensity and a low ecosystem functioning), light blue symbols stand for the subset of Ecological Strategies (ES, low spraying intensity and a high ecosystem functioning).

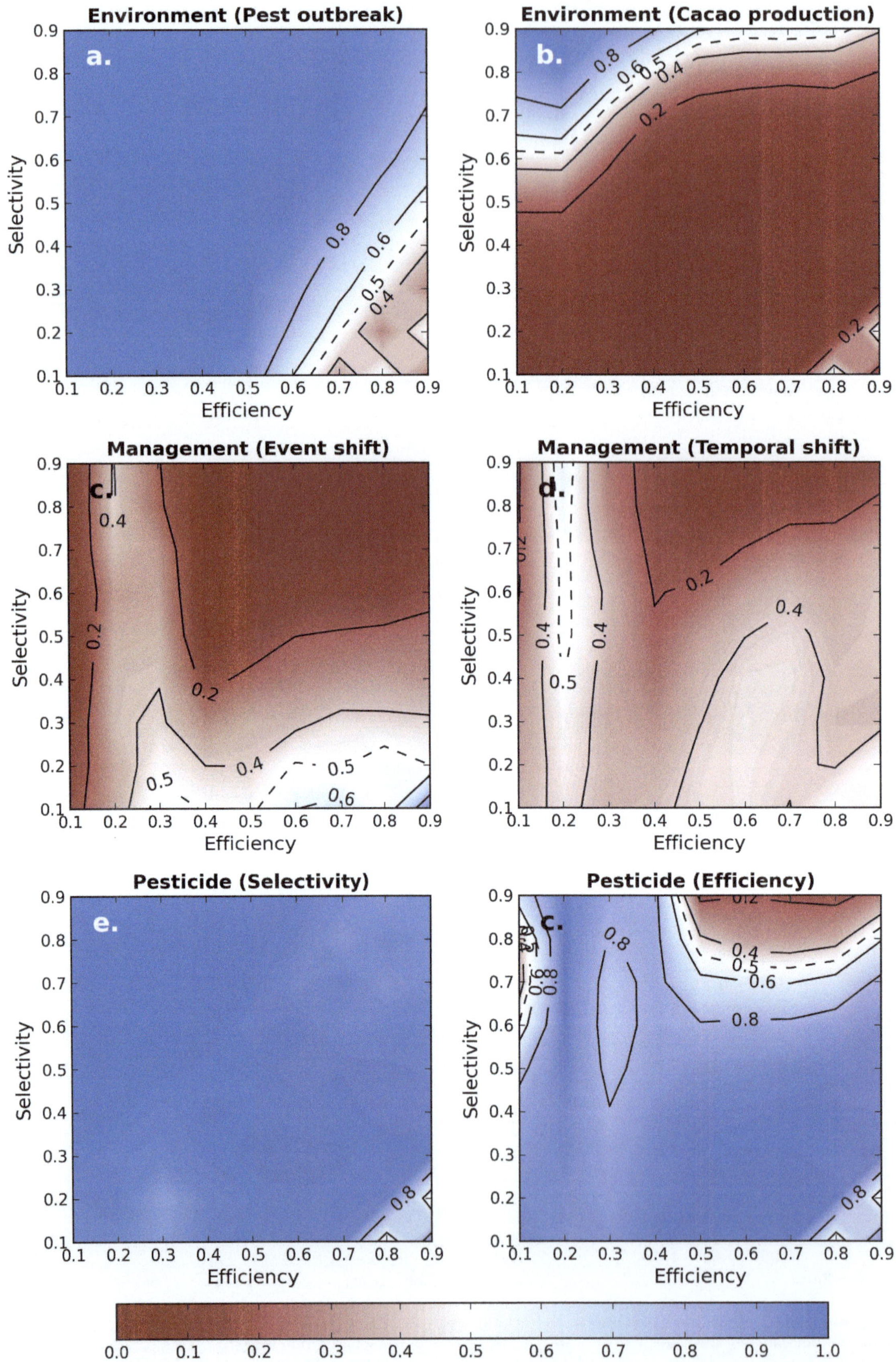

Figure 5. Comparison of the robustness of the agroecosystem managed by Chemical Based Strategies (CBS) or Ecological Strategies (ES) under 6 different perturbations. Figures a and b correspond to an environmental perturbation (a: pest outbreak, b: production perturbation). Figures c and d correspond to perturbation of the management (c: shift of one spraying/non spraying event, d: temporal shift of the

management strategy). Figures e and f correspond to perturbations of the pesticide characteristics (e: selectivity; f: efficiency). Color indicates the comparison index calculated, i.e. the percentage of pairs of strategies in which the ES performs better than the CBS.

smooth but general conclusions can still be drawn. For most pesticides, CBS were more robust than ES to Cacao production-related perturbations (average *CInd* = 0.22) as well as to management-related perturbations (event shift, average *CInd* = 0.24, temporal shift, average *CInd* = 0.32) and ES were more robust than CBS to pest outbreaks (average *CInd* = 0.88) as well as to pesticide-related perturbations (efficiency, average *CInd* = 0.80, selectivity, average *CInd* = 0.96).

However, yield showed a more balanced pattern (Figure 6, average *CInd* = 0.51). ES showed higher yields with pesticides of high efficiency and low selectivity while CBS showed higher yields with pesticides of high selectivity and low efficiency. This illustrates the non-trivial role of pesticides in a complex agroecosystem.

Discussion

Our results showed two opposite behaviors of the Cacao agroecosystem that we modeled depending on their management: First, Cacao agroecosystems relying on a high input of pesticides and showing low ecosystem functioning that fit to the conventional model of chemical intensification inspired by the Green Revolution; second, Cacao agroecosystems relying on ecosystem functioning with a low level of chemical input that fit to the model of ecological intensification inspired by agroecology-like paradigms. These two types of management led to different levels of production as well as different robustness. From the production point of view, no system clearly outclassed the other and their ranking depended on the type of pesticide used. From the robustness point of view, the two types of systems performed differently when subjected to different types of perturbations. Ecologically intensive systems were more robust to pest outbreaks and pesticide-related perturbation while chemically intensive systems were more robust to management perturbation and production-related perturbations.

Generality of the results

Our results were obtained with a simplified model of a Cacao agroecosystem and the generality of our results should be discussed. The agroecosystem that we modeled only includes three insects, which is far less than what can be found in the Cacao agroecosystem of the study area [26] and in agroecosystems in general (e.g. [27]). However, ecosystem properties depend much more on functional diversity than on species richness *per se* [28] and agroecosystems are no exception [29]. Choosing these three species, we focused on the three main above-ground ecosystem services and disservices commonly found in agroecosystems (pest

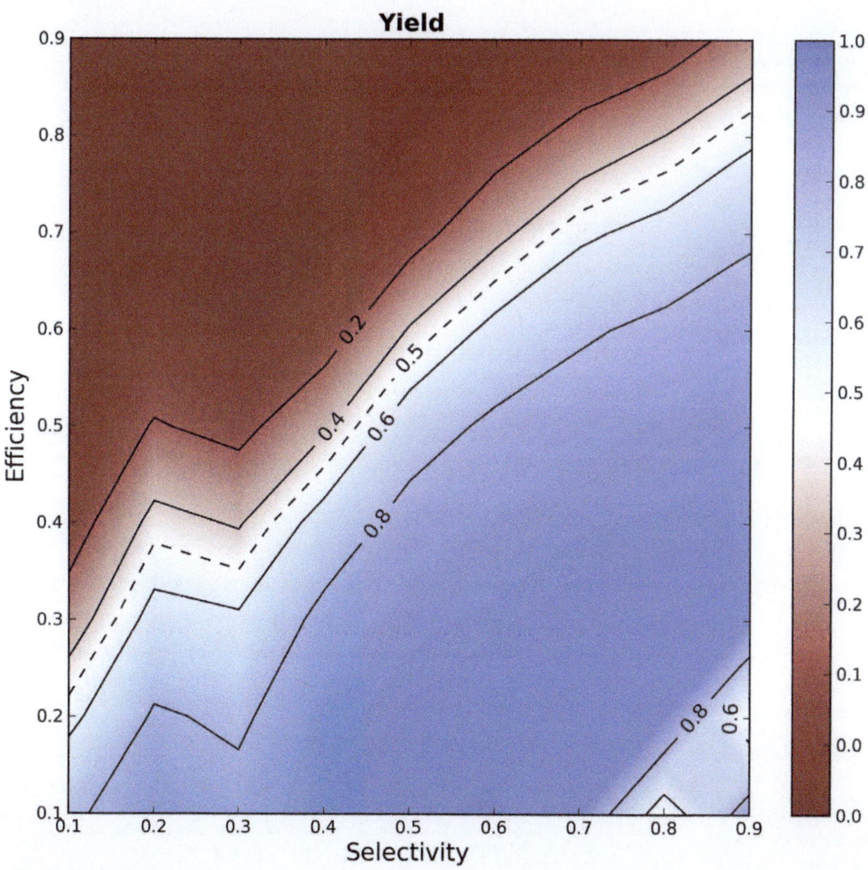

Figure 6. Comparison of the yield obtained in agroecosystems managed by Chemical Based Strategies (CBS) or Ecological Strategies (ES). Color indicates the comparison index calculated, i.e. the percentage of pairs of strategies in which the ES performs better than the CBS.

damage, pest regulation and pollination; [10]). In this sense, even though the quantitative outputs of our models may differ when applied to other agroecosystems, the following qualitative results should remain valid to a broad range of agroecosystems:

- There is a strong negative correlation between the frequency of spraying events and the functioning of the agroecosystem.
- Higher yields can be reached with farming practices based on ecosystem functioning when broad-spectrum pesticides (high efficiency and low selectivity) are used.
- Farming practices based on ecosystem functioning are more robust to pest outbreak perturbations than farming practices based on chemical inputs.
- Farming practices based on ecosystem functioning are more sensitive to management perturbation than farming practices based on chemical inputs.

We see two ways of validating these general findings through modelling. These general results could be tested by applying our modeling framework to other types of agroecosystems (e.g. orchards, oil palm plantations,...). The application of other models of agroecosystems (e.g. [30–33]) to the question raised in this article would also be a way of testing the generality of these results. However, most of these models would first have to be extended to explicitly include crop dynamics and the effects of insects on production. The use of such models developed in different scientific contexts would contribute to the cross-validation of our results.

Limits and perspectives

In this article, we only focused on Cacao agroecosystem management at the field scale through the use of pesticides. Therefore, our model could be developed in two main directions to improve our understanding of the mechanisms of ecological intensification: adding new dimensions to management and increasing spatial scale.

With respect to new dimensions, the dynamics of shade trees could be added to the model. In the specific case of agroforestry systems such as Cacao, the management of shade trees is a strong driver of ecosystem services and disservices [34,35]. Including this aspect in our model would improve its predictive power and allow us to address the paradox raised by [36]: on the one hand, several studies highlight the strong importance of shade trees for the sustainability of the production system, but on the other hand, farmers tend to remove shade trees to improve yields and do not notice any major drawback.

With respect to scaling-up, several studies have emphasized the importance of the landscape scale when considering ecological dynamics of agroecosystems [37]. Especially, the spatial distribution of insects involved in ecosystem services such as pest control has a strong effect on pest populations [38,39]. In our study area, parasitism rate, for instance, has been shown to depend strongly on the distance to forest [40] and pollination also depends strongly on landscape structure [41]. Transferring our model to the landscape scale would make it possible to consider both ecological and economic interactions between different fields. Refining the management component of our model and transferring it to greater spatial scales would increase the number of tools available to balance between ecosystem services and disservices and give a better understanding of how ecological intensification could be put into practice.

Conclusion

We modeled a Cacao agroecosystem under two management scenarios. The Cacao agroecosystem managed in an ecologically intensive way strongly differed from the one managed in a more conventional way using high quantities of chemical inputs. The ecologically intensive Cacao agroecosystem was more robust to pest outbreaks. It also showed higher yields when broad spectrum pesticides were used. However, the ecologically intensive Cacao agroecosystem was more sensitive to management-related perturbations and Cacao production perturbations, which confirms the high level of expertise needed to conduct such a management.

Supporting Information

Appendix S1 Description of the model.

Acknowledgments

We thank Andreas Spangenberg for farm surveys and Gerald Moser for providing us the Cacao yield databases.

Author Contributions

Conceived and designed the experiments: RS. Performed the experiments: RS. Analyzed the data: RS. Wrote the paper: RS KW KM.

References

1. Fukuoka M (2010) The one-straw revolution: An introduction to natural farming. New York Review Books. 229 p.
2. Altieri MA, Nicholis CI (2005) Agroecology and the search for a truly sustainable agriculture. Mexico D.F., Mexico: United Nations Environmental Programme, Environmental Training Network for Latin America and the Caribbean.
3. Gliessman SR (2012) Agroecology: Researching the ecological basis for sustainable agriculture. Springer London, Limited. 400 p.
4. Swaminathan MS (1996) Sustainable Agriculture: Towards an evergreen revolution. Konark Publishers. 219 p.
5. Conway G (1998) The doubly green revolution: food for all in the twenty-first century. Comstock Pub Associates. 356 p.
6. Kremen C, Williams NM, Aizen MA, Gemmill-Herren B, LeBuhn G, et al. (2007) Pollination and other ecosystem services produced by mobile organisms: a conceptual framework for the effects of land-use change. Ecol Lett 10: 299–314. doi:10.1111/j.1461-0248.2007.01018.x.
7. Bommarco R, Miranda F, Bylund H, Björkman C (2011) Insecticides suppress natural enemies and increase pest damage in cabbage. J Econ Entomol 104: 782–791.
8. Altieri MA, Ponti L, Nicholls CI (2005) Manipulating vineyard biodiversity for improved insect pest management: case studies from northern California. Int J Biodivers Sci Manag 1: 191–203.
9. Ives AR, Carpenter SR (2007) Stability and diversity of ecosystems. Science 317: 58–62. doi:10.1126/science.1133258.
10. Zhang W, Ricketts TH, Kremen C, Carney K, Swinton SM (2007) Ecosystem services and dis-services to agriculture. Ecol Econ 64: 253–260. doi:10.1016/j.ecolecon.2007.02.024.
11. Tilman D, Downing JA (1994) Biodiversity and stability in grasslands. Nature 367: 363–365. doi:10.1038/367363a0.
12. Holling CS (1973) Resilience and stability of ecological systems. Annu Rev Ecol Syst 4: 1–23. doi:10.1146/annurev.es.04.110173.000245.
13. Stelling J, Sauer U, Szallasi Z, Doyle FJ 3rd, Doyle J (2004) Robustness of cellular functions. Cell 118: 675–685. doi:10.1016/j.cell.2004.09.008.
14. Carpenter S, Walker B, Anderies JM, Abel N (2001) From metaphor to measurement: resilience of what to what? Ecosystems 4: 765–781. doi:10.1007/s10021-001-0045-9.
15. Ruf F (1995) Booms et crises du cacao: les vertiges de l'or brun. KARTHALA Editions. 504 p.
16. FAO website (2013) Available: www.faostat.fao.org, Accessed 2013 July 17.
17. Entwistle PF, Philip F (1972) Pests of cocoa. [London]: Longman.
18. Groeneveld JH, Tscharntke T, Moser G, Clough Y (2010) Experimental evidence for stronger Cacao yield limitation by pollination than by plant resources. Perspect Plant Ecol Evol Syst 12: 183–191. doi:10.1016/j.ppees.2010.02.005.
19. Sabatier R, Meyer K, Wiegand K, Clough Y (2013) Non-linear effects of pesticide application on biodiversity-driven ecosystem services and disservices in

a Cacao agroecosystem: A modeling study. Basic Appl Ecol 14: 115–125. doi:10.1016/j.baae.2012.12.006.
20. Day R (1985) Control of the cocoa pod borer (Conopomorpha cramerella) London: University of London.
21. Hassell MP (1978) The spatial and temporal dynamics of host-parasitoid interactions. Oxford; New York: Oxford University Press.
22. Séchan Y, Faaruia M, Tetuanui A (1996) Lutte contre le nono blanc des plages: rapport d'exécution et de fin des travaux. Papeete: ITRMLM. Available: http://www.documentation.ird.fr/hor/fdi:010013638. Accessed 8 August 2013.
23. Moser G, Leuschner C, Hertel D, Hölscher D, Köhler M, et al. (2010) Response of cocoa trees (Theobroma cacao) to a 13-month desiccation period in Sulawesi, Indonesia. Agrofor Syst 79: 171–187. doi:10.1007/s10457-010-9303-1.
24. Schwendenmann L, Veldkamp E, Moser G, Hölscher D, Köhler M, et al. (2010) Effects of an experimental drought on the functioning of a Cacao agroforestry system, Sulawesi, Indonesia. Glob Change Biol 16: 1515–1530. doi:10.1111/j.1365-2486.2009.02034.x.
25. Wielgoss A, Clough Y, Fiala B, Rumede A, Tscharntke T (2012) A minor pest reduces yield losses by a major pest: plant-mediated herbivore interactions in Indonesian Cacao: Indirect interaction between Cacao pests. J Appl Ecol 49: 465–473. doi:10.1111/j.1365-2664.2012.02122.x.
26. Kessler M, Abrahamczyk S, Bos M, Buchori D, Putra DD, et al. (2011) Cost-effectiveness of plant and animal biodiversity indicators in tropical forest and agroforest habitats. J Appl Ecol 48: 330–339. doi:10.1111/j.1365-2664.2010.01932.x.
27. Duelli P, Obrist MK, Schmatz DR (1999) Biodiversity evaluation in agricultural landscapes: above-ground insects. Agric Ecosyst Environ 74: 33–64. doi:10.1016/S0167-8809(99)00029-8.
28. Hooper DU, Chapin Iii FS, Ewel JJ, Hector A, Inchausti P, et al. (2005) Effects of biodiversity on ecosystem functioning: a consensus of current knowledge. Ecol Monogr 75: 3–35.
29. Moonen A-C, Bàrberi P (2008) Functional biodiversity: An agroecosystem approach. Agric Ecosyst Environ 127: 7–21. doi:10.1016/j.agee.2008.02.013.
30. Chatterjee S, Isaia M, Venturino E (2009) Spiders as biological controllers in the agroecosystem. J Theor Biol 258: 352–362. doi:10.1016/j.jtbi.2008.11.029.
31. Ives AR, Settle WH (1997) Metapopulation dynamics and pest control in agricultural systems. Am Nat 149: 220–246.
32. Drechsler M, Settele J (2001) Predator–prey interactions in rice ecosystems: effects of guild composition, trophic relationships, and land use changes—a model study exemplified for Philippine rice terraces. Ecol Model 137: 135–159.
33. Bambaradeniya CNB, Edirisinghe JP (2008) Composition, structure and dynamics of arthropod communities in a rice agro-ecosystem. Ceylon J Sci Biol Sci 37: 23–48.
34. Steffan-Dewenter I, Kessler M, Barkmann J, Bos MM, Buchori D, et al. (2007) Tradeoffs between income, biodiversity, and ecosystem functioning during tropical rainforest conversion and agroforestry intensification. Proc Natl Acad Sci 104: 4973–4978. doi:10.1073/pnas.0608409104.
35. Tscharntke T, Clough Y, Bhagwat SA, Buchori D, Faust H, et al. (2011) Multifunctional shade-tree management in tropical agroforestry landscapes – a review. J Appl Ecol 48: 619–629. doi:10.1111/j.1365-2664.2010.01939.x.
36. Ruf FO (2011) The myth of complex cocoa agroforests: the case of Ghana. Hum Ecol 39: 373–388. doi:10.1007/s10745-011-9392-0.
37. Tscharntke T, Klein AM (2005) Landscape perspectives on agricultural intensification and biodiversity – ecosystem service management. Ecol Lett 8: 857–874.
38. Bianchi F, Schellhorn NA, Buckley YM, Possingham HP (2010) Spatial variability in ecosystem services: simple rules for predator-mediated pest suppression. Ecol Appl 20: 2322–2333.
39. Ricci B, Franck P, Toubon J-F, Bouvier J-C, Sauphanor B, et al. (2009) The influence of landscape on insect pest dynamics: a case study in southeastern France. Landsc Ecol 24: 337–349. doi:10.1007/s10980-008-9308-6.
40. Klein A-M, Steffan-Dewenter I, Tscharntke T (2006) Rain forest promotes trophic interactions and diversity of trap-nesting Hymenoptera in adjacent agroforestry. J Anim Ecol 75: 315–323. doi:10.1111/j.1365-2656.2006.01042.x.
41. Priess JA, Mimler M, Klein AM, Schwarze S, Tscharntke T, et al. (2007) Linking deforestation scenarios to pollination services and economic returns in coffee agroforestry systems. Ecol Appl Publ Ecol Soc Am 17: 407–417.
42. Juhrbandt J (2011) Economic valuation of of land use change-A case study on rainforest conversion and agroforestry intensification in Central Sulawesi, Indonesia. Available: https://ediss.uni-goettingen.de/handle/11858/00-1735-0000-0006-AB32-C. Accessed 7 August 2013.

The Soil Biota Composition along a Progressive Succession of Secondary Vegetation in a Karst Area

Jie Zhao[1,2], Shengping Li[1,2,3]**, Xunyang He[1,2], Lu Liu[1,2,3], Kelin Wang[1,2]***

1 Key Laboratory of Agro-ecological Processes in Subtropical Region, Institute of Subtropical Agriculture, Chinese Academy of Sciences, Changsha, Hunan, China, 2 Huanjiang Observation and Research Station for Karst Ecosystems, Chinese Academy of Sciences, Huanjiang, Guangxi, China, 3 Graduate University of Chinese Academy of Sciences, Beijing, China

Abstract

Karst ecosystems are fragile and are in many regions degraded by anthropogenic activities. Current management of degraded karst areas focuses on aboveground vegetation succession or recovery and aims at establishing a forest ecosystem. Whether progressive succession of vegetation in karst areas is accompanied by establishment of soil biota is poorly understood. In the present study, soil microbial and nematode communities, as well as soil physico-chemical properties were studied along a progressive succession of secondary vegetation (from grassland to shrubland to forest) in a karst area in southwest China. Microbial biomass, nematode density, ratio of fungal to bacterial biomass, nematode structure index, and nematode enrichment index decreased with the secondary succession in the plant community. Overall, the results indicated a pattern of declines in soil biota abundance and food web complexity that was associated with a decrease in soil pH and a decrease in soil organic carbon content with the progressive secondary succession of the plant community. Our findings suggest that soil biota amendment is necessary during karst ecosystem restoration and establishment and management of grasslands may be feasible in karst areas.

Editor: Dafeng Hui, Tennessee State University, United States of America

Funding: This study was supported by the Chinese Academy of Sciences Action Plan for the Development of Western China (KZCX2-XB3-10), the National Key Technology Research and Development Program of the Ministry of Science and Technology of China (2010BAE00739-02) given to Kelin Wang. The study was also supported by a Western Light Program from CAS and a NSFC program (31300448) given to Jie Zhao. The funders had no role in study design, data collection and analysis, decision to publish, or preparation of the manuscript.

Competing Interests: The authors have declared that no competing interests exist.

* Email: kelin@isa.ac.cn

These authors contributed equally to this work.

Introduction

Karst landscapes account for about 12% of the Earth's land surface [1]. Ecosystems in karst areas are fragile and suffering extreme degradation in many regions of the world, such as southwest China, Southeast Asia (e.g., Vietnam and Thailand), and Central America (e.g., Mexico, Puerto Rico, Dominican Republic, and Cuba) [2–7]. Karst areas in southern and central Europe (e.g., Austria, Bosnia, Herzegovina, Croatia, Germany, Italy, Russia, and Spain [not all land cover of these countries are karst but there are karst areas in them]) also are facing significant land degradations [3]. One important example of this degradation is "karst rocky desertification", which is characterized in part by shallow soils and exposed bedrock [1]. Karst rocky desertification results largely from anthropogenic activities such as irrational cultivation and extensive logging. Since the 1980s, karst rocky desertification has been recognized as a critical factor limiting the social and economic development of the karst region in southwest China [8–9].

The Chinese central and local governments have promoted the ecological restoration of karst areas in southwest China, and as a consequence many farmlands and plantations have been abandoned and allowed to recover naturally. Several studies have documented the succession of plant communities in the karst region in southwest China. For example, Zeng et al. (2007) identified the following six plant communities occurring during succession following the natural revegetation of areas suffering from rocky desertification: sparse grass, grass, shrub, liana-shrub, deciduous broad-leaved forest, and mixed evergreen and deciduous broad-leaved forest [8]. An earlier study had also identified six vegetation types: grass, grass-shrub, shrub, shrub-tree, tree forest, and climax mixed evergreen and deciduous broad-leaved forest [10]. Although the classification of the plant communities is somewhat different in these two studies, the succession from grass to shrub to forest is considered typical in this area [11–12].

The restoration of degraded karst areas in southwest China is currently evaluated based on the degree of forest vegetation establishment but the restoration of ecosystem services will also depend on the recovery of other organisms including soil organisms. Whether a secondary vegetation succession can promote the increment of biomass and abundance of soil biota in degraded karst areas is largely unknown, and this lack of knowledge limits our ability to understand and predict the functioning of such ecosystems.

In studying the succession of plant communities and soil biota, researchers frequently use space-for-time substitution, i.e., chronosequences [13]. In addition, researchers recognize two general types of succession, which are primary and secondary. Secondary succession is widespread and occurs at a high rate relative to primary succession [14]. In the past two decades, a number of studies have explored the relationship between secondary vegetation succession and soil biota. For example, vegetation succession can influence the composition of soil biota through resource and habitat changes [15–17], and the soil biota can affect the vegetation succession by selective feeding and by altering nutrient availability [18–19]. Few studies, however, have explored the changes in soil biota (especially nematodes) during secondary succession from grassland to other ecosystem types, and these studies have yielded inconsistent results concerning the relationship between vegetation succession and soil biota succession. For example, soil microbial biomass increased during the secondary succession of abandoned fields ranging in age from 1 to 60 years in east-central Minnesota [17] but declined across a secondary vegetation succession from short grassland to tall grassland to shrub-land in northwest France [20]. In addition, soil microbial biomass did not differ between early and late stages in a secondary succession from shrubland to forest in New Zealand [16].

In this study, we determined whether the progressive succession of the plant community is accompanied by the establishments of soil microbial and nematode communities in karst shallow soils; by progressive succession, we mean that the community becomes more complex, with increased species richness, community biomass, and food web trophic links. We explored the compositions of soil microbial and nematode communities along a secondary vegetation succession from grassland to shrubland to forest in the karst region in southwest China, where ecosystems have been severely disrupted and where the soils are very shallow. We test the hypothesis that progressive succession of the plant community should improve the soil habitat (because of increases in soil organic matter and nutrients) and thereby support the establishments of the soil microbial and nematode communities.

Materials and Methods

No specific permits were required for the described field studies. No specific permissions were required for these locations/activities. The location is not privately owned or protected in any way and the field studies did not involve endangered or protected species.

Site description

This study was carried out at the Huanjiang Observation and Research Station for Karst Ecosystems (107°51′–108°43′E, 24°44′–25°33′N), Chinese Academy of Sciences (CAS), Guangxi Province, China. The climate is subtropical monsoon with a distinct wet (from April to September) and dry season (from October to March). The mean annual temperature and precipitation are 18.5°C and 1,389 mm, respectively. The soil developed from a dolostone base and is calcareous.

Three plant communities (or ecosystems) were identified on the hillsides around the Huanjiang Observation and Research Station: grassland, shrubland, and forest. The main criteria used in selecting and designating these vegetation types were as follows: the plant community composition should be different, the disturbance history of the site should be known (based on information obtained from local farmers), and the environmental conditions (soil origin, altitude, slope orientation, and slope gradient) should be as similar as possible. All three vegetation types were situated in the middle of south-facing slopes with gradients ranging from 30 to 50°. The altitude ranged from 280 to 464 m. The grasslands were developed from managed forests

Table 1. Soil physico-chemical properties at two depths in a grassland, shrubland, and forest in the degraded karst region of southwest China.

Property	Depth (cm)	Vegetation type		
		Grassland	Shrubland	Forest
SWC (%)	0–5	47.3 (1.4)	54.9 (4.4)	44.7 (8.3)
	5–10	43.2 (2.9)	45.8 (2.6)	40.1 (6.3)
pH_{WATER}	0–5	8.17 (0.02) a[a]	7.53 (0.32) b	6.89 (0.42) c
	5–10	8.20 (0.03)	7.60 (0.31)	6.85 (0.33)
TN (g/kg)	0–5	4.79 (0.22)	4.29 (0.20)	5.25 (0.79)
	5–10	4.33 (0.07)	3.56 (0.14)	4.29 (0.73)
SOC (g/kg)	0–5	96.9 (5.0) a	67.8 (9.6) b	56.5 (8.8) b
	5–10	87.9 (6.1)	50.6 (13.6)	39.4 (6.9)
C:N	0–5	20.4 (1.9) a	16.1 (3.1) a	10.8 (0.5) b
	5–10	20.4 (1.7)	14.5 (4.3)	9.2 (0.5)

[a]Means for a property (averaged over both soil depths) followed by different letters are significantly different ($p<0.05$) according to the LSD test.
SWC (%), pH, TN (g kg^{-1} dry soil), SOC (g kg^{-1} dry soil), and C:N stand for soil water content, soil pH, soil total nitrogen, soil organic carbon, and ratio of soil organic carbon to total nitrogen, respectively.

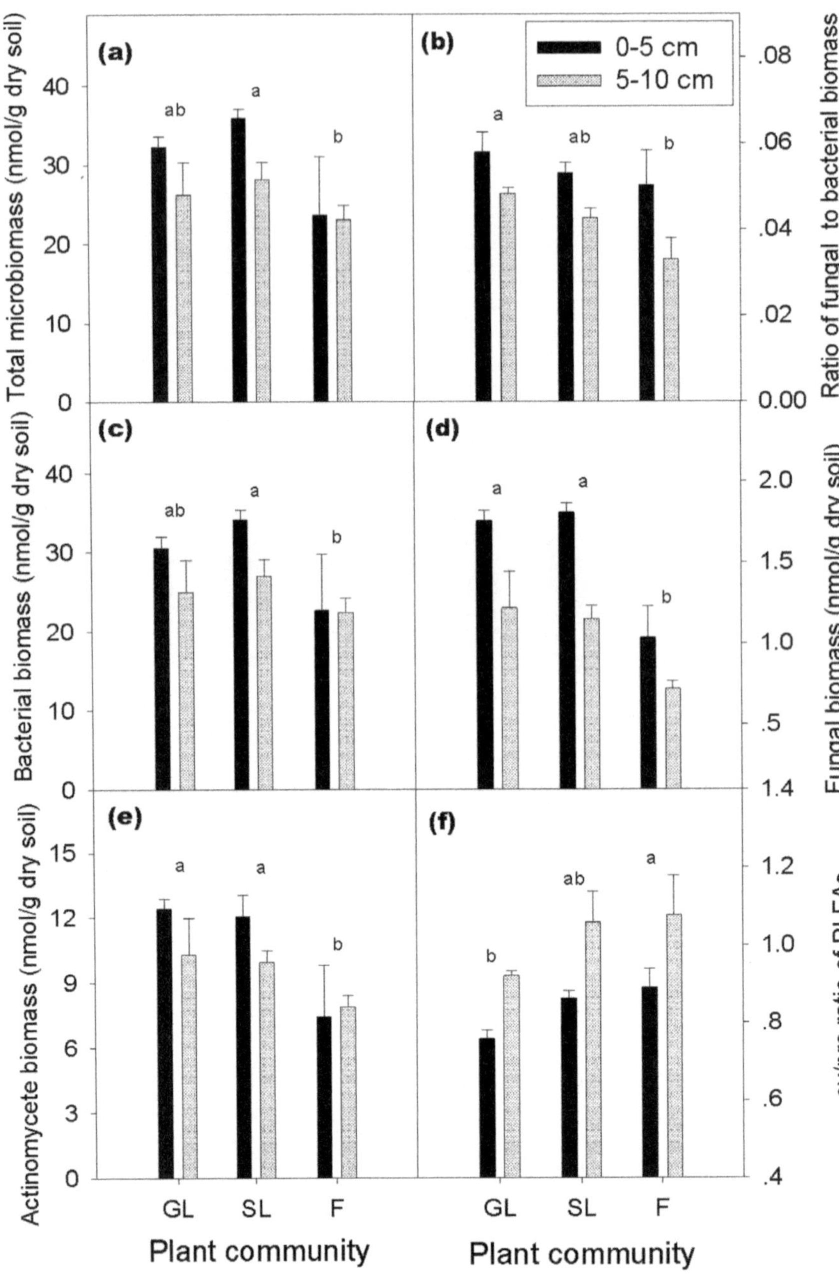

Figure 1. Characteristics of microbial communities (as determined by PLFA analysis) in the grassland (GL), shrubland (SL), and forest (F) at 0–5 and 5–10 cm soil depths. (a) Microbial biomass; (b) Ratio of fungal biomass to bacterial biomass; (c) Bacterial biomass; (d) Fungal biomass; (e) Actinomycetes biomass; and (f) cy/pre ratio. Bars indicate standard errors of means. Within each vegetation type, values with the same letters are not significantly different ($p>0.05$) according to the LSD test.

which had been abandoned for 9 to 10 years, and wildfire had occurred three times since all of the trees had been harvested and replaced by grasses. The dominant plant in the grassland was *Miscanthus floridulu*. The shrublands were developed from managed forests which had been abandoned for 25 to 30 years, at which time all of the trees were removed (i.e., all of the wood was harvested). The common shrub species were *Loropetalum chinensis*, *Litsea coreana* lvl. var. *Sinensis*, and *Pittosporum tonkinense Gagnep*. The forest sites had previously been managed forests and had been abandoned 50 to 55 earlier. The dominant overstory species in the forest was *Cyclobalanopsis glauca*; the common understory species included *Litsea coreana* levl. var. *Sinensis* and *Loropetalum chinensis*.

In 2011, the three identified plant communities at nine sites were selected, among which each plant community including three sites (replicates). A 10×10 m plot was established at each of the grassland and shrubland site, while a 20×30 m plot was established at each of the forest site. Because of anthropogenic disturbances in this area over the past decades, no primary forest ecosystem on calcareous soil that developed from a dolomite base existed near the research station. Therefore, no primary forest ecosystems were included in the present study.

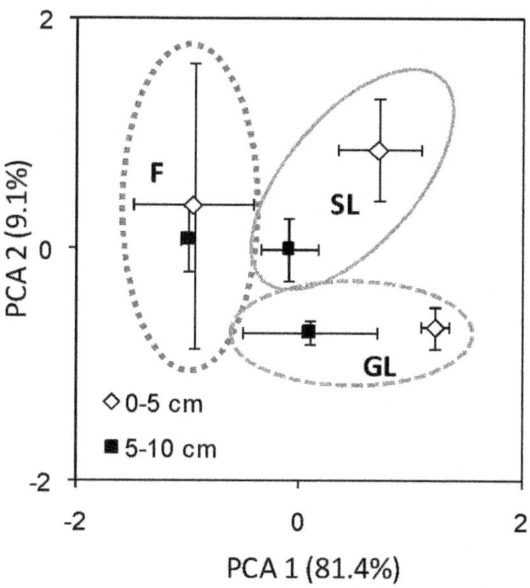

Figure 2. Principal component analysis (PCA) of microbial PLFA biomarkers of the grassland (GL), shrubland (SL), and forest (F) at 0–5 and 5–10 cm soil depth.

Soil sampling and analysis

Soil was sampled in September 2012. Soil cores (2.5 cm in diameter, 5 cm in length) were taken at 0–5 cm and 5–10 cm depths from 15 random locations within each of the nine plots. The 15 cores of the same depth from each plot were combined to form one composite sample, giving three replicate samples for each combination of vegetation type and depth. The surface litter was carefully removed before soil cores were collected. Samples were brought to the laboratory in insulated boxes. Then, the soil was sieved to pass through a 2 mm sieve to remove the roots. The soil water content, microbial and nematode communities were determined on a fresh subsample. Another subsample was thereafter air-dried and used for determination of soil pH, soil organic C and total nitrogen.

Soil pH was determined in 1:2.5 (w/v) soil: water suspensions, and soil water content (SWC %, g of water per 100 g dry soil) was measured by oven-drying the soil for 48 h at 105°C. Soil organic C (SOC, g kg^{-1} dry soil) was determined by dichromate oxidation, and total N (TN, g kg^{-1} dry soil) was measured with an ultraviolet spectrophotometer after Kjeldahl digestion [21].

Phospholipid fatty acids (PLFA) were extracted from 8 g fresh soil and were analyzed according to Bossio and Scow (1998) [22]. Concentrations of each PLFA were calculated relative to 19:0 internal standard concentrations. Bacterial biomass was considered to be represented by 10 PLFAs (i15:0, a15:0, 15:0, i16:0, 16:1ω7, i17:0, a17:0, 17:0, cy17:0, and cy19:0), fungal biomass was considered to be represented by the PLFA 18:2ω6,9, and actinomycete biomass was considered to be represented by three PLFAs (10 Me 16:0, 10 Me 17:0, and 10 Me 18:0) [23–25]. Microbial biomass was considered to be represented by the 10 bacterial PLFAs and the one fungal PLFA. Other PLFAs such as 16:1ω5, 18:1ω7, and 18:1ω9 were also used to analyze the soil microbial community.

Nematodes were extracted from 50 g of fresh soil using the Baermann funnel method [26]. After fixation in a 4% formalin solution, nematodes were counted with a differential interference contrast (DIC) microscope (ECLIPSE 80i, Nikon), and the first 100 individuals encountered were identified to genus. All nematodes were identified to genus when the sample contained fewer than 100 individuals. Nematodes were assigned to main trophic groups (bacterivores, fungivores, herbivores, omnivores, and predators) [27] and colonizer-persister guilds [28].

The microbial data were used to calculate a ratio of fungal biomass to bacterial biomass (F:B) and a cy/pre ratio of PLFAs [(cy17:0+cy19:0)/(16:1ω7c+18:1ω7c)]. The nematode data were used to calculate an enrichment index (EI), a structure index (SI), a maturity index (MI), and a plant-parasitic index (PPI) for each sample [29–30].

Data analysis

Two-way ANOVAs were used to examine the main and interaction effects of plant community and soil depth on dependent variables describing soil physico-chemical properties and soil microbial and nematode communities. Soil physico-chemical properties and microbial community composition were also analyzed by transforming the data to their principal components (PCA) and analyzing these using ANOVAs [31–32]. Data were transformed (natural log, square root, or rank) to meet assumptions of normality and homogeneity of variance. Statistical significance was determined at $p<0.05$. ANOVAs and PCAs were performed using SPSS software (SPSS Inc., Chicago, IL). LSD was used to test differences among treatment means. Tamhane's T2 was used to test differences among treatments when variances of transformed data were unequal. Redundancy analysis (RDA) was used to determine the relationship between soil physico-chemical properties and microbial and nematode community indices, microbial biomass, or nematode abundance. Soil layer was converted to a nominal variable (0, 1) and was treated as a covariable [33–35]. Monte Carlo permutation tests were used to compute statistical significance. RDA was performed with CANOCO 4.5 software [33].

Results

Soil physico-chemical properities

Soil pH was highest in the grassland, lowest in the forest, and intermediate in the shrubland (Table 1). Soil organic carbon (SOC) was greater in the grassland than in the shrubland and forest. Soil C/N was greater in the grassland and shrubland than in the forest. Soil water content (SWC) and total nitrogen (TN) content, however, did not differ among the plant communities (Table 1).

Soil microbial community

Microbial biomass and bacterial biomass were higher in the shrubland than in the forest (Fig. 1a and c). Fungal biomass and actinomycete biomass were higher in the grassland and shrubland than in the forest (Fig. 1d and e). The ratio of fungal to bacterial biomass (F:B) decreased gradually from grassland to shrubland to forest and was significantly higher in the grassland than in the forest (Fig. 1b). The cy/pre ratio of PLFAs increased from grassland to shrubland to forest (Fig. 1f) and was significantly higher in the forest than in the grassland ($p<0.05$) and tended to be higher in the shrubland than in the grassland ($p=0.058$). In addition, ANOVA results of the transformed principal components of the soil microbial community showed that the microbial community structure of the forest differed from that of the grassland and shrubland ($p<0.01$) (Fig. 2). Moreover, the soil microbial community structure tended to differ between the grassland and shrubland ($p=0.069$) (Fig. 2).

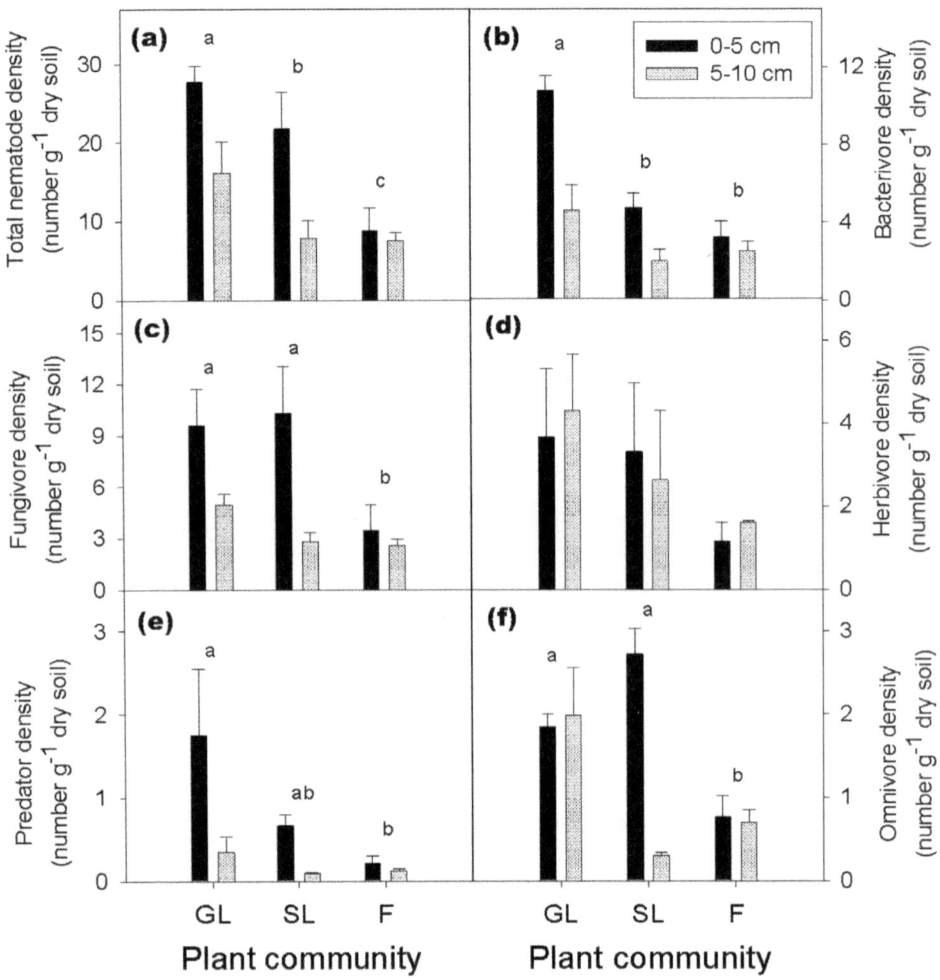

Figure 3. Densities of total nematodes and different nematode trophic groups in the grassland (GL), shrubland (SL), and forest (F) at 0–5 and 5–10 cm soil depth. (a) Total nematodes; (b) Bacterivores; (c) Fungivores; (d) Herbivores; (e) Predators; and (f) Omnivores. Bars indicate standard errors of means. Within each panel, values (averaged across both depths for each plant community) with the same or no letters are not significantly different ($p>0.05$) according to the LSD test.

Soil nematode community

Fungivores and bacterivores were the most abundant trophic groups (Fig. 3b and c). Total nematode density declined gradually from grassland to shrubland to forest at both 0–5 and 5–10 cm soil depths (Fig. 3a). Bacterivore density was greater in the grassland than in the shrubland and forest (Fig. 3b). Fungivore density and omnivore density were greater in the grassland and shrubland than in the forest (Fig. 3c and f); omnivore density in the shrubland, however, was very low at 5–10 cm depth (Fig. 3f). Predator density declined gradually from grassland to shrubland and forest and was significantly greater in the grassland than in the forest ($p=0.023$) (Fig. 3e). Herbivore density declined gradually from grassland to shrubland to forest but did not significantly differ among the three ecosystems (Fig. 3d).

The weighted soil nematode fauna analysis showed that the nematode communities of all three ecosystems at both 0–5 and 5–10 cm soil depths projected onto quadrat III of the nematode fauna plot (Fig. 4a). The composition of the soil nematode community differed among the three ecosystems, and the value of the nematode structure index (SI) was greater in the grassland than in the forest ($p=0.003$) (Fig. 4b). The SI value of the shrubland was intermediate to that of the grassland and forest, and tended to differ from that of the grassland ($p=0.092$) and of the forest ($p=0.086$). The nematode enrichment index (EI) was not significantly influenced by ecosystem type.

Relationships between soil physico-chemical properties and soil biota

The first two canonical axes explained about 71.6%, 51.9%, and 62.1% of the total variance of microbial and nematode community indices (Fig. 5a), microbial community biomass (Fig. 5b), and nematode community density (Fig. 5c), respectively. The microbial and nematode community indices were related to soil water content ($p=0.012$, $F=7.99$), SOC ($p=0.004$, $F=7.46$), and soil pH ($p=0.040$, $F=3.44$) (Fig. 5a). For each index, SI and MI were mainly and positively correlated with SOC; EI and F:B were mainly and positively correlated with pH; cy/pre was mainly and negatively correlated with pH; and PPI was mainly and positively correlated with TN (Fig. 5a). The microbial community biomass tended to be correlated with soil pH ($p=0.052$, $F=4.02$) (Fig. 5b). Total microbial biomass, fungal biomass, bacterial biomass, and actinomycetes biomass were mainly and negatively correlated with TN (Fig. 5b). Nematode community density was significantly related to SOC ($p=0.010$, $F=8.19$) (Fig. 5c). Total

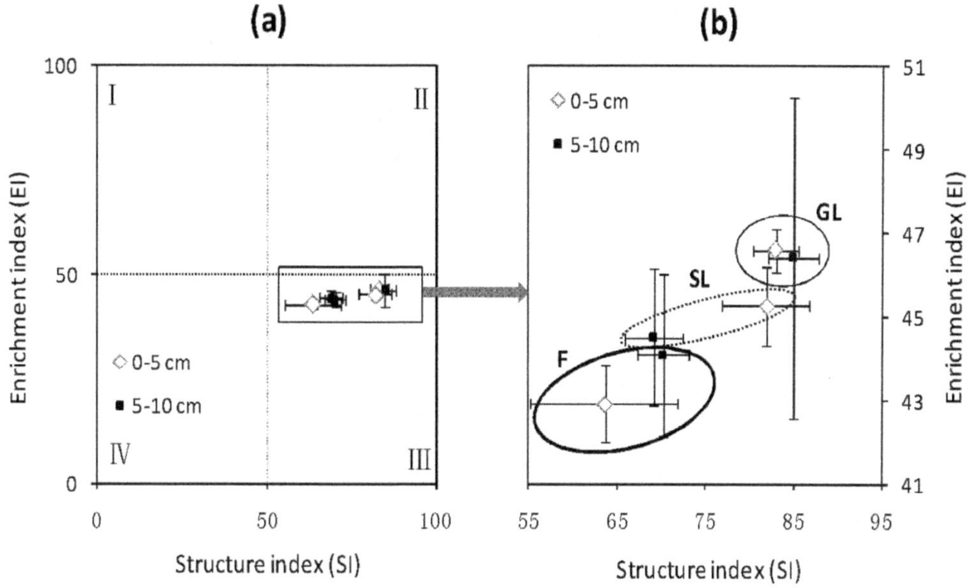

Figure 4. Weighted soil nematode faunal analysis at 0–5 depth and 5–10 cm depth in three vegetation types (GL, grassland; SL, shrubland; and F, forest). (a) Bi-plot with four quadrats; (b) Magnified quadrate-area of bi-plot A. Bars indicate the standard errors of means.

nematode density, omnivore density, and predator density were mainly and positively correlated with pH and SOC; fungivore density was mainly and positively correlated with soil water content and C/N; and herbivore density and bacterivore density were positively correlated with TN, pH, and SOC (Fig. 5c).

Discussion

Succession of soil microbial communities

In this study of a degraded karst region, the progressive secondary succession in vegetation was generally accompanied by a decline of the soil microbial community in that microbial biomass and other microbial variables decreased as the plant community succeeded from grassland to shrubland to forest. Consistent with our results, microbial biomass declined along a secondary succession from grassland to early stages of forest in calcareous soils in northwestern France [20]. In contrast, a previous study reported that succession from grassland to forest increased the soil microbial biomass and biodiversity in karst calcareous soils in southwest China [36]. Another study reported that bacterial phylogenetic diversity was higher but that bacterial metabolic diversity was lower in a grassland than in a forest in karst calcareous soils in southwest China [11]. Conversely, soil microbial biomass decreased with intensity of land degradation from forest to shrubland and grassland in karst calcareous soils in southwest China [37]. In these previous studies in karst areas in southwest China, the responses of soil microbial communities (mainly total microbial biomass) to vegetation succession could be ascribed to the shifts in soil organic carbon, i.e., microbial biomass was positively correlated with soil organic carbon. Although this was also the case in this study, soil organic carbon declined with plant succession. It has been well established that soil organic carbon is usually greater in grasslands than in forests [20,38], and land use changes from grassland to forest have decreased soil carbon stocks worldwide [39]. It is also possible that the relationship between microbial communities and soil organic carbon differs among ecosystems. For instance, the relationship between the F/B ratio and soil organic carbon was positive in agricultural soils but negative in prairie soils at Fermilab, Batavia, in Illinois [40].

In the present study, the increase in the cy/pre ratio with secondary succession might indicate that the soil habitat was becoming more harsh with plant succession. Many previous studies have documented that increasing cy/pre ratios is associated with stress and starvation (e.g., anaerobic conditions, nutrient stress, and water stress) [22,41–42]. Moreover, soil microbial communities in the current study were primarily influenced by soil pH, and the cy/pre ratio was negatively correlated with soil pH. Therefore, soil acidification and the deterioration of the soil habitat occurring with secondary vegetation succession might be the primary environmental factors affecting soil microbial communities. Previous studies had demonstrated that soil acidification significantly reduced the soil microbial biomasses and/or activities and soil nematode abundances [43–45]. In addition, many studies have reported that succession from grassland to forest decreased the soil pH [15,46–48]. Soil pH in the present study was not only negatively associated with the cy/pre ratio but was also positively associated with the F:B ratio. In agreement with our results, the F:B ratio increased slightly with increasing pH as reported by [49]. However, the relationship between the F:B ratio and soil pH is inconsistent [24,50]. Although soil total nitrogen was not influenced by vegetation type, total microbial biomass, bacterial biomass, fungal biomass, and actinomycetes biomass tended to be negatively correlated with soil TN. The reasons for this unexpected result are unclear.

Succession of soil nematode communities

Like soil microbial biomass, soil nematode abundance showed declined patterns with respect to the progressive successional patterns of the plant communities in this study. Few studies have explored the soil nematode communities associated with plant community succession from grassland to forest or from forest to grassland. One study reported that soil nematode densities and biodiversity (i.e., plant feeders and non-plant feeders) increased with secondary succession from savanna to forest in the tropical semi-arid zone of West Africa (Senegal) [51]. Although many

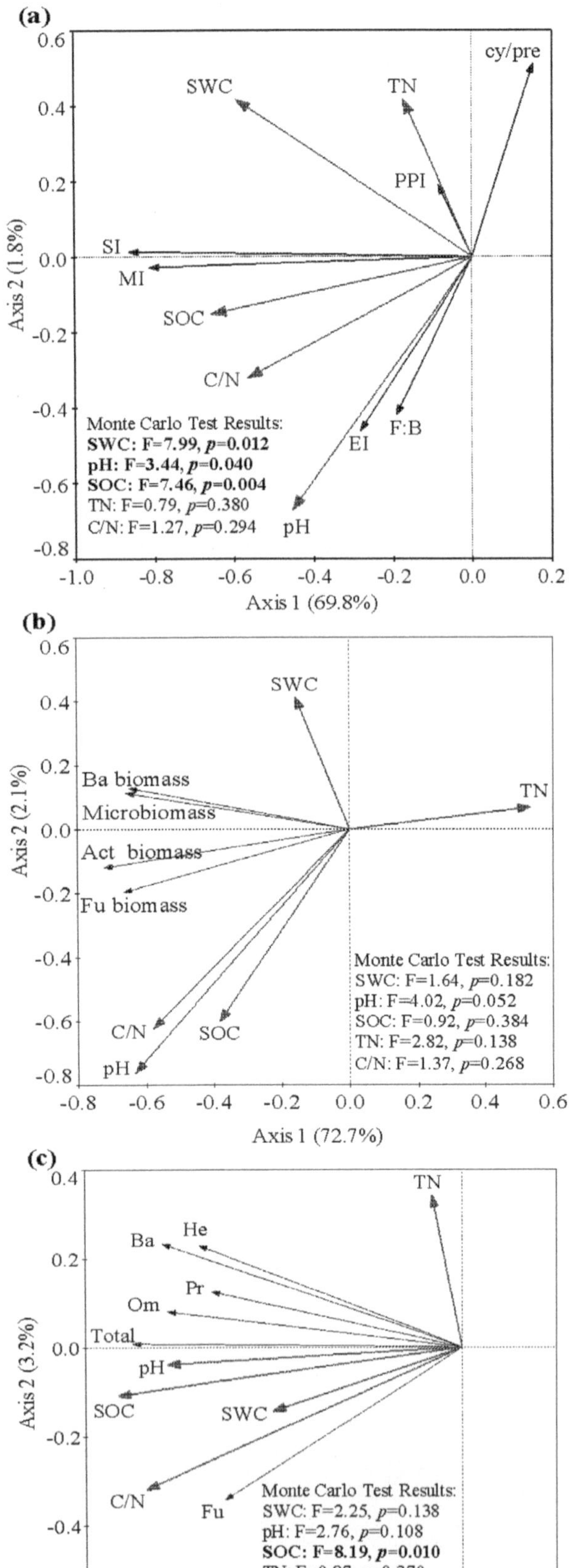

Figure 5. Redundancy analysis (RDA) of microbial and nematode community indices (a), microbial community biomass (b), and nematode community density (c). Ordination diagrams present species scores and environmental factor scores (vectors). Environmental factors include soil water content (SWC), soil pH, soil total nitrogen (TN), soil organic carbon (SOC), and ratio of soil organic carbon to total nitrogen (C:N). Microbial and nematode community indices include ratio of fungal biomass to bacterial biomass (F:B), cy/pre ratio of PLFAs (cy/pre), nematode maturity index (MI), plant parasite index (PPI), structure index (SI), and enrichment index (EI). Variables of microbial community biomass include total microbial biomass (Microbiomass), fungal biomass (Fu biomass), bacterial biomass (Ba biomass), and actinomycetes biomass (Act biomass). Variables of nematode community density include densities of total nematodes (Total), bacterivores (Ba), fungivores (Fu), herbivores (He), omnivores (Om), and predators (Pr).

previous studies have compared the compositions of soil nematode communities of grasslands, shrublands, and forests, it is unclear how inferences based on ecosystem type can be applied to ecosystem succession. For example, Wasilewska (1979) reported that densities of herbivorous, omnivorous, carnivorous, and total nematodes and number of nematode genera were higher in grassland than in forest ecosystems but that densities of bacterivorous and fungivorous nematodes were similar in grassland and forest ecosystems [52]. In another study, land-use change from grassland to shrub heathland suppressed densities of nematode trophic groups (i.e., herbivores, bacterivores, fungivores, omnivores, and predators) in northwest Europe [47].

The weighted soil nematode fauna analysis showed that the nematode communities of all three ecosystems projected onto quadrat III in the nematode fauna plot, which demonstrated that the structures of the soil food webs of all three ecosystems were relatively similar [29]. However, the values of the nematode structure index (SI) decreased from grassland to shrubland to forest ecosystems, which demonstrated that the complexity of the soil food webs decreased in the order grassland > shrubland > forest. In addition, soil organic carbon was the main factor associated with total nematode density and the density of each trophic group. In agreement with our results, many studies have reported that soil organic carbon affects nematode communities [53–55]. Soil nematodes are also affected by spatial heterogeneity, plant community composition, and the soil environment [56]. Although the nematode plant-parasitic index (PPI) was positively correlated with soil total nitrogen, and although the enrichment index (EI) was positively correlated with pH, soil total nitrogen and EI did not significantly change with the secondary vegetation succession. Therefore, soil nematode communities might be more affected by the main soil resource (i.e., soil organic carbon) than by pH or other components of the soil habitat.

Potential implications for land conservation and vegetation recovery in karst areas

The aim of traditional ecosystem restoration in degraded karst areas has usually been to restore forest ecosystems [4–5,8,10]. Our study clearly shows that ecosystem restoration in degraded karst areas might not necessarily be favored by secondary succession to forests. Based on the results of the current study, soil carbon storage and soil biodiversity are greater in grasslands and shrublands than in forests in degraded karst areas. The low soil organic carbon content and biota biomass associated with the forest might be explained in part by karst soils, which are generally very shallow and which can provide only low levels of nutrients and water [57]. The limitations of nutrient and water supply can be exacerbated when the live plant biomass increases during succession [58]. In addition, because, the live biomass is greater for forests than for grasslands or shrublands, natural and/or anthropogenic disruptions (e.g., disease and pest outbreak, drought, harvest, and contamination) may cause greater material loss from forests than from grasslands or shrublands. Consequently, natural and/or anthropogenic disruptions may be more damaging in forests than in grassland or shrubland ecosystems. Therefore, soil carbon and biota ought to be amended in degraded karst areas during vegetation succession or vegetation recovery.

Karst areas of southwest China, Southeast Asian, and Central America have large human populations [1–2]. The challenge is to provide the basic needs of these populations while protecting the fragile karst ecosystems. Agricultural development is the cause of soil erosion and rocky desertification in karst areas [1,12]. As noted in the previous paragraph, disruptions of karst forest ecosystems might be especially destructive. Thus, the cultivation of crops and forestry does not seem to be suitable for maintaining karst ecosystems. In contrast, herbivorous animal husbandry in karst areas of southwest China has provided economic benefits to local farmers while reducing soil erosion [9]. In addition, the perennial C4 grass *Miscanthus* (i.e., the dominant genus in the grassland of the current study) has produced substantial biomass crops worldwide [59]. Another perennial C4 grass, hybrid napiergrass (*Pennisetum hydridum*), has also been widely cultivated and is mainly used for raising livestock in karst areas in southwest China and in other subtropical and tropical areas [60].

Most restoration studies have focused on soil nutrient recovery rather than on soil biota recovery, and many previous studies have used a "space-for-time substitution approach" to explore the soil nutrient status during succession from grassland to forest in karst areas. Most of these studies have reported that soil nutrients accumulate with secondary succession [11,36–37] but that was not the case in the current study. This difference might be explained by the differences in the grasslands. In contrast to grasslands in the earlier studies, the studied grassland was much older and had experienced repeated wildfires.

Conclusions

Contrary to our hypothesis, the progressive succession of vegetation occurring in karst sites in southwest China was accompanied by declines in soil biota abundances and food web complexity. The declines of the soil microbial and nematode communities might be explained by changes in soil habitat (i.e., reduced pH) and resource availability (i.e., reduced soil organic carbon). Because sustainable ecosystem management should integrate aboveground and belowground ecosystem properties and functions [61–62], our results indicate that soil biota amendment is necessary in karst areas during ecosystem restoration and management and establishment of grasslands in degraded karst areas are feasible. This finding provides a scientific basis for the development of herbivorous animal husbandry in karst areas that are under population pressure.

Acknowledgments

We thank Bruce Jaffee for his help in preparing the manuscript. We also thank the editor and two anonymous reviewers for their comments.

Author Contributions

Conceived and designed the experiments: JZ XH KW. Performed the experiments: JZ SL LL. Analyzed the data: JZ SL. Contributed reagents/materials/analysis tools: XH LL. Wrote the paper: JZ SL KW.

References

1. Yuan D (2002) Geology and Geohydrology of Karst and its Relevance to Society; Paris. UNESCO. pp. 15–18.
2. Tuyet D (2001) Characteristics of karst ecosystems of Vietnam and their vulnerability to human impact. Acta Geologica Sinica 75: 325–329.
3. Parise M, De Waele J, Gutierrez F (2009) Current perspectives on the environmental impacts and hazards in karst. Environmental Geology 58: 235–237.
4. Rivera LW, Aide TM (1998) Forest recovery in the karst region of Puerto Rico. Forest Ecology and Management 108: 63–75.
5. Rivera LW, Zimmerman JK, Aide TM (2000) Forest recovery in abandoned agricultural lands in a karst region of the Dominican Republic. Plant Ecology 148: 115–125.
6. Querejeta JI, Estrada-Medina H, Allen MF, Jiménez-Osornio JJ (2007) Water source partitioning among trees growing on shallow karst soils in a seasonally dry tropical climate. Oecologia 152: 26–36.
7. Latinne A, Waengsothorn S, Herbreteau V, Michaux J (2011) Thai limestone karsts: an impending biodiversity crisis. The 1st Environment Asia International Conference. Bangkok, Thailand. pp. 176–187.
8. Zeng F, Peng W, Song T, Wang K, Wu H, et al. (2007) Changes in vegetation after 22 years' natural restoration in the Karst disturbed area in northwestern Guangxi, China. Acta Ecologica Sinica 27: 5110–5119.
9. Wang K, Su Y, Zeng F, Chen H, Xiao R (2008) Ecological process and vegetation restoration in karst region of southwest China. Research of Agricultural Modernization 29: 641–645 (in Chinese with English abstract).
10. Yu L, Zhu S, Wei L, Chen Z, Xiong Z (1998) Study on the natural restoration process of degraded Karst communities–Successional sere. Journal of Mountain Agriculture and Biology 2: 71–77. In Chinese with English abstract.
11. Xiong Y, Xia H, Li Za, Cai Xa, Fu S (2008) Impacts of litter and understory removal on soil properties in a subtropical *Acacia mangium* plantation in China. Plant and Soil 304: 179–188.
12. Qi X, Wang K, Zhang C (2013) Effectiveness of ecological restoration projects in a karst region of southwest China assessed using vegetation succession mapping. Ecological Engineering 54: 245–253.
13. Walker LR, Wardle DA, Bardgett RD, Clarkson BD (2010) The use of chronosequences in studies of ecological succession and soil development. Journal of Ecology 98: 725–736.
14. Walker LR (1999) Ecosystems of disturbed ground. Amsterdam: Elsevier.
15. Schipper LA, Degens BP, Sparling GP, Duncan LC (2001) Changes in microbial heterotrophic diversity along five plant successional sequences. Soil Biology and Biochemistry 33: 2093–2103.
16. Wardle D (1993) Changes in the microbial biomass and metabolic quotient during leaf litter succession in some New Zealand forest and scrubland ecosystems. Functional Ecology 7: 346–355.
17. Zak D, Grigal D, Gleeson S, Tilman D (1990) Carbon and nitrogen cycling during old-field succession: Constraints on plant and microbial biomass. Biogeochemistry 11: 111–129.
18. van der Heijden MGA, Bardgett RD, van Straalen NM (2008) The unseen majority: soil microbes as drivers of plant diversity and productivity in terrestrial ecosystems. Ecology Letters 11: 296–310.
19. De Deyn GB, Raaijmakers CE, Zoomer HR, Berg MP, de Ruiter PC, et al. (2003) Soil invertebrate fauna enhances grassland succession and diversity. Nature 422: 711–713.
20. Chabrerie O, Laval K, Puget P, Desaire S, Alard D (2003) Relationship between plant and soil microbial communities along a successional gradient in a chalk grassland in north-western France. Applied Soil Ecology 24: 43–56.
21. Liu G (1996) Analysis of soil physical and chemical properties and description of soil profiles. Beijing: China Standard.
22. Bossio DA, Scow KM (1998) Impacts of carbon and flooding on soil microbial communities: Phospholipid fatty acid profiles and substrate utilization patterns. Microbial Ecology 35: 265–278.
23. Joergensen RG, Wichern F (2008) Quantitative assessment of the fungal contribution to microbial tissue in soil. Soil Biology and Biochemistry 40: 2977–2991.
24. Frostegård A, Bååth E (1996) The use of phospholipid fatty acid analysis to estimate bacterial and fungal biomass in soil. Biology and Fertility of Soils 22: 59–65.
25. Frostegård Å, Tunlid A, Bååth E (2011) Use and misuse of PLFA measurements in soils. Soil Biology and Biochemistry 43: 1621–1625.
26. Barker KR, Carter CC, Sasser JN (1985) An advanced treatise on Meloidogyne. Volume II: Methodology. Raleigh: United States Agency for International Development. vi + 223 pp. p.
27. Yeates GW, Bongers T, Degoede RGM, Freckman DW, Georgieva SS (1993) Feeding habits in soil nematode families and genera–an outline for soil ecologists. Journal of Nematology 25: 315–331.
28. Bongers T, Bongers M (1998) Functional diversity of nematodes. Applied Soil Ecology 10: 239–251.

29. Ferris H, Bongers T, de Goede RGM (2001) A framework for soil food web diagnostics: extension of the nematode faunal analysis concept. Applied Soil Ecology 18: 13–29.
30. Neher DA (2001) Role of nematodes in soil health and their use as indicators. Journal of Nematology 33: 161–168.
31. Zhao J, Wan S, Li Z, Shao Y, Xu G, et al. (2012) *Dicranopteris*-dominated understory as major driver of intensive forest ecosystem in humid subtropical and tropical region. Soil Biology and Biochemistry 49: 78–87.
32. Zhao J, Wan S, Fu S, Wang X, Wang M, et al. (2013) Effects of understory removal and nitrogen fertilization on soil microbial communities in *Eucalyptus* plantations. Forest Ecology and Management 310: 80–86.
33. Lepš J, Šmilauer P (2003) Multivariate analysis of ecological data using CANOCO. New York: Cambridge University Press. xii + 269 pp. p.
34. Zhao J, Neher D (2013) Soil nematode genera that predict specific types of disturbance. Applied Soil Ecology 64: 135–141.
35. Zhao J, Shao Y, Wang X, Neher DA, Xu G, et al. (2013) Sentinel soil invertebrate taxa as bioindicators for forest management practices. Ecological Indicators 24: 236–239.
36. Zhu H, He X, Wang K, Su Y, Wu J (2012) Interactions of vegetation succession, soil bio-chemical properties and microbial communities in a Karst ecosystem. European Journal of Soil Biology 51: 1–7.
37. Zhang P, Li L, Pan G, Ren J (2006) Soil quality changes in land degradation as indicated by soil chemical, biochemical and microbiological properties in a karst area of southwest Guizhou, China. Environmental Geology 51: 609–619.
38. Evrendilek F, Celik I, Kilic S (2004) Changes in soil organic carbon and other physical soil properties along adjacent Mediterranean forest, grassland, and cropland ecosystems in Turkey. Journal of Arid Environments 59: 743–752.
39. Guo LB, Gifford RM (2002) Soil carbon stocks and land use change: a meta analysis. Global Change Biology 8: 345–360.
40. Allison VJ, Miller RM, Jastrow JD, Matamala R, Zak DR (2005) Changes in soil microbial community structure in a tallgrass prairie chronosequence. Soil Science Society of America Journal 69: 1412–1421.
41. Schmitt A, Glaser B (2011) Organic matter dynamics in a temperate forest soil following enhanced drying. Soil Biology and Biochemistry 43: 478–489.
42. Kieft TL, Ringelberg DB, White DC (1994) Changes in ester-linked phospholipid fatty acid profiles of subsurface bacteria during starvation and desiccation in a porous medium. Applied and Environmental Microbiology 60: 3292–3299.
43. Räty M, Huhta V (2003) Earthworms and pH affect communities of nematodes and enchytraeids in forest soil. Biology and Fertility of Soils 38: 52–58.
44. Chen D, Lan Z, Bai X, Grace JB, Bai Y (2013) Evidence that acidification-induced declines in plant diversity and productivity are mediated by changes in below-ground communities and soil properties in a semi-arid steppe. Journal of Ecology.
45. Rousk J, Brookes PC, Baath E (2010) The microbial PLFA composition as affected by pH in an arable soil. Soil Biology & Biochemistry 42: 516–520.
46. Dmowska E, Ilieva-Makulec K (2006) Secondary succession of nematodes in power plant ash dumps reclaimed by covering with turf. European Journal of Soil Biology 42: S164–S170.
47. Holtkamp R, Kardol P, van der Wal A, Dekker SC, van der Putten WH, et al. (2008) Soil food web structure during ecosystem development after land abandonment. Applied Soil Ecology 39: 23–34.
48. Robertson GP, Vitousek PM (1981) Nitrification potentials in primary and secondary succession. Ecology: 376–386.
49. Bååth E, Anderson T-H (2003) Comparison of soil fungal/bacterial ratios in a pH gradient using physiological and PLFA-based techniques. Soil Biology and Biochemistry 35: 955–963.
50. Rousk J, Brookes PC, Bååth E (2009) Contrasting soil pH effects on fungal and bacterial growth suggest functional redundancy in carbon mineralization. Applied and Environmental Microbiology 75: 1589–1596.
51. Pate E, Ndiaye-Faye N, Thioulouse J, Villenave C, Bongers T, et al. (2000) Successional trends in the characteristics of soil nematode communities in cropped and fallow lands in Senegal (Sonkorong). Applied Soil Ecology 14: 5–15.
52. Wasilewska L (1979) The structure and function of soil nematode communities in natural ecosystems and agrocenoses. Polish Ecological Studies 5: 97–145.
53. Yeates G, Bongers T (1999) Nematode diversity in agroecosystems. Agriculture, Ecosystems & Environment 74: 113–135.
54. Liang W, Lou Y, Li Q, Zhong S, Zhang X, et al. (2009) Nematode faunal response to long-term application of nitrogen fertilizer and organic manure in Northeast China. Soil Biology and Biochemistry 41: 883–890.
55. Freckman D (1988) Bacterivorous nematodes and organic-matter decomposition. Agriculture, Ecosystems & Environment 24: 195–217.
56. Viketoft M (2013) Determinants of small-scale spatial patterns: Importance of space, plants and abiotics for soil nematodes. Soil Biology and Biochemistry 62: 92–98.
57. Zhang X, Wang K (2009) Ponderation on the shortage of mineral nutrients in the soil-vegetation ecosystem in carbonate rock-distributed mountain regions in Southwest China. Earth and Environment 37: 337–341 (in Chinese with English abstract).
58. Guo K, Liu C, Dong M (2011) Ecological adaptation of plants and control of rocky-desertification on karst region of Southwest China. Chinese Journal of Plant Ecology 35: 991–999 (in Chinese with English abstract).
59. Matlaga DP, Davis AS (2013) Minimizing invasive potential of Miscanthus × giganteus grown for bioenergy: identifying demographic thresholds for population growth and spread. Journal of Applied Ecology 50: 479–487.
60. Zhao J, Zhang W, Wang K, Song T, Du H (2014) Responses of the soil nematode community to management of hybrid napiergrass: The trade-off between positive and negative effects. Applied Soil Ecology 74: 134–144.
61. Bardgett RD, Bowman WD, Kaufmann R, Schmidt SK (2005) A temporal approach to linking aboveground and belowground ecology. Trends in Ecology & Evolution 20: 634–641.
62. Smith RS, Shiel RS, Bardgett RD, Millward D, Corkhill P, et al. (2003) Soil microbial community, fertility, vegetation and diversity as targets in the restoration management of a meadow grassland. Journal of Applied Ecology 40: 51–64.

Inorganic Nitrogen Supply and Dissolved Organic Nitrogen Abundance across the US Great Plains

Megan L. Mobley[1,2]*, Matthew J. Cleary[1], Ingrid C. Burke[1,2,3]

1 Department of Botany, University of Wyoming, Laramie, Wyoming, United States of America, **2** Helga Otto Haub School of Environment and Natural Resources, University of Wyoming, Laramie, Wyoming, United States of America, **3** Program in Ecology, University of Wyoming, Laramie, Wyoming, United States of America

Abstract

Across US Great Plains grasslands, a gradient of increasing mean annual precipitation from west to east corresponds to increasing aboveground net primary productivity (ANPP) and increasing N-limitation. Previous work has shown that there is no increase in net N mineralization rates across this gradient, leading to the question of where eastern prairie grasses obtain the nitrogen to support production. One as-yet unexamined source is soil organic N, despite abundant literature from other ecosystems showing that plants take up dissolved soil organic N. This study measured KCl-extractable dissolved organic N (DON) in surface soils across the grassland productivity gradient. We found that KCl-extractable DON pools increased from west to east. If available to and used by plants, this DON may help explain the high ANPP in the eastern Great Plains. These results suggest a need for future research to determine whether, in what quantities, and in what forms prairie grasses use organic N to support primary production.

Editor: Hormoz BassiriRad, University of Illinois at Chicago, United States of America

Funding: This research was funded by the University of Wyoming Excellence Funds. The funders had no role in study design, data collection and analysis, decision to publish, or preparation of the manuscript.

Competing Interests: The authors have declared that no competing interests exist.

* Email: meg.mobley@alumni.duke.edu

Introduction

It has in the past been widely accepted that nitrogen compounds must undergo mineralization – that is, transformation from organic to inorganic forms - to become available to plants [1]. Yet, in a variety of ecosystems, plants have been shown to take up dissolved organic nitrogen (DON) directly from the soil, as reviewed by [2,3]. These plant-accessible organic N compounds include amino acids, amino sugars, amino-polysaccharides, nucleic acids, and various purine and pyrimidine compounds [2]. A new paradigm concerning N cycling has gained acceptance [2], in which the rate of depolymerization of organic N polymers into monomers by microbial enzymes, i.e., the production of DON, regulates ecosystem N cycling rather than the rate of N mineralization.

The ecological importance of DON acquisition is still unclear in many ecosystems [4], including temperate grasslands. Jones et al. [5,6] suggest that temperate grassland species can take up DON directly from the soil, although it is unclear how important DON is relative to other sources. Knowledge about the magnitude and importance of plant uptake of DON is important for our understanding and predictions of N cycling using models, and for clarifying the dynamics of plant-microbial competition. Understanding the potential for grassland plant use of DON could lead to management strategies that reduce fertilizer use and associated nutrient runoff and pollution while still supporting productive and N-retentive cereal croplands and biofuel-producing grasslands.

For grasslands of the United States, the combined results of previous research do not frame a coherent understanding of grassland N cycling. A number of studies have shown grassland plant production to respond to experimental N additions, suggesting that these ecosystems are N–limited [7–10]; a literature review concluded that although water was an important co-limiter of the semi-arid western grasslands, N-limitation was the rule across the Great Plains [11]. Aboveground net primary production (ANPP) increases with mean annual precipitation from semiarid shortgrass steppe in the west through mixed grass prairie to the subhumid tallgrass prairie in the east [12]. Both the concentration of N in biomass and the litter decomposition rate decrease, and nutrient use efficiency increases, along this precipitation gradient [13]. Nevertheless, plant N uptake, as indicated by the quantity of N in aboveground production (ANPP-N), increases along this precipitation and production gradient [14,15]. These lines of evidence suggest that the severity of N-limitation increases from west to east.

To explain observed patterns of ANPP, simulations of the Century model predict a linear increase of net N mineralization rates with precipitation [16], yet field studies have reported no west-to-east trend in net N mineralization [14,15]. Total soil N, on the other hand, has been shown to be higher in tallgrass than shortgrass prairie [13–15,17], suggesting that organic N may play a role in supplying the additional N needed to support eastern tallgrass prairie production. This could explain the discrepancy between model predictions and field studies, as commonly used ecosystem models do not consider the rate of organic N depolymerization or plant uptake of organic N, focusing only on inorganic N mineralization [18–21].

This exploratory study examined the spatial pattern in abundances and rates of production of KCl-extractable dissolved inorganic N (DIN) and DON along the west to east precipitation gradient across US grasslands.

Materials and Methods

Site Description

The Nature Conservancy, Fort Hays State University, USDA-ARS, and the Long-Term Ecological Research programs at Konza Prairie and Shortgrass Steppe provided access to field sites spanning from eastern Colorado to eastern Kansas [14,15,17]. The entire gradient spans precipitation and temperature ranges of 500 mm and 4.4°C, respectively. The two westernmost sites, Shortgrass Steppe (SGS) Long-Term Ecological Research (LTER) and Arikaree River (ARI) Nature Conservancy site, are located in the drier shortgrass steppe. Two sites, Smokey River (SVR) Nature Conservancy site and a property of Ft. Hays State University (HAYS), were located in the northern mixed grass prairie. The easternmost site was located in the wetter tallgrass prairie, Konza Prairie LTER (KON) (Table 1). Each sampling location was located on native upland grasslands with moderate grazing by cattle in the summer and minimal disturbance. The soils at all sites ranged from loam to sandy loam. The vegetation of the shortgrass steppe was dominated by *Bouteloua gracilis* (Willd. ex Kunth) Lag. ex Griffiths and *Buchloë dactyloides* (Nutt.) J.T. Columbus, whereas *Andropogon gerardii* Vitman and *Schizachyrium scoparium* (Michx.) Nash dominated the tallgrass prairie. The mixed-grass prairie comprised a combination of all of the species above.

Field Sampling

At each site, we set up three 25 m long transects within a 100 m radius of each other in undisturbed areas representative of the surrounding vegetation. From June 6–15, 2011, we collected soil cores (10 cm length ×5 cm diameter) from two microsites (under plant canopies and interspaces between plants) and at two depths (0–5 cm and 5–10 cm) at three random locations on each transect, for a total of 180 soil samples (2 microsites ×2 depths ×3 locations ×3 transects ×5 sites). Soils were immediately sieved through a 2 mm mesh and homogenized in a plastic bag.

We extracted soils immediately for dissolved N pools, in order to minimize disturbance-induced N transformations associated with delays in processing, and to obtain the most realistic estimates of DON and DIN concentrations possible [22]. We extracted dissolved pools of N following the procedure outlined by Robertson et al. [23] and Bardgett et al. [24]. A 10 g subsample of field-fresh, sieved, homogenized soil was placed in an extraction cup with 100 mL 2M KCl solution and shaken for 10 minutes. The extraction cups were returned to the laboratory at the University of Wyoming, Laramie, WY, to be filtered and frozen within two days of sampling. The remaining soil was air-dried and stored in paper bags until further analysis.

We installed one pair of Plant Root Simulator (PRS) resin probes (Western Ag Innovations Inc., Saskatoon, Saskatchewan) per microsite at 10 random locations along each transect (2 microsites ×10 locations ×3 transects ×5 sites = 300 pairs), to estimate inorganic N supply rate. Each pair consisted of one anion and one cation probe to collect mobile NO_3^- and NH_4^+, respectively. Probes were inserted into the soil with the adsorption membrane extending from the surface to 5.5 cm depth, and were buried for 30 days from early May to early June 2011. At each site, the probes were removed on the same day as the soil sampling. Probes were sorted into two groups of five pairs of probes from each microsite type. This resulted in two replicate sets of probes per each microsite type per each transect for a total of 60 (2 replicates ×2 microsites ×3 transects ×5 sites) probe sets. Probes were washed in the laboratory with DI water within two days of unearthing and then sent to Western Ag Innovations, Inc. (Saskatoon, Saskatchewan) for inorganic N analysis.

Laboratory Analyses

The eluate from the extraction cups was shaken for 1 hour and filtered in the laboratory using Whatman #1 filter paper and frozen immediately thereafter. For measurement of total KCl-extractable dissolved total N and inorganic N, we thawed the eluate and diluted a 2.5 mL subsample with 17.5 mL DI water. We analyzed total dissolved N on a Shimadzu TOC-V Analyzer (Shimadzu Scientific Instruments, Wood Dale, IL, USA). Dissolved inorganic N was analyzed at the Natural Resource Ecology Laboratory (Colorado State University, Fort Collins, CO) using an Alpkem Flow Solution IV Automated wet chemistry system (O.I. Analytical, College

Table 1. Locations and attributes of grassland sites.

	SGS:	ARI:	SVR:	HAYS:	KON:
	Shortgrass Steppe LTER	Arikaree River TNC	Smokey River TNC	Ft. Hays State Univ.	Konza Prairie LTER
Latitude, Longitude	40°48.8′N, 104°42.8′W	39°45.4′N, 102°26.9′W	38°53.6′N, 100°58.7′W	38°52.4′N, 99°23.3′W	39°4.8′N, 96°33′W
MAP (mm)	320	450	502	575	835
MAT (°C)	8.6	9.7	10.8	11.9	13.0
Grassland Type	Shortgrass Steppe	Shortgrass Steppe	Mixed Grass Prairie	Mixed Grass Prairie	Tallgrass Prairie
ANPP-N [15] (g N m^{-2} y^{-1})	1.1	1.7	1.5	1.6	2.4
ANPP-C [15] (g C m^{-2} y^{-1})	143	123	221	211	337
Net N min. [14] (g N m^{-2} y^{-1})	1.5 (0.49)	3.0 (1.04)	1.5 (0.53)	3.0 (0.63)	2.4 (0.26)
Soil C (g C m^{-2})	1367 (67.9)	1440 (68.6)	1851 (52.9)	5204 (155.4)	2373 (151.2)
Soil N (g N m^{-2})	287.2 (13.1)	274.2 (13.0)	214.6 (13.3)	285.1 (5.9)	210.7 (11.0)
Soil C:N	4.91 (0.31)	5.33 (0.25)	9.05 (0.50)	18.3 (0.44)	11.2 (0.32)
Volumetric Soil Moisture (%)	13.8 (0.40)	7.3 (0.53)	3.7 (0.17)	3.4 (0.36)	6.3 (0.30)

MAP = mean annual precipitation (mm), MAT = mean annual temperature (°C), TNC = The Nature Conservancy, LTER = National Science Foundation Long-Term Ecological Research site. Data in lower panel is from this study; values in parentheses represent one standard error of the mean.

Station, TX, USA). We estimated KCl-extractable DON as the difference between total dissolved N and dissolved inorganic N.

We mixed and subsampled remaining air-dried soils for analysis of total C and N. Subsamples were ground on a rotary soil grinder (Planetary Micro Mill Pulverisette 7, Fritsch GmbH, Idar-Oberstein, Germany), and 2 mg of ground soil were analyzed on a Perkin Elmer 2400 Elemental Analyzer (Perkin Elmer, Inc., Waltham, MA, USA) for total C and N. Volumetric water content was estimated after drying a 100 g subsample of field moist soil at 105°C for one day [23].

Statistical Analyses

We performed all analyses in the statistical program R, version 2.14.2 [25]. Extractable DIN and DON concentrations were multiplied by the soil sample depth (5 cm) and bulk density (mean $=0.877$ g cm^{-3}) to obtain DIN and DON pools on a per-area basis. Data for the two soil depth increments (0–5 cm, 5–10 cm) were aggregated such that all analyses were of soil properties and constituents to 10 cm depth, with the exception of PRS probe data that characterized the top 5 cm of soil. We employed two-way ANOVAs coupled to Tukey multiple comparison tests [26] to test differences in the dependent variables total soil C (g C m^{-2}), N (g N m^{-2}), C:N (mass C:N), and 0–5 cm soil inorganic N supply rate (µg N 10 cm^{-2} month^{-1}) among sites and between microsites. Two-way ANCOVAs tested differences in the dependent variables total soil KCl-extractable DIN (g DIN m^{-2}) and KCl-extractable DON (g DON m^{-2}) among sites and between microsites with volumetric soil moisture as a continuous independent variable. One extreme outlier value for DON with strong influence on the ANCOVA (DON>17 g N m^{-2}; outlier test Bonferonni $p=4.4\times10^{-26}$) was excluded from DON analyses. ANOVAs were called using aov(), with drop1(model, test = "F") to extract type III sums of squares. Multiple comparisons were performed using TukeyHSD(). We performed linear regressions using lm() to assess the effect of soil moisture on DIN and DON measurements, and to test for a relationship between KCl-extractable DON pools and MAP. A paired t-test tested differences in DON and DIN abundance (g N m^{-2}) across all sites and microsites. Significance was assessed at $\alpha=0.05$. Data for this study can be found in the Supporting Information.

Results

Soil C differed significantly among sites (among-sites $F=212$, d.f. = 4, $p<0.0001$) and was significantly higher in the tallgrass prairie than shortgrass steppe ($p<0.0001$, Table 1). Total soil N (g N m^{-2}) decreased from west to east across the grassland gradient (among-sites $F=10.7$, d.f. = 4, $p<0.0001$, Table 1). Soil C:N increased significantly from west to east across the gradient (among-sites $F=211$, d.f. = 4, $p<0.0001$). There were no significant differences in soil C ($p=0.439$), N ($p=0.537$), or C:N between under-plant and plant interspace microsites ($p=0.101$).

The 30-day inorganic N supply rate estimated with PRS probes was highly variable and showed a slight decline across the gradient, with the supply rate at the easternmost site significantly lower than that of the westernmost site (among-sites $F=8.07$, d.f. = 4, $p<0.0001$; Figure 1a). There were no significant differences in inorganic N supply rates between microsites.

The pool of KCl-extractable dissolved inorganic nitrogen (DIN) was greater in the eastern tallgrass prairie compared to the western shortgrass steppe (among-sites $F=44.9$, d.f. = 4, $p<0.0001$; Figure 1b). KCl-extractable dissolved organic N was significantly more abundant in the tallgrass region compared to the shortgrass steppe (among-sites F = 10.6, d.f. = 4, p<0.0001; Figure 1b).

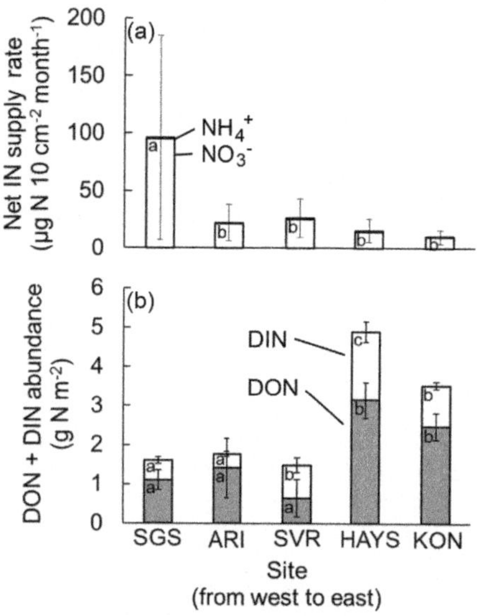

Figure 1. Inorganic N supply rates and extractable dissolved N pools at grassland sites. a) 0–5 cm soil net inorganic N supply rates (NH_4^+ in black bars, NO_3^- in white bars) at five grasslands of the US Great Plains, arranged from west to east. Error bars represent 95% confidence intervals on total (NH_4^++NO_3^-) inorganic N supply rate. Lowercase letters represent significant site differences at $\alpha=0.05$. b) Total extractable dissolved soil N (DIN+DON) to 10 cm soil depth across the US Great Plains. Lowercase letters in the gray columns represent significant differences in DON among sites; lowercase letters in the white columns indicate significant differences in DIN among sites. Multiple comparisons were performed on output of ANOVAs that did not include soil moisture as a covariate. Error bars represent 95% confidence intervals. Significance assessed at $\alpha=0.05$. One extreme outlier (DON>17 g N m^{-2} at SVR) was excluded from the analysis.

There were no significant differences in DIN ($p=0.128$) or DON ($p=0.461$) between microsites. Across all sites, abundances of KCl-extractable DON were significantly higher than DIN (paired $t=6.648$, d.f. = 82, $p<0.0001$). Both DIN (F = 10.4, d.f. = 79, $R^2=0.106$, p = 0.0018) and DON (F = 21.1, d.f. = 78, $R^2=0.202$, p<0.0001) increased with soil moisture (Figure 2), but soil moisture was not significant as a covariate of DIN ($p=0.277$) or DON ($p=0.241$) in the ANCOVAs. There was a positive relationship between DON measurements and site-level ANPP (F = 6.257, d.f. = 42, $R^2=0.130$, $p=0.0164$; Figure S1).

Discussion

This study tested the hypothesis that extractable dissolved organic N pools would increase along a precipitation and production gradient from west to east across US grasslands. Unlike two previous studies at the same sites [14,15], we did not find a west to east increase in total soil N; soil C and C:N did increase from west to east as in previous studies. Also like previous studies, we found no increase in inorganic N supply rates from west to east, even though plant uptake does increase (ANPP-N, Table 1). This supports previous conclusions that N becomes more limiting, and the N cycle grows tighter, from west to east. We did observe, however, that pools of KCl-extractable dissolved inorganic and organic N, representing exchangeable + free soil N [27], increased from west to east.

These data must be considered as suggestive, rather than conclusive, given a number of limitations to this small study. Our one-time measurements of KCl-extractable DIN and DON and

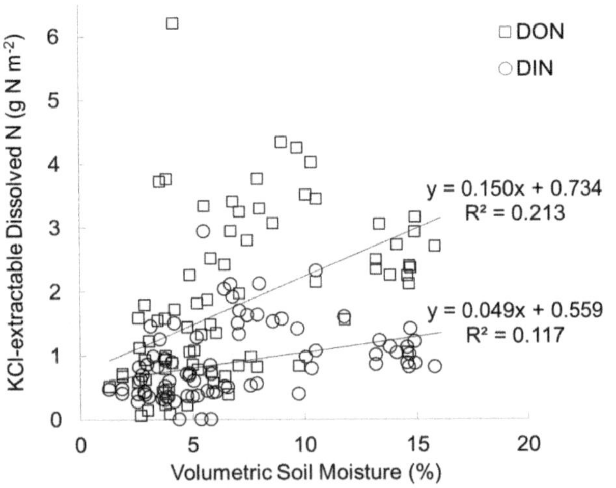

Figure 2. Simple linear regressions of KCl-extractable soil dissolved inorganic (DIN; circles) and organic nitrogen (DON; squares) against volumetric soil moisture for 0–10 cm soil samples from five sites across the Great Plains.

our one resin probe incubation give an incomplete characterization of the long-term mean pools and rates for these grassland sites. Pool size is not necessarily indicative of cycling rate or ecological importance, as a large pool might result from high production or low demand. We did not characterize the chemical composition or molecular weight distribution of DON, thus we cannot speculate on how much of the DON is potentially plant-available. There are artifacts associated with each of the field methods that could reduce the sensitivity of N pool and production estimates. The soil and root disturbance associated with soil sampling may release DON that would not naturally be available in the soil [4], and use of KCl as an extractant may result in a pool of extractable dissolved N that is quite different in composition, quantity, and quality than that found under natural conditions [28,29]. As evidenced by the significant relationships between KCl-extractable N pools and soil moisture (Figure 2), the KCl-extraction likely overestimates what is actually plant-accessible under normal dry soil conditions. Because less nitrate is likely to reach the resin probes when soil moisture is limiting, regardless of actual mineralization rates, we consider resin probes to be better indicators than extractions of the available N that is accessible by plants [30]. These methodological artifacts should not affect comparisons among sites within a given study if consistent methods are used at all sites, as was the case in this study and previous studies at these sites [14,15].

Alternative or additional sources of ANPP-N may include biological N fixation and atmospheric N deposition. N fixation, largely by free-living organisms, has been estimated at <0.1 g N m^{-2} y^{-1} in shortgrass steppe [31] and ~0.8 g N m^{-2} y^{-1} in tallgrass prairie [32], although it is unclear how much of that fixed N enters the plant-soil N cycle versus that volatilized or denitrified [33]. Total wet+dry inorganic N deposition increases from approximately 0.35 g N m^{-2} y^{-1} at SGS to approximately 0.71 g N m^{-2} y^{-1} at KON [34,35], and so could also provide a portion of the additional N required to support ANPP.

This study provides one piece of support for the broader hypothesis that organic N may serve as a nitrogen source for grassland production—a source that would explain how plant N uptake (ANPP-N) is able to increase along the precipitation gradient without a concomitant increase in N mineralization rates. Our results show that a sizeable pool of KCl-extractable dissolved organic N resides in these grassland soils, and that this N pool increases from west to east across the Great Plains precipitation gradient. The larger KCl-extractable DON pools at the eastern end of the precipitation gradient open the possibility that DON as a source of plant N may help resolve discrepancies in between model predictions and measurements of available N. If so, ecosystem models such as Biome-BGC, Century, PnET, and TEM may be missing a source of N for grasslands. These models represent the N required for primary productivity as "mineral N" generated by N mineralization [18–21]. If organic N, which these results indicate is abundant at these sites, is important for plant growth, then an important nutrient source is being left out of model simulations.

Conclusions

We found that the abundance of DON increased across the west-to-east along the precipitation and productivity gradient of US Great Plains grasslands, suggesting that it could be a significant source of N if plants are able to utilize it. If organic N is an important component of available N in US grasslands, it may explain why ecosystem models that do not include plant use of organic N are unable to predict N supply rates across the US grasslands. Additional studies must be conducted to determine if grass species dominating the US Great Plains assimilate biologically-relevant quantities of organic N.

Supporting Information

Figure S1 The relationship between KCl-extractable soil DON and ANPP. Simple linear regression of KCl-extractable soil dissolved organic nitrogen (DON; microsites and depths for each sample location mathematically combined to yield $n = 9$ replicate measurements per site) against site-level mean aboveground net primary production (ANPP; [15]) for five grassland sites across the US Great Plains.

Data S1 Bulk soil C and N, KCl-extractable N, and inorganic N supply rate data.

Acknowledgments

We wish to thank The Nature Conservancy, Fort Hays State University, USDA-ARS, and the Long-Term Ecological Research Sites at Konza Prairie and Shortgrass Steppe for access to field sites. We thank Colin Pinney (EcoCore Analytical Services, a Colorado State University Core Facility) and Leann Naughton for assistance with nitrogen analysis. This manuscript was improved by comments from William Lauenroth, Urszula Norton, David Bell, Daniel Schlaepfer, Otgonsuren Avirmed, Ariuna Lkhagva, Lynn Moore, Cait Rottler, and four anonymous reviewers.

Author Contributions

Conceived and designed the experiments: MJC ICB. Performed the experiments: MJC. Analyzed the data: MJC MLM. Contributed reagents/materials/analysis tools: ICB. Wrote the paper: MJC MLM ICB.

References

1. Stanford G, Smith S (1972) Nitrogen mineralization potentials of soils. Soil Science Society of America Journal 36: 465–472.
2. Schimel JP, Bennett J (2004) Nitrogen mineralization: challenges of a changing paradigm. Ecology 85: 591–602.
3. Neff JC, Chapin FS, Vitousek PM (2003) Breaks in the cycle: dissolved organic nitrogen in terrestrial ecosystems. Frontiers in Ecology and the Environment 1: 205–211.
4. Hobbie JE, Hobbie EA (2013) Microbes in nature are limited by carbon and energy: the starving-survival lifestyle in soil and consequences for estimating microbial rates. Frontiers in Microbiology 4: Article 324.
5. Jones DL, Healey JR, Willett VB, Farrar JF, Hodge A (2005) Dissolved organic nitrogen uptake by plants–an important N uptake pathway? Soil Biology and Biochemistry 37: 413–423.
6. Jones DL, Shannon D, Murphy DV, Farrar J (2004) Role of dissolved organic nitrogen (DON) in soil N cycling in grassland soils. Soil Biology and Biochemistry 36: 749–756.
7. Dodd JL, Lauenroth WK (1979) Analysis of the Response of a Grassland Ecosystem to Stress. In: French NR, editor.Perspectives in Grassland Ecology: Springer New York.pp. 43–58.
8. Lauenroth WK, Dodd JL, Sims PL (1978) The effects of water- and nitrogen-induced stresses on plant community structure in a semiarid grassland. Oecologia 36: 211–222.
9. Risser PG, Parton WJ (1982) Ecosystem Analysis of the Tallgrass Prairie: Nitrogen Cycle. Ecology 63: 1342–1351.
10. Clenton EO, Hyde RM, Anderson KL (1970) Effects of Clipping and Supplemental Nitrogen and Water on Loamy Upland Bluestem Range. Journal of Range Management 23: 341–346.
11. Hooper D, Johnson L (1999) Nitrogen limitation in dryland ecosystems: Responses to geographical and temporal variation in precipitation. Biogeochemistry 46: 247–293.
12. Sala OE, Parton W, Joyce L, Lauenroth W (1988) Primary production of the central grassland region of the United States. Ecology 69: 40–45.
13. Murphy KL, Burke IC, Vinton MA, Lauenroth WK, Aguiar MR, et al. (2002) Regional analysis of litter quality in the central grassland region of North America. Journal of Vegetation Science 13: 395–402.
14. McCulley RL, Burke IC, Lauenroth WK (2009) Conservation of nitrogen increases with precipitation across a major grassland gradient in the Central Great Plains of North America. Oecologia 159: 571–581.
15. Barrett J, McCulley R, Lane D, Burke I, Lauenroth W (2002) Influence of climate variability on plant production and N-mineralization in Central US grasslands. Journal of Vegetation Science 13: 383–394.
16. Burke IC, Lauenroth WK, Parton WJ (1997) Regional and temporal variation in net primary production and nitrogen mineralization in grasslands. Ecology 78: 1330–1340.
17. Vinton MA, Burke IC (1997) Contingent effects of plant species on soils along a regional moisture gradient in the Great Plains. Oecologia 110: 393–402.
18. Parton W, Schimel DS, Cole C, Ojima D (1987) Analysis of factors controlling soil organic matter levels in Great Plains grasslands. Soil Science Society of America Journal 51: 1173–1179.
19. McGuire AD, Melillo JM, Kicklighter DW, Joyce LA (1995) Equilibrium responses of soil carbon to climate change: empirical and process-based estimates. Journal of Biogeography 22: 785–796.
20. Running SW, Hunt ER Jr (1993) Generalization of a forest ecosystem process model for other biomes, BIOME-BGC, and an application for global-scale models. In: Ehleringer JR, Field CB, editors.Scaling Physiological Processes: Leaf to Globe.New York: Academic Press, Inc. pp. 141–158.
21. Aber JD, Ollinger SV, Driscoll CT (1997) Modeling nitrogen saturation in forest ecosystems in response to land use and atmospheric deposition. Ecological Modelling 101: 61–78.
22. Van Miegroet H (1995) Inorganic nitrogen determined by laboratory and field extractions of two forest soils. Soil Science Society of America Journal 59: 549–553.
23. Robertson GP, Wedin DA, Groffman PM, Blair JM, Holland EA, et al. (1999) Soil carbon and nitrogen availability. Nitrogen mineralization, nitrification and soil respiration potentials. In: Robertson GP, Coleman DC, Bledsoe CS, Sollins P, editors.Standard Soil Methods for Long-Term Ecological Research.New York: Oxford University Press. pp. 258–271.
24. Bardgett RD, Streeter TC, Bol R (2003) Soil microbes compete effectively with plants for organic-nitrogen inputs to temperate grasslands. Ecology 84: 1277–1287.
25. R Development Core Team (2012) R: A Language and Environment for Statistical Computing. 2.14.2 ed. Vienna, Austria: R Foundation for Statistical Computing.
26. Hsu JC (1996) Multiple comparisons: theory and methods. London: Chapman & Hall. xiv, 277 p., 274 p. of plates p.
27. Näsholm T, Kielland K, Ganeteg U (2009) Uptake of organic nitrogen by plants. New Phytologist 182: 31–48.
28. Inselsbacher E, Näsholm T (2012) The below-ground perspective of forest plants: soil provides mainly organic nitrogen for plants and mycorrhizal fungi. New Phytologist 195: 329–334.
29. Jones DL, Willett VB (2006) Experimental evaluation of methods to quantify dissolved organic nitrogen (DON) and dissolved organic carbon (DOC) in soil. Soil Biology and Biochemistry 38: 991–999.
30. Hook PB, Burke IC (1995) Evaluation of methods for estimating net nitrogen mineralization in a semiarid grassland. Soil Science Society of America Journal 59: 831–837.
31. Copley PW (1972) Evaluation of biological N_2 fixation in a grassland ecosystem. Fort Collins, CO: Colorado State University. 90 p.
32. DuBois JD, Kapustka LA (1983) Biological Nitrogen Influx in an Ohio Relict Prairie. American Journal of Botany 70: 8–16.
33. West NE, Skujins J (1977) The nitrogen cycle in North American cold-winter semi-desert ecosystems. Oecologia plantarum 12: 45–53.
34. National Atmospheric Deposition Program (NRSP-3) (2011) NADP Program Office, Illinois State Water Survey, 2204 Griffith Dr., Champaign, IL 61820.
35. U.S. Environmental Protection Agency Clean Air Markets Division (2011) Clean Air Status and Trends Network (CASTNET) Annual Total Deposition.

PERMISSIONS

All chapters in this book were first published in PLOS ONE, by The Public Library of Science; hereby published with permission under the Creative Commons Attribution License or equivalent. Every chapter published in this book has been scrutinized by our experts. Their significance has been extensively debated. The topics covered herein carry significant findings which will fuel the growth of the discipline. They may even be implemented as practical applications or may be referred to as a beginning point for another development.

The contributors of this book come from diverse backgrounds, making this book a truly international effort. This book will bring forth new frontiers with its revolutionizing research information and detailed analysis of the nascent developments around the world.

We would like to thank all the contributing authors for lending their expertise to make the book truly unique. They have played a crucial role in the development of this book. Without their invaluable contributions this book wouldn't have been possible. They have made vital efforts to compile up to date information on the varied aspects of this subject to make this book a valuable addition to the collection of many professionals and students.

This book was conceptualized with the vision of imparting up-to-date information and advanced data in this field. To ensure the same, a matchless editorial board was set up. Every individual on the board went through rigorous rounds of assessment to prove their worth. After which they invested a large part of their time researching and compiling the most relevant data for our readers.

The editorial board has been involved in producing this book since its inception. They have spent rigorous hours researching and exploring the diverse topics which have resulted in the successful publishing of this book. They have passed on their knowledge of decades through this book. To expedite this challenging task, the publisher supported the team at every step. A small team of assistant editors was also appointed to further simplify the editing procedure and attain best results for the readers.

Apart from the editorial board, the designing team has also invested a significant amount of their time in understanding the subject and creating the most relevant covers. They scrutinized every image to scout for the most suitable representation of the subject and create an appropriate cover for the book.

The publishing team has been an ardent support to the editorial, designing and production team. Their endless efforts to recruit the best for this project, has resulted in the accomplishment of this book. They are a veteran in the field of academics and their pool of knowledge is as vast as their experience in printing. Their expertise and guidance has proved useful at every step. Their uncompromising quality standards have made this book an exceptional effort. Their encouragement from time to time has been an inspiration for everyone.

The publisher and the editorial board hope that this book will prove to be a valuable piece of knowledge for researchers, students, practitioners and scholars across the globe.

LIST OF CONTRIBUTORS

Alistair W. R. Seddon
Long-Term Ecology Laboratory, Biodiversity Institute at the Oxford Martin School, Department of Zoology, University of Oxford, Oxford, United Kingdom

Cynthia A. Froyd
Long-Term Ecology Laboratory, Biodiversity Institute at the Oxford Martin School, Department of Zoology, University of Oxford, Oxford, United Kingdom
Department of Geography, Swansea University, Swansea, United Kingdom

Melanie J. Leng
NERC Isotope Geosciences Laboratory, British Geological Survey, Keyworth, Nottingham, United Kingdom

Glenn A. Milne
Department of Earth Sciences, University of Ottawa, Marion Hall, Ottawa, Ontario, Canada

Katherine J. Willis
Long-Term Ecology Laboratory, Biodiversity Institute at the Oxford Martin School, Department of Zoology, University of Oxford, Oxford, United Kingdom
Department of Biology, University of Bergen, Bergen, Norway

Katherine A. Dafforn, Graeme F. Clark and Emma L. Johnston
Evolution and Ecology Research Centre, School of Biological, Earth and Environmental Sciences, University of New South Wales, Sydney, NSW, Australia
Sydney Institute of Marine Science, Mosman, NSW, Australia

Brendan P. Kelaher
National Marine Science Centre, Southern Cross University, Coffs Harbour, NSW, Australia

Stuart L. Simpson
Commonwealth Scientific and Industrial Research Organisation, Land and Water, Kirrawee, NSW, Australia

Melinda A. Coleman and Nathan A. Knott
Aquatic Ecosystem Research, New South Wales Department of Primary Industries, Nelson Bay, NSW, Australia

Pat A. Hutchings
Marine Invertebrates, Australian Museum, Sydney, NSW, Australia

Martina A. Doblin
Sydney Institute of Marine Science, Mosman, NSW, Australia
University of Technology Sydney, Broadway, NSW, Australia

Matthew Tighe, Rebecca E. Haling, Richard J. Flavel and Iain M. Young
School of Environmental and Rural Science, University of New England, Armidale, New South Wales, Australia

Jason Kreitler
Western Geographic Science Center, U.S. Geological Survey, Boise, Idaho, United States of America

Michael Papenfus
Woods Institute for the Environment, Stanford University, Stanford, California, United States of America

Kristin Byrd
Western Geographic Science Center, U.S. Geological Survey, Menlo Park, California, United States of America

William Labiosa
Western Geographic Science Center, U.S. Geological Survey, Seattle, Washington, United States of America

Megan P. Keville and Cory C. Cleveland
Department of Ecosystem and Conservation Sciences, University of Montana, Missoula, Montana, United States of America

Sasha C. Reed
U.S. Geological Survey, Southwest Biological Science Center, Moab, Utah, United States of America

James H. Hunt
Departments of Biology and Entomology and W. M. Keck Center for Behavioral Biology, North Carolina State University, Raleigh, North Carolina, United States of America

Navdeep S. Mutti
School of Life Sciences, Arizona State University, Tempe, Arizona, United States of America

Gro V. Amdam
School of Life Sciences, Arizona State University, Tempe, Arizona, United States of America
Department of Chemistry, Biotechnology and Food Science, Norwegian University of Life Sciences, Aas, Norway

Heli Havukainen
Department of Chemistry, Biotechnology and Food Science, Norwegian University of Life Sciences, Aas, Norway
Department of Biomedicine, University of Bergen, Bergen, Norway

Michael T. Henshaw
Department of Biology, Grand Valley State University, Allendale, Michigan, United States of America

Luciano M. Chiaverano
Kewalo Marine Laboratory, Pacific Biosciences Research Center, University of Hawaii at Mānoa, Honolulu, Hawaii, United States of America

Brenden S. Holland
Center for Conservation Research & Training, Pacific Biosciences Research Center, University of Hawaii at Mānoa, Honolulu, Hawaii, United States of America

Gerald L. Crow
Waikiki Aquarium, University of Hawaii at Mānoa, Honolulu, Hawaii, United States of America

Landy Blair
Ocean Safety and Lifeguard Services, City and County of Honolulu, Honolulu, Hawaii, United States of America

Angel A. Yanagihara
Bekesy Laboratory, Pacific Biosciences Research Center, and Department of Tropical Medicine, John A. Burns School of Medicine, University of Hawaii at Mānoa, Honolulu, Hawaii, United States of America

Corey B. Wakefield, Paul D. Lewis and David V. Fairclough
Western Australian Fisheries and Marine Research Laboratories, Department of Fisheries, Government of Western Australia, Perth, Western Australia, Australia.

Teresa B. Coutts
Western Australian Fisheries and Marine Research Laboratories, Department of Fisheries, Government of Western Australia, Perth, Western Australia, Australia.
Department of Environment and Conservation, Government of Western Australia, Broome, Western Australia, Australia

Timothy J. Langlois
Western Australian Fisheries and Marine Research Laboratories, Department of Fisheries, Government of Western Australia, Perth, Western Australia, Australia.
School of Plant Biology, The University of Western Australia, Perth, Western Australia, Australia

Walter R. Tschinkel
Department of Biological Science, Florida State University, Tallahassee, Florida, United States of America

Joshua R. King
Biology Department, University of Central Florida, Orlando, Florida, United States of America

C. Anela Choy and Jeffrey C. Drazen
University of Hawaii, Department of Oceanography, Honolulu, Hawaii, United States of America

Peter C. Davison
Scripps Institution of Oceanography at University of California, San Diego, California, United States of America

Adrian Flynn
The University of Queensland, School of Biomedical Sciences, Brisbane, Australia
Commonwealth Scientific and Industrial Research Organization, Marine and Atmospheric Research, Hobart, Australia

Elizabeth J. Gier and Brian N. Popp
University of Hawaii, Department of Geology & Geophysics, Honolulu, Hawaii, United States of America

Joel C. Hoffman
U.S. Environmental Protection Agency Office of Research and Development, National Health and Environmental Effects Research Laboratory, Mid-Continent Ecology Division, Duluth, Minnesota, United States of America

Jennifer P. McClain-Counts
United States Geological Survey, Southeastern Ecological Science Center, Gainesville, Florida, United States of America

Todd W. Miller
Ehime University, Global Center of Excellence, Center for Marine Environmental Studies, Matsuyama, Japan

Steve W. Ross
University of North Carolina at Wilmington, Center for Marine Sciences, Wilmington, North Carolina, United States of America

Tracey T. Sutton
College of William & Mary, Virginia Institute of Marine Science, Gloucester Point, Virginia, United States of America

Jennifer L. Kelley
School of Animal Biology, University of Western Australia, Perth, Western Australia, Australia

Lesley J. Morrell
Department of Biological Sciences, University of Hull, Hull, United Kingdom

Chloe Inskip
Durrell Institute of Conservation and Ecology, University of Kent, Kent, United Kingdom

Jens Krause
Leibniz-Institute of Biology and Ecology of Fishes, Berlin, Germany
Institute of Integrative and Comparative Biology, University of Leeds, Leeds, United Kingdom

Darren P. Croft
School of Psychology, University of Exeter, Exeter, Devon, United Kingdom

Václav Pižl, Ladislav Háněl, Josef Starý, Karel Tajovský and Alena Nováková
Institute of Soil Biology, Biology Centre Academy of Sciences of the Czech Cepublic, České Budě jovice, Czech Republic

Jan Frouz
Institute of Soil Biology, Biology Centre Academy of Sciences of the Czech Cepublic, České Budě jovice, Czech Republic
Institute for Environmental Studies, Charles University, Prague, Czech Republic

Elisa Thébault
Laboratoire Bioge´ochimie et Ecologie des Milieux Continentaux Unité mixte de recherche 7618, Centre National de la Recherche Scientifique, Paris, France

Sina Adl
Department of Soil Science, University of Saskatchewan, Saskatoon, Canada

Tomáš Cajthaml and Petr Baldrián
Institute of Microbiology Academy of Sciences of the Czech Cepublic, Prague , Czech Republic

Jan Materna
Krkonoše National Park, Museum of Krkonoše in Vrchlabí, Vrchlabí, Czech Republic

Peter C. de Ruiter
Biometris, Wageningen University, Wageningen, The Netherlands

Alexander S. Flecker
Department of Ecology and Evolutionary Biology, Cornell University, Ithaca, New York, United States of America

Krista A. Capps
Department of Ecology and Evolutionary Biology, Cornell University, Ithaca, New York, United States of America
Sustainability Solutions Initiative, University of Maine, Orono, Maine, United States of America

Matthew Gordon and Jamie Seymour
School of Marine and Tropical Biology, James Cook University, Cairns, Queensland, Australia

David Raffaelli and Piran C. L. White
Environment Department, University of York, York, United Kingdom

Tsuyuko Yamanaka
Environment Department, University of York, York, United Kingdom
School of Environmental Sciences, University of Liverpool, Liverpool, United Kingdom

Mariano Lastra
Departamento de Ecología y Biología Animal, Universidad de Vigo, Vigo, Spain

Iván F. Rodil
Departamento de Ecología y Biología Animal, Universidad de Vigo, Vigo, Spain
National Institute of Water and Atmospheric Research, Hamilton, New Zealand

Tanya J. Compton
National Institute of Water and Atmospheric Research, Hamilton, New Zealand

Michael P. Lesser
Department of Molecular, Cellular and Biomedical Sciences, University of New Hampshire, Durham, New Hampshire, United States of America

Marc Slattery
Department of Pharmacognosy, University of Mississippi, University, Mississippi, United States of America

Garth Herring and Collin A. Eagles-Smith
United States Geological Survey, Forest and Rangeland Ecosystem Science Center, Corvallis, Oregon, United States of America

Dale E. Gawlik and James M. Beerens
Department of Biological Sciences, Florida Atlantic University, Boca Raton, Florida, United States of America

Joshua T. Ackerman
United States Geological Survey, Western Ecological Research Center, Dixon Field Station, Dixon, California, United States of America

Eckhard W. Heymann
Abt. Verhaltensö kologie & Soziobiologie, Deutsches Primatenzentrum, Göttingen, Germany

Dietmar Zinner
Abt. Kognitive Ethologie, Deutsches Primatenzentrum, Göttingen, Germany

Jörg U. Ganzhorn
Abt. Tierökologie, Zoologisches Institut, Universität Hamburg, Hamburg, Germany

Mayuko Nakamaru and Kanan Miura
Department of Value and Decision Science, Tokyo Institute of Technology, Tokyo, Japan

Takenori Takada
Graduate School of Environmental Earth Science, Hokkaido University, Sapporo, Japan

Akiko Ohtsuki
Department of Value and Decision Science, Tokyo Institute of Technology, Tokyo, Japan
The Graduate University for Advanced Studies, Kanagawa, Japan

Sayaki U. Suzuki
Department of Value and Decision Science, Tokyo Institute of Technology, Tokyo, Japan
The Graduate University for Advanced Studies, Kanagawa, Japan
National Agricultural Research Center, Ibaraki, Japan

Kazuki Tsuji
Faculty of Agriculture, University of the Ryukyus, Okinawa, Japan

Maija Karala and Tapio Mappes
Division of Ecology and Evolutionary Biology, Department of Biological and Environmental Science, University of Jyväskylä, Jyväskylä, Finland

Zbyszek Boratyński
Division of Ecology and Evolutionary Biology, Department of Biological and Environmental Science, University of Jyväskylä, Jyväskylä, Finland
CIBIO/InBIO, Centro de Investigação em Biodiversidade e Recursos Genéticos, Universidade do Porto, Vairão, Portugal

José Carlos Brito and João Carlos Campos
CIBIO/InBIO, Centro de Investigação em Biodiversidade e Recursos Genéticos, Universidade do Porto, Vairão, Portugal
Departamento de Biologia, Faculdade de Ciências da Universidade do Porto, Porto, Portugal

Rodolphe Sabatier
Department of Ecosystem Modelling, Büsgen-Institute, Georg-August-University of Göttingen, Göttingen, Germany
INRA, UMR 1048 SADAPT, Paris, France

Kerstin Wiegand and Katrin Meyer
Department of Ecosystem Modelling, Büsgen-Institute, Georg-August-University of Göttingen, Göttingen, Germany

Jie Zhao, Kelin Wang and Xunyang He
Key Laboratory of Agro-ecological Processes in Subtropical Region, Institute of Subtropical Agriculture, Chinese Academy of Sciences, Changsha, Hunan, China
Huanjiang Observation and Research Station for Karst Ecosystems, Chinese Academy of Sciences, Huanjiang, Guangxi, China

Shengping Li and Lu Liu
Key Laboratory of Agro-ecological Processes in Subtropical Region, Institute of Subtropical Agriculture, Chinese Academy of Sciences, Changsha, Hunan, China
Huanjiang Observation and Research Station for Karst Ecosystems, Chinese Academy of Sciences, Huanjiang, Guangxi, China
Graduate University of Chinese Academy of Sciences, Beijing, China

Matthew J. Cleary
Department of Botany, University of Wyoming, Laramie, Wyoming, United States of America

Megan L. Mobley
Department of Botany, University of Wyoming, Laramie, Wyoming, United States of America
Helga Otto Haub School of Environment and Natural Resources, University of Wyoming, Laramie, Wyoming, United States of America

Ingrid C. Burke
Department of Botany, University of Wyoming, Laramie, Wyoming, United States of America
Helga Otto Haub School of Environment and Natural Resources, University of Wyoming, Laramie, Wyoming, United States of America
Program in Ecology, University of Wyoming, Laramie, Wyoming, United States of America

Index